Introduction to Cell Biology

Stephen L. Wolfe
University of California, Davis

Wadsworth Publishing Company
Belmont, California
A Division of Wadsworth, Inc.

Biology Editor: Jack Carey

Production Editor: Sally Schuman

Managing Designer: MaryEllen Podgorski

Copy Editor: Margo Quinto

Technical Illustrators: Florence Fujimoto, Virginia Mickelson, Catherine Brandel, Darwen Hennings, Vally Hennings, Victor Royer

Printed in the United States of America

2 3 4 5 6 7 8 9 10—87 86 85 84 83

Library of Congress Cataloging in Publication Data
Wolfe, Stephen L., 1932–
    Introduction to cell biology.
    Includes index.
    1. Cytology. I. Title.
QH581.2.W645 1983   574.87   82-13594
ISBN 0-534-01270-1

ISBN 0-534-01270-1 NB2I

Preface

Introduction to Cell Biology offers the same integration of cell structure, biochemistry, and physiology as my more advanced textbook, *Biology of the Cell*, but places greater emphasis on concepts and is less detailed than *Biology of the Cell*.

Like the more advanced text, *Introduction to Cell Biology* presents the most important hypotheses and concepts of cell biology in language that is as simple and direct as possible. After an introduction to cell biology, biological molecules, energy, and enzymes, the discussion moves to the structure and function of membranes and the cell surface in cellular organization, transport of molecules between cells and their environment, cell-to-cell recognition, and cell adhesion. The role of membranes in the capture of energy and the conversion of energy in photosynthesis and respiration to forms that can be utilized by cells follows this introduction to membrane structure and function. Later units describe the use of cellular energy in motility and cell synthesis, with particular attention to the activities of the nucleus and cytoplasm in the synthesis of proteins. The discussion then turns to the primary result of cellular synthesis, cell reproduction in meiotic and mitotic division. The book also includes a chapter relating nuclear structure and function to the major findings and concepts of genetics. The text closes with a description of the current hypotheses of the origin of cellular life. The major techniques and methods of research in cell biology are described in an extensive appendix.

As in *Biology of the Cell*, the major concepts emphasized in this book are supported and explained in terms of the results of scientific research with animals, plants, and microorganisms. However, the experimental details and exhaustive literature references included in the more advanced book are omitted in this text to make the presentation as readable as possible. Although this approach has been used throughout the book, the parallel organization of the two textbooks allows *Biology of the Cell* to be used directly as a source of additional information and supporting references from the scientific literature if these are needed by instructors or students.

The topics presented in each chapter of this book are amplified by end-of-chapter supplements, as in the more advanced text. Some of the supplements give additional information at more advanced levels; others provide more specialized topics or describe key experiments that support the material presented in the main text. By choosing the appropriate supplements, the book can be tailored to meet the interests and aims of a variety of students and courses. As an additional aid, each chapter concludes with questions to facilitate review of chapter topics.

The material included in this book from *Biology of the Cell* has been completely rewritten to suit the needs and interests of the introductory student in cell biology. The text has been reviewed both by scientists with expert knowledge and by teachers with experience and expertise in cell biology to further improve the accuracy, readability, and pertinence of the book. *Introduction to Cell Biology* also benefits from the extensive reviews carried out during preparation of the more advanced book from which its text is derived.

I am indebted to these reviewers, whose efforts greatly improved the accuracy and completeness of this book. I am also indebted to the many scientists and authors who willingly and generously supplied photographs, diagrams, and tables. Thanks are also due to Jack C. Carey, my subject editor at Wadsworth, who provided enthusiastic support and made many valuable suggestions concerning the form and substance of the text. Special thanks also go to Sally Schuman for her very competent efforts and assistance as production editor, to MaryEllen Podgorski, who ably designed the book, and to Virginia Mickelson and Florence Fujimoto, who provided new artwork and redrew illustrations from *Biology of the Cell* to suit the aims and approach of this book.

Stephen L. Wolfe

Reviewers

June Aprille
Tufts University

William Battin
SUNY at Binghamton

J. D. Brammer
North Dakota State University

P. R. Burton
University of Kansas

C. H. Chen
South Dakota State University

William L. Downing
Hamline University

Sharyn A. Endow
Duke University

Gordon G. Evans
Tufts University

Richard Falk
University of California, Davis

Knute Fisher
University of California

Susan Germeraad
University of Santa Clara

Donald L. Hybertson
Mankato State University

Bruce Haggard
Hendrix College

William R. Hargreaves
Harvard University

Ian B. Heath
York University

Peter Horton
State University of New York

Harold Kasinsky
University of British Columbia

Virginia Latta
Jeff State Junior College

Robert L. Macey
University of California

Peter B. Moens
York University

Brian Mulloney
University of California, Davis

Alan R. Orr
University of Northern Iowa

Jane Overton
University of Chicago

R. E. Pacha
Central Washington University

Arthur B. Pardee
Sidney Farber Cancer Institute

Lee Peachey
University of Pennsylvania

Kenneth R. Skjegstad
Moorhead State University

Frank B. Salisbury
Utah State University

Patricia Schulz
University of San Francisco

Rose Sheinin
University of Toronto

William R. Sistrom
University of Oregon

Gary Stein
University of Florida School of Medicine

Joan Steitz
Yale University

D. Lansing Taylor
Harvard University

Terrence Trivett
Pacific Union College

Fred D. Warner
Syracuse University

C. D. Watters
Middlebury College

Robert A. Weinberg
Massachusetts Institute of Technology

Christopher L. F. Woodcock
University of Massachusetts

Contents

SEVEN

Cell Origins   361

ONE

Introduction

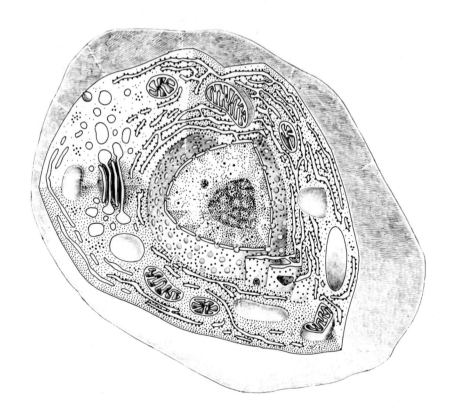

1

An Introduction to Cells and Cell Structure

Cells are the basic functional units of living organisms. All the complex activities of animals and plants depend ultimately on the activities of their individual cells. A single cell removed from an organism, if kept under the proper conditions, may remain alive indefinitely, and may grow and reproduce. If cells are separated into their constituent parts this quality of life is lost. Thus the organization of matter that we recognize as life does not exist in units smaller than cells.

Cells take highly varied forms in different plants, animals, and microorganisms. Figures 1-1 and 1-2 give an idea of the wide range of cell types. They may exist singly, as in the protozoa and bacteria, or packed together by the billions as in larger plants and animals. In size, cells range from the smallest bacteria, just barely visible in the light microscope, to units as large as the hen's egg: the yolk of a hen's egg is a single cell, several centimeters in diameter. Although most cells are roughly spherical in shape, some, like the nerve cells of larger animals, may carry long extensions that are microscopic in diameter but more than a meter in length.

Cell Structure

In spite of their varied sizes, shapes, and activities, all cells have two major functional regions. The *nuclear region* contains the molecules that store and transmit the hereditary information required for cell growth and reproduction. The second region, the *cytoplasm*, uses the nuclear information to make most of the molecules required for growth and reproduction (some are also made in the nuclear region), and also provides the energy needed to carry out these activities. The total living matter of cells, including both the nuclear region and the cytoplasm, is collectively called the *protoplasm.*

Cells are maintained as distinct compartments by *membranes* formed by thin layers of fat- or oil-like molecules called *lipids* in combination with proteins. The lipid layers, which form the basic framework of membranes, are only two molecules in thickness (see Fig. 4-3). These layers, with their associated proteins, control the movement of molecules to and from cells, and between regions within cells. The result is an effective separation of the cell contents from the outside world, and the subdivision of the cell interior into regions with specialized functions.

The cells of all organisms fall into one of two major subdivisions according to the organization of their membranes and the complexity of the nuclear region. The smaller and more primitive subdivision includes only two major groups, the bacteria and blue-green algae (the blue-green algae are also called *cyanobacteria*). In these organisms, called *prokaryotes*, the cellular membrane systems are limited to the surface or *plasma membrane* and relatively simple inner membranes derived from it. No membranes separate the nuclear region from the surrounding cytoplasm in the prokaryotes. (The name *prokaryotes*, from *pro* = before and *karyon* = nucleus, refers to the primitive organization of the nuclear region in these organisms.) Because of its primitive organization, the region containing the nuclear material in prokaryotes is termed the *nucleoid*. In the second major division, the *eukaryotes* (from *eu* = typical, *karyon* = nucleus), which includes all of the remaining plants, animals, and microorganisms on the earth, cells are divided by a large number of independent membrane systems into separate interior compartments. These separate membrane-bound interior structures, called *organelles*, are specialized to carry out the various cellular functions of eukaryotes. The largest and most conspicuous of the internal organelles is the *nucleus*, which contains the nuclear material. In contrast to the nucleoid in prokaryotes, the nucleus in eukaryotes is completely separated from the surrounding cytoplasm by a continuous system of membranes.

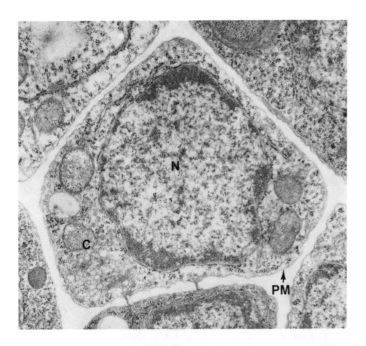

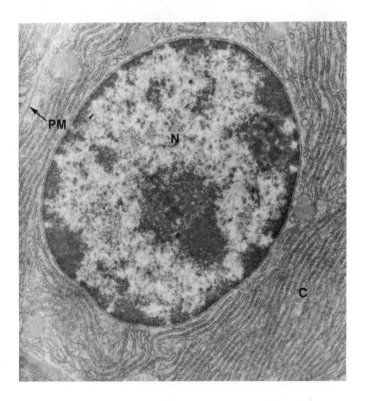

Figure 1-1 Electron micrographs of plant and animal cells. *N*, nucleus; *C*, cytoplasm; *PM*, plasma membrane. (**a**) An embryonic plant cell of *Sorghum bicolor*, a type of grass. × 12,800. Courtesy of Chin Ho Lin. (**b**) A cell from the pancreas of a rat. × 12,800. Photograph by the author.

Prokaryotic Cells

Prokaryotic cells (Figs. 1-3 and 1-4) are comparatively small, usually not much more than a few micrometers long and a micrometer or slightly less wide (the units of measurement most commonly used in cell biology are listed in Information Box 1-1). The boundary membrane of the cell, the plasma membrane, is surrounded in almost all prokaryotes by a rigid external layer of material, the *cell wall*. This external layer may range in thickness from 15 to 100 nanometers or more and may itself be coated with a thick, jellylike *capsule*. The cell wall provides rigidity to prokaryotic cells, and, with the capsule, protects the cell within the wall.

The plasma membrane lining the inner surface of the cell wall may be smooth or may include folds or pockets that extend from the surface into the cell interior (as in Fig. 1-4). This membrane system carries out a variety of vital functions in prokaryotes in addition to controlling the movement of substances to and from the cell. Most of the molecular systems that break down fuel substances to release energy for cell activities are linked to the plasma membrane in prokaryotes. In the photosynthetic bacteria and in the blue-green algae, the molecules carrying out the absorption and conversion of light to chemical energy are also associated with the plasma membrane and its interior extensions or derivatives. In addition, the plasma membrane is thought likely to play a part in replication and division of the nuclear material in prokaryotes. Thus many functions that are located in separate membrane-bound organelles in eukaryotes are associated with the plasma membrane or its derivatives in prokaryotic cells.

The nucleoid containing the nuclear material in prokaryotes is suspended directly in the cytoplasm without boundary membranes of any kind (see Fig. 1-3). The nucleoid appears as a structure of indefinite outline, less dense than the surrounding cytoplasm, containing tangled masses of fibers 3–5 nanometers in thickness. In the bacteria, which have been studied more intensively than the blue-green algae, the material of the nucleoid has been shown to consist of a single, long molecule of *DNA* (*deoxyribonucleic acid*), irregularly folded into a compact mass. DNA, a *nucleic acid*, is an informational molecule that stores the directions required for synthesizing cellular proteins. When isolated, the DNA of a bacterial nucleoid proves to form a single closed circle. While the nucleoids of blue-green algae also contain DNA, the number of DNA molecules per nucleus and the structure of the DNA are still unknown. The DNA of both bacteria and the blue-green algae is found primarily without large quantities of associated proteins.

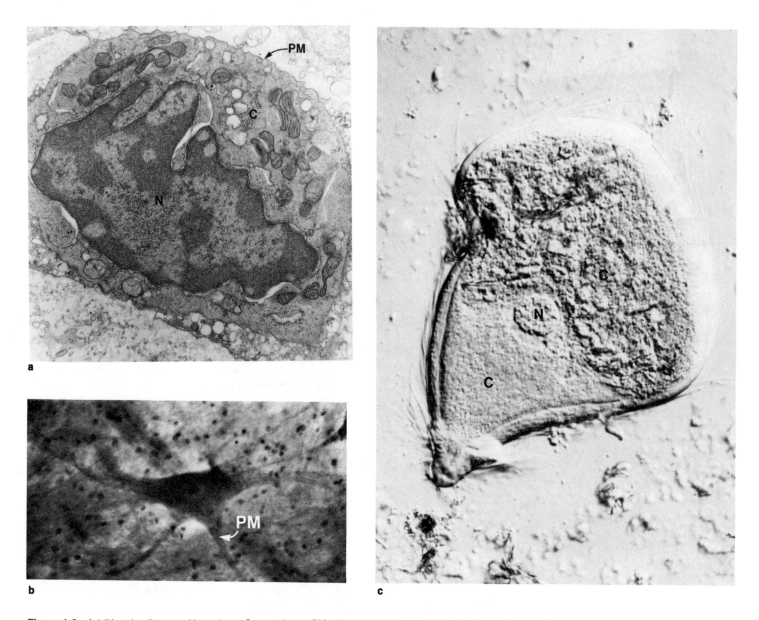

Figure 1-2 Additional cell types. *N*, nucleus; *C*, cytoplasm; *PM*, plasma membrane. (a) An electron micrograph of a human leucocyte (white blood cell). × 13,600. Courtesy of S. Brecher. (b) A light micrograph of a nerve cell from bovine spinal cord. × 1,750. Photograph by the author. (c) A light micrograph of a protozoan cell from the gut of a termite. × 400. Courtesy of T. K. Golder.

The cytoplasm surrounding the nucleoid usually appears darkly stained in electron micrographs. Most of this density is due to the presence of large numbers of small, roughly spherical particles about 20–30 nanometers in diameter, the *ribosomes*. In bacteria, ribosomes are complex structures containing more than 50 different proteins in combination with several types of a second nucleic acid,

RNA *(ribonucleic acid)*. These complex spherical bodies are the sites of protein synthesis in both prokaryotic and eukaryotic cells (the ribosomes of eukaryotes are slightly larger and contain more protein and RNA molecules).

Other structures may be present in the cytoplasm of the more complex prokaryotes. In some, particularly the photosynthetic bacteria and blue-green algae, the cyto-

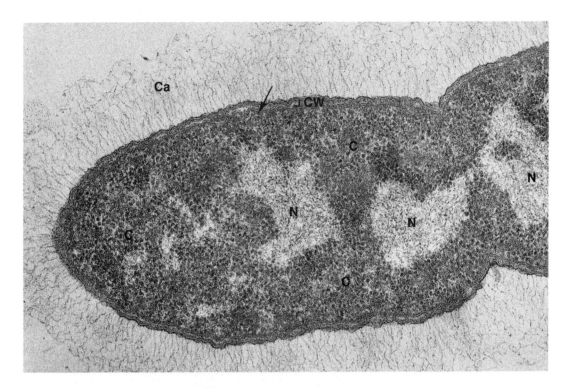

Figure 1-3 A prokaryotic cell, the bacterium *Klebsiella pneumoniae*. The nucleoid (*N*), the prokaryote equivalent of a nucleus, occupies the center of the cell. The cytoplasm surrounding the nucleoid is packed with ribosomes. The cell is surrounded by the cell wall (*CW*); the plasma membrane (arrow) lies just beneath the cell wall. The capsule material (*Ca*) is visible just outside the cell wall. × 62,000. Courtesy of E. N. Schmid, from *Journal of Ultrastructure Research* 75 (1981): 41.

plasm contains numerous baglike, closed sacs with walls formed by a single, continuous membrane. In these cells, the molecules carrying out photosynthesis are associated with the internal sacs as well as with the plasma membrane. No ribosomes are present inside the sacs; in the photosynthetic bacteria and blue-green algae, some may be filled with gas. Prokaryotes may also contain deposits of polysaccharides or inorganic phosphates that appear as small, dense, spherical bodies scattered in the cytoplasm. Many of these structures can be identified in the bacterial and blue-green algal cells in Figures 1-3 and 1-4.

Many types of bacteria are capable of rapid movement, generated by the action of threadlike structures covering the cell surface. These threads (Fig. 1-5), called *flagella* (singular = *flagellum*), are long, individual chains of protein molecules that extend from the cell surface, usually with no boundary membranes of any kind. The flagella, even though they may be as much as five times longer than the cell, are only 12–18 nanometers in diame-

ter. Usually, the flagella of a single bacterial species contain a single type of protein.

The apparent simplicity of prokaryotic cells is deceptive. Most bacteria and blue-green algae contain all of the biochemical mechanisms required to make the complex organic substances needed for life. In most cases, they are able to make these substances from simple inorganic precursors. In many respects, in fact, prokaryotes are more versatile in their synthetic activities than eukaryotes.

Eukaryotic Cells

In contrast to the prokaryotes, eukaryotic cells (Figs. 1-6 and 1-7) are divided into distinct interior compartments by systems of internal membranes. The plasma membrane completely encloses eukaryotic cells as in prokaryotes; all substances entering and leaving cells must pass through

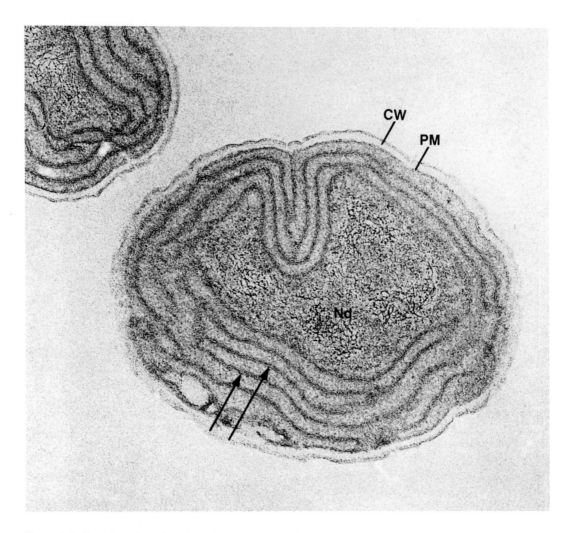

Figure 1-4 The blue-green alga, *Synechococcus lividus*. The center of the cell is occupied by the nucleoid (*Nd*), in which DNA fibers are clearly visible. The cell wall (*CW*), plasma membrane (*PM*), and cytoplasmic membranes associated with photosynthesis (arrows) are visible in the cytoplasm. × 72,000. Courtesy of M. R. Edwards, New York State Department of Health, from *Journal of Phycology* 4 (1968): 283.

this barrier. Although the molecules that break down fuel substances or convert light to chemical energy are not associated with the eukaryotic plasma membrane as they are in prokaryotes, molecular systems with other functions may be present. Among the most important of these, as in prokaryotes, are the molecules that transport substances through the plasma membrane.

The protoplasm of eukaryotic cells is divided into two distinct regions, nucleus and cytoplasm, by another membrane system, the *nuclear envelope*. The nuclear envelope, in contrast to the plasma membrane, consists of *two* concentric membranes, one layered just outside the other. The nuclear envelope is crowded with large numbers of *pore complexes* that perforate the membranes and evidently serve as transport channels between the nucleus and cytoplasm (arrows, Fig. 1-6; see also Figs. 9-19 to 9-22). The pores in the envelope, which average about 70 nanometers in diameter, are filled in by a ring of dense material, the *annulus*. The molecules of the annulus, which have so far defied analysis, act in some way to control the movement of large molecules such as RNA and protein through the nuclear envelope.

The Nucleus Most of the space inside the nucleus is occupied by masses of very fine, irregularly coiled and

Information Box 1-1

Units of Measurement Used in Cell Biology and Their Equivalents

Units	Equivalence in Millimeters	Equivalence in Micrometers	Equivalence in Nanometers	Equivalence in Angstroms
Millimeter (mm)	1	1,000	1,000,000	10,000,000
Micrometer (μm)	0.001	1	1,000	10,000
Nanometer (nm)	0.000001	0.001	1	10
Angstrom (A)	0.0000001	0.0001	0.1	1

The micrometer, equivalent to 1/1000 millimeter, is convenient for describing the dimensions of whole cells or larger cell structures such as the nucleus, and is much used in this book. Most cells are between 5 and 200 micrometers in diameter, although some animal eggs may be much larger. Objects about 200 micrometers (0.2 millimeter) in diameter are just visible to the unaided eye.

Particles roughly the size of mitochondria and chloroplasts (objects in this range of dimensions are called *ultrastructures* of the cell) are most often measured in nanometers or Angstroms, units that are useful for descriptions from this level down to particles as small as molecules and atoms. The nanometer, equivalent to 1/1000 micrometer, is difficult to visualize, but with experience a relative appreciation can be made of the size of objects measured in this unit. Lipid molecules, for example, are about 2 nanometers long, and amino acids are about 1 nanometer long. Protein molecules may be 10 nanometers or so in diameter. On the level of cell organelles, membranes are 7.5 to 10 nanometers thick, and ribosomes are about 25 to 30 nanometers in diameter. The electron microscope, incidentally, can "see" objects with diameters as small as 0.7 to 0.8 nanometer. The Angstrom unit, equal to 0.1 nanometer or 1/10,000 micrometer, is employed for measurements in the same size range.

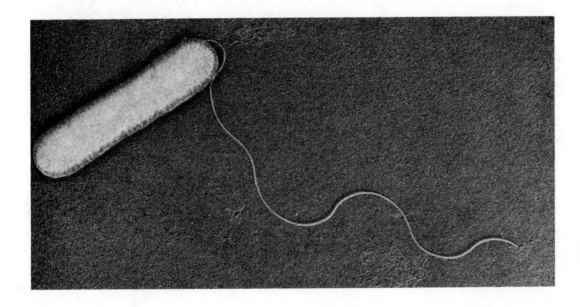

Figure 1-5 A bacterial cell with a single flagellum. × 30,000. Courtesy of J. Pangborn.

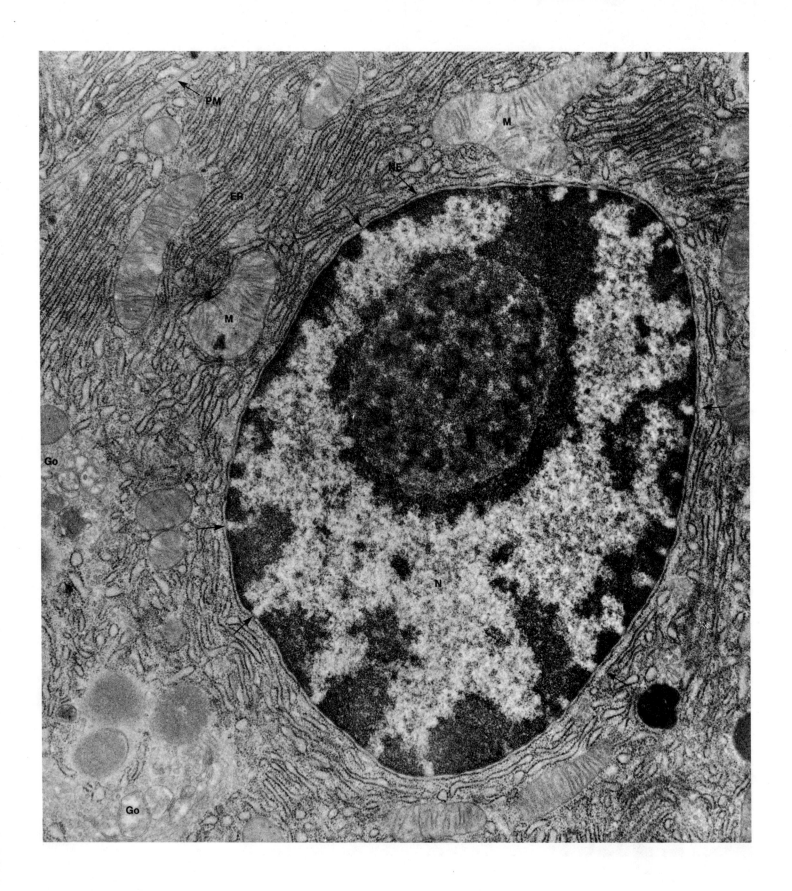

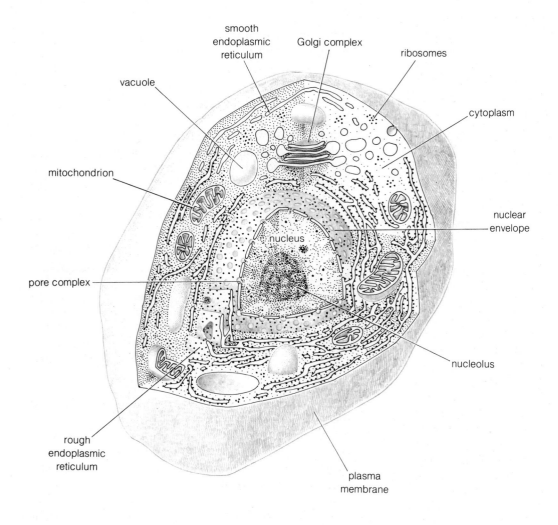

smooth
endoplasmic
reticulum

Golgi complex

ribosomes

vacuole

cytoplasm

mitochondrion

nuclear
envelope

nucleus

pore complex

nucleolus

rough
endoplasmic
reticulum

plasma
membrane

Figure 1-7 Structures typically seen in electron micrographs of eukaryotic cells.

folded fibers. These fibers, which average about 10–20 nanometers in diameter, have been shown to contain the DNA of the nucleus, along with two types of protein specifically associated with DNA in eukaryotes, the *histone* and *nonhistone* chromosomal proteins. The DNA molecules, in combination with the histone and nonhistone chromosomal proteins, are collectively called the chromatin of the eukaryotic nucleus. The chromosomal proteins found with the DNA maintain the structure of the chromatin fibers, and also regulate the activity of the DNA.

Suspended within the chromatin of the nucleus are one or more irregularly shaped bodies, the *nucleoli* (singular = *nucleolus*). The nucleolar material is so densely packed that its boundaries are easily traced, even though no membranes separate the nucleolus from the surrounding chromatin. Two major components, called simply the

Figure 1-6 A eukaryotic animal cell, from the pancreas of a rat. *N*, nucleus; *Nu*, nucleolus; *M*, mitochondria; *ER*, endoplasmic reticulum; *Go*, Golgi complex; *PM*, plasma membrane; *NE*, nuclear envelope; arrows, nuclear pore complexes. × 24,000. Photograph by the author.

nucleolar fibers and *nucleolar granules,* are visible inside the nucleolus. The nucleolar fibers are indistinct and difficult to trace in electron micrographs; generally they appear to be somewhat thinner than chromatin fibers. The nucleolar granules are about half the size of cytoplasmic ribosomes. Frequently, spaces containing chromatin fibers extend into the nucleolus, giving parts of the structure a more open, coarsely coiled appearance. The overall size and shape of the nucleolus and the distribution of fibrils and granules inside it change as cells go through their cycles of growth and division.

The nucleus functions as the ultimate control center for cell activities. Within the chromatin the information required to synthesize cellular proteins is coded into the DNA molecules. This information is copied into "messenger" RNA molecules that move to the cytoplasm through the pore complexes of the nuclear envelope. The nucleolus synthesizes the RNA of ribosomes, and assembles this RNA with ribosomal proteins into the subunits of ribosomes, which also pass to the cytoplasm through the pores of the nuclear envelope. In the cytoplasm, the subunits assemble by twos into complete ribosomes. Messenger RNA molecules are then used by the ribosomes as the directions for assembling proteins.

The nucleus also duplicates the chromatin as a part of cell reproduction. Just before cell division, all of the components of chromatin, including both DNA and chromosomal proteins, are precisely doubled. During cell division, the two copies of the chromatin are separated and exactly divided so that the two cells resulting from the division each receive a complete set of the directions for cell synthesis.

The Cytoplasm *Mitochondria* Eukaryotic cytoplasm, the portion of the cell outside the nucleus, is packed with ribosomes and a variety of membrane-bound organelles. Among the most conspicuous of these organelles are *mitochondria* (singular = *mitochondrion,* from *mitos* = thread, and *chondros* = grain). The Greek roots for *thread* and *grain* refer to the fact that mitochondria may be long and filamentous or compact and granulelike, or may change between these forms. Mitochondria are formed from two separate membrane systems, one enclosed within the other (Fig. 1-8; see also Figs. 7-2 and 7-3). The outer membrane is smooth and continuous; the inner membrane is thrown into numerous folds or tubular projections called *cristae* (singular = *crista*). Mitochondria carry out most of the reactions that break down fuel substances and release energy for cellular activities. Because they carry out this function mitochondria are frequently termed the "powerhouses" of the cell. Mitochondria are partially autono-

mous and contain their own DNA and ribosomes, which in structure and activity resemble the DNA and ribosomes of bacteria.

Ribosomes, Endoplasmic Reticulum, and the Golgi Complex Eukaryotic cytoplasm also contains large numbers of ribosomes and a variety of membrane-bound sacs or *vesicles.* Eukaryotic ribosomes, at diameters of about 25–35 nanometers, are somewhat larger than the ribosomes of prokaryotes and contain more protein and RNA. These ribosomes may be either freely suspended in clusters or attached to the surface of membranous sacs in the cytoplasm. The freely suspended ribosomes make proteins that primarily become part of the soluble, background substance of the cytoplasm. The ribosomes attached to membranous sacs form extensive channels in the cytoplasm known as the *rough endoplasmic reticulum* or *rough ER* (Fig. 1-6; see also Figs. 10-6 to 10-9). The rough ER forms a system that synthesizes and transports proteins. Proteins synthesized on the ribosomes penetrate into the membranes of the ER, forming a part of the membrane structure, or into the enclosed ER channels. The newly made proteins may have various fates in cells. The proteins becoming a part of the ER membranes eventually move from the ER to other cell structures to become permanent membrane proteins. The proteins entering the ER channels become enclosed in spherical, membrane-bound sacs or vesicles that pinch off from the ER. These sacs of protein may follow a variety of routes through cells. Some remain in the cytoplasm as storage vesicles; others migrate to the plasma membrane to expel their contents to the cell exterior. Some merge with another membranous complex, the *Golgi complex,* where the enclosed proteins undergo further processing before storage or release from the cell.

The Golgi complex or apparatus, named for its discoverer, Camillo Golgi, is a collection of membranous vesicles without attached ribosomes. The complex often takes the form of closely stacked, flattened sacs (Fig. 1-9; see also Figs. 10-11 to 10-13). Within the complex, proteins are modified by the attachment of other chemical groups, including lipids and sugars. Following these modifications, sacs containing the proteins pinch off from the Golgi complex and either remain suspended in storage in the cytoplasm or are secreted to the cell exterior.

Not all of the interconnected membranous vesicles collectively identified as endoplasmic reticulum are associated with ribosomes. The ribosome-free membranes, known as *smooth endoplasmic reticulum* (*smooth ER;* see Fig. 10-10), have various functions in the cytoplasm. One is the transport of proteins synthesized in the rough ER. Other smooth ER membranes, probably unrelated to the ones

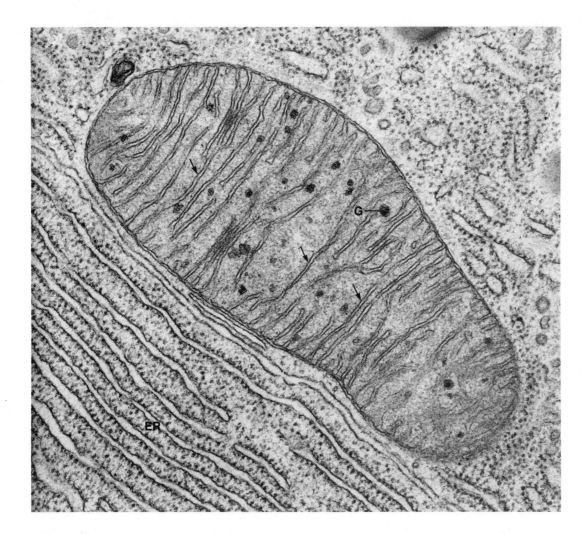

Figure 1-8 A mitochondrion from bat pancreas. Cristae (arrows) extend into the interior of the mitochondrion as folds from the inner boundary membrane. The darkly stained granules (G) are believed to be deposits of lipids in the mitochondrial interior. Courtesy of K. R. Porter.

transporting proteins, are involved in the initial breakdown of fats to release energy for cellular activities. Some segments of smooth ER have been identified with the synthesis of lipids, or with reactions that break down toxic substances absorbed by cells.

Microtubules, Microfilaments, and Cell Movement Almost all cell movements in eukaryotes are generated by one or both of two cytoplasmic structures, *microtubules* and *microfilaments*. Microtubules (Fig. 1-10) are long, pipelike cylinders, somewhat smaller in diameter than a ribosome and varying in length from a few nanometers to many micro-

meters. The cylindrical walls of microtubules enclose a central channel that is distinctly less dense than the walls. Microfilaments (Fig. 1-11) are extremely fine fibers, with thickness no greater than the wall of a microtubule. Each of the two motile structures are built up from proteins with a motile function—microtubules from a protein called *tubulin,* and microfilaments from a protein called *actin.*

Certain kinds of cellular movement, such as the movement of the chromosomes during cell division, depend on the activity of microtubules. Microfilaments are responsible for other types of movement, including the active flowing motion of cytoplasm called *cytoplasmic*

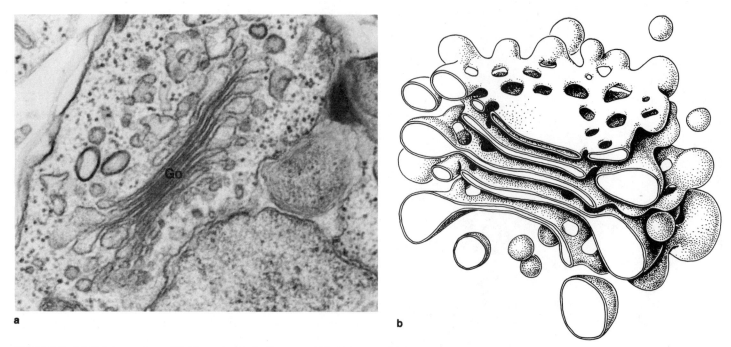

Figure 1-9 (**a**) Golgi complexes (*Go*) in a plant cell. Courtesy of W. A. Jensen. (**b**) A three-dimensional reconstruction of a Golgi complex.

streaming. More organized movements, such as the contraction of muscle cells, also depend on the activity of microfilaments.

Both microtubules and microfilaments produce the force for cell movement by an active sliding mechanism. The force for the sliding is believed to depend on chemical bonds that are alternately made and broken between the proteins of adjacent microtubules and microfilaments. Although microtubules and microfilaments may coordinate to produce a complex cell movement, they apparently do not interact directly. That is, microtubules do not produce motion by sliding over microfilaments, or vice versa.

Many types of plant and animal cells move by means of flagella that are fundamentally different from the much smaller and simpler flagella of bacteria. Within a flagellum is a remarkably complex system of microtubules consisting of a circle of nine peripheral double microtubules (the *doublets*) surrounding a central pair of single microtubules (the *central singlets;* see Fig. 1-12 and Figs. 8-1 to 8-3). With rare exceptions, the same 9 + 2 arrangement of microtubules is found inside flagella of all types, including the tails of plant and animal sperm cells, the flagella found on many animal cells, such as the cells lining the respiratory tract of mammals, and the flagella of protozoa and algae.

In all of these systems, the microtubules within each flagellum are believed to generate the force for movement by sliding actively over each other.

The Cytoskeleton or Cellular Matrix Microtubules and microfilaments, in addition to their functions in cell motility, also form supportive networks in the cytoplasm called the *cytoskeleton* or *cellular matrix* (Fig. 1-13). Supportive structures are also formed by another class of fibers somewhat larger than microfilaments called *intermediate filaments* or *fibers* (Fig. 1-14). These intermediate filaments, with diameters of about 10 nanometers, are built up from various nonmotile structural proteins, among them *keratin, desmin,* and *vimentin*. As the name suggests, the cytoskeletal networks of microtubules, microfilaments, or intermediate filaments form frameworks providing structural support to the cell. These networks are particularly highly developed in cells forming connective tissues in vertebrate animals.

Specialized Cytoplasmic Structures of Plant Cells All of the nuclear and cytoplasmic structures described up to this point, with the possible exception of intermediate filaments, occur in both plant and animal cells. Plant cells also have a number of organelles and components not

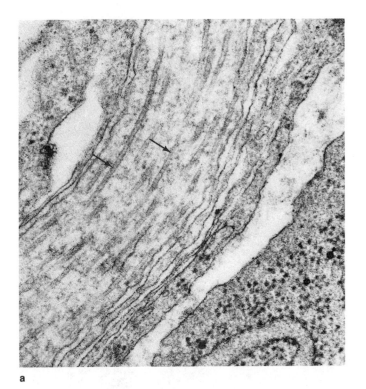

a

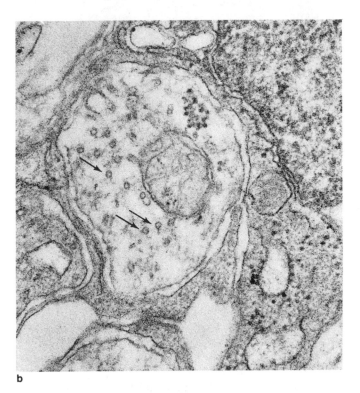

b

Figure 1-10 Microtubules (arrows) in a longitudinal section (**a**) and cross section (**b**) of a developing nerve cell. × 65,000. Courtesy of M. P. Daniels and the New York Academy of Sciences.

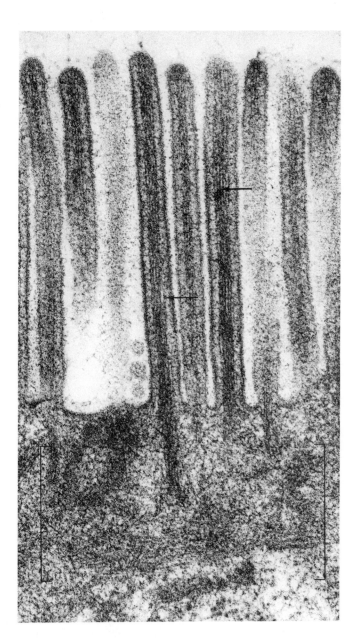

Figure 1-11 Microfilaments (arrows) in the fingerlike surface extensions of a cell from the intestine of a chick. The surface extensions, called *microvilli,* are capable of movement; the microfilaments inside them are believed to provide the force for this movement. Other microfilaments (brackets) are present in the cytoplasm under the microvilli. While microvilli can move, their primary function seems to be absorption of nutrients rather than cell motility. × 80,000. Courtesy of C. Chambers.

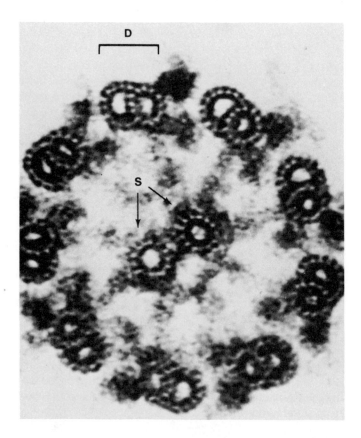

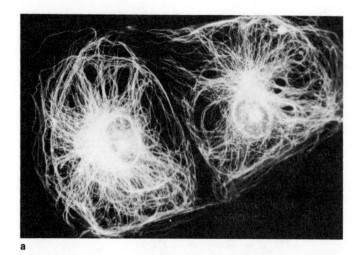

a

Figure 1-12 The 9 + 2 system of microtubules in a eukaryotic flagellum in cross section. *D*, peripheral doublet; *S*, central singlets. Subunits in the microtubule walls have been made visible by tannic acid staining. Courtesy of K. Fujiwara and the New York Academy of Sciences. Figures 1-10 and 1-12 are from *The Biology of Cytoplasmic Microtubules, Annals of the New York Academy of Sciences* 253 (1975).

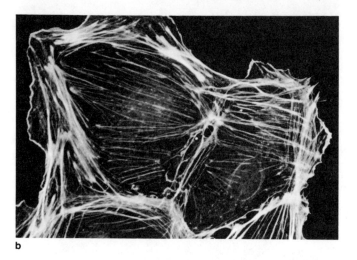

b

Figure 1-13 The cytoskeleton or cellular matrix in cultured kangaroo rat cells. The staining technique used to prepare these light micrographs makes the cytoskeleton appear white against a dark background. (**a**) The microtubule network of the cytoskeleton. (**b**) The microfilament network of the cytoskeleton. × 600. Courtesy of M. Osborn, reprinted from M. Osborn, W. W. Franke, and K. Weber, *Proceedings of the National Academy of Sciences* 74 (1977): 2490.

found in animal cells (Figs. 1-15 and 1-16). The most conspicuous of these are the *plastids,* large *vacuoles,* and the *cell wall.*

Plastids constitute a family of organelles with a variety of functions in plants. In green, photosynthesizing plant tissues the characteristic plastid is the *chloroplast* (Fig. 1-17; see also Figs. 6-1 and 6-2), a membranous organelle somewhat like mitochondria in structure. The chloroplast surface is formed by two concentric membranes. The outer membrane, as in mitochondria, is smooth, and the inner one is highly convoluted. Most of the chloroplast interior is filled with a third system of saclike membranes called *thylakoids.* Within the thylakoid membranes, light is converted to chemical energy. This energy is then used by molecules suspended in solution inside the chloroplast to

synthesize complex organic molecules such as sugars from water, carbon dioxide, and other simple inorganic substances. Like mitochondria, chloroplasts also contain DNA and ribosomes that resemble the equivalent prokaryotic systems in structure and function.

The plastids of embryonic plant tissues are small and contain only a few inner membranes. These plastids, called *proplastids,* develop into chloroplasts if the tissue is

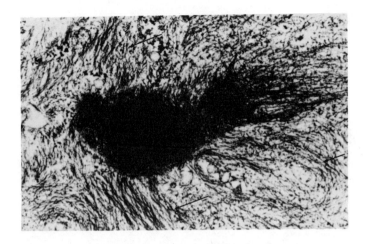

Figure 1-14 Intermediate filaments (arrows) in the cytoplasm of a mammalian cell. × 16,000. Courtesy of S. Brecher.

exposed to light. Other plastids, packed with stored lipid, protein, or starch rather than photosynthetic membranes, are called *leucoplasts* (*leukos* = colorless or white). In ripening fruit or in leaves undergoing the change to fall colors, leucoplasts or chloroplasts are transformed into *chromoplasts*, colored plastids in which red and yellow lipid pigments predominate.

The various types of plastids observed in plants are dependent upon the tissue type and developmental stage of the cells containing them. The usual sequence of development is from proplastid to leucoplast or chloroplast, and in some tissues subsequently to chromoplast. This sequence indicates that the several plastid types are different developmental forms of the same basic organelle.

Vacuoles are identified as separate, distinct organelles of plant cells because they reach much larger dimensions than any of the membrane-bound sacs or vesicles of animal cells. Often most of the volume of a mature plant cell, up to nine-tenths or more, is occupied by one or more large vacuoles. These vacuoles, consisting of a single, continuous membrane enclosing an inner, fluid-filled space, are formed, coalesce and increase in size during the maturation of plant cells; much of the growth of the developing cells is due to the enlargement of their vacuoles. Plant vacuoles contain dilute solutions of a variety of substances, including sugars and various pigments. Frequently the colors of flowers are due to pigments concentrated in cell vacuoles.

Plant vacuoles also function in the support of plant cells and tissues through the development of internal pressure that forces the cells against the rigid cell walls. In plants with small amounts of woody tissue, support of the stems and leaves depends primarily on pressure exerted by cell vacuoles, which may be as much as 20 or more times greater than atmospheric pressure (the pressure is developed by a pattern of water movement called *osmosis*; see p. 78).

Cell walls (Figs. 1-15 and 1-16; see also Figs. 5-14 to 5-20) are layered structures that provide rigidity to plant cells. These structures are outside the plasma membrane and are considered to be external to the cell. The *primary* cell walls of embryonic plant cells are flexible and capable of extension during growth, but the walls of mature cells, called *secondary* walls, are usually rigid and inextensible. At intervals, cell walls in higher plants are perforated by minute channels, the *plasmodesmata* (see Figs. 1-16, 5-19, and 5-20). These openings, sometimes as narrow as a few nanometers, contain cytoplasmic bridges that directly connect the protoplasm of adjacent cells. Both primary and secondary walls are built up primarily from fibers of *cellulose*—long, chainlike molecules consisting of repeating subunits of a sugar, *glucose*. The cellulose fibers of primary and secondary walls are held in place by a matrix consisting of a variety of complex branched molecules formed from sugars and other substances.

A Summary of Prokaryotic and Eukaryotic Cell Structure

Prokaryotes are relatively small cells with a single membrane system based on the plasma membrane and its derivatives. The prokaryotic nuclear material, contained in the nucleoid, is suspended directly in the cytoplasm with no boundary membranes marking the division between the two regions. The nucleoid contains little more than DNA; no nucleolus is present, and no proteins equivalent to the histone and nonhistone proteins of eukaryotes occur in quantity in association with the DNA. Prokaryotic cytoplasm consists primarily of masses of ribosomes. Although small membranous vesicles may occur in addition in the cytoplasm of the more complex prokaryotes, none of the discrete, membrane-bound organelles of eukaryotes such as mitochondria or chloroplasts are present.

Eukaryotic cells are divided into specialized compartments by systems of internal membranes. The nucleus is separated from the cytoplasm by a double layer of membranes, the nuclear envelope. Within the cytoplasm other membrane systems form mitochondria, chloroplasts, the endoplasmic reticulum, the Golgi complex, and a variety of vesicles and vacuoles. A plasma membrane, separate

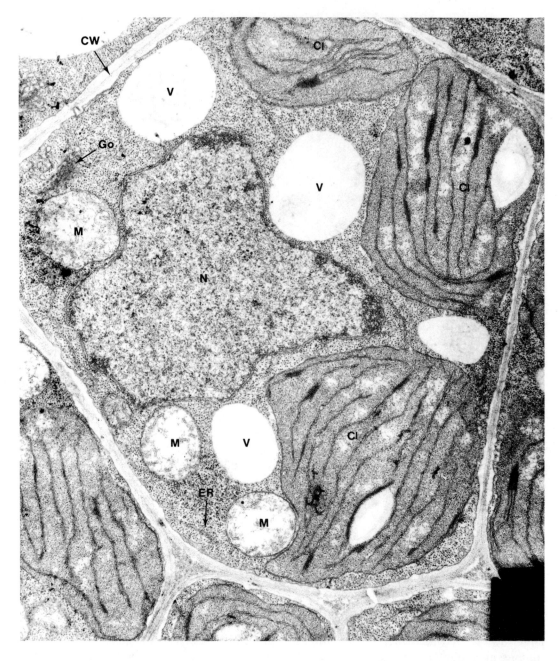

Figure 1-15 A maturing plant cell from a bean seedling (*Phaseolus vulgaris*). A number of large chloroplasts (*Cl*) and a few mitochondria (*M*) are visible in the cytoplasm. A prominent nucleus (*N*) is present; the nucleolus has not been caught in the plane of section. Several vacuoles (*V*) are developing in the cytoplasm. Elements of the endoplasmic reticulum (*ER*) and Golgi complex (*Go*) are also present. The cell is surrounded by a cell wall (*CW*). As the cell matures, the vacuoles will grow in size and coalesce into a single, large, central vacuole. Courtesy of Chin Ho Lin.

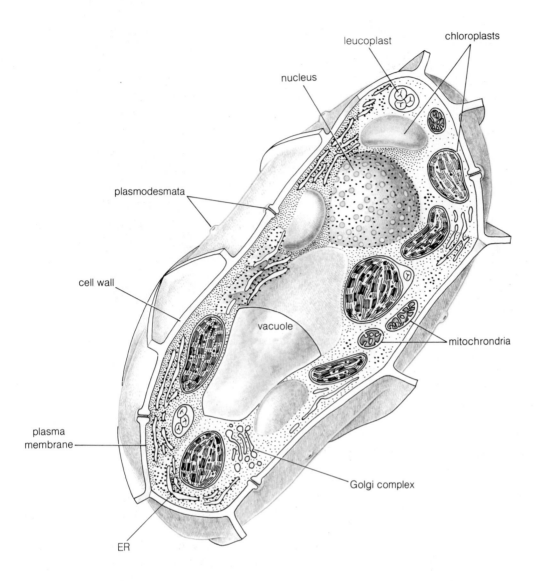

leucoplast

chloroplasts

nucleus

plasmodesmata

cell wall

vacuole

mitochrondria

plasma
membrane

Golgi complex

ER

Figure 1-16 The structures typically visible in electron micrographs of plant cells.

and distinct from the internal membranes, forms the outer boundary of the cell. The nucleus contains chromatin fibers, consisting of DNA in association with the histone and nonhistone proteins, and a prominent nucleolus. Both microtubules and microfilaments are present in eukaryotic cells as motile elements. Microtubules and microfilaments, along with intermediate filaments, also form a cytoskeleton or cytoplasmic matrix that supports regions of the cell. The major differences between prokaryotic and eukaryotic cells are summarized in Table 1-1.

Viruses

Viruses are minute particles that are capable of infecting both prokaryotic and eukaryotic cells and converting them to the production of more virus particles. Outside their host cells, virus particles consist of a *core* of nucleic acid, either DNA or RNA, surrounded by a *coat* of protein, all concentrated into a particle smaller than a ribosome. All degrees of complexity exist in viruses, from simple parti-

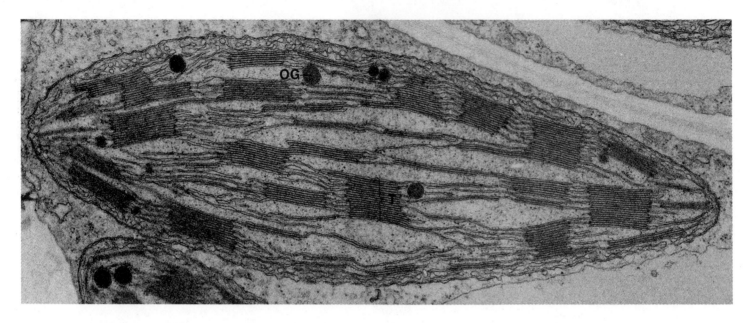

Figure 1-17 Chloroplast in a tobacco leaf cell. The reactions of photosynthesis converting light to chemical energy are concentrated in the thylakoid membranes (*T*) in the chloroplast interior. *OG*, osmiophilic granule, a darkly staining granule commonly observed inside chloroplasts. Courtesy of W. M. Laetsch.

cles consisting of a core with a single nucleic acid molecule and a coat of protein molecules of a single type, to more complex particles with coats made up of more than 50 different kinds of protein. A few viruses infecting animal cells, such as the *influenza* and *herpes* viruses, are surrounded by an outer membrane derived from the plasma membranes of their hosts.

The nucleic acid molecules of virus particles may be either linear or circular. The viruses infecting plant cells usually contain linear RNA molecules. The viruses infecting animal and bacterial cells may contain either RNA or DNA molecules, in either linear or circular form.

The protein coat enclosing the nucleic acid core, depending on the virus, may be rod- or lollipop-shaped, or spherical. The coat of the much studied *tobacco mosaic virus* (Fig. 1-18), for example, is a rod-shaped structure about 15 by 300 nanometers long, built up from more than 2000 identical protein units. The linear RNA molecule of this virus winds into a regular spiral extending through the axis of the rod. The viruses infecting bacteria, called *bacteriophages*, are among the most complex viral particles known. Best studied of these are the bacteriophages infecting the human intestinal bacterium *Escherichia coli* (Fig. 1-19 shows one of the *E. coli* bacteriophages). The "T-even" *E. coli* bacteriophages (T_2, T_4, and T_6) have a polyhedral *head* enclosing a DNA core, and a *tail* containing

several different proteins. The tail is complex and consists of a *collar*, at the point of attachment to the head, a cylindrical *sheath*, and a *baseplate*. The baseplate carries six long, hairlike extensions, the *tail fibers*.

The general pattern by which virus particles infect their host cells, illustrated by the life cycle of a T-even bacteriophage (Fig. 1-20), provides clues to the status of viruses as living or nonliving material. Free bacteriophage particles come into contact with bacterial cells by random collisions. If a virus particle collides with a bacterial cell, the tail fibers bind to sites on the bacterial cell wall (Fig. 1-20a). The head and tail sheath then contract and inject the DNA core into the cell (Fig. 1-20b). The proteins of the virus remain outside.

Once inside the bacterial wall, the bacteriophage DNA immediately takes over the synthetic machinery of the infected cell and directs it to make from hundreds to thousands of new copies of the viral DNA and protein (Fig. 1-20c and d). As the head and tail segments accumulate in the bacterial cytoplasm, the newly synthesized bacteriophage DNA condenses into the heads, and the heads and tails assemble into completed particles (Fig. 1-20e). After synthesis of the virus particles is complete, a final viral protein causes breakdown of the bacterial cell wall (Fig. 1-20f). Rupture of the cell wall releases the newly completed bacteriophage particles to the surrounding medium,

Table 1-1 Major Cell Structures of Prokaryotes and Eukaryotes

Structure	Prokaryotes	Eukaryotes Plants	Eukaryotes Animals
Plasma Membrane	+	+	+
Nucleus	−	+	+
Nucleoid	+	−	−
Nucleolus	−	+	+
Nuclear Envelope	−	+	+
Mitochondria	−	+	+
Chloroplasts	−	+	−
Ribosomes	+	+	+
Endoplasmic Reticulum	−	+	+
Golgi Complex	−	+	+
Microtubules	−	+	+
Microfilaments	−	+	+
Intermediate Filaments	−	±	+
Cell Wall	+	+	−

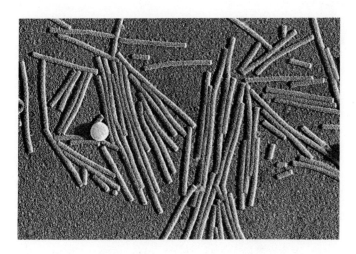

Figure 1-18 Tobacco mosaic virus, an example of a rod-shaped virus. The white sphere is a plastic particle of known size added to the preparation as a reference standard for determining the dimensions of the specimen. × 52,000. Courtesy of R. B. Park.

where chance collisions with additional bacterial cells may lead to another cycle of infection and release of virus particles.

This pattern indicates that viruses are best classified as nonliving matter when they are outside their host cells. In this form, they carry out none of the activities of life and are inert except for the capacity to attach to their host cells. They can be purified and crystallized in this form, and stored indefinitely without change or damage. The viral nucleic acid molecule carries only the information required to direct the host cell machinery to make more viral particles, and is active in this function only when inside a host cell. Thus a virus particle probably represents nothing more or less than a fragment of a nucleoid or chromosome derived from a once-living cell, surrounded by a layer of protein with protective and cell recognition functions. The information of the DNA contained within the virus is reduced to a set of directions coding for little more than production of more virus particles of the same kind.

The Origins of Cell Biology

The concept that cells are the basic functional units of life developed gradually over a period of almost two centuries, beginning with the first description of cells by Robert Hooke in 1665. During the earliest period of its development, investigations in cell biology were almost entirely descriptions of cell morphology. During the eighteenth and nineteenth centuries, the study of cell chemistry began, largely as an effort that proceeded independently from the morphological studies of cells being carried out at the time. As a result, the early investigations into cell chemistry had little effect on cell biology. By the arrival of the twentieth century, however, the rapidly developing field of biochemistry began to influence cell biology; by the 1930s the emphasis in cell biology started its shift from primarily morphological investigations to its present emphasis on biochemistry. In recent years the biochemical and molecular approach has dominated most of the work carried out in the study of cells.

The earliest developments in cell biology were closely tied to the invention and gradual improvement of the light microscope. Soon after invention of the light microscope in the seventeenth century, Hooke, an Englishman, made the first descriptions of cells. In *Micrographia*, published in 1665, Hooke reported his observations of small compartmentlike units in the woody tissues of plants; he called these compartments "pores" or "cells." It is frequently

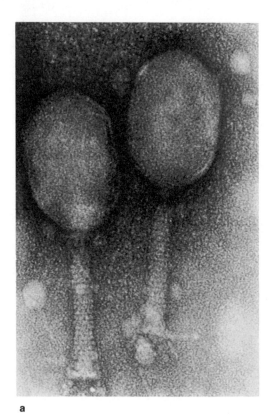

a

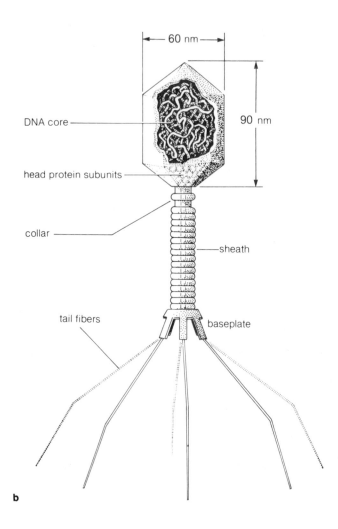

b

Figure 1-19 (a) T-even bacteriophages infecting *E. coli* cells. × 480,000. Courtesy of the Perkin–Elmer Corporation. (b) The structures of a T-even bacteriophage.

claimed that Hooke did not actually observe living cells because in some of the tissues he examined the cells were simply empty spaces outlined by residual cell walls. But in other tissue, such as the "inner pulp or pith of an Elder, or almost any other tree," Hooke "plainly enough discover'd these cells or Pores fill'd with juices" and thus saw living cells. Hooke's observations were extended by several investigators toward the end of the 1600s, most notably by Anton van Leeuwenhoek, a Dutchman and amateur microscopist who made remarkably accurate observations of the microscopic structure of protozoa, sperm cells, and a variety of other "animalcules," as he called them. Leeuwenhoek reported his findings in letters to the British Royal Society that arrived over a period of about 50 years.

After the last of Leeuwenhoek's letters, written in the early 1700s, cell biology entered a period of quiescence that continued for a hundred years, well into the nineteenth century. This delay in the evolution of the field was largely due to the imperfections of the microscopes available in the seventeenth and eighteenth centuries.

Practical methods for correcting the lens defects of light microscopes were introduced in the early nineteenth century. These innovations made investigations of the inner details of cell structure technically possible, and new discoveries in cell biology quickly followed.

The early observations of Hooke and others, made primarily in plant tissues, had given the impression that cells were units of living matter outlined by conspicuous walls. Because the cells of animal tissues lack distinct walls, structures equivalent to plant cell walls could not at first be resolved in animals, and the basic similarities in microscopic structure between plants and animals escaped notice.

The parallels in structure between animal and plant tissue were finally drawn in 1839, when Theodor Schwann observed that cartilaginous tissues of animals contain a

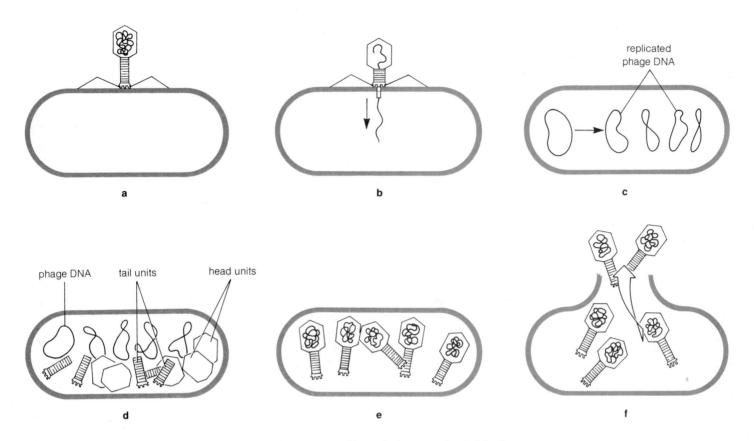

Figure 1-20 Life cycle of a T-even bacteriophage. (**a**) Attachment of bacteriophage to a bacterial cell; (**b**) injection of viral DNA; (**c**) replication of viral DNA; (**d**) synthesis of head and tail proteins; (**e**) assembly of DNA, head, and tail units into finished virus particles; (**f**) release of completed bacteriophage particles by rupture of the infected cell.

microscopic structure that "exactly resembles . . . [the] cellular tissues of plants." Schwann's work was aided by the fact that the extracellular material in cartilage occupies a position analogous to the cell wall in plants. Thus, while not possessing walls as an integral structural component, the cells of cartilage are separated by partitions that are more easily resolved in the light microscope than the boundaries of cells in most other animal tissues. This fact enabled Schwann to recognize the cellular nature of animal tissue. Schwann also remarked on the presence of nuclei in the cartilage cells, but was at first not as impressed by the cell contents as by the walls, which seemed to provide the main basis for the structural parallel between plant and animal tissues.

The improvements in light optics soon allowed recognition of the much thinner cell boundaries in other types of animal tissues. As the details of structure in the cell interior became discernible, emphasis gradually shifted from the walls to the contents, and the term *cell* began to take on modern connotations. At this time, it was recognized that the fluid contents of the cell make up the primary substance of living organisms. J. E. Purkinje adapted the term *protoplasm* for this substance in a publication appearing in 1840.

Development of the Cell Theory

In 1833, a few years before the definition of protoplasm by Purkinje, Robert Brown published a paper describing the microscopic structure of the reproductive organs of plants. In this paper, Brown drew attention to the nucleus as a constant feature of the plant cells investigated in his studies. This work, which subsequently received wide notice, established that the nucleated cell was the unit of living tissue in plants. Brown's study laid the foundations for the later development of the concept that cells are the fundamental units of all living organisms, now known as the *cell theory.*

The first two of the three hypotheses forming the cell theory were developed by Theodor Schwann and an eminent botanist of the day, M. J. Schleiden. During his work with cartilage in the 1830s, Schwann's attention was drawn to the cell nucleus by Schleiden, who had developed a series of hypotheses about the growth of cells in plants. As visualized by Schleiden, the nucleus took on a central role in the reproduction of cells. Under Schleiden's influence, Schwann realized the universality of nucleated cells as the structural units of living matter, and developed two postulates that were to provide the basis for the cell theory: that all living organisms are composed of one or more nucleated cells, and that the cell is the functional unit of living organisms. Historians usually attribute the origins of the two hypotheses jointly to Schwann and Schleiden.

These conclusions were soon supplemented by a third hypothesis that completed the cell theory. This hypothesis was supplied by scientists investigating cell origins, who observed that cells arise in both plants and animals by the division of a parent cell into two daughter cells. By 1855 this work had progressed far enough for Rudolf Virchow to state that all cells arise only from preexisting cells by a process of division. Virchow's famous statement of this concept, *"Omnis cellula e cellula,"* completed the cell theory:

1. All living organisms are composed of nucleated cells.

2. Cells are the functional units of life.

3. All cells arise only from preexisting cells.

Further work established that the nucleus is the repository of hereditary information, and that the essential feature of cell division is transmission of the hereditary information to daughter cells. The physical continuity of the internal matter of the nucleus through cycles of division was confirmed by Eduard Strasburger and Walter Flemming, who discovered that the substance of the nucleus, the *chromatin,* is transformed into compact rodlets, the *chromosomes* (see Fig. 11-8) during division. When fully formed, the chromosomes are clearly seen to be double. As the stages of division progress, the two halves of each chromosome split apart and pass into separate daughter cells, where they form the daughter nuclei. Later work by Flemming and others showed that the numbers of chromosomes appearing during division are constant for all members of a given species.

As a result of the discovery of the cellular mechanisms that preserve the physical continuity of the chromosomes during cell division, enthusiasm grew for the hypothesis that the factors controlling cell heredity are probably located in the nucleus. In 1884, Strasburger declared that the physical basis of heredity resides in the nucleus; in 1885, August Weismann concluded that "the complex mechanism for cell division exists practically for the sole purpose of dividing the chromatin, and . . . thus the [chromatin] is without doubt the most important part of the nucleus."

Some years before these developments took place, an Austrian monk, Gregor Mendel, analyzed the inheritance of traits such as seed shape and color in garden peas. (Mendel's experiments are described in Chap. 13.) Through his mathematical analysis of the distribution of these traits in parents and offspring, Mendel discovered genes and their patterns of inheritance. Among other observations, Mendel noted that the factors controlling heredity occur in pairs in an individual, and that the pairs separate independently in the formation of gametes. The fact that the inheritance of Mendel's factors, the genes, exactly parallels the distribution of chromosomes in cell reproduction was not realized in 1865 when Mendel published his results. At the time, cell biology had not progressed far enough for the connection to be made. As a result, Mendel's results were unnoticed and lay forgotten until just after the turn of the century when they were discovered anew by Hugo De Vries, C. Correns, and E. von Tschermak. By this time, the behavior of the chromosomes in cell division and reproduction were sufficiently well known for the correlation between genes and chromosomes to become apparent. In 1903 the American cell biologist Walter S. Sutton made the connection and pointed out the exact equivalence between the behavior of genes and chromosomes during reproduction. These findings and conclusions fully established the concept that the genes are carried on the chromosomes, and that the nucleus and chromosomes function in the storage and transmission of the hereditary information of the cell.

The Beginnings of Chemical Investigation of Cells

The application of physics and chemistry to the study of living organisms also has its roots in the eighteenth and nineteenth centuries. The beginnings in this field were made in 1772 when Joseph Priestley discovered that oxygen is released by green plants exposed to light. At almost the same time, Antoine Lavoisier recognized that "respiration is a . . . combustion, slow, it is true, but otherwise perfectly similar to that of charcoal." Because of the philosophical climate of the time, however, the fundamental importance of these discoveries was not appreciated. During the latter part of the eighteenth century, it was still believed that the substances and processes in living organisms were basically different from those of the inorganic world and that the techniques used for studying inanimate objects could not be applied to life.

The first significant movement from these attitudes came in 1828 when Friedrich Wöhler synthesized urea, at about the same time that cell biologists were beginning to formulate the cell theory. Wöhler was able to convert the inorganic chemical ammonium cyanate to urea, an organic substance commonly excreted by animals. This synthesis of a chemical found in living organisms, and the other organic molecules subsequently synthesized, gradually established that the same elements occur in both living and inanimate objects and are governed by the same chemical and physical laws. By the end of the nineteenth century, investigators had synthesized many of the chemicals found in plants and animals. Most successful in this work was Emil Fischer, who extracted, degraded, and resynthesized many substances from living organisms and laid the foundation for the chemical description of proteins, fats, and sugars.

This descriptive chemical work was complemented in the early nineteenth century by the beginnings of functional biochemical studies. Before this time, the substances and the reactions occurring in living systems were generally thought to be moved by a mysterious "vital force." The discovery of catalysts, the substances that speed reactions to completion, provided clues to the real nature of the chemical interactions in living organisms. In 1836 Jons Jakob Berzelius, a Swedish chemist, wrote that it is "justifiable . . . to suppose that, in living plants and animals, thousands of catalytic processes take place . . . and result in the formation of the great number of dissimilar chemical compounds, for whose formation out of the common raw material . . . no probable cause could be assigned."

Berzelius's intuitions about the role of catalysts in living organisms were later proved correct as a result of investigations into the nature of alcoholic fermentation, which has been of interest to mankind since ancient times. In the 1850s Louis Pasteur began his efforts to determine the cause of fermentation. Pasteur found that fermentation occurs only if particular microorganisms are present—if the microorganisms are eliminated or killed, sugar is not converted into alcohol. The first insights into the catalytic nature of the reactions occurring in fermentation came toward the end of the nineteenth century. In 1897, Hans and Eduard Buchner were working with the problem of isolating and preserving extracts from yeast cells for medicinal purposes. In order to preserve their extract, made by grinding and pressing yeast cells, the Buchners added sugar, commonly used then as today as a method for preserving food. To their surprise, the sugar was rapidly fermented by the cell-free yeast extract. Intensive studies of the yeast extracts led to the discovery of *enzymes*, the proteinaceous catalysts of living systems (the term *enzyme* was coined from a Greek word that means "in yeast").

Integration of the Chemical and Morphological Study of Cells

Significantly enough, investigations into the chemical nature of the nucleus were among the first efforts to integrate the chemical and morphological approaches to the study of cellular life. This work stemmed directly from the growing realization in the latter half of the nineteenth century that the nucleus is of central importance in cell function and heredity.

Friedrich Miescher, a physician and physiological chemist, became interested in the chemical composition of cell nuclei and developed a method for isolating nuclei in quantity for analysis. From his preparations, he isolated a previously unknown substance with properties then considered unusual for organic matter, including high acidity and large phosphorus content. Miescher, in announcing his discovery in 1871, called the new substance "nuclein." Soon after the announcement of Miescher's discovery, Walter Flemming concluded that if the chromatin of the nucleus and nuclein (later to be called nucleic acid) were not one and the same substance, "one carries the other."

A method for purifying nucleic acids was worked out by R. Altman in 1889, and the major chemical constituents of the nucleic acids were identified. One type of nucleic acid, DNA, was subsequently established to be characteristic of all cell nuclei. Later work in the 1920s and 1930s confirmed that DNA is located in the chromosomes, and many cell biologists began to suspect direct involvement of this substance in heredity. Finally, in the 1940s and 1950s, a series of experiments by Oswald Avery, Alfred D. Hershey and Martha Chase (see p. 214) confirmed that DNA provides the molecular basis of heredity.

Similar explorations into the integration of morphology and biochemistry were made with other cell organelles, beginning just before the turn of the century with investigations into the function of chloroplasts and mito-chondria. Work in this area proceeded slowly until the 1930s, when Albert Claude developed a technique for iso-lating and purifying mitochondria. This work culminated in the 1940s with the identification of mitochondria by Claude and others as the major site of the oxidative re-actions that release energy for cellular use. The same techniques later revealed that photosynthesis occurs in chloroplasts and that protein synthesis takes place on ribosomes.

This integrated approach to the study of cell mor-phology and function, aided in the 1950s by the develop-ment of powerful new methods such as electron microscopy and the use of radioactive tracers, led directly to the pres-ent-day appreciation of cells, in which structure and bio-chemistry are understood to be inseparable. The recent developments in cell biology and the unity of structure and function revealed by this approach are the subjects of this book.

The Plan of This Book

The introductory unit of this book continues with a de-scription of the major types of molecules occurring in liv-ing cells (Chap. 2), and a survey of the principles of energy flow and enzymatic catalysis (Chap. 3). Following this introduction, the second unit takes up membrane structure and function (Chap. 4), and specializations of the cell surface (Chap. 5), including cell junctions and cell walls. The structure and function of mitochondria (Chap. 6) and chloroplasts (Chap. 7) in the provision of energy for cellular activities are taken up in Unit Three. The use of this energy by microtubule-based and microfilament-based systems in the generation of cell motility is de-scribed in Unit Four. Unit Five, which surveys cellular syn-thesis, describes the structure and function of cell nuclei in RNA synthesis (Chap. 9), and of the cytoplasm in pro-tein synthesis (Chap. 10). The final unit of the book con-siders the division and origins of cells, including DNA replication and mitosis (Chap. 11), meiosis (Chap. 12), ge-netics (Chap. 13), and the origins of cellular life (Chap. 14). Major techniques used in cell biology are described in the Appendix.

Suggestions for Further Reading

Claude, A. 1975. The coming of age of the cell. *Science* 189:433–35.

Fruton, J. S. 1976. The emergence of biochemistry. *Science* 192:327–34.

Hughes, A. 1959. *A history of cytology.* Abelard-Schuman: New York.

Wolfe, S. L. 1981. *Biology of the cell.* 2nd ed. Wadsworth: Belmont, Calif.

2

A Chemical Background for Cell Biology

In its early years, cell biology was concerned primarily with the description of newly discovered cell structures. This concentration on descriptive morphology began to change near the close of the nineteenth century, when new techniques for studying the chemistry of life were discovered. These techniques developed rapidly and were so successfully applied that by the mid-twentieth century much of cell biology was transformed into a biochemical and molecular science. As a result, it is impossible to understand the conclusions of cell biology without an introduction to chemistry and to the major types of molecules carrying out the activities of life.

Although many of the molecules of living systems are highly complex, most of them fall into one or more of only four classes: (1) *carbohydrates*, (2) *lipids*, (3) *proteins*, and (4) *nucleic acids*. These four classes of organic molecules, along with water, form almost the entire substance of living organisms. Understanding the structure of these biological molecules, and the chemical bonds holding them together, will provide the basic information needed to follow the biochemical and molecular systems described in this book.

Atomic Structure and Chemical Bonds

Atomic Structure

Molecules are formed from *atoms* linked together in definite numbers and ratios by chemical bonds. The atoms of a molecule may be the same, as in a molecule of oxygen (O_2) or hydrogen (H_2), or different, as in a molecule of water, which contains two hydrogen atoms linked to a single oxygen atom (H_2O).

An atom is the smallest unit possessing the chemical and physical properties of an element. Although nearly a hundred different kinds of atoms occur naturally on the earth and link in various ways to form the molecules of both living and nonliving systems, all are basically similar in structure. Each atom consists of an atomic *nucleus* surrounded by one or more smaller, fast-moving particles, the *electrons* (Fig. 2-1). Most of the space occupied by an atom contains the electrons; the nucleus represents only about 1/10,000 of the total volume of an atom. The nucleus, however, is much heavier than all of the electrons put together; it makes up more than 99 percent of the total mass of an atom.[1]

The Atomic Nucleus All atomic nuclei contain one or more positively charged particles called *protons*. The number of protons in the nucleus of each type of atom is always the same. Hydrogen, the smallest atom, has a single proton in its nucleus; the nucleus of uranium, the heaviest naturally occurring atom, has 92 protons. Since the atoms of a given type always contain the same number of protons, this number, called the *atomic number*, specifically identifies an atom. Hydrogen, for example, has the atomic number 1. Similarly, nitrogen with seven protons, and oxygen with eight, have atomic numbers of 7 and 8 respectively (Fig. 2-2).

Each type of atom, except hydrogen, also has a number of uncharged particles in its nucleus. These particles, called *neutrons*, occur in variable numbers approximately equal to the number of protons. For example, carbon in its

---

[1]Mass is defined as the tendency of a body or particle to resist an accelerating force. Consider the difference between kicking (applying an accelerating force to) a soccer ball filled with air and one filled with lead. The difference in resistance to the accelerating force would be immediately obvious.

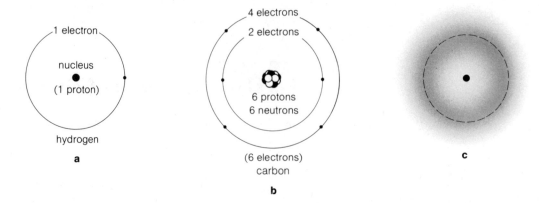

Figure 2-1 Atoms consist of a nucleus surrounded by fast-moving electrons. (**a**) In hydrogen, the simplest atom, the nucleus is surrounded by the orbital of a single electron. (**b**) Carbon, a more complex atom, has a nucleus surrounded by two successive shells of electrons. (**c**) The orbital (dotted line over region of deepest shade) represents the most probable location for an electron to occupy. Less probable locations are shown as regions of lighter shade.

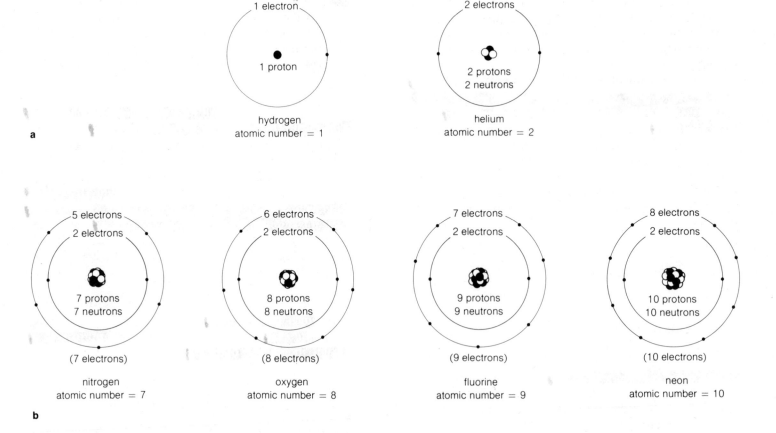

Figure 2-2 (**a**) Hydrogen and helium, with one and two protons respectively, have atomic numbers of 1 and 2. (The atomic number is equal to the number of protons in the nucleus of an atom.) (**b**) Atomic numbers of more complex atoms.

most common form has six protons and six neutrons in its nucleus. About one percent of naturally occurring carbon atoms have six protons and seven neutrons in their nuclei. Other carbon atoms, with six protons and eight neutrons, are found in natural carbons in even smaller proportions. These different forms of an atom, all with the same number of protons but varying numbers of neutrons, are called *isotopes*. The isotopes of an atom have essentially the same chemical properties, but differ in mass and other physical characteristics.

A neutron and a proton have the same mass. This mass is given an arbitrary value of 1, and atoms are assigned a *mass number* based on the total number of protons and neutrons in the atomic nucleus (the mass of electrons is so small that it can be ignored in determinations of atomic mass). The three isotopes of carbon with mass numbers of 12 (six protons plus six neutrons), 13 (six protons plus seven neutrons), and 14 (six protons plus eight neutrons) are identified as ^{12}C, ^{13}C, and ^{14}C.

Some isotopes of an atom are unstable and undergo radioactive decay. For example, the carbon isotope ^{14}C is unstable and slowly breaks down into an atom of nitrogen. In this breakdown, one neutron in the ^{14}C nucleus splits into a proton and an electron. The proton is retained in the nucleus, giving a new total of seven protons and seven neutrons, and the electron is ejected from the atom. When ejected, the electron is called a *beta particle*:

^{14}C (6 protons + 8 neutrons) → ^{14}N (7 protons + 7 neutrons) + 1 electron (beta particle)

An unstable isotope of hydrogen, ^{3}H, called *tritium*, which has one proton and two neutrons in its nucleus, also undergoes radioactive decay with ejection of a beta particle from its nucleus. The electrons ejected as beta particles from ^{14}C and ^{3}H have considerable energy and can be detected by various means.

The radioactive isotopes of carbon and hydrogen, along with unstable isotopes of sulfur and phosphorus, among others, have been of great value in biological research because molecules containing them can be traced by their radioactivity as they go through chemical reactions in living organisms. A number of stable, nonradioactive isotopes, such as ^{15}N (heavy nitrogen), can be detected by their mass differences and have also proved to be valuable as tracers in biological experiments.

The Electrons of an Atom The electrons surrounding an atomic nucleus carry a negative charge and normally occur in numbers equal to the positively charged protons. However, electrons have a mass equivalent to only 1/1800 of the mass of a proton.

Electrons are in constant, rapid motion around the atomic nucleus. Until the 1920s, electrons were believed to follow definite paths in their movement around a nucleus, much as the planets follow a defined orbit around the sun. More accurate analysis of the behavior of electrons has shown that an orbiting electron may actually be found in almost any location, ranging from the immediate vicinity of an atomic nucleus to practically infinite space. In moving through these locations, an electron travels so fast that it can almost be regarded as being at all of these points at the same time. However, if an electron could be tracked in its movements around the nucleus, it would be found to pass through some locations more frequently than others. These most probable locations surround the atomic nucleus in layers of different shapes called *orbitals*. In the hydrogen atom shown in Figure 2-1c, the orbital of the single electron moving around the nucleus is depicted as a shaded region. In this region, the most probable locations of the electron at any given time are shown as areas of deepest shade. Although either one or two electrons may occupy a given orbital, the most stable and balanced conditions are provided when an orbital is occupied by a *pair* of electrons.

The orbitals followed by electrons occur in successive layers or *shells* around an atomic nucleus. The innermost layer, the one closest to the nucleus, contains a maximum of two electrons and is thus filled by a single orbital. Hydrogen has one electron in this orbital; helium has two (see Fig. 2-2a). Atoms with total numbers of electrons between 3 (lithium) and 10 (neon) have two layers or shells of orbitals. The innermost layer contains two electrons in a single orbital. The second layer of orbitals, which may contain as many as eight electrons traveling in four orbitals, surrounds the nucleus at a greater distance. As this shell is filled, the atoms from lithium to neon are formed, including beryllium (with four electrons), boron (with five), carbon (six), nitrogen (seven), oxygen (eight), fluorine (nine) and neon (with ten electrons). Carbon is shown in Figure 2-1; the series from nitrogen to neon is shown in Figure 2-2b. Larger atoms have successive layers of orbitals that may contain more than eight electrons, up to a maximum of 32 electrons for any single shell. However, no matter how many electrons occur in the intermediate shells of large atoms, the most stable conditions are reached when the outermost shell contains a total of eight.

Electrons and the Chemical Activity of an Atom The chemical properties of an atom are determined largely by the number of electrons in the outermost shell. Most significant is the difference between this number and the stable number of either two electrons, for atoms near helium in size, or eight electrons, for larger atoms. Helium, with two electrons in its single shell, and atoms such as neon and argon, with eight electrons in their outer shells, are stable and essentially inert chemically. Atoms with outer shells containing electrons near these numbers tend to gain or lose electrons to approximate these stable configurations. For example, sodium, with two electrons in its first shell, eight in the second, and one in the third and outermost shell, readily loses its single outer electron to leave a stable second shell with eight electrons. Chlorine, with seven electrons in its outermost shell, tends to attract an electron from another atom to attain the stable configuration of eight. Other atoms, such as oxygen or calcium, which differ from a stable configuration by two electrons, tend to gain or lose electrons in pairs. Atoms differing from the stable, eight-electron outer shell by more than two electrons are inclined to share electrons with other atoms rather than gain or lose electrons completely. This sharing pattern is characteristic of carbon, which has four electrons in its outer shell and thus falls at the midpoint between the tendency to gain or lose electrons. The relative tendency to gain, lose, or share electrons underlies the formation of the chemical bonds holding the atoms of molecules together. Most important of these bonds in biological molecules are *electrostatic bonds, covalent bonds,* and *hydrogen bonds.*

Chemical Bonds

Electrostatic Bonds   Electrostatic bonds form through the tendency of some atoms to gain or to lose electrons entirely in order to form stable outer shells. A bond of this type may form if sodium, which has a tendency to lose an electron, is brought together with chlorine, which has a tendency to gain an electron (Fig. 2-3a and b). Under the correct conditions, the sodium atom readily loses an electron to the chlorine atom. After the transfer, the sodium atom, now with 11 protons in its nucleus and only 10 electrons in surrounding orbitals, carries a single positive charge. The chlorine, now with 17 protons and 18 electrons, carries a single negative charge. In this charged condition, the sodium and chlorine atoms are called *ions,* and are identified as Na$^+$ and Cl$^-$. The difference in charge sets up a strong attraction that tends to hold the atoms close together (Fig. 2-3c). This attraction is termed an *electrostatic* or *ionic bond.*

Electrostatic bonds are readily broken and remade. In solid sodium chloride, Na$^+$ and Cl$^-$ ions are held together by their opposite charges. If the sodium chloride is placed

in water, molecules of water take up positions between the Na$^+$ and Cl$^-$ ions, greatly reducing the attraction of the ions for each other. The ions, each surrounded by a layer of water molecules, may then separate and diffuse through the solution as free ions. If the water molecules are removed by evaporation, the Na$^+$ and Cl$^-$ ions will reassociate into a solid crystal as the solution dries. Water molecules are effective in causing the transition from bound to free ions because they have relatively positive and negative ends (see Unequal Electron Sharing and Polarity below). The ends of the water molecules are attracted to the surfaces of the ions, surrounding them and thus reducing their effective charge.

Electrostatic bonds are common among the forces holding ions, atoms, and molecules together in biological systems. Because these bonds are easily made and broken under the influence of water, electrostatic bonds are frequently important in rapidly changing systems where molecules or ions are brought together briefly during their interactions.

Covalent Bonds   We have noted that atoms such as carbon, with 4 electrons in their outer shells, have little tendency to gain or lose electrons completely. Instead, in forming chemical bonds, these atoms tend to share electrons to form stable outer shells. The hydrogen atom also shares its single electron readily to form a stable outer shell of two electrons.

The sharing of electrons by two atoms of hydrogen to form molecular hydrogen, H$_2$, provides the simplest example of the sharing mechanism. Each hydrogen atom has a single electron occupying an orbital around its nucleus. If two hydrogen atoms approach closely enough, the single electron of each atom may join in a new, combined orbital that surrounds both nuclei. The new orbital, containing two electrons, satisfies the conditions for stability in atoms with a single shell of electrons. As a consequence, the hydrogen atoms tend to remain linked together, forming a stable molecule. This linking force, set up by shared electrons, is the *covalent bond.*

Covalent bonds are relatively stable and can be broken only through the expenditure of considerable amounts of energy. In molecular diagrams, the covalent bond is designated by a pair of dots or a single line to represent the shared electrons:

$$H^. + .H \rightarrow H{:}H \text{ or } H{-}H$$

Carbon, with four unpaired outer electrons, forms four separate covalent bonds to complete its outermost shell. The formation of methane (CH$_4$) by electron sharing

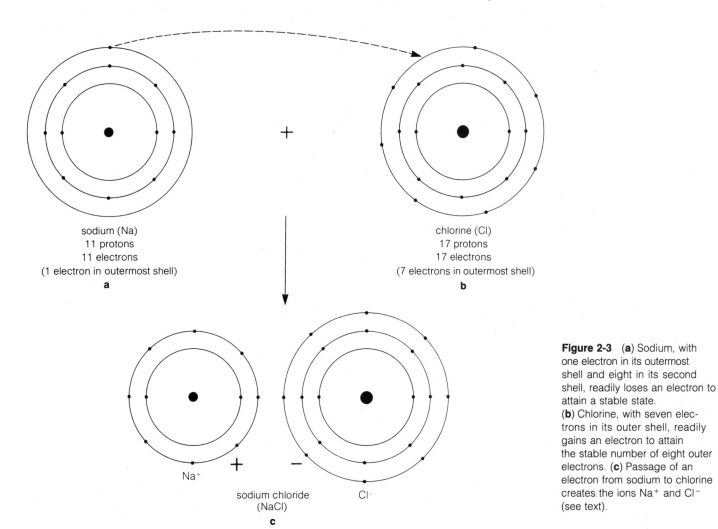

sodium (Na)
11 protons
11 electrons
(1 electron in outermost shell)
a

chlorine (Cl)
17 protons
17 electrons
(7 electrons in outermost shell)
b

Na$^+$

sodium chloride
(NaCl)
c

Cl$^-$

Figure 2-3 (**a**) Sodium, with one electron in its outermost shell and eight in its second shell, readily loses an electron to attain a stable state. (**b**) Chlorine, with seven electrons in its outer shell, readily gains an electron to attain the stable number of eight outer electrons. (**c**) Passage of an electron from sodium to chlorine creates the ions Na$^+$ and Cl$^-$ (see text).

between one carbon and four hydrogen atoms illustrates this process:

$$H \cdot + \overset{\displaystyle \cdot}{\underset{\displaystyle \cdot}{C}} \cdot + \cdot H \rightarrow H \overset{\displaystyle H}{\underset{\displaystyle H}{:C:}} H \text{ or } H{-}\overset{\displaystyle H}{\underset{\displaystyle H}{C}}{-}H$$

Because carbon can form four separate covalent bonds, carbon atoms link together readily to form highly branched or chainlike structures. These carbon chains form the "backbone" of an almost unlimited variety of molecules. In the molecules shown in Figure 2-6, each covalent bond representing a pair of shared electrons is depicted as a single line between linked atoms. Note that each carbon atom is linked to its neighbor by a total of four covalent bonds.

The other atoms present in the molecule—oxygen and hydrogen—also form covalent linkages readily and are commonly found with carbon in biological molecules. In these linkages, oxygen forms two covalent bonds and hydrogen one. Double bonds indicate that two pairs of electrons are shared between the atoms involved in the linkage.

Unequal Electron Sharing and Polarity Electrons are not always shared equally between two atoms held together by a covalent bond. In traveling their orbitals, the electrons may pass near one of the two atomic nuclei more frequently than the other. If this is the case, the atom retaining the electrons a greater part of the time will tend to carry a relatively negative charge. The deprived atom will become relatively positive. A covalent bond of this type has some of the properties of an electrostatic linkage and is said to be *polar*. If electrons are equally shared, so that

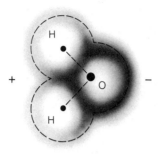

Figure 2-4 Polarity in the water molecule, created by unequal electron sharing in the covalent bonds between hydrogen and oxygen. The most probable locations of the shared electrons are indicated by regions of deepest shade. At any instant, the shared electrons are likely to be closer to the oxygen nucleus, making this end of the molecule relatively negative and the hydrogen end relatively positive.

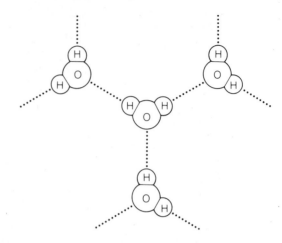

Figure 2-5 Hydrogen bonding (dotted lines) between water molecules. The bonds are formed through an attraction between the relatively negative oxygen atoms and relatively positive hydrogen atoms of adjacent molecules.

there is no greater probability that they will be closer to one atom than another in the linkage, the covalent bond is *nonpolar.*

The covalent bonds linking two atoms of hydrogen to oxygen to form water are polar (Fig. 2-4). The shared electrons in this molecule are more likely at any instant to be nearest the oxygen nucleus, leaving the hydrogen atoms with a relatively positive charge and giving the oxygen a relatively negative charge. Since the two hydrogen atoms in water are located asymmetrically, to one side of the oxygen, the whole molecule has positive and negative ends and is polar in character.

Polar molecules attract and align themselves with other polar or charged ions and molecules. The polar environments created by this association tend to exclude nonpolar substances, which also "prefer" their own company. This can be illustrated easily by mixing a polar substance, such as water, with a nonpolar substance such as a fat or oil. No matter how vigorously the oil and water are shaken, they will quickly separate into their respective polar and nonpolar environments when placed at rest. Because of their reaction to water, polar substances are often identified as *hydrophilic* (hydro = water, *philic* = preferring) and nonpolar substances as *hydrophobic* (hydro = water, *phobic* = avoiding).

Polar and nonpolar regions are important in the organization of cells. The plasma membrane at the cell surface, for example, is primarily nonpolar in character, but both the cell interior and the watery medium bathing the cell are polar. The nonpolar surface membrane tends to exclude the polar molecules of both the inside and outside mediums and thus acts as a barrier to the free movement of polar substances to and from cells. The tendency of po-

lar and nonpolar groups to associate in mutually exclusive domains or regions also provides a part of the force folding many complex biological molecules into their three-dimensional forms, or holding molecules together in molecular assemblies.

Hydrogen Bonds Hydrogen bonds also result from unequal electron sharing in covalent linkages. In the example given above, the hydrogen atoms in a water molecule are relatively positive because the electrons shared with oxygen tend at any instant to be nearer the oxygen nucleus. Because of this positive charge, a hydrogen nucleus in a water molecule is attracted to other atoms with a relatively negative charge, such as the oxygen atom of an adjacent water molecule (Fig. 2-5). This type of attraction, the *hydrogen bond*, is comparatively weak (about 1/10 as strong as a covalent bond) and is easily disturbed and broken, particularly by elevated temperatures. Hydrogen bonds are readily broken at temperatures above 50 to 60° C and become practically nonexistent at 100° C. Hydrogen bonds are frequently depicted by a dashed or dotted line between the hydrogen and the adjacent atom involved in the linkage.

Hydrogen bonds are important in biological molecules because opportunities for hydrogen bonding are frequently so great that many such links may form. The individual bonds, although relatively weak, are collectively strong when numerous and lend stability to the three-dimensional structure of complex molecules such as nucleic acids and proteins. In these large biological molecules, hydrogen bonds form between a hydrogen nucleus linked covalently to nitrogen or oxygen (both of these atomic nuclei attract electrons strongly) and another oxygen or nitrogen nucleus located nearby. Hydrogen bonds of this type are shown by the dotted lines in the protein structure diagrammed in Figure 2-19.

The Major Classes of Biological Molecules

The Organic Subunits

Many of the substances in the four major classes of biological molecules—the carbohydrates, lipids, proteins, and nucleic acids—are large molecules with complex structures and activities. However, all are based on different combinations of a few relatively simple organic subunits, primarily *organic acids, alcohols, aldehydes* and *ketones*. The amino acids and the fatty acids, which are components of proteins and lipids respectively, are examples of organic acids that often appear as building blocks of biological molecules (acids, bases, and pH are reviewed in Supplement 2-1). These subunit molecules owe their acidic properties to the —COOH (carboxyl) group, which ionizes in solution to release an H+ ion:

$$-C\overset{\displaystyle O}{\underset{\displaystyle OH}{\big\langle}} \;\rightleftharpoons\; -C\overset{\displaystyle O}{\underset{\displaystyle O^-}{\big\langle}} \;+\; H^+$$

Alcohols, which serve as important building blocks of fats and many carbohydrates, have in common the reactive *alcoholic* group:

$$\overset{\displaystyle H}{\underset{\displaystyle \vert}{-\overset{\vert}{C}-OH}}$$

Linkages formed by the alcohols in building more complex substances are based primarily on the reactivity of the —OH segment of this group of atoms. Carbohydrates also contain a reactive *carbonyl* (C=O) group, consisting of an oxygen double-bonded to a carbon of the chain. The car-

bonyl group occurs in either of two reactive forms. In one, called an *aldehyde*, the oxygen is carried on a terminal carbon of the chain:

$$\overset{\displaystyle H}{\underset{\displaystyle \vert}{-\overset{\vert}{C}-\overset{\vert}{C}=O}}$$

In the other reactive group, called a *ketone*, the oxygen is carried on a carbon at a point in the interior of the chain:

$$-\overset{\vert}{C}-\overset{\Vert}{\underset{\displaystyle O}{C}}-\overset{\vert}{C}-$$

These reactive groups and some other arrangements of atoms important in biological molecules are summarized in Table 2-1.

Carbohydrates

Carbohydrates are important both as fuel substances and structural molecules in cells. Carbohydrate fuels are stored in cells as *starches*, long-chain molecules made up of repeating carbohydrate units linked end-to-end. Structural carbohydrates are also long-chain molecules assembled from repeating carbohydrate subunits. The main constituent of plant cell walls, *cellulose*, a long-chain molecule of this type, is probably the most abundant organic molecule on the earth. Carbohydrate units also link to lipids to form *glycolipids*, and to proteins to form *glycoproteins*. Both of these complex molecular types are found in quantity at cell surfaces, where they are active in the recognition of stimuli reaching the cell from outside.

Carbohydrate Structure Carbohydrates contain carbon, hydrogen, and oxygen in the approximate ratio 1C:2H:1O, or (CH_2O). The basic building blocks of the carbohydrate family are short carbon chains from three to seven carbons long. Each carbon atom in the chains forming the units, except one, carries an —OH (hydroxyl) group. The remaining carbon carries an oxygen attached by a double bond, forming a C=O (carbonyl) group (see Fig. 2-6a). In most of these carbohydrate subunits all of the other available binding sites of the carbons are occupied by hydrogen atoms. The carbonyl oxygen may be at the end of the chain, forming an aldehyde ($\overset{\displaystyle H}{\underset{\displaystyle -C=O}{\vert}}$) group, or in the in-

Table 2-1 Some Important Reactive Groups of Organic Molecules

Radical	Structure	Reactivity
Alcohol	$\overset{\displaystyle H}{\underset{\displaystyle \vert}{-\overset{\vert}{C}-OH}}$	Reacts with organic acids to form many important biological substances; part of many carbohydrate molecules
Carboxyl (acid group)	$-C\overset{\displaystyle O}{\underset{\displaystyle OH}{\diagdown}}$	Ionizes in solution to release H^+ ions and thus acts as acid; part of amino acids, fatty acids
Carbonyl	C=O in	
Aldehydes	$-\overset{\displaystyle H}{\underset{}{\overset{\vert}{C}=O}}$	Important components of carbohydrates and intermediate compounds formed in synthesis and breakdown of carbohydrates
Ketones	$\overset{}{\underset{\displaystyle \vert}{\overset{\vert}{C}=O}}$	
Amino	$-N\overset{\diagup H}{\diagdown H}$	Component of amino acids and other important biological molecules; can combine with H^+ ions in solution to produce an $-NH_3^+$ group, thus acting as base by reducing H^+ ion concentration
Phosphate	$-O-\overset{\displaystyle O^-}{\underset{\displaystyle O}{\overset{\vert}{\underset{\Vert}{P}}}}-O^-$	An acidic group; forms links between organic groups in complex biological molecules; enters in energy reactions as energy carrier
Sulfhydryl	$-S-H$	Reactive group on amino acid cysteine that enters into disulfide ($-S-S-$) linkages and stabilizes protein structure

terior of the chain, forming a ketone ($-\overset{\displaystyle C}{\underset{\displaystyle O}{\Vert}}-$) group (see Table 2-1).

One unit of carbohydrate of this type, with a chain of three to seven carbons linked to —OH groups and the single carbonyl oxygen, is known as a *monosaccharide*. Monosaccharides are named according to the number of carbons present. Of the various possibilities, only *trioses* (three carbons), *pentoses* (five carbons), and *hexoses* (six carbons) are common in nature. Table 2-2 lists a number of the monosaccharides and gives some of their functions in living organisms.

Monosaccharides with five or more carbons can form stable ring structures in water solutions. The 6-carbon monosaccharide, glucose, for example, occurs in either of two closely related ring structures formed by an interaction between the aldehyde ($-\overset{\displaystyle H}{\underset{\displaystyle C=O}{\vert}}$) group at one end of the chain and the hydroxyl group at the next-to-last position at the other end of the chain (Fig. 2-6b). The ring formed is not flat or in one plane in space, as suggested in Figure 2-6b, but commonly has the three-dimensional "chair" conformation shown in Figure 2-6c.

The two closely related ring forms of glucose differ only in the direction pointed by the hydrogen attached to the 1-carbon of the sugar (see shaded arrows in Fig. 2-6b). The relatively small difference between these two forms, called α- and β-glucose, has great significance for the

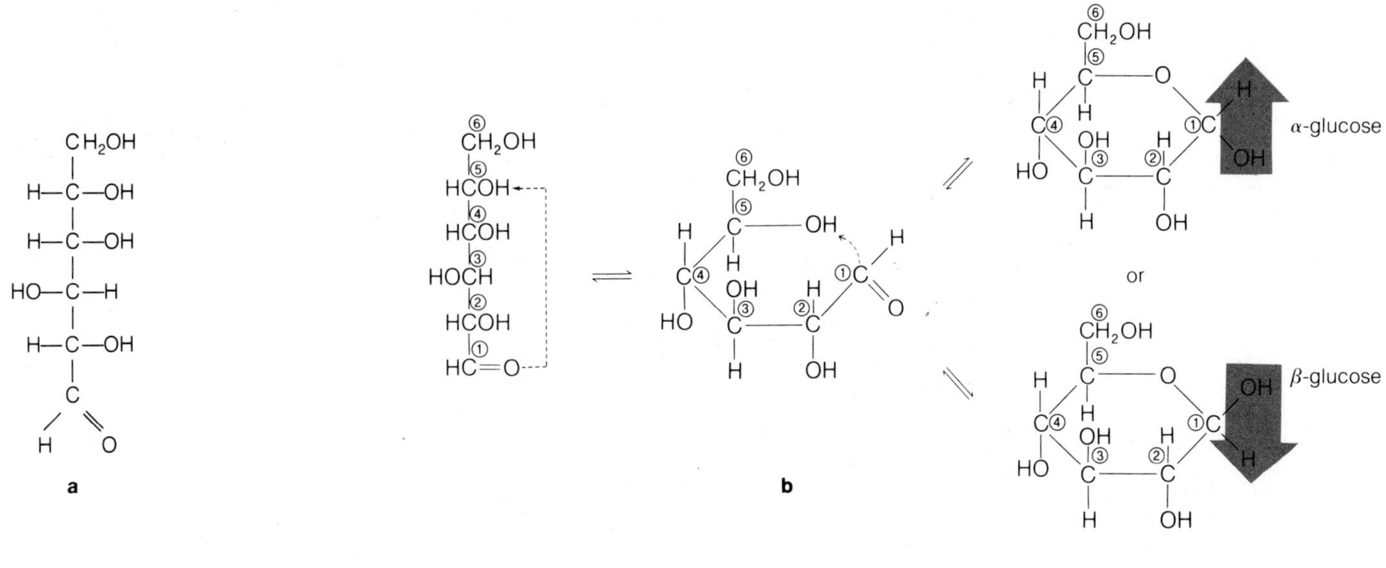

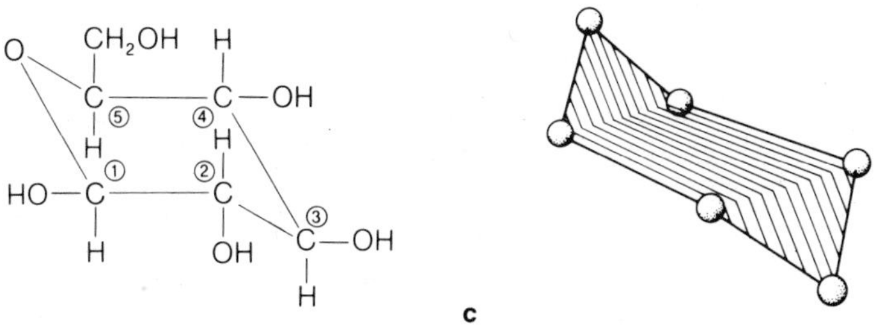

a

b

c

Figure 2-6 (**a**) A monosaccharide, glucose, in the extended form. (**b**) Ring formation by glucose, and the two alternate conformations of glucose, α and β, determined by the orientation of the hydrogen linked to the 1-carbon (shaded). (**c**) The "chair" conformation, the most common three-dimensional arrangement of the glucose carbon ring. The carbons of monosaccharides are numbered in order beginning with the carbon of the aldehyde (—CHO) group as number 1.

chemical properties of polysaccharides formed from glucose units. For example, cellulose, built up from glucose links in the β-form, is insoluble and cannot be digested as a food source by most animals. Starches, however, which contain glucose links in the α-form, are soluble and easily digested. Other sugars with a general structural plan resembling glucose also have equivalent α- and β-forms.

Linkage of Monosaccharides into Disaccharides and Polysaccharides Monosaccharides may link together by twos to form *disaccharides* or in greater numbers to form *polysaccharides*. Some polysaccharides, such as starch, cellulose, and glycogen, consist of very long chains of monosaccharides linked in either an unbranched chain, as in cellulose, or a chain with forks or branches, as in gly-

cogen (see Fig. 2-8c). Polysaccharide molecules most commonly contain a single type of monosaccharide building block.

Linkage of two glucose molecules to form the disaccharide *maltose* illustrates the general pattern of the reactions forming disaccharides and polysaccharides (Fig. 2-7). The reaction shown in Figure 2-7 also illustrates how water frequently participates in reactions involving assembly or disassembly of biological molecules (see Information Box 2-1). Note that the linkage formed between the two glucose molecules extends between the 1-carbon of the first glucose molecule and the 4-carbon of the second glucose. Bonds of this type between monosaccharides are designated as 1→4 linkages; 1→2, 1→3, and 1→6 linkages are also common in disaccharides and polysaccharides in na-

Table 2-2 Carbohydrate Units Found in Nature

Number of Carbons	Type	Examples	Importance
3	Triose	Glyceraldehyde, dihydroxyacetone	Intermediates in energy-yielding reactions and photosynthesis
4	Tetrose	Erythrose	An intermediate in photosynthesis and energy-releasing reactions
5	Pentose	Ribose, deoxyribose, ribulose	Intermediates in photosynthesis; components of molecules carrying energy; components of the informational nucleic acids DNA and RNA; structural molecules in cell walls of plants
6	Hexose	Glucose, fructose, galactose, mannose	Fuel substances; products of photosynthesis; building blocks of starches and cellulose
7	Heptose	Sedoheptulose	Intermediate in photosynthesis and energy-releasing reactions

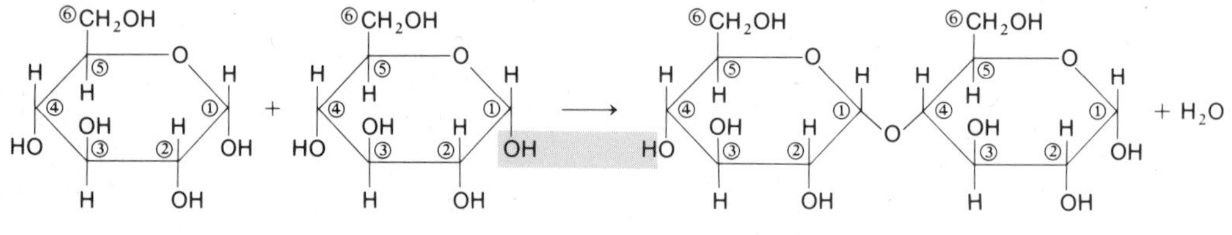

Figure 2-7 Combination of two glucose molecules to form the disaccharide maltose. The interaction involves removal of the elements of a molecule of water from the monosaccharides (shaded groups).

ture. These linkages are designated as *alpha* (α) or *beta* (β) depending on the orientation of the hydrogen at the 1-carbon forming the bond. In the maltose molecule, the hydrogen at the 1-carbon forming the linkage is in the α position; therefore the link between the two glucose subunits of maltose is called an α(1→4) linkage. As noted, α linkages are characteristic of sugars and starches that can be metabolized and used as an energy source by higher organisms; β linkages occur in structural molecules such as cellulose that cannot be digested by most eukaryotic organisms.

Cellulose (Fig. 2-8a) is formed by a series of reactions of the type shown in Figure 2-7, except that the bonds holding the glucose subunits together are β(1→4) linkages. Plant starch molecules, which also consist of long chains of glucose links, are similar (Fig. 2-8b), except that the bonds holding the glucose subunits together are α(1→4)

linkages. Animal starch (glycogen) molecules also contain glucose subunits held in chains by α(1→4) linkages. In contrast to plant starch, glycogen molecules form branched structures in which the side branches are linked to main chains by α(1→6) linkages (shaded in Fig. 2-8c).

Linkage of glucose units to form long chains in cellulose or starch molecules illustrates the general pattern followed by a *polymerization* reaction, a common assembly mechanism in which large, complex molecules are built up from repeats of a building block unit. In a polymerization reaction, the individual units, called *monomers*, join together like links in a chain to form the long, complex molecule, called the *polymer*. Thus in cellulose formation, the individual glucose molecules are the monomers of cellulose before they are joined together; the finished cellulose molecule is the polymer.

Much of the organic matter on earth consists of

Condensation and Hydrolysis

Many types of biological reactions in which molecular subunits are assembled or disassembled involve addition or removal of water. As glucose molecules link together to form a disaccharide, for example, a molecule of water appears as an additional product (see Fig. 2-10.) In this type of assembly reaction, called a *condensation*, the components of water, H^+ and OH^-, are split from the reacting chemical groups of the combining molecules. (The atoms contributing to the formation of water in Fig. 2-7 are shaded.) Most of the biological reactions in which complex molecules are assembled from smaller subunits are condensation reactions.

In the reverse reaction, chemical building blocks are broken off from a larger molecule with the *addition* of a molecule of water. This type of reaction, called a *hydrolysis*, is also common in biological systems:

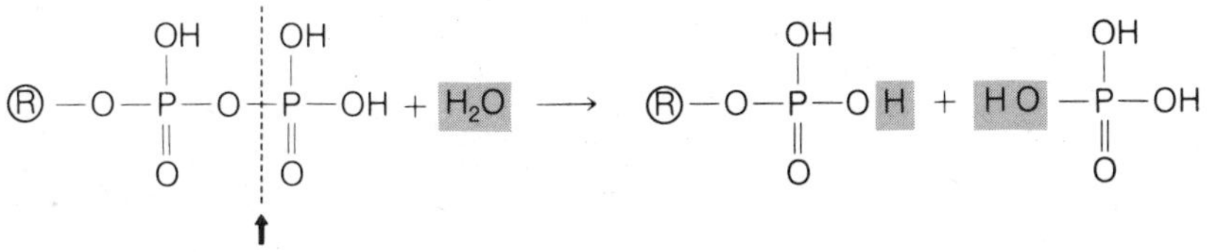

The R in the chemical structures stands for a complex organic group that forms the rest of the molecule. Hydrolysis is a part of all biological reactions in which larger molecules are broken into their subunits. The reactions of digestion, in which proteins, fats, carbohydrates, and nucleic acids are broken into smaller units before being absorbed in the intestine, are typical hydrolysis reactions.

(1) carbohydrates in the form of monosaccharides, (2) polysaccharides containing glucose units as the sole building block, or (3) polysaccharides containing glucose in combination with other hexoses. Most of this material serves as a reservoir of stored chemical energy or provides structural support for cells. In either of these roles, carbohydrates occur in all living organisms.

Lipids

Lipids form a mixed group of biological molecules so diverse that they defy description in simple, all-inclusive terms. Perhaps the best definition is based on solubility: lipids are biological substances that dissolve more readily in nonpolar solvents such as acetone, ether, chloroform, and benzene than in water, a strongly polar solvent. This solubility property reflects the presence of the nonpolar groups that make up all or part of lipid molecules, giving them a strongly hydrophobic character. This characteristic is of extreme importance in the interactions of lipids in cells because lipids tend to set up water-excluding domains and barriers, as in the cell membranes that form boundaries between and within cells. In addition to their structural and functional roles in membranes, lipids are important as an energy source and are stored and utilized in many cell types for this purpose. Three types of lipids— *neutral lipids*, *phospholipids*, and *steroids*—are especially important in biological systems and are found in quantity in different cells and tissues.

Neutral Lipids The neutral lipids, commonly found in cells as storage *fats* and *oils*, illustrate the molecular pattern

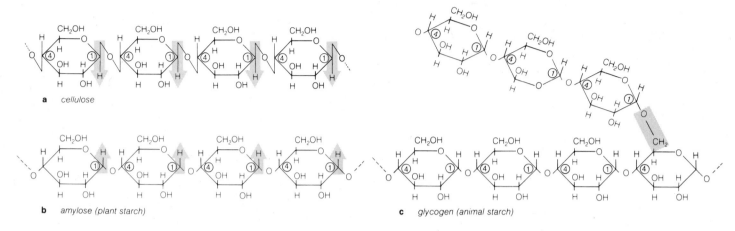

a cellulose

b amylose (plant starch)

c glycogen (animal starch)

Figure 2-8 (**a**) Cellulose, formed by end-to-end linkages of glucose molecules in the β form, or β(1→4) linkages. (**b**) Plant starch, formed by end-to-end linkages of glucose molecules in the α form, or α(1→4) linkages. (**c**) Glycogen, a branched molecule formed by glucose units joined by α(1→4) and α(1→6) linkages; the α(1→6) linkage is shaded in (**c**).

of many cell lipids. The neutral lipids are so called because at the degrees of acidity in living cells they bear no charged groups. Almost all these lipids are formed by a combination between *fatty acids* and the alcohol *glycerol*.

A fatty acid molecule consists of a long, unbranched chain of carbon atoms with attached hydrogens and other groups (Fig. 2-9a and b). A carboxyl (—COOH) group at one end of the chain gives the molecule its acidic properties. Almost all the fatty acids in neutral lipids contain an even number of carbon atoms. Although a few with odd numbers are found in all organisms, these make up only a minor fraction of the total. The carbon chains of fatty acids vary in length from as few as four carbons to much longer structures containing 24 or more carbons. Most commonly, the fatty acids linked to glycerol in neutral lipids have chains of carbon atoms in even numbers from 14 to 22; most have either 16 or 18 carbons. Within the chains, hydrogen atoms are bound to the carbon atoms forming the backbone of the molecules. If the maximum possible number of hydrogen atoms are bound to a fatty acid chain, the chain is said to be *saturated* (Fig. 2-9a). If hydrogen atoms are absent from adjacent carbon atoms in the interior of the chain, the fatty acid is said to be *unsaturated* (Fig. 2-9b). At points where hydrogen atoms are missing from adjacent carbon atoms, the carbons share a double instead of a single bond. This arrangement may occur at multiple points in the interior of the chain, forming the so-called *polyunsaturated* fatty acids. Unsaturated fatty acids have lower melting points than saturated fatty acids and are more abundant in living organisms (some of

the common saturated and unsaturated fatty acids are listed in Table 2-3).

Fully saturated fatty acids extend as more or less straight chains without major bends or kinks (see Fig. 2-9a). Unsaturated fatty acids usually have a kink at each double bond (Fig. 2-9b). The kinks affect the packing of unsaturated fatty acid chains in cell structures, making the chains more disordered and consequently more fluid at biological temperatures. This property is important in maintaining the flexibility and fluidity of cell membranes at lower temperatures.

Glycerol has three hydroxyl (—OH) sites at which fatty acids may attach (shaded in Fig. 2-9c). If a fatty acid binds to each of the three sites, the resulting compound is known as a *triglyceride*. The neutral lipids in living systems are primarily of this type.

In the formation of triglycerides, the three —OH sites of a glycerol molecule react with the carboxyl (—COOH) groups of three fatty acids (Fig. 2-10). One molecule of water is released as a byproduct for each linkage formed. In the resulting triglycerides (Fig. 2-11), the three fatty acid chains may occur in any combination: they may be all the same, as in the example shown in Figure 2-11, or all different. Different animal species usually have distinctive mixtures of fatty acids in their triglycerides; these may change to some degree depending on the diet of the organism.

Generally, the fluidity of triglycerides decreases as the length of their fatty acid chains increases. Triglycerides that are liquid at biological temperatures are called oils,

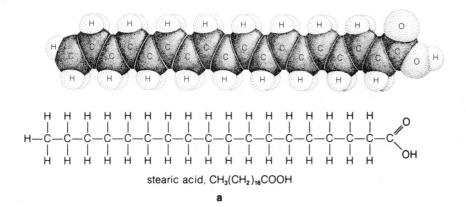

stearic acid, CH₃(CH₂)₁₆COOH

$$\text{stearic acid, } CH_3(CH_2)_{16}COOH$$

a

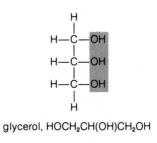

glycerol, $HOCH_2CH(OH)CH_2OH$

c

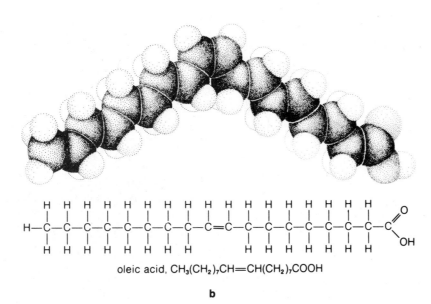

oleic acid, $CH_3(CH_2)_7CH{=}CH(CH_2)_7COOH$

b

Figure 2-9 The components of a neutral lipid or triglyceride. (**a**) Stearic acid, a saturated fatty acid. (**b**) Oleic acid, an unsaturated fatty acid. The molecular model shows the "kink" introduced in the fatty acid chain by the double bond. (**c**) Glycerol. The shaded —OH groups react with the —COOH groups of fatty acids to form triglycerides.

and those that are semisolid or solid are called fats. Most of the oils and fats inside cells are stored as energy reserves. Waxes, which contain very long-chain fatty acids in combination with glycerol or other alcohols, are solid at biological temperatures. Waxes occur most frequently on the exterior surfaces of cells or cell walls, particularly in plants, where they form a protective covering that resists water loss and invasion by infective agents.

Phospholipids Triglycerides rarely occur as functional parts of cell structures such as membranes. Of much greater importance in this role are the *phospholipids*, a group of phosphate-containing molecules with structures

basically similar to the triglycerides. In phospholipids, a phosphate group:

$$\begin{array}{c} O^- \\ | \\ -O-P-OH \\ \| \\ O \end{array}$$

is substituted for one of the fatty acids bound to glycerol (Fig. 2-12). In natural phospholipids, the phosphate group binds to either of the outside —OH groups on glycerol, never to the middle one. In membrane phospholipids, the phosphate group is linked in turn to one of a group of

Table 2-3 Saturated and Unsaturated Fatty Acids Occuring in Lipids

Saturated Fatty Acids		Unsaturated Fatty Acids	
Butyric acid	$CH_3(CH_2)_2CO_2H$	Crotonic acid	$CH_3CH—CHCO_2H$
Caproic acid	$CH_3(CH_2)_4CO_2H$	Palmitoleic acid*	$CH_3(CH_2)_5CH—CH(CH_2)_7CO_2H$
Caprylic acid	$CH_3(CH_2)_6CO_2H$	Oleic acid*	$CH_3(CH_2)_7CH—CH(CH_2)_7CO_2H$
Capric acid	$CH_3(CH_2)_8CO_2H$	Linoleic acid*	$CH_3(CH_2)_3(CH_2CH—CH)_2(CH_2)_7CO_2H$
Lauric acid	$CH_3(CH_2)_{10}CO_2H$	Linolenic acid	$CH_3(CH_2CH—CH)_3(CH_2)_7CO_2H$
Myristic acid	$CH_3(CH_2)_{12}CO_2H$	Arachidonic acid	$CH_3(CH_2)_3(CH_2CH—CH)_4(CH_2)_3CO_2H$
Palmitic acid	$CH_3(CH_2)_{14}CO_2H$	Nervonic acid	$CH_3(CH_2)_7CH—CH(CH_2)_{13}CO_2H$
Stearic acid	$CH_3(CH_2)_{16}CO_2H$		
Arachidic acid	$CH_3(CH_2)_{18}CO_2H$		
Lignoceric acid	$CH_3(CH_2)_{22}CO_2H$		

*Occur as major fatty acids in human storage fats.

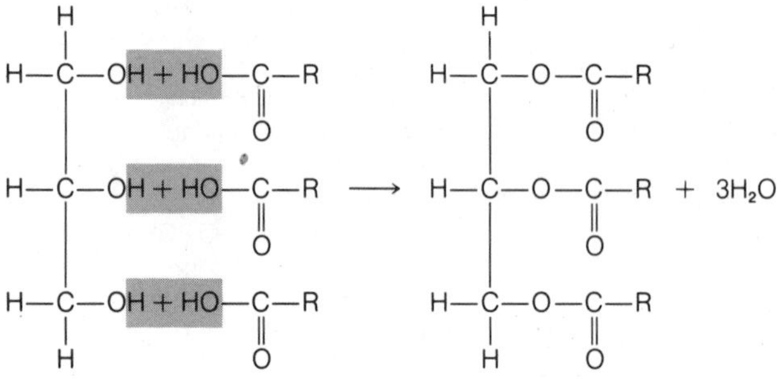

Figure 2-10 Combination of glycerol with three fatty acids to form a triglyceride (see text). The "R" groups represent the carbon chains of the fatty acids.

alcohols, usually containing one or more atoms of nitrogen (Fig. 2-13). In these more complex phospholipids, the phosphate group thus forms a linking bridge between the glycerol and the alcohol (see Fig. 2-12).

Phospholipids have dual solubility properties because the end of the molecule containing the alcohol and phosphate group is polar and hydrophilic. The fatty acid chains, in contrast, retain their strongly nonpolar and hydrophobic character. As a result, one end of the molecule is hydrophilic, and the other hydrophobic.

When placed in solution, phospholipids take up arrangements that satisfy their dual solubility properties. For example, if introduced into the interface formed when a nonpolar solvent such as benzene is layered on top of water, phospholipids orient so that their nonpolar fatty acid chains extend into the benzene and their polar phosphate-alcohol groups extend into the water (Fig. 2-14a). Phospholipids placed under the surface of water (or any strongly polar liquid) meet their dual affinities in an interesting and highly significant way, by forming layers just two molecules thick called *bilayers*. In a bilayer, the phospholipid molecules orient so that the fatty acid chains of the double layer associate together in a nonpolar, hydrophobic region in the interior of the layer. The phosphate-alcohol groups face the surrounding water (Fig. 2-14b). Bilayers of this type form the basic structural framework of membranes (see Fig. 4-3).

Steroids The steroids comprise a class of lipids based on a complex framework of four interconnected rings of carbon atoms (Fig. 2-15a). The various steroids differ in the position and number of double bonds linking the carbons in the rings, and in the side groups attached to the rings. The most abundant group of steroids, the *sterols*, have a hydroxyl group linked at one end and a complex, nonpolar carbon chain linked at the opposite end of the ring

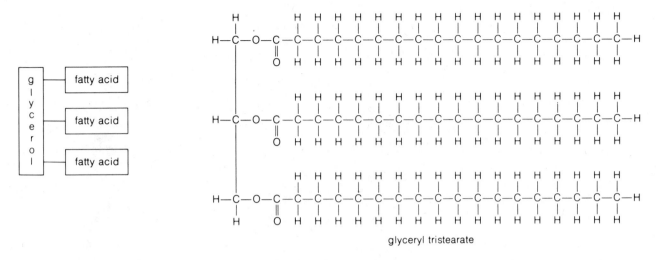

glyceryl tristearate

Figure 2-11 The basic structure of triglycerides (fats, oils, and some waxes), formed by interaction between glycerol and three fatty acids. The three fatty acids in the interaction may be the same or all different.

structure (shaded groups, Fig. 2-15b). Although sterols are almost completely hydrophobic, the single hydroxyl group gives the end containing it a slightly polar or hydrophilic character. As a result, the sterols also tend to orient in arrangements that satisfy their dual solubility properties.

The sterol shown in Figure 2-15b and c, *cholesterol*, is an important part of the surface membranes of all animal cells; similar sterols occur in plant cell membranes. Deposits derived from cholesterol (a normal component of blood) also form a major part of the material deposited inside arteries in the disease *atherosclerosis* (hardening of the arteries).

Other steroids are important as *hormones* in animals. Although steroid hormones occur in only trace amounts in animal tissues, they have regulatory effects on cells far out of proportion to their low concentrations. The male and female sex hormones of humans and other animals are steroid hormones; so are the hormones of the adrenal cortex that regulate cell growth and activity. Some toxic steroids also occur as cell poisons in the venoms of toads and other animals.

Proteins

Proteins are large, complex molecules that carry out three vital functions in living organisms: (1) as structural mole-cules, providing much of the framework of cells; (2) as enzymes, the biological catalysts that speed the rate of cellular reactions; and (3) as molecules that provide movement to cells and cell parts. Proteins also function in less central roles, as in the protein-based hormones of animals (insulin is a prime example) and as antibodies.

The protein molecules carrying out these diverse functions are all basically similar in structure. All consist of one or more long, unbranched chains of subunits. The individual subunits or links in the chains are the *amino acids*. Although only 20 different amino acids link initially in different combinations to make proteins, these may be modified to other forms after proteins are synthesized. As a result, as many as 150 different amino acids may occur in natural proteins. The total number of amino acids in different proteins may range from a minimum of 30 or so to giant molecules containing more than 50,000 amino acids. For any given protein, the amino acid sequence shows little or no variation; this sequence determines the properties and chemical activity of a particular type of protein molecule.

An almost endless variety of proteins is possible. At any point in an amino acid chain, any one of the 20 amino acids may be present as a link in either modified or unmodified form. In even the smallest proteins, with roughly 50 amino acids, this allows 20^{50} different sequences to be made without modification of individual amino acids. This incomprehensibly huge number is equivalent to one unique protein for every gram of matter in the universe!

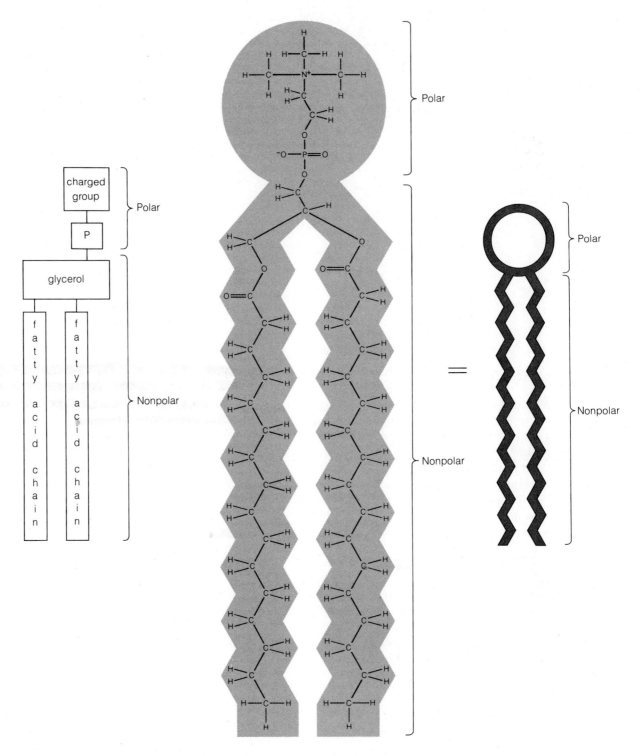

Figure 2-12 An important membrane phospholipid, phosphatidyl choline. The diagram at the right is used to depict a phospholipid molecule. The circle represents the polar end of the molecule, and the zigzag lines the nonpolar carbon chains of the fatty acid residues.

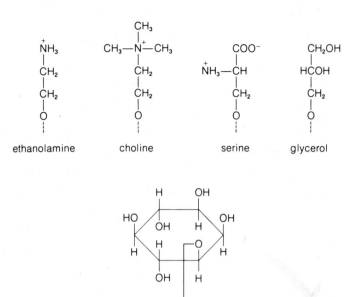

ethanolamine   choline   serine   glycerol

inositol

Figure 2-13 Chemical groups linked to glycerol by the phosphate group of phospholipids.

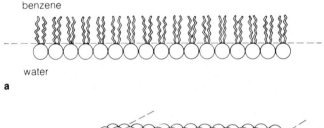

benzene

water

a

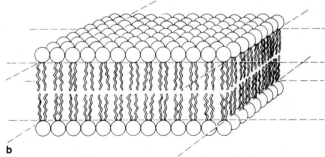

b

Figure 2-14 (**a**) Phospholipid molecules introduced between the layers formed when benzene is layered on water orient with the nonpolar tails extending into the benzene and their polar heads facing the water. (**b**) Phospholipid molecules surrounded by water assemble into bilayers, with their nonpolar tails associated together in the bilayer interior, and their polar heads facing the surrounding water molecules.

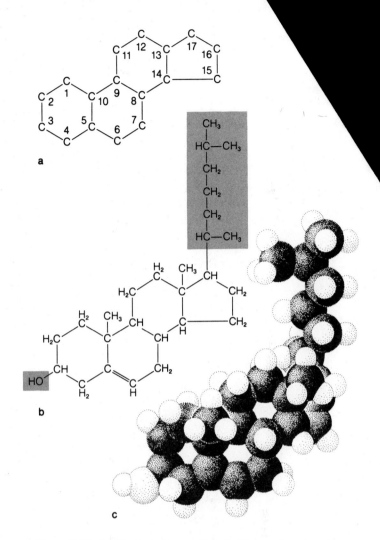

a

b

c

Figure 2-15 (**a**) The typical arrangement of four carbon rings in a steroid molecule. (**b**) A sterol, cholesterol. Sterols have a long nonpolar side chain at one end of the molecule and a single —OH group at the other end (shaded). The —OH group makes its end of the cholesterol molecule slightly polar. The remaining parts of the molecule are nonpolar.

The Amino Acid Subunits of Proteins Nineteen of the 20 amino acids initially assembled into proteins are based on the same structural plan (Fig. 2-16). These amino acids have a central carbon atom, called the *alpha* (α) carbon, to which are attached, one on either side, an amino (—NH₂) group and a carboxyl (—COOH) group. One of the remaining bonds of the central carbon is linked to a hydrogen atom, giving the structure:

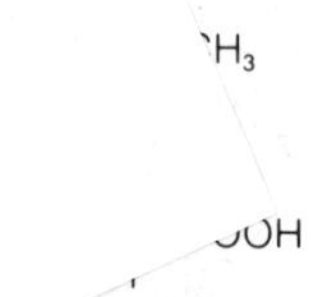

Figure 2-16 The 20 amino acids used by cells to assemble proteins. The side chains are shown above the dashed line, and the alpha carbon, with its attached amino (—NH$_2$) and carboxyl (—COOH) groups below. One amino acid, proline, differs from this structural plan (see text).

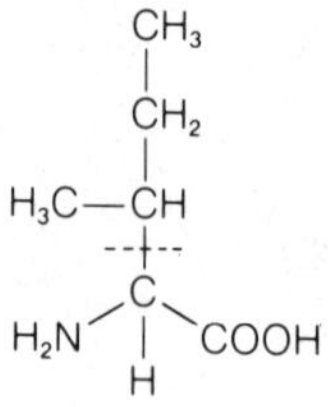

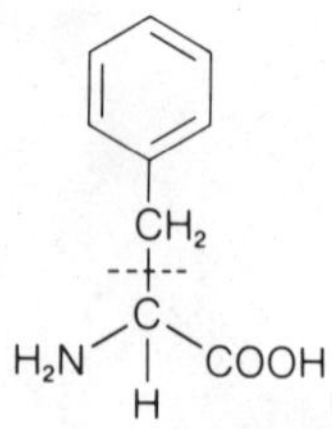

valine
Val

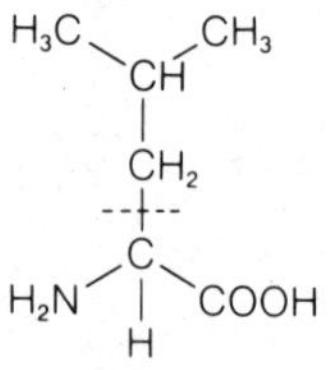

leucine
Leu

isoleucine
Ile

phenylalanine
Phe

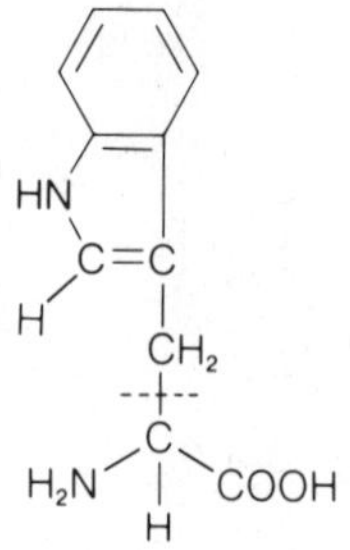

tryptophan
Trp

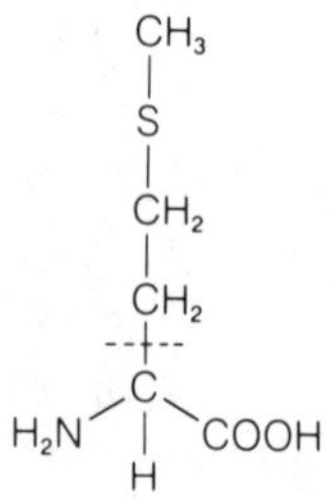

methionine
Met

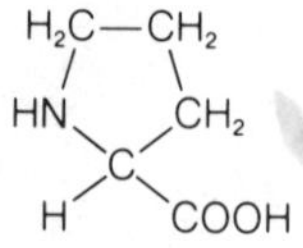

proline
Pro

Uncharged Polar Amino Acids

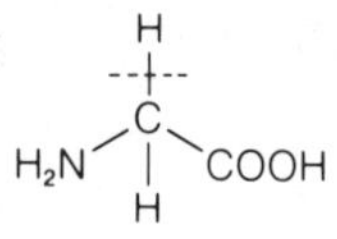

glycine
Gly

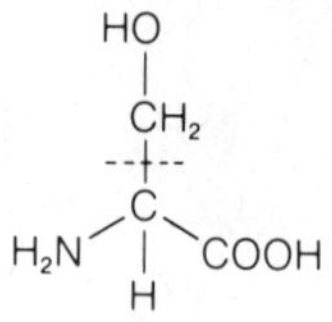

serine
Ser

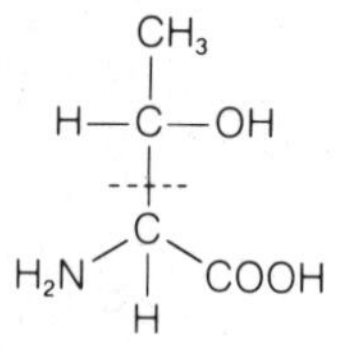

threonine
Thr

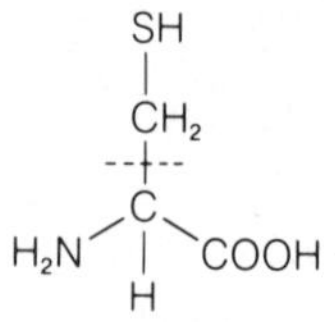

cysteine
Cys

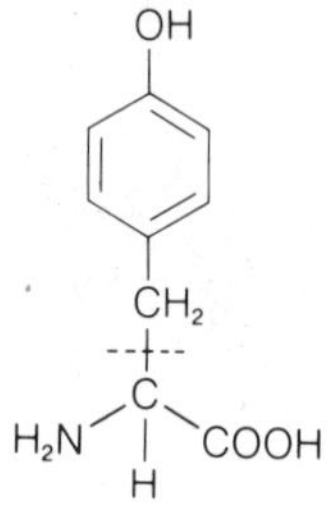

tyrosine
Tyr

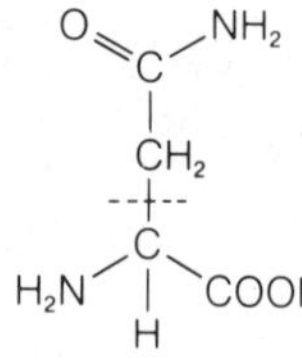

asparagine
Asn

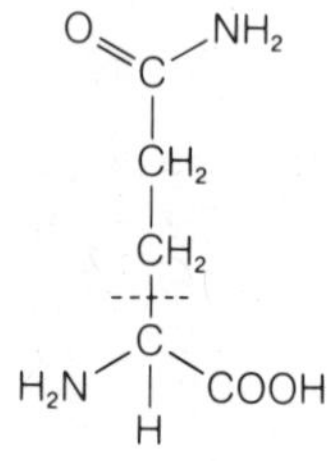

glutamine
Glu

Negatively Charged (Acidic) Amino Acids

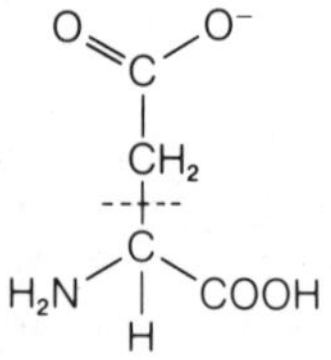

aspartic acid
Asp

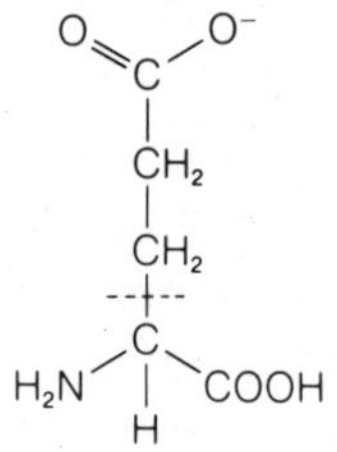

glutamic acid
Glu

Positively Charged (Basic) Amino Acids

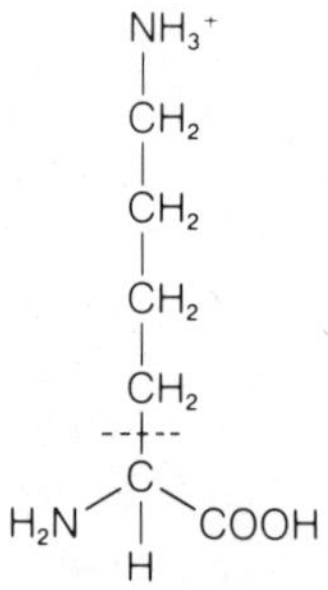

lysine
Lys

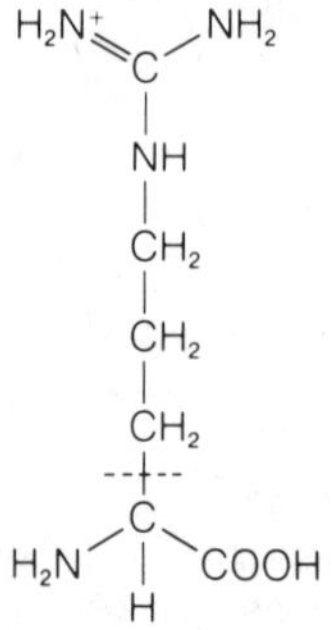

arginine
Arg

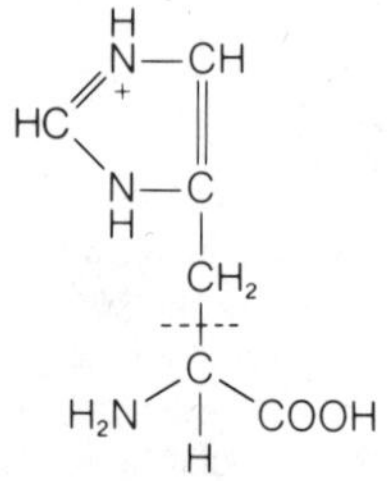

histidine
His

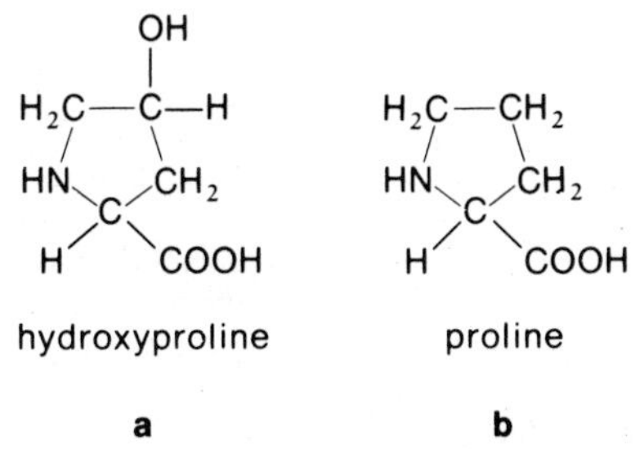

Figure 2-17 (**a**) Hydroxyproline, a modified amino acid occurring in the protein collagen. Modified amino acids are produced in cells by chemical conversion of the 20 amino acids used in initial assembly of proteins. (**b**) The unmodified form of proline.

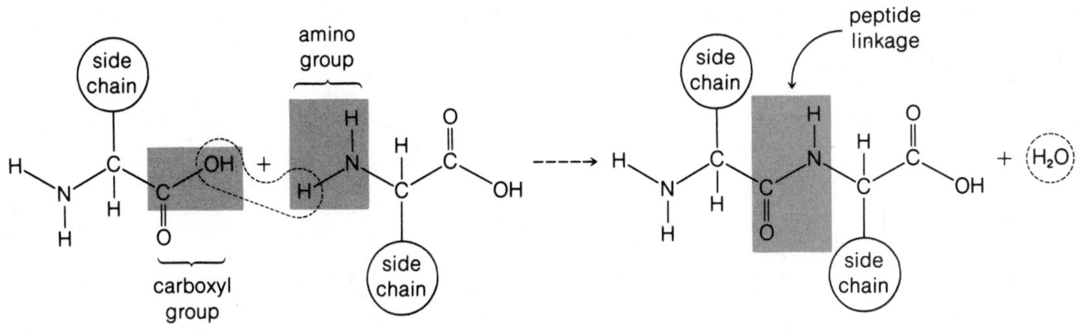

Figure 2-18 Reaction between the carboxyl and amino groups of two amino acids to form a dipeptide. The reaction takes place on ribosomes in cells.

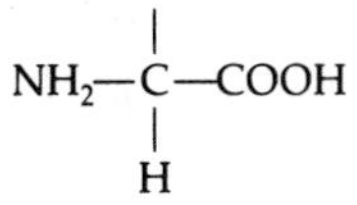

common to these amino acids. The fourth bond of the alpha carbon may be attached to any one of 19 different side chains, ranging in complexity from a single hydrogen atom in the simplest amino acid, *glycine*, to long carbon chains or rings in other amino acids. Some of the more complex side chains contain oxygen, nitrogen, or sulfur atoms in addition to carbon and hydrogen. The remaining amino acid, *proline*, an exception to these general rules, is based on a ring structure including the central carbon atom (see Fig. 2-16).

The types and configuration of the atoms in the side chain give each amino acid special properties, including differences in polarity, charge, and acidic or basic reaction. A number of amino acid side chains have groups capable of readily entering into reactions with atoms and molecules located outside the protein or elsewhere on an amino acid chain. These reactive groups include amino (—NH₂),

hydroxyl (—OH), carboxyl (—COOH), and sulfhydryl (—SH) groups.

Modified amino acids are common in natural proteins. For example, the protein *collagen*, a structural molecule that is abundant in animals, contains quantities of the amino acid *hydroxyproline* (Fig. 2-17), a form modified from the original proline by addition of a hydroxyl group. These additional amino acids, as noted, are produced by chemical modifications of the original 20 amino acids after the initial synthesis of a protein.

The Peptide Bond and Primary Structure Amino acids can be readily linked together into chains of two or more units. The covalent bond linking the units together, the *peptide bond* or *peptide linkage*, is produced by a reaction between the amino group of one amino acid and the carboxyl group of a second amino acid (Fig. 2-18). For each peptide bond formed, the equivalent of one molecule of water is released as a byproduct.

The linkage of two amino acids into a short chain produces a *dipeptide*. Chains of more than two units are called *polypeptides*. The dipeptide shown in Figure 2-18 retains an

Table 2-4 Protein Structural Nomenclature

Structural Level	Definition	Primary Bonds Holding Structure Together
Primary structure	Sequence of amino acids in a protein	Covalent bonds of the peptide linkages
Secondary structure	Folding of the amino acid chain into patterns such as the alpha helix, random coil, and pleated sheet, which are subparts of a whole protein molecule	Hydrogen bonds
Tertiary structure	Complete folding pattern of an entire protein molecule containing a single polypeptide chain	Hydrogen bonds, covalent bonds of disulfide linkages, polar and nonpolar associations
Quaternary structure	Complete folding pattern when a protein molecule contains two or more peptide chains	Same as tertiary structure

amino group at one end and a carboxyl group at the other. Thus the end groups of a dipeptide are the same as those of a single amino acid and can enter into the formation of additional peptide linkages at either the carboxyl or the amino end. In nature, additional amino acids are added only to the —COOH end of a growing polypeptide chain. This means that a finished protein has an —NH₂ group exposed at its "front" end and a —COOH group at its "rear" end. The particular sequence in which amino acids occur in a finished chain is called the *primary structure* of a protein (see Table 2-4). By virtue of the pattern in which proteins are assembled, the primary structure of a protein begins with the first amino acid at the —NH₂ end and progresses in sequence to the last amino acid at the —COOH end.

Secondary Structure in Proteins

The Alpha Helix and the Random Coil The amino acid chain inside proteins often twists into a regular spiral called the *alpha helix* (Fig. 2-19). The spiral, discovered by Linus Pauling and Robert Corey of the California Institute of Technology, is a relatively rigid, rodlike structure containing 3.6 amino acids in each of its turns. The amino acids are placed so that the side chains extend outward from the spiral. The amino acid chain is held in the alpha helix by a regularly spaced series of hydrogen bonds formed between hydrogen and oxygen atoms along the chain.

The alpha-helical portions of protein molecules resist bending and deformation; where sharp bends occur in folded backbone chains the alpha helix gives way to a much less regular arrangement known as a *random coil*.

Formation of an alpha helix or a random coil in any particular region of the chain is largely dependent on the amino acids present. Some amino acids allow a stable helix to form; others reduce the stability or even interrupt the helix. Most significant of the latter kind of amino acid is proline, which has no hydrogen available for the formation of stabilizing hydrogen bonds. The unusual ring structure of proline also prevents the chain from twisting into the alpha helix at points containing this amino acid. The helix is therefore interrupted at any position occupied by proline. The myoglobin molecule diagrammed in Figure 2-22 shows the distribution of rigid alpha-helical spirals and the sharply bent, flexible regions containing random coils in this protein.

The Pleated Sheet Pauling and Corey also discovered a second regular arrangement of the amino acid chains of proteins which they termed the *pleated sheet* (Fig. 2-20). In the pleated sheet arrangement, the amino acid chain is extended in a zigzag conformation instead of a coil as in the alpha helix. Further, the zigzag chains twist back (dotted lines in Fig. 2-20) so that a region is formed in which segments of the chain are arranged parallel to each other. The parallel regions are held together in a plane by hydrogen bonds formed between adjacent segments. The polypeptide chain may twist back on itself so that the parallel segments run in opposite directions, as in Figure 2-20, or in the same direction. Pleated sheet conformations occur in many natural proteins; they are particularly important in the structure of antibody molecules, proteins that can recognize and bind to foreign substances entering the body in vertebrate animals.

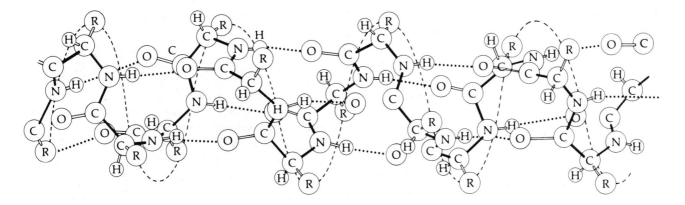

Figure 2-19 The alpha helix. The backbone of the amino acid chain is held in a spiral (dashed line) by hydrogen bonds (short dotted lines) formed at regular intervals. The small spheres labeled "R" represent the side chains of amino acids in the alpha helix. Redrawn from "Proteins" by Paul Doty, *Scientific American* 197 (1957): 173. Copyright © 1957 by Scientific American, Inc. All rights reserved.

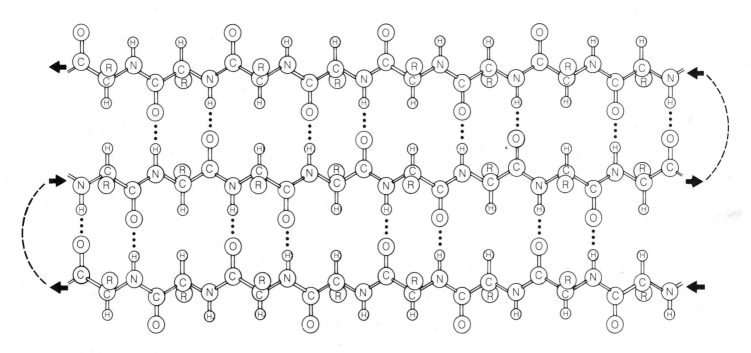

Figure 2-20 The arrangement of amino acids in a pleated sheet. The small spheres labeled "R" represent the amino acid side chains, which in this view extend above and below the page.

The amino acid chain of different proteins, depending on the sequence of amino acids present, may twist into the alpha helix, form segments of random coil, or assemble into pleated sheets. The amount and arrangement of these three folding conformations is termed the *secondary structure* of proteins.

Tertiary and Quaternary Structure Primary and secondary structure provide only a part of the total folding arrangement of protein molecules. The distribution of alpha helix, random coil, and pleated sheet arrangements causes the polypeptide chain of a protein to fold into a three-dimensional arrangement that is as significant for the function of the protein as the sequence of amino acids itself. Some proteins, when folded in their natural three-dimensional form, are spherical or globular; others are ellipsoids or long fibers. This overall three-dimensional folding conformation taken on by the polypeptide chain is called the *tertiary structure* of a protein.

A number of complex proteins, such as the hemoglobin molecule that carries oxygen in vertebrate blood, are built up from several polypeptide chains held together by combinations of hydrogen bonds, covalent bonds, and polar and nonpolar associations. Hemoglobin consists of four individual polypeptide chains; some proteins may have as many as ten or twelve. The pattern in which the individual chains of a multichain protein are held together is called the *quaternary structure* of a protein.

Proteins thus have four levels of structure. The sequence of amino acids in a protein constitutes its primary structure. The degree to which the amino acid chain is twisted into arrangements such as the alpha helix, random coil, or pleated sheet defines the secondary structure of a protein. Folding of the chain upon itself produces the tertiary or three-dimensional structure of a protein. Most commonly the three-dimensional structure of proteins is roughly spherical or elliptical. Quaternary structure refers to the number and arrangement of multiple polypeptide chains within a complex protein containing more than one chain.

Proteins are held in their final globular or fibrous shapes by a number of bonds and attractions in addition to the peptide linkages between amino acids. Among these are hydrogen bonds, electrostatic attractions between positively and negatively charged groups, associations between hydrophilic and hydrophobic amino acid side chains, and an interchain covalent bond called a *disulfide (—S—S—) linkage*.

Although hydrogen bonds are individually relatively weak, many hydrogen bonds can be formed in proteins since each peptide linkage has hydrogen and oxygen atoms capable of forming them. The total effect of hydrogen bonding all along the chain is to restrict the twisting of the amino acid backbone and produce a limited number of ways in which a protein of given sequence may fold.

Disulfide linkages (Fig. 2-21) are produced by covalent bonding between the —SH (sulfhydryl) groups of cysteine amino acid side groups located at different points in the backbone chain. Where these links form, they anchor the amino acid chain of a protein in a folded position.

The net effect of hydrogen and disulfide bonds, electrostatic attractions, and hydrophilic and hydrophobic associations is to establish a unique folding conformation for each protein with a particular amino acid sequence. This total three-dimensional shape is called the *folding conformation* of a protein. The three-dimensional folding conformation of a relatively simple protein, *myoglobin*, is shown in Figure 2-22.

The folding conformation of a protein is flexible and changes readily in response to variations in temperature, pH, and the binding of other molecules, ions, or chemical groups. The resulting changes in shape, called *conformational changes*, are vital to the activities of proteins as enzymes and motile molecules. Extreme changes in temperature or pH seriously disturb the internal bonds and associations holding proteins together, causing the molecules to unfold or refold into random shapes. This effect, called *denaturation*, results in loss of the functional activity of the protein. For example, temperatures above about 50°C break most or all of the hydrogen bonds within proteins, causing their polypeptide chains to unfold into random, extended shapes, with total loss of functional activity. Denaturation of proteins at higher temperatures provides the primary reason why almost no living organisms can tolerate temperatures above 45 to 55°C.

Combinations Between Proteins and Other Substances By virtue of the different reactive groups carried on their amino acid side chains, proteins may bind to a variety of ions, chemical groups, and organic molecules within cells. Inorganic ions, particularly sodium, potassium, chloride, calcium, magnesium, phosphate, copper, and iron atoms are often found in combination with enzymes. Vitamins also combine with proteins, particularly with enzymes active in the reactions capturing energy for cellular use. Proteins in combination with carbohydrates form glycoproteins; these complexes are frequently found at the surfaces of cells or in the gelatinous or sticky secretions of many cell types. Nonpolar amino acids are important in the hydrophobic associations between proteins and lipids that form lipoproteins; these complexes are basic structural units in membranes. Last but not least is the electrostatic binding between proteins and nucleic acids that forms the *nucleo-*

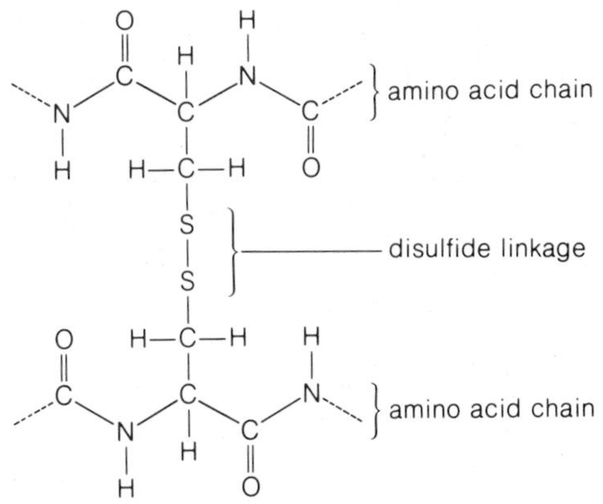

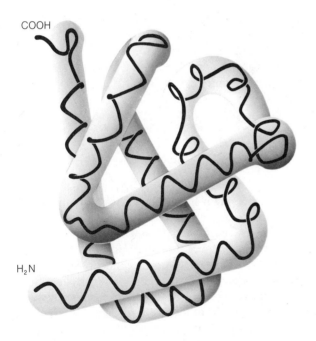

Figure 2-21 A disulfide linkage, formed by an interaction between sulfhydryl (—SH) groups on cysteine amino acids located in different parts of an amino acid chain. Disulfide linkages provide one of the bonds holding protein molecules in a three-dimensional folding conformation.

Figure 2-22 The three-dimensional folding of the amino acid chain in the protein myoglobin. The solid line shows the folding arrangement of the amino acid backbone of the molecule.

proteins, the major constituents of ribosomes and the chromatin of eukaryotic cell nuclei.

Nucleotides and the Nucleic Acids, DNA and RNA

The two nucleic acids, *deoxyribonucleic acid (DNA)* and *ribonucleic acid (RNA)*, are the informational molecules of all living organisms. Both of these polymer-type molecules are long structures built up from chains of repeating monomers or building blocks, the *nucleotides.* The sequence of nucleotides in nucleic acid molecules forms a code that stores and transmits the cellular information required for cell growth and reproduction. Individual nucleotides also transfer units of energy or important reactants from one system to another in cells.

Nucleotides Each nucleotide (Fig. 2-23) consists of (1) a nitrogen-containing base, (2) a 5-carbon sugar, and (3) one or more phosphate groups, all linked together by covalent bonds. The nitrogenous bases, called *pyrimidines* and *purines* (Fig. 2-24), are ring-shaped molecules containing both carbon and nitrogen atoms. (Pyrimidines contain one carbon-nitrogen ring, and purines contain two rings.) Three pyrimidine bases—uracil (U), thymine (T), and cytosine (C)—and two purine bases—adenine (A) and guanine (G)—are the most common bases in nucleotides.

These nitrogenous bases link covalently in nucleotides to one of two 5-carbon sugars, either *ribose* or *deoxyribose.* The two sugars differ only in the chemical group bound to the carbon at the shaded position in Figure 2-23. Ribose has an —OH group at this position, and deoxyribose a single hydrogen. The ribose or deoxyribose sugars of nucleotides are also linked to a chain of 1, 2, or 3 phosphate groups, forming the mono-, di-, and triphosphates of the sugar-base unit as in the AMP, ADP, and ATP family of molecules shown in Figure 2-23. The two forms of the ATP nucleotide, one with the ribose and one with the deoxyribose sugar, are written as ATP (with ribose) and dATP (with deoxyribose), to indicate which form is meant (see also Information Box 9-1).

The names used for the nucleotides can be confusing. The term nucleo*tide* refers to a unit containing all three subunits: a nitrogenous base, a 5-carbon sugar, and one or more phosphates. Cells also contain units consisting of only the base and sugar without any phosphates. These units are called nucleo*sides,* and are given names derived from the nitrogenous base present. The base-sugar complex containing adenine and ribose, for example, is called

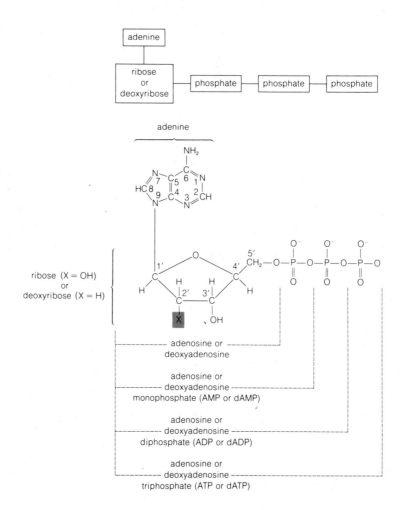

Figure 2-23 Structural plan of the nucleotides (see text).

adenosine; if deoxyribose is the sugar in the complex, it is called *deoxyadenosine* (the names derived for the remaining base-sugar complexes are listed in Fig. 2-23). The nucleotides are named by considering them as nucleosides with added phosphates. The adenine-ribose complex with one phosphate, for example, is called *adenosine monophosphate;* with two phosphates, the complex is called *adenosine diphosphate;* with three, *adenosine triphosphate.* One further convention is worth noting. Frequently, the carbons in the sugars are written with primes (1', 2', 3', and so on) to distinguish them from the carbons in the bases, which are written without primes (see Fig. 2-23).

DNA and RNA   Nucleotides link together into long chains to form the two nucleic acids DNA and RNA (Fig. 2-25). The chains of nucleotides in DNA contain the sugar deoxyribose linked to one of four bases—adenine, thymine, guanine, or cytosine (A, T, G, C; see Fig. 2-24). Each nu-

cleotide in the chain making up RNA contains the sugar ribose and one of the four bases adenine, uracil, guanine, or cytosine (A, U, G, C). Thus DNA and RNA differ in the sugar present (ribose or deoxyribose) and in the presence of either uracil in RNA or thymine in DNA. As in the amino acid side groups of proteins, the nitrogenous bases of the nucleotides may be modified chemically to other forms after initial synthesis of a nucleic acid. These *modified bases* are especially common in some types of RNA molecules.

DNA exists in cells in the form of a *double helix* containing two nucleotide chains twisted around each other in a regular, double spiral (see Fig. 9-4). The double helix is held together by hydrogen bonds and by hydrophobic associations between the nitrogenous bases of opposite chains in the spiral. RNA exists as single, rather than double nucleotide chains in living cells. However, RNA molecules may fold back on themselves to form extensive

25. What is the difference bet

26. How do DNA and RNA

27. What are the primary fu
acids?

Supplement 2-1:
Acids, Bases, and pH

Acids and Bases

Most inorganic substances
act as either *acids* or *base*
bases are substances that
drogen ions (H$^+$) and h
Water always contains bo
proportion of the water m
arate to produce H$^+$ and (

$$H_2O \rightarrow$$

Acids are substances that
lease additional hydrogen
tive concentration of H$^+$
bind H$^+$ ions, or release
dissolved in water, and tl
centration of H$^+$ ions.

Acidity and pH   The rel
OH$^-$ ions in a water solu
solution. The degree of aci
reactivity of many organi
solved in water and modif
protein molecules.

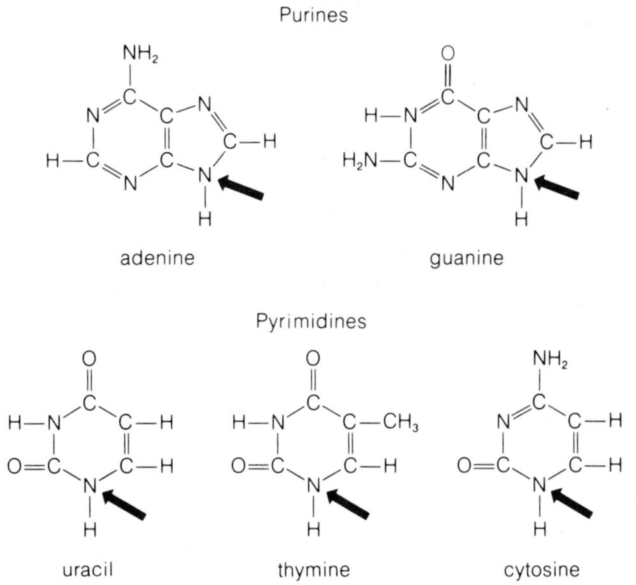

Purines

adenine   guanine

Pyrimidines

uracil   thymine   cytosine

Figure 2-24   The purine and pyrimidine bases of nucleic acids and nucleotides. The arrows indicate where the base links to ribose or deoxyribose sugars in the formation of nucleotides.

regions with double-helical structure (see Fig. 9-11). *Hybrid double helices*, containing one DNA and one RNA nucleotide chain wound into a double spiral, also occur as temporary intermediates during RNA synthesis in living cells.

Suggestions for Further Reading

Davidson, J. N. 1972. *The biochemistry of nucleic acids.* Academic Press: New York.

Dickerson, R. E. and I. Geis. 1969. *The structure and action of proteins.* Harper & Row: New York.

Dickerson, R. E. and I. Geis. 1976. *Chemistry, matter and the universe.* W. A. Benjamin: Menlo Park, Calif.

Lehninger, A. L. 1975. *Biochemistry.* 2nd ed. Worth: New York.

Stryer, L. 1981. *Biochemistry.* 2nd ed. W. H. Freeman: San Francisco, Calif.

White, E. H. 1964. *Chemical background for the biological sciences.* Prentice-Hall: New York.

For Further Information

Questions

1. What determines the chemical activity of an atom?

2. What is an electrostatic bond? How are electrostatic bonds formed?

3. What is a covalent bond? How are covalent bonds formed? How are covalent bonds represented in molecular diagrams?

4. What are polar and nonpolar molecules? What condition produces polarity?

5. What do the terms hydrophobic and hydrophilic mean? How is the property of being hydrophobic or hydrophilic related to polarity?

6. How is polarity important in molecular and cellular structure?

7. What is a hydrogen bond? What conditions are necessary for the establishment of a hydrogen bond? In what ways are hydrogen bonds related to polarity?

8. Diagram a carboxyl, hydroxyl, sulfhydryl, aldehyde, ketone, and phosphate group.

9. What structural features do carbohydrates have in common? What is a monosaccharide? A disaccharide? A polysaccharide?

10. What is a polymerization? List several polymerization reactions of importance in biological systems.

11. What is the difference between α- and β-glucose? Does this difference have any biological significance?

12. What are the primary differences between cellulose, plant starch, and animal starch (glycogen)?

13. Diagram a saturated and an unsaturated fatty acid. What are the characteristics of cellular fatty acids?

14. What is the basic structural plan of a neutral lipid? Of a phospholipid?

3

Energy and Enzymes in Biological Reactions

The assembly of and interactions between the carbohydrates, lipids, proteins, nucleic acids and other organic molecules of living organisms that are responsible for the activities of life require a constant input of energy. These life processes also require enzymes, the protein molecules that speed and regulate the rate of biological reactions and allow life to proceed at ordinary temperatures. The right enzymes are made in cells at the right place and time, and in the right numbers, to increase the rate of the biochemical interactions required for a given type of growth or life activity.

The enzyme-catalyzed reactions making up the activities of life are not unique to cells or to the living condition. They can proceed outside cells, in a test tube, provided the necessary physical and chemical conditions are duplicated. Not even the enzymes normally catalyzing the reactions are necessary for the reactions to proceed. The primary role of enzymes is simply to increase the rate of reactions that would take place, however slowly, without the presence of an enzyme. All of this is because biochemical reactions must obey the same chemical and physical laws operating anywhere in the universe, whether inside cells or in the outside world.

To understand how enzymes increase the rate of biological reactions we must first understand the basic, universal rules governing the probability and direction of chemical changes. These rules, or laws, are principles drawn from *thermodynamics*, the science dealing with energy changes in all collections of matter.

The Nature of Energy and Its Changes in the Cell and Elsewhere

What is energy? Although everyone has an intuitive grasp of the meaning of the word, energy is difficult to define precisely. We cannot perceive energy itself; we can only observe its effects on physical objects in the environment. Because these effects often involve movement of physical objects—whether electrons, atoms, molecules, or larger pieces of matter—energy is often defined as "the capacity to do work." Although not entirely adequate, this definition is useful and for most purposes does not violate the truth.

There are two basic principles, called the *first* and *second laws of thermodynamics*, that describe energy changes in reactions of all kinds and specify whether such changes are probable and will proceed on their own. Reactions that are probable are called *spontaneous* reactions in thermodynamics. This usage of the word *spontaneous* does not carry its more common connotation that the change is instantaneous: spontaneous reactions, in thermodynamics, are simply reactions that will "go." However, they may take place at any rate from unmeasurably slow to practically instantaneous.

Spontaneous Reactions and the First and Second Laws of Thermodynamics

The first law of thermodynamics is a formal statement of the common observation that "you can't get something for nothing." The law was developed partly through studies of the many unsuccessful attempts to build perpetual motion machines. Machines of this type, once set in motion, would run indefinitely without any input of energy. However, all such machines, no matter how ingeniously they are constructed, eventually run down because of friction. Some of the initial push of energy used to start the machine is lost as heat due to friction between the working parts. This heat flows from the machine into its surround-

glutamic acid $+ NH_3 \rightarrow$ glutamine $+ H_2O$ (3-10)

requires an energy input of 3400 cal/mole to make it proceed to the right. In cells, this uphill reaction is coupled to the breakdown of ATP to ADP, to produce an overall downhill reaction that releases free energy and proceeds spontaneously to the right:

glutamic acid $+$ ATP $+ NH_3 \rightarrow$ glutamine $+$ ADP $+ HPO_4^{2-}$ (3-11)

Now the total products, glutamine, ADP, and phosphate, have less energy content and greater entropy than the total reactants, which include glutamic acid, ATP, and NH_3. Free energy amounting to 3900 cal/mole is released as each gram molecular weight of reactants is converted to products.

In coupled reactions, therefore, an uphill reaction requiring energy is coupled to the breakdown of ATP to produce an overall reaction that proceeds downhill, and the reacting system becomes thermodynamically probable. Almost all of the chemical, electrical, and mechanical work of the cell, and all the other energy-requiring activities of growth, reproduction, movement, and responsiveness are made energetically favorable in this way (Fig. 3-2).

The Role of Enzymes in Biological Reactions

Enzymes and Enzymatic Catalysis

The laws of thermodynamics, as described up to this point, apply to all chemical reactions, whether they occur inside cells or not. None can violate the rule that they must proceed to a level of minimum energy and maximum entropy, or a favorable balance between the two factors. What effects do enzymes have, then, on the biochemical reactions taking place inside cells? The answer is simply that enzymes greatly increase the *rate* at which spontaneous reactions take place. (Rate means the number of reactant molecules that are converted to products per unit of time under constant conditions of concentration, temperature, and pressure.) Enzymes cannot make a reaction go if it will not already proceed spontaneously without the enzyme.

Many reactions, although spontaneous, proceed so slowly that their rate at room temperature is essentially

and products, a situation producing r maximum entropy with respect to ences. This balance point is called the the reaction. At the equilibrium point ture, equal numbers of molecules from reactants to products, and ba reactants. If, at this point, more reac reaction will proceed to the right u point is reached again. Conversely, added, the reaction will move in the until the system is again balanced. termed *reversible* and are written with

reactants $\rightleftharpoons$ produ

Most biological reactions are of this t

The position of the equilibrium reaction depends on the difference in entropy between reactants and produ is large, so that the products contain ergy and greater entropy than the re a large release of free energy as the r equilibrium point will lie far to the which most of the reaction mixture at sist of products. If the energy and e tween reactants and products is sm point will lie further to the left, and of reactants will be present in the mi

Since most biological reactions ar be written either way, with either pr the left hand side of the reaction. Fo ological reaction of importance, an a is broken down by removal of an a verted to glutamic acid:

glutamine $+ H_2O \rightleftharpoons$ glutami

The reaction proceeds far in the dire releases free energy as it takes plac also synthesize glutamine from glut nia by running this reaction in rever

glutamic acid $+ NH_3 \rightleftharpoons$ gluta

As written, very little product is exp librium, when the reaction is written far to the left. But cells need glutam regular basis. How do cells accompl tion without violating the first thermodynamics?

ings and is lost. To keep the machine running, the lost energy must be constantly replaced by additional pushes of energy from the outside. Otherwise, the operator would be getting something for nothing, the "something" in this example being the energy needed to replace the energy lost as frictional heat.

Careful measurements of the heat flowing from such machines show that as the machines come to a stop, the total amount of heat lost due to friction is exactly equivalent to the amount of energy added in the initial push used to start them. These observations, and similar observations made in other energy changes, led to the more formal statement of the first law of thermodynamics. This law affirms that in any process involving an energy change (as in the change from mechanical to heat energy in a so-called perpetual motion machine) the total amount of energy remains constant. Or in other words, in such changes, *energy can neither be created nor destroyed.* The first law means that the energy in the universe remains constant. Although it may be transformed from one form to another, the total quantity remains the same.

In thermodynamics, a collection of molecules undergoing change, such as a perpetual motion machine in the process of running down or a chemical reaction in progress, is called a *system.* A system can be defined to include any reacting molecules of interest. Everything outside the defined system is called the *surroundings.* In undergoing a change, systems go from an *initial state* to a *final state.* In these terms, the first law of thermodynamics states that in undergoing a change from the initial state at the beginning of a reaction to the final state at the end of a reaction the *total amount of energy of the system and its surroundings remains constant.*

It is useful to consider what total energy means when it applies to a collection of molecules. Part of the energy content of any collection of molecules at temperatures above absolute zero ($-273°C$) is reflected in their constant rotation, vibration, and movement from one place to another. This part of the total energy, reflected in molecular movement, is the *kinetic energy* of the collection of molecules. The second part depends on the energy contained in the arrangement of chemical bonds in the molecules. This second component is called *potential energy.*

The two energy components are interchangeable. For example, an unlit match held at room temperature contains considerable potential energy in the form of the arrangement of chemical bonds in its wood and the phosphorus of its tip, but only moderate kinetic energy. Once struck, much of the potential energy is transformed into increases in the kinetic motion of molecules of the match and its immediate surroundings, which we can de-

tect as an increase in temperature. Some energy also flows from the burning match as heat and light.

What does the first law governing energy content have to do with the probability and direction taken by chemical reactions? From our own intuition, based on observations of ordinary events within our common experience, we would expect that reactions in which the energy content of the system in its initial state is higher than the final state would be spontaneous. This is illustrated, for example, by a system represented by a rock placed on the side of a steep hill. We know that that rock will roll spontaneously downhill until it comes to a stop at the bottom. At the initial state, when placed on the side of the hill, the rock has a quantity of potential energy equivalent to its distance from the center of the earth. On reaching its position of rest at the final state at the bottom of the hill, the rock is closer to the center of the earth and has a smaller amount of potential energy. The difference between the initial and final states is lost as heat energy to the surroundings. From this type of observation, we would therefore expect that spontaneous chemical reactions would be those in which the total chemical energy of the products is smaller than that of the reactants. That is, we expect chemical reactions to proceed spontaneously to a state in which the molecules of the system have *minimum energy content.*

For many spontaneous processes, including chemical reactions, the assumption that the molecules of a system always change toward a state of minimum energy content (potential + kinetic) is correct. However, it is not difficult to find examples in which the opposite turns out to be the case. For example, consider the system represented by the collection of water molecules in a block of ice. As long as the temperature of the surroundings is above 0°C, the ice will melt. This occurs spontaneously, even though the ice absorbs heat and the energy content of the system at the final state is higher than at the initial state. The energy required for this increase is absorbed from the surroundings, so that the total energy of the system and its surroundings still remains constant.

From this it is obvious that the energy content of the system is insufficient by itself to predict whether a process is spontaneous. What is missing? Once again, common experience comes to the rescue. It is our common observation that, as spontaneous changes take place, the objects around us generally tend to get out of order rather than assuming more ordered arrangements. This simple observation is the basis for the second law of thermodynamics, which states the same thing in more formal terms: In any process involving a spontaneous change from an initial to a final state *the disorder or randomness of the system and its*

surroundings always increases. If the syst
entire universe, the second law mea
take place anywhere, the disorder o
stantly increases. In thermodynamics
ness is given the special name *entropy*

Combining together the first and
modynamics allows us to state the con
to determine whether a reaction is lik
tions that go to a condition of minimum
disorder (maximum entropy) are mo
tend to proceed on their own.

In proceeding to the final state,
acting system may give off energy to i
ergy given off in this way is called *free*
meaning simply that the energy is ava
living organisms, the primary work a
energy is the chemical work of mol
biochemical interactions.

It is possible for reactions to pro
toward conditions in which, althoug
of the system decreases, randomnes
conversely, although randomness in
content of the system may actually r
initial to final states. In such reaction
ergy content and entropy in the sys
tendencies, one tending to favor th
other opposing it. In these cases, th
ing tendencies within the system det
reaction will proceed.

This is the case with the water m
melting ice. When ice melts, the ener
tem increases, which is an unfavo
spontaneous reaction. However, the
tropy increase, reflecting the chang
cules from the highly ordered arrang
to the more random condition in liqu
between the two opposing factors is
to make the change proceed spontar

The first and second laws still a
in which the change in energy conte
system act as opposing tendencies
summarize the total change expected
surroundings. If energy flows from
the system, as in a block of melting
content of the system and the surro
the same, since energy can neithe
stroyed. The same thing is true in
amount of order increases (entro
spontaneous changes. This is the ca
tems undergoing changes such as
into a tree, or a fertilized egg into an

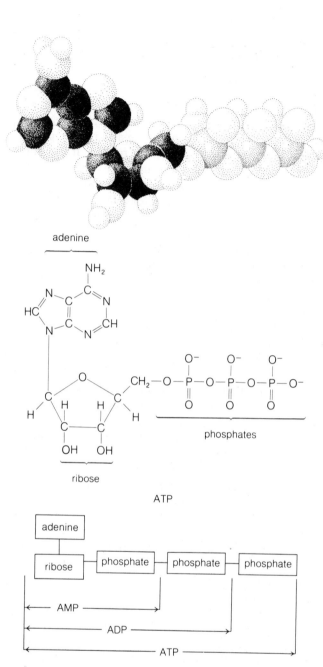

Figure 3-1 Top: The ATP molecule, adenosine triphosphate.
Bottom: Sequential removal of the phosphate groups produces the
closely related molecules adenosine diphosphate (ADP) and adenosine monophosphate (AMP).

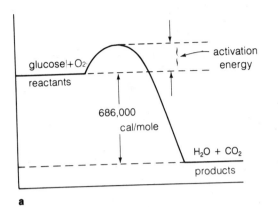

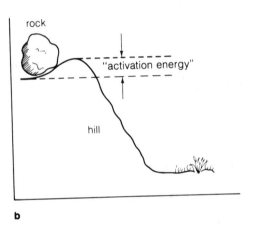

Figure 3-3 (a) The energy of activation for the oxidation of glucose,
an energy "barrier" over which glucose molecules must be raised
before they can be oxidized to CO_2 and H_2O. (b) An analogous
physical situation in which a rock is poised in a depression at the top
of a hill. The rock will not move downward unless enough "activating
energy" is added to raise it over the lip of the depression.

depression at the top of a hill (Fig. 3-3b). As long as the
rock is undisturbed, it will not spontaneously begin its
travel downhill, even though the total "reaction," the pro-
gression of the rock downward, is energetically favorable.
In this physical example, the activation energy may be regarded as the effort that would be required to raise the
rock over the lip of the depression holding it in place.

The requirement for activation energy raises the question of why reactions proceed spontaneously at all. In
other words, how can any molecules in a system go over
the energy barrier? Where does the energy come from?

Movement over the barrier is possible because, instead of
being at rest as in the example of the rock at the top of a
hill, molecules at temperatures above absolute zero are in
constant motion. The amount of motion, or kinetic energy,
is not the same in all of the molecules present, although
the average energy is below the amount required for activation. Some molecules in the population are below this
average, and some are above. The distribution occurs
partly as a result of random collisions in which energy is
gained and lost by individual molecules, depending on
factors such as the angle and number of collisions per unit
time. Depending on the height of the activation barrier,
these random collisions may raise a number of molecules
to the energy level required for the reaction to proceed.
This happens when molecules strike each other at exactly
the right place, with sufficient force for interaction to occur. A high activating barrier indicates that successful collisions of this type are not very likely to occur.

One way to increase the probability that molecules in
the reacting system will pass over the energy barrier is to
raise the temperature. At elevated temperatures, both the
speed of travel of individual molecules and the frequency
of collisions increase, raising the probability that sufficiently forceful collisions, at the correct angle and place,
will occur. Once large numbers of molecules begin to pass
over the barrier, the reaction is frequently self-sustaining:
sufficient heat energy is released by molecules converted
to products to keep the reaction temperature high. Glucose may be ignited in this way by heating it over a flame.
Once ignited, the glucose will continue to burn in a self-
sustaining reaction until conversion to products is complete.

For obvious reasons, ignition is not a satisfactory approach for pushing reacting molecules over the activation
barrier in biological systems. Enzymes speed reactions by
a different mechanism, by *lowering the activation energy re-*
quired for a reaction to proceed (Fig. 3-4). Exactly how enzymes lower the activation energy is not completely
understood. However, it is known that enzymes combine
with the reactants during a reaction; this combination
places the reactants in a state in which their interaction is
favored instead of unlikely. Inorganic catalysts work in the
same way, by combining with reacting molecules and
greatly increasing the probability of favorable interaction
at reduced temperatures.

The Characteristics of Enzymes The enzyme molecules
lowering the energy barrier of biological reactions have
several important characteristics. All known enzymes are
proteins. In catalyzing biological reactions, enzymes combine only very briefly with the reacting molecules and are
released unchanged on completion of the reaction. As a

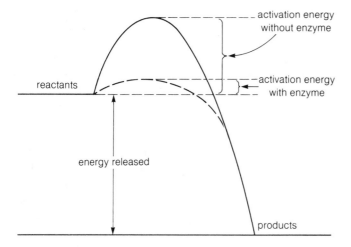

Figure 3-4 Enzymes increase the rate of a spontaneous reaction by reducing the energy of activation (see text).

result, a single enzyme molecule may cycle repeatedly through its reaction sequence, carrying many reactant molecules in succession over the reduced activation barrier. The number of reactions catalyzed by a single enzyme molecule, depending on the enzyme, may vary from 100 to more than 30,000,000 per minute. The magnitude of these numbers means that only a relatively small number of enzyme molecules are required to catalyze a large quantity of reactants. Finally, enzymes are *specific* in their catalytic activity. Usually they are tailored to catalyze only a single type of biochemical reaction, and combine only with specific, single molecular types or closely related groups of molecules. The specific molecule or molecular group whose reaction is catalyzed is known as the *substrate* of the enzyme. (The characteristics of enzymes are summarized in Table 3-1.)

Thousands of different enzymes have been detected and described. Of these, several hundred have been purified to the extent that they can be crystallized and chemically characterized. These enzymes vary from relatively small molecules having about 100 amino acids to large complexes containing several polypeptide chains totalling thousands of amino acids. Many enzymes are linked, in addition, to an inorganic ion or nonprotein organic group that contributes to their catalytic function. These nonprotein groups, called *cofactors*, may be covalently linked to the enzyme, or may be held by weaker linkages such as electrostatic attractions or hydrogen bonds.

The inorganic cofactors are all metallic ions, including iron, copper, magnesium, zinc, potassium, manganese,

Table 3-1 A Summary of Enzyme Characteristics

1. All known enzymes are proteins.

2. Enzymes combine briefly with the reactants during an enzyme-catalyzed reaction.

3. Enzymes are released unchanged after catalyzing the conversion of reactants to products.

4. Enzymes are specific in their activity: Each enzyme catalyzes the reaction of a single molecular type or of groups of closely related molecules.

5. Many enzymes are combined with nonprotein groups called *cofactors* that contribute to their activity. The inorganic cofactors are all metallic ions. The organic cofactors, called *coenzymes*, are complex groups derived from vitamins.

molybdenum, and cobalt ions. These metal ions, when present as a part of the enzyme, contribute directly to the reduction of activation energy by the enzyme. That is, they form a functional part of the site on the enzyme that combines with substrates and speeds their conversion to products. The organic cofactors, called *coenzymes*, are all complex chemical groups containing segments derived from vitamins (the biological role of vitamins, in fact, is apparently in forming parts of coenzymes of various kinds). When present as cofactors, coenzymes act as *carriers* of chemical groups, atoms, or electrons removed from substrates during reactions (representative coenzymes are shown in Figs. 6-12 and 6-13).

Enzymes are named by adding the suffix *ase* to their substrates or the type of reaction they catalyze. For example, an enzyme catalyzing the polymerization of nucleotides into DNA is called *DNA polymerase*. Certain enzymes, named before the current rules of nomenclature were agreed upon, are also known by their original or "trivial" names. The digestive enzymes *trypsin* and *pepsin* are examples of this group.

How Enzymes Lower the Energy of Activation

The Role of the Active Site The mechanisms by which enzymes lower the energy of activation have been the subject of much research and speculation and are still incompletely understood. Interest in this work has centered on the *active site* of the enzyme, the part of the enzyme molecule that actually combines with the substrate during a biological reaction.

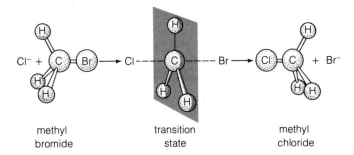

methyl bromide   transition state   methyl chloride

Figure 3-5 Formation of the transition state in the interaction between methyl bromide and chloride ion to produce methyl chloride. The transition state is unstable and easily pushed in the direction of either reactants or products (see text). Redrawn from an original courtesy from W. Ferdinand, *The enzyme molecule*, John Wiley and Sons, Ltd., 1976. Reprinted by permission.

Part of the mechanism lowering the energy of activation probably depends simply on the fact that, by virtue of their combination with the active site on the surface of the enzyme, reacting molecules are brought close together, in a position favoring their collision and interaction. Combination with the active site may also align the reacting molecules so that collisions between them are in the most favorable direction and place for reaction to occur. However, there are indications that the active site of enzymes may contribute more to reducing the energy of activation than simply placing the reactants in favorable positions. This is in line with the contemporary view that active sites are flexible and can undergo conformational changes (see p. 46) that may actually "push" reactants toward an interaction.

The Active Site and the Transition State of the Reactants
The contribution of conformational changes in the active site to enzymatic catalysis, according to the hypotheses of Daniel E. Koshland of the University of California at Berkeley and other workers in the field of enzymology, is in induction of an intermediate condition of the reactants known as the *transition state*. During any chemical reaction, the reactants briefly enter a state in which the "old" chemical bonds are incompletely broken and the "new" ones are incompletely formed. In response, atoms in the reactant molecules, not directly connected with the bonds being broken and remade, assume intermediate positions between their locations in the reactants and their new positions in the products. This altered arrangement, the transition state, is highly unstable and persists only briefly.

The transition state is most easily understood in a relatively simple inorganic reaction such as the interaction between methyl bromide and chloride ion to produce methyl chloride (Fig. 3-5). In this reaction, a transition state is formed in which both the chlorine and bromine atoms are linked to the central carbon atom for a fraction of a second. The transitory bond formed may be visualized as a new electron orbital that extends, for a brief instant, over all three atoms (the transitory bond is shown in dotted lines in Fig. 3-5). During the transition state, the hydrogens attached to the carbon are pushed to a new position, intermediate between the old methyl bromide and new methyl chloride conformations. This transition state is highly unstable and readily changes in the direction of either products or unchanged reactants.

According to the current hypotheses of enzymatic catalysis, the active site, rather than fitting only the reactant molecules in a rigid lock-and-key fashion, has additional conformational arrangements that also fit the transition state and the products. Of these, the tightest binding, and the most precise fit, is to the transition state. Substrate binding and enzymatic catalysis then proceed as follows. The initial conformation of the active site exposes chemical groups tailored to fit the reactants closely, but not precisely. Binding of the reactants induces a change in the conformation of the enzyme molecule, changing the active site to a conformation that precisely matches the transition state. This tight fit "bends" or "warps" the substrate into the transition state. Once in this unstable state, the reacting molecules are easily pushed by collisions in the direction of products. The collisions may occur between the reacting molecules, or between the enzyme-substrate complex and other molecules in the surrounding solution. The energy of these collisions, transmitted through the flexible structure of the enzyme molecule, is easily sufficient to push the unstable transition state in the direction of products.

The idea that the active site may also contain a conformation that fits the products, which at first might seem surprising, helps explain the role of enzymes in catalyzing reversible reactions. Depending on the relative concentrations of reactants and products in a reversible reaction, enzymes may either combine with reactants, speeding their conversion to products, or the reverse: at high concentrations of products, an enzyme may combine with the products and speed their conversion to reactants. This means that the active site must contain arrangements of chemical groups that can bind to either the products or reactants. Which of the two molecular groups, either reactants or products, actually binds to the active site (since it can fit both) depends on their relative concentrations in the medium. If the reactants are present in highest concentration, collisions between the enzyme and reactant molecules will be more likely and the reactants will bind most frequently. If the products are present in highest concentration, the reverse will be true. Once either is bound, the enzyme undergoes its conformational change in response, altering the active site to the conformation that fits the transition state. Once in the unstable transition state, the reaction can easily proceed in either direction, toward reactants or products.

Regulatory Changes in the Active Site: Allosteric Enzymes
Some enzymes contain active sites that can be turned on or off as a means for regulating or "fine-tuning" enzyme activity to the needs of the cell. Typically, these enzymes catalyze key reactions forming the first step in complex pathways leading to end products required by an organism. The final product of the pathway acts as a regulator of the enzyme catalyzing the first step in the sequence, in a process called *end product* or *feedback inhibition*.

The sequence converting the amino acid threonine to another amino acid, isoleucine (Fig. 3-6), serves as the classic example of enzyme regulation by feedback inhibition of the active site. Five enzymes catalyze the pathway, acting in sequence. The final product of the pathway, isoleucine, can act as an inhibitor of the first enzyme of the pathway, an enzyme called *threonine deaminase*. As isoleucine is produced by the pathway, any molecules produced in excess of cell requirements combine reversibly with threonine deaminase enzyme molecules, inhibiting their ability to catalyze the first step in the pathway leading to isoleucine. If the final product is present in great excess, the pathway is completely turned off by inhibition of all threonine deaminase molecules present. If the concentration of isoleucine later falls as a result of its use in cell synthesis, inhibition of the threonine deaminase system is released, opening the pathway to synthesis of isoleucine again.

Examination of the effects of isoleucine on threonine deaminase reveals that inhibition of the enzyme by the end product follows an unusual pattern. Most commonly, enzymes are inhibited by substances resembling the normal substrate; these substances combine with the active site of the enzyme, blocking access by the normal substrate (see Saturation and Inhibition of Enzymes, below). However, isoleucine combines with a site elsewhere in the enzyme, somehow reducing the ability of the active site to combine with the substrate without directly blocking it.

From observations of this and other examples of feedback inhibition, a French scientist, Jacques Monod, proposed a hypothesis in 1965 to account for enzyme regulation

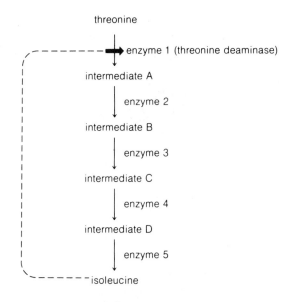

Figure 3-6 Feedback inhibition of the pathway leading from the amino acid threonine to isoleucine. Five enzymes catalyze successive steps in the pathway. If the product of the pathway, isoleucine, is present in excess, it inhibits the enzyme catalyzing the first step in the sequence (dotted line).

of this type. According to Monod's idea, the end product of the pathway combines with the key enzyme at a site separate from the active site (Fig. 3-7a). The separate site is called the *allosteric site*, meaning the "other location." Combination with the end product causes the enzyme to undergo a conformational change, closing or altering the active site so that it no longer matches and combines with the substrate (Fig. 3-7b). Release of the inhibitor (called the *allosteric effector* in the model) returns the enzyme to its active form (Fig. 3-7c). Enzymes containing the additional regulatory site are called *allosteric enzymes;* many examples of cellular enzymes following this pattern are known. In some cases, combination with the allosteric effector activates, rather than inhibits, activity of the enzyme.

Other Factors That Affect Enzyme Activity

Enzymatic catalysis of biochemical reactions depends on the three-dimensional structure of the active site. As might

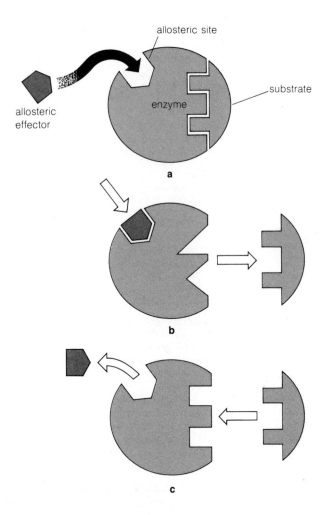

allosteric site

allosteric
effector

enzyme

substrate

a

b

c

Figure 3-7 (**a**) Reversible combination of an allosteric effector with the allosteric site causes a conformational change in the enzyme, altering the active site so that it no longer fits the substrate (**b**). (**c**) The active site returns to its original conformation upon release of the allosteric effector, enabling it to bind the substrate again (see text).

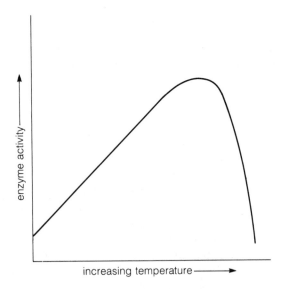

enzyme activity

increasing temperature

Figure 3-8 The effects of increasing temperature on enzymatic activity. Typically, the rate of the reaction reaches an optimum, usually at 40 to 50°C. Once past the optimum, the reaction rate falls rapidly due to unfolding of the enzyme at elevated temperatures. *Activity* refers to the rate at which the same number of enzyme molecules converts reactants to products.

be expected, exposing enzymes to conditions that seriously alter their folding conformation interferes with their ability to increase biological reaction rates.

Proteins are held in their final structure by interactions between their amino acid side groups (see p. 46). The pattern and strength of two types of these interactions, hydrogen bonding and electrostatic attractions, are highly sensitive to changes in temperature and pH of the medium surrounding the enzyme.

The Effects of Temperature Gradual increases in temperature increase the kinetic motion of both the amino acid chains and side groups forming an enzyme molecule, and the molecules colliding with the enzyme in the surrounding solution. At elevated temperatures, these disturbances become strong enough to overcome the attraction of hydrogen bonds, which are individually relatively weak. As its internal hydrogen bonds are broken, an enzyme gradually unfolds and loses its "correct" three-dimensional folding conformation.

These changes affect enzymatic activity in a characteristic way (Fig. 3-8). Initially, rises in temperature over the range from 0° to about 40°C increase the activity of most enzymes along the lines followed by all chemical reactions: each 10°C rise in temperature approximately doubles the reaction rate. This effect is due to increases in the force and frequency of collisions of enzymes and reactant molecules, reflecting the heightened kinetic motion of all molecules in the solution. The collisions at these temperatures are forceful enough to increase the reaction rate, but do not seriously disturb enzyme structure. However, in most enzymatically catalyzed reactions, rates begin to fall off at temperatures above 40°C as collisions become violent enough to begin unfolding the enzyme. The drop in activ-

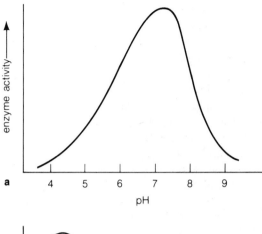

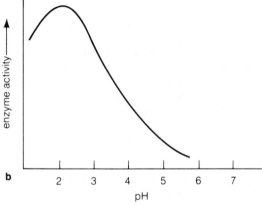

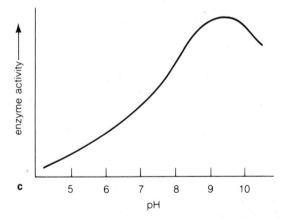

Figure 3-9 The effects of pH changes on enzyme activity. (a) Enzymes with maximum activity at neutral pH. (b) Enzymes with maximum activity at acid pH. (c) Enzymes with maximum activity at alkaline pH (see text).

ity becomes steep at 55°C and falls to zero at 60°C. At the highest temperature, the disturbance in hydrogen bonding causes the enzyme to unfold into a completely inactive form, totally counteracting the positive effects of increased kinetic motion at elevated temperatures.

As a result of these two opposing effects of increased temperature, all enzymes exhibit an optimum activity representing the highest temperature at which reaction rates are increased with no significant disturbance of internal hydrogen bonding and three-dimensional folding arrangement of the enzyme. For most enzymes, this optimum lies between 40 and 50°C. A few organisms, such as the bacteria living in hot springs, possess enzymes with structures so resistant to disturbance that they remain active at temperatures of 85°C or more.

The Effects of pH on Enzyme Activity Changes in the pH of the medium surrounding enzymes affect enzyme structure and activity by altering the charge of groups carried on amino acid side chains. Particularly important among the affected groups are those, such as —COOH and —NH$_2$ groups, that are capable of releasing or accepting an H$^+$ ion and converting to a charged form. Each of these groups, depending on their location in a protein molecule, undergoes conversion from uncharged to charged form at a characteristic pH. Changes of this type affect the charge of groups holding enzymes in their final three-dimensional shape, and also the activity of charged groups that may be tailored in the active site to fit the reactants, products, or transition state of the reaction catalyzed by the enzyme.

As a result, changes in pH also affect the activity of enzymes in characteristic ways (Fig. 3-9). Typically, each enzyme has a pH optimum at which the charge taken on by its amino acid side groups is exactly "correct" and the enzyme is most efficient in speeding the rate of its specific biochemical reaction. On either side of this optimum pH, the charges of these groups change; the resulting alterations in the folding pattern of the enzyme and the charge of the active site causes the rate of the reaction to drop off. The effects become more extreme as pH values farther from the optimum are reached until the rate drops to zero. Some enzymes reach their optimum activity level at intermediate pH (Fig. 3-9a); in most such cases, the optimum lies in the vicinity of neutrality, near pH 7. Others (Figs. 3-9b and c) have their optimum catalytic activity at higher or lower pH, and drop in activity as the pH rises or falls from the optimum value. For example, the digestive enzyme pepsin, which speeds the breakdown of proteins being digested in the stomach, has its optimum activity at the acid pH characteristic of the stomach contents.

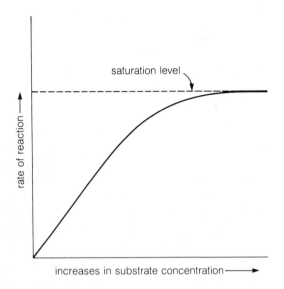

 — Figure contains y-axis label "rate of reaction →", x-axis label "increases in substrate concentration →", a curve rising and leveling off at a horizontal dashed line labeled "saturation level".

Figure 3-10 Saturation of an enzyme by increases in substrate concentration (see text).

Saturation and Inhibition of Enzymes The fact that enzymes combine briefly with their substrate molecules to speed biological reactions has two important effects on the rate of a reaction being speeded by enzymatic activity. One effect is related to the concentration of substrate molecules. At very low substrate concentrations, the enzyme molecules do not collide frequently with the substrate molecules and the reaction proceeds slowly. As the concentration of substrate molecules increases, the reaction rate increases at first because collisions between the enzyme molecules and the reactants become more frequent (Fig. 3-10). Continued increases in the concentration of the substrate molecules cause proportionate increases in the reaction until the enzymes colliding with the substrate molecules are cycling through the reaction as fast as they can (each enzyme has a maximum rate, the *turnover number*, at which it can combine with the reactants, speed the reaction, and release the products). Since the enzymes are cycling as rapidly as they can, further increases in substrate concentration have no effect on the reaction rate. At this point, the enzyme is said to be *saturated*, and the reaction remains at the saturation level represented by the horizontal dotted line in Figure 3-10.

The characteristic saturation curve shown in Figure 3-10 provides a valuable biochemical tool for determining whether a given reaction is speeded by an enzyme in a biological system. In order to determine whether an enzyme is involved, the concentration of reactants is increased experimentally and the rate of the reaction is followed. If a point is reached at which further increases in reactants have no effect in increasing the reaction rate, indications are good that an enzyme is involved (uncatalyzed reactions will increase in rate almost indefinitely with increases in concentration of the reactants).

The second effect of the combination of enzymes with their reactants is susceptibility to *inhibition* by nonreactant molecules that resemble the substrate molecules in structure. These molecules, by virtue of their resemblance to the substrate molecules, can combine with the reactive site of the enzyme. Since they cannot be converted to the products of the reaction normally catalyzed by the enzyme, they tend to remain attached to the active site without change and block access by the normal substrate. As a result, the rate of the reaction slows; if the concentration of the inhibitor becomes high enough, the reaction may stop completely. Many poisons have their toxic effects by inhibiting enzymes in this way. For example, the action of cyanide and carbon monoxide as poisons depends on their ability to act as inhibitors of enzymes important in the utilization of oxygen in cellular respiration.

Many examples of enzymes and enzymatic activity are included in the subsequent chapters of this book. In many respects, the story of cellular life is the story of enzymes, so much so that cells have been described as "bags of enzymes." While this is obviously an oversimplification, it is true that each of the thousands of individual biochemical reactions coordinated to achieve cellular life is speeded by its own enzyme, tailored by evolution to fit its normal substrates with specificity approaching perfection. This specificity permits cells to control their activities with precision and sensitivity by regulating which enzymes are present in the various cellular compartments at different times.

Suggestions for Further Reading

Bender, M. L. and L. J. Brubacher. 1973. *Catalysis and enzyme action.* McGraw-Hill: New York.

Dickerson, R. E. and I. Geis. 1969. *The structure and action of proteins.* Harper & Row: New York.

Ferdinand, W. 1976. *The enzyme molecule.* John Wiley: New York.

Koshland, D. E., Jr. 1973. Protein shape and biological control. *Scientific American* 229: 52–64 (October).

Lehninger, A. L. 1975. *Biochemistry.* 2nd ed. Worth: New York.

Miller, G. Tyler, Jr. 1971. *Energy, kinetics and life.* Wadsworth: Belmont, Calif.

Wolfe, S. L. 1981. *Biology of the cell.* 2nd ed. Wadsworth: Belmont, Calif.

Questions

1. State the first and second laws of thermodynamics. Give examples from your own experience that illustrate the operation of the laws.

2. What does the word *spontaneous* mean in thermodynamics?

3. Define a system, the surroundings, and initial and final states.

4. What does energy content mean with reference to molecules? Define kinetic and potential energy.

5. What is entropy?

6. How do energy content and entropy interact in determining whether chemical reactions will proceed to completion? What is a reversible reaction?

7. What happens at the equilibrium point of a reversible reaction? What happens if more reactants are added to a reaction at equilibrium? If more products are added?

8. How do biological organisms run reactions that require energy? What does coupling mean?

9. It has sometimes been claimed that organisms can violate the second law of thermodynamics because they become more complex as they develop from a fertilized egg or seed. Why is this statement incorrect?

10. Which of the following reactions is spontaneous? Why?
    a. Fertilized egg → adult animal
    b. Fertilized egg + raw materials (glucose, fats, etc.) → adult animal + waste products

11. How does ATP act as the agent coupling uphill biological reactions to other reactions that release energy?

12. Draw the structures of ATP, ADP, and AMP.

13. What are calories? Cal/mole?

14. What effects do enzymes have on spontaneous reactions?

15. What is activation energy? How do enzymes affect activation energy?

16. What are the characteristics of enzymes?

17. Define cofactors and coenzymes.

18. What is the active site of an enzyme?

19. What is the transition state for a reaction?

20. How does the active site of an enzyme affect the transition state? How does this effect on the transition state change the activation energy for a reaction?

21. Why is it likely that the active site of an enzyme can fit and bind both the reactants and products of a reaction?

22. What effects do changes in temperature have on enzyme-catalyzed reactions? Why?

23. What effects do changes in pH have on enzyme-catalyzed reactions? Why?

24. What is enzyme saturation? Enzyme inhibition? Feedback inhibition? What are allosteric enzymes?

Membranes and the Cell Surface

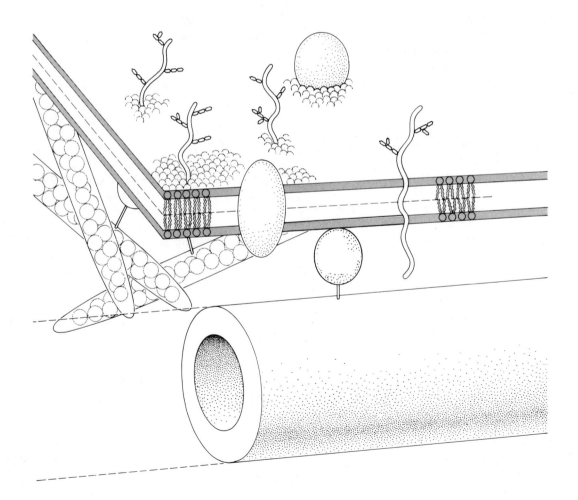

4

Structure and Transport in Cellular Membranes

The survey of prokaryotic and eukaryotic cell structure presented in Chapter 1 makes it obvious that cells are organized by systems of membranes. The plasma membrane surrounds cells and separates them from the outside world; the membranes of internal organelles divide the interior of cells into compartments with distinct environments and functions. Regions within membranes also provide hydrophobic environments in which certain reaction systems, such as some of the reactions producing cellular energy, take place exclusively.

The plasma membrane also serves as the zone of contact between cells and the outside world. All of the fuel substances and raw materials necessary for life enter cells from the outside through this membrane. Waste materials and cell secretions travel in the opposite direction, from inside to outside. These movements maintain the concentrations of molecules inside cells at the levels required for life and cellular functions. Within eukaryotic cells, additional membranes, such as those surrounding mitochondria, chloroplasts, and the nucleus, maintain the interior of these organelles as separate compartments with their own distinct molecular makeup.

The movement of substances to and from cells and their internal compartments reflects the properties of both the lipid and protein molecules of membranes. Nonpolar or hydrophobic molecules pass through the lipid regions of membranes. Membrane proteins, in contrast, are responsible for the transport of polar, hydrophilic, or charged substances. Thus structure and function in cellular membranes are clearly and closely interrelated.

The Structure of Cellular Membranes

The chemical analysis of isolated membranes has confirmed that they contain lipids and proteins in quantity. The proportions of lipids and proteins in eukaryotic membranes vary from the extremes of myelin (a membrane type found in nerve tissue), with 90 percent lipids and 10 percent proteins by weight, to the inner membrane of mitochondria with 25 percent lipids and 75 percent proteins by weight. Most plant and animal cell membranes fall about midway between these extremes, approaching a distribution in which lipids make up from 30 to 50 percent of the total membrane molecules by weight, with proteins forming the remainder. Carbohydrates, as noted, are also found in membranes as chemical groups linked to the lipids and proteins. Membrane carbohydrates occur in greatest abundance in eukaryotic plasma membranes, where they are found in large quantities on the surface facing the cell exterior.

How Proteins and Lipids Are Arranged in Cellular Membranes

Early Models Leading to the Modern Concept of Membrane Structure The importance of membranes in cells was recognized early in the development of cell biology.

From the late 1800s onward, investigations into the nature of cell boundaries gradually revealed the molecules that are present in membranes, and how these molecules are put together to form the thin surface coats of cells and internal cell structures.

The first significant experiments, carried out by the Swiss investigator Ernst Overton in the late 1890s, showed that in general cells absorb small, nonpolar lipid molecules much more rapidly than polar molecules. These experiments suggested for the first time that the surface of cells is probably covered by a layer of lipids.

The first clues about the physical arrangement of the lipid molecules in the surface layer were developed in Holland in the 1920s by E. Gorter and F. Grendel. For their experiments, these investigators used the red blood cells of animals, a cell type that is of uniform size and shape, and relatively easy to break open. Disruption releases the cell contents, leaving empty plasma membranes called red cell "ghosts." By measuring the ghosts under a light microscope, Gorter and Grendel were able to estimate the total area occupied by a single red cell membrane. They then extracted the lipids from the membranes and calculated the lipid quantity per red blood cell ghost. Comparing this result with the surface area per cell revealed that just enough lipid was present to make a layer exactly two molecules in thickness around the cell. On this basis, Gorter and Grendel proposed for the first time that the lipid molecules in cell membranes occur in *bilayers* (see Fig. 2-14 and p. 38).

Proteins were first implicated as components of membranes through cooperative research carried out in the 1930s and 1940s by two Englishmen, James Danielli and Hugh Davson. These workers made measurements of the surface tension of oil droplets in water and compared the values obtained with the surface tension of living cells. They reasoned that if cells are actually covered with a layer of lipids, as the earlier work suggested, then the surface tension of oil droplets and cells ought to be the same. The results they obtained turned out to be quite different—the surface tension of cells was consistently much lower than that of oil droplets. Danielli and Davson found, however, that they could mimic the surface properties of living cells by adding proteins to the oil droplets. The proteins formed a hydrophilic film over the surface of the droplets and reduced their surface tension to levels much like living cells.

From this information, Danielli and Davson proposed a model for membrane structure that was to shape biological thinking for many years to come. They proposed, in agreement with Gorter and Grendel, that the lipid mole-

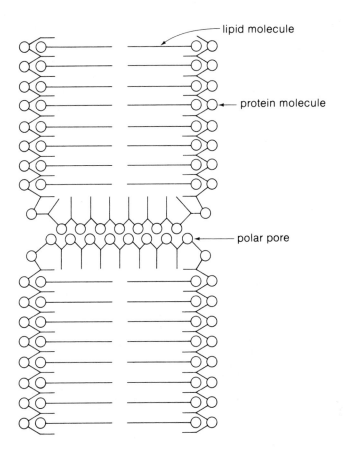

Figure 4-1 The Danielli-Davson model for membrane structure. In the model, both surfaces of the bilayer are coated with a layer of protein molecules unwound into extended amino acid chains.

cules of membranes are arranged in bilayers, with the nonpolar fatty acid chains of the membrane lipids facing each other in the interior of the bilayer (Fig. 4-1). Noting that the surface tension of living cells is much lower than the values obtained for pure oil droplets, they proposed in addition that the lipid bilayer is coated on both membrane surfaces by a layer of proteins that reduces its surface tension. To accommodate a complete layer of protein on both surfaces within the very thin dimensions actually observed for biological membranes, Danielli and Davson proposed that the proteins extend over the lipid bilayer as completely unfolded chains only one amino acid in thickness.

Development of the Modern Concept of Membrane Structure: The Fluid Mosaic Model The Danielli-Davson model attracted wide attention and provided the conceptual framework for essentially all the thought and experimentation concerning membrane structure until the late 1960s. During these years a variety of experiments confirmed that phospholipids (see p. 37) and proteins are important constituents of membranes. Particularly important in these years was the work of J. D. Robertson of Duke University, who drew attention to the almost uniform appearance of cellular membranes in the electron microscope. On the basis of his research, Robertson proposed, in his *unit membrane* concept, that all biological membranes are fundamentally similar in structure.

Work during the 1960s began to reveal inadequacies in the Danielli-Davson model. In 1966, S. J. Singer, working at the University of California at San Diego, found that membrane proteins transmit light in a pattern indicating that as much as 30 percent of their amino acid chains is folded into an alpha helix (see Fig. 2-19 and p. 44). Thus at least this proportion of the membrane proteins is extensively coiled rather than spread into a layer only one amino acid in thickness as Davson and Danielli had proposed. Singer noted that the total content of alpha helix in membranes, in fact, is typical of proteins with a spherical rather than flattened shape. However, spherical proteins seemed incompatible with the Danielli-Davson model, because spreading proteins in this form on the two sides of a bilayer would build a structure several times thicker than the dimensions actually observed for biological membranes.

Further problems for the Danielli-Davson model came from Singer's observation that unfolding membrane proteins into thin, essentially two-dimensional layers on both sides of a membrane would inevitably expose some hydrophobic amino acids, which are present in quantity in all membrane proteins, to the watery medium at the cell surface. This condition is extremely unlikely, since considerable energy must be expended to maintain hydrophobic groups in a hydrophilic environment. The energy required to maintain proteins in thinly spread layers would be so high, in fact, that membranes would be unstable and very unlikely to remain intact for any length of time.

Observations of the actual behavior of phospholipids in water solved some of these problems. These observations made it clear that it is unnecessary to assume that phospholipids must be coated with proteins to reduce their surface tension when they are placed in water. The substances used by Danielli and Davson in their experiments were pure oils that are totally hydrophobic in character. Substances of this type round up into spheres when placed in water. Since the spherical shape exposes the smallest possible surface area to the surrounding water molecules, the spheres resist deformation to more flattened shapes. This resistance to deformation is reflected in the relatively high surface tension of oil droplets in water. Phospholipid droplets, in contrast, can take up the bilayer arrangement in which all of the exposed surfaces are hydrophilic. This allows them to take on essentially any shape in water, and reduces the surface tension of phospholipids with no requirement for surface coats of protein. Thus, as often happens in science, Danielli and Davson arrived at a correct conclusion, that proteins are important in membrane structure, for what turned out to be the wrong reasons.

A new technique for preparing cells for electron microscopy, developed during the 1960s, provided clues to the actual arrangement of protein molecules in membranes. In this method, called the *freeze-fracture* technique, a tissue sample is rapidly frozen by placing it in liquid nitrogen. The sample is then fractured by striking it with a sharp knife edge. Since cell membranes are hydrophobic, nonpolar regions, they produce weakly frozen "faults" in the preparations that fracture more readily than the surrounding polar regions. As a consequence, the fracture tends to follow membranes, often splitting them into bilayer halves and revealing the hydrophobic membrane interior (Fig. 4-2; see the Appendix for further information on preparative techniques for electron microscopy).

The freeze-fracture technique preserves membrane proteins and makes their distribution in membranes directly visible. It shows that proteins are spaced as discrete, individual particles both on and within membranes (see Fig. 4-2). Some proteins are attached at scattered points to the outside surface of membranes and some to the inner surface. Other proteins are embedded within membranes, some extending from either surface, and others extending all the way through, from the inside to the outside membrane surfaces.

The Fluid Mosiac Model In 1972, Singer and a colleague, Garth Nicolson, combined all of this information into a new idea for membrane structure, the *fluid mosaic model* (Fig. 4-3). In simplest terms, their model retains the phospholipid bilayer as the framework of membranes, and proposes that proteins, rather than being spread in a continuous layer on both membrane surfaces, are suspended as individual units in or on the bilayer.

As in the Danielli-Davson hypothesis, the Singer-Nicolson model proposes that phospholipid molecules orient in the membrane bilayers with their hydrophilic ends fac-

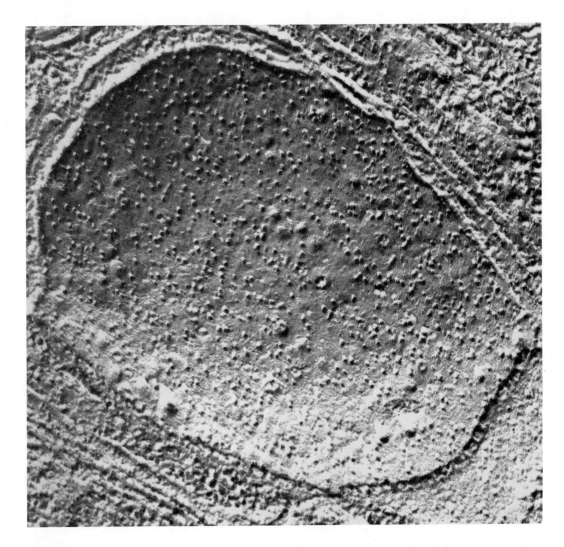

Figure 4-2 A segment of a cell containing several layers of membranes prepared for electron microscopy by the freeze-fracture technique. In the central region of the micrograph the fracture has split the bilayer in half and exposed the membrane interior. The particles are protein molecules suspended as individual units in the membrane. × 110,000. Courtesy of D. W. Deamer.

ing the aqueous medium and their hydrophobic ends associated together in the membrane interior. The model proposes in addition that the membrane lipids are in a mobile, fluid state, and that individual lipid molecules are free to exchange places and move through the bilayer. This is the "fluid" part of the fluid mosaic model.

Singer and Nicolson proposed that membrane proteins float as spherical or globular units in the fluid bilayer like "icebergs in the sea." Some, according to the model,

penetrate entirely through the membrane and are exposed on both sides, and some penetrate only part way through. The distribution of proteins in membranes, in dispersed units rather than as continuous sheets covering the membrane surfaces, is the "mosaic" part of the fluid mosaic model. Singer and Nicolson called the proteins embedded partly or completely in the phospholipid bilayer *integral* membrane proteins. These proteins, according to the fluid mosaic model, are functional parts of membranes. Other

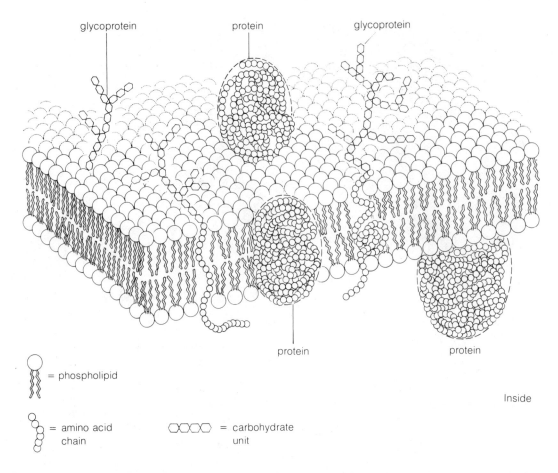

glycoprotein   protein   glycoprotein

protein   protein

Inside

= phospholipid

= amino acid
chain

= carbohydrate
unit

Figure 4-3 The fluid mosaic model for membrane structure, as proposed by Singer and Nicolson. Proteins forming an integral part of the membrane are shown deeply embedded in the phospholipid bilayer; more loosely bound proteins attached to the membrane surface, called peripheral proteins, are shown at the arrows.

proteins attached to membrane surfaces were termed *peripheral* membrane proteins (see Fig. 4-3). The peripheral proteins form parts of structures that are associated with the membrane surfaces.

According to the model, the integral membrane proteins are held in suspension in the fluid lipid bilayer by their solubility properties. The proteins that pass entirely through the membrane have two hydrophilic ends and a hydrophobic middle region. The hydrophobic middle regions are held in association with the membrane interior, and the hydrophilic ends extend into the watery medium at the membrane surfaces. Integral proteins that extend only partly through the membrane have a hydrophobic end suspended in the membrane interior, and a hydro-

philic end facing the surrounding watery medium. The proteins remain in stable suspension in the bilayer because any change in orientation would expose their hydrophobic regions to the watery surroundings. Within these limitations, the proteins are free to displace phospholipid molecules and move laterally through the fluid bilayer. Most of these integral proteins are transport molecules or enzymes that catalyze the specialized functions of membranes in different regions of the cell.

Evidence Supporting the Fluid Mosaic Model The best evidence that integral proteins are embedded as particulate units in membranes comes from a source already described, the freeze-fracture technique. The particles exposed in mem-

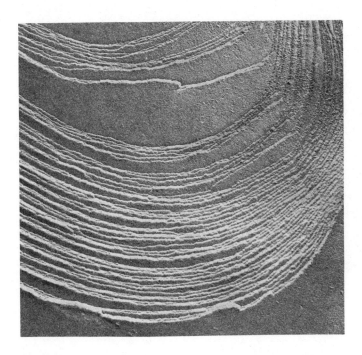

Figure 4-4 Freeze-fracture preparation of the membrane sheath surrounding a nerve cell extension. The fracture has exposed successive layers of the membrane sheath. The exposed surfaces are smooth and contain no evidence of particulate structure, reflecting the fact that these membranes contain few proteins (compare with Fig. 4-2). Courtesy of D. Branton and Academic Press, Inc., from *Experimental Cell Research* 45 (1967): 703.

branes by the technique approximate the dimensions expected for protein molecules, and are distributed in membranes according to expectations from the model.

The exposed particles have been identified as proteins by several experiments. The long extensions of nerve cells in higher animals are surrounded by a sheath formed from layered membranes. The membranes of the sheath, by analysis, have been shown to contain almost no proteins. These membranes, as expected from the model, show almost no globular particles in freeze-fracture preparations (Fig. 4-4). The membranes of cell structures known to contain proteins in greater quantity are crowded with particulate units when fractured (as in Fig. 4-2).

More direct evidence for the conclusion that the membrane particles visible in freeze-fracture preparations are integral proteins comes from studies of artificial bilayers carried out by David W. Deamer of the University of California at Davis. Artificial bilayers created by suspending pure phospholipids in water show smooth surfaces without particles when they are fractured (Fig. 4-5a). Addition of proteins to the phospholipids results in the appearance of particles in the artificial bilayers (Fig. 4-5b). These particles, in dimensions and structure, closely resemble the particles visible in fractured natural membranes. All of this evidence supports the proposal that integral proteins are embedded as particulate units in biological membranes, as proposed in the fluid mosaic model.

Singer and others showed that the integral membrane proteins, as expected from the model, could be removed from membranes only by chemicals such as detergents that totally destroy the lipid bilayer framework. The peripheral proteins, in contrast, could be released by relatively mild treatments such as adjustments in the salt concentrations of the fluids surrounding the membrane. These findings gave further support to the idea that integral membrane proteins are deeply embedded in the hydrophobic membrane interior, and that peripheral proteins are attached to the polar membrane surfaces.

Evidence that membrane bilayers are fluid, and that integral proteins are free to move laterally in membranes comes from an interesting and graphic experiment carried out by L. David Frye and Michael A. Edidin at the Johns Hopkins University. Frye and Edidin worked with mouse and human cells fused together by exposing them to a virus, the *Sendai* virus. When two cells are fused in this way, their plasma membranes join together into one continuous surface membrane covering the combined cytoplasm and nuclei of both cells.

Frye and Edidin used the cell fusion technique in combination with molecules that could bind specifically to membrane proteins of either cell type. The molecules were marked for identification with a dye that glows, or fluoresces, in ultraviolet light. The molecules specific for mouse and human membrane proteins were marked respectively with dyes that fluoresce green and red. When mouse and human cells were fused together and reacted with the fluorescent molecules, the green and red colors were at first separated on the surface of the combined cells (Fig. 4-6a), with one half of the fused membranes glowing green, and one half red. This showed that membrane proteins derived from the mouse cells were initially restricted to one half of the fused plasma membranes, and the human membrane proteins to the other. After about 40 minutes, most of the cells showed complete intermixing of the two colors, indicating that the membrane proteins could move readily and freely through the membrane bilayer (Fig. 4-6b). This would be unlikely unless the membrane bilayer is in a fluid state, as proposed in the fluid mosaic model.

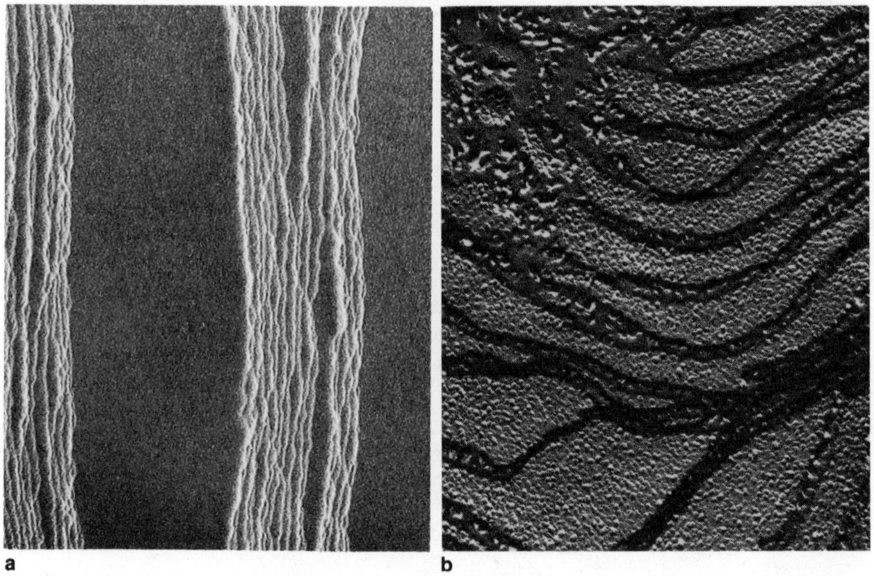

Figure 4-5 (a) An artificial bilayer without added proteins. The fractured bilayer surfaces are smooth and show no particles. (b) An artificial bilayer with added proteins. Particles similar to the particulate units visible in natural membranes are now present in the fractured bilayer surfaces. Courtesy of D. W. Deamer.

Other evidence for the fluidity of membranes comes from experiments in which individual lipid molecules are "tagged" by chemical groups that allow their distribution in membranes to be detected. These tagged molecules are introduced into membranes in groups so that they are at first concentrated into a localized region. The rate at which they separate and reach an equal distribution throughout the membrane is then followed. The rates of movement, about 1 to 2×10^{-8} square centimeters per second, are so rapid that a phospholipid molecule moving at this rate could pass from one end of a bacterium to the other in about one second! Rates at this level mean that the phospholipid bilayers of membranes have about the same fluidity as a light household lubricating oil.

Membrane Carbohydrates in the Fluid Mosaic Model

The carbohydrate part of cellular membranes is detected primarily at the surfaces of cells, on the side of the plasma membrane facing the cell exterior. In accordance with this observation the carbohydrate groups of membrane glycolipids and glycoproteins are considered in the fluid mosaic model to extend only from the outside surfaces of plasma membranes, or from the surfaces of internal cell membranes that line spaces that communicate with the cell exterior (see Fig. 4-3).

One of the experiments showing that the carbohydrate groups of membrane glycolipids and glycoproteins are actually restricted to the exterior surfaces of membranes comes from the Singer laboratory. Singer and Nicolson developed a technique for marking the carbohydrate groups of membranes by using *concanavalin A*, a plant protein that binds specifically to the sugar groups occurring on glycoproteins. In order to make concanavalin A visible in the electron microscope, they attached a large, dense marker molecule containing iron atoms to it. They noted that the marked concanavalin A molecules attached only to the outer surfaces of cellular membranes, supporting the hypothesis that the carbohydrate groups extend only from the side of the membrane facing the cell exterior. Several carbohydrate groups are shown in this position in Figure 4-3.

All the major parts of the fluid mosaic model are thus strongly supported by experimental evidence, and the model has come to be accepted as the basis for membrane structure in living cells. Membranes are held in this arrangement by the same forces that keep the hydrophobic portions of membrane proteins in suspension in the membrane interior. When surrounded on both sides by water molecules, as they normally are in cells, the phospholipid and integral protein molecules remain stable in membranes because any disturbance of the structure would expose their hydrophobic portions to the surrounding water molecules. This resistance to mixing, equivalent to the stable layering of oil on water when the two are mixed, keeps

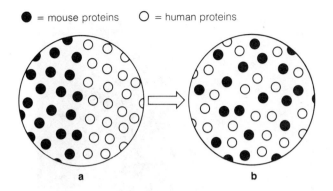

● = mouse proteins   ○ = human proteins

Figure 4-6 Evidence for the mobility of membrane proteins from fused mouse-human cells. (**a**) Immediately after fusion of the cells, membrane proteins of the two cell layers are separated into distinct regions. (**b**) After 40 minutes, the mouse and human membrane proteins are completely intermixed (proteins not drawn to scale).

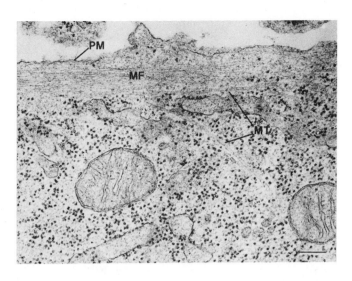

Figure 4-7 Membrane-associated microtubules (*MT*) and microfilaments (*MF*) located just inside the plasma membrane (*PM*) of a mammalian cell. × 37,000. Courtesy of G. L. Nicolson, from *Biochimica et Biophysica Acta* 457 (1976): 57.

membranes intact and provides them with their unique structural properties as stable cell boundaries.

Microtubules and Microfilaments in Cell Membranes

After Singer and Nicolson first proposed their model in 1972, investigations into membrane structure revealed that both microtubules and microfilaments are associated with cell membranes, particularly the plasma membrane. Both of these motile and supportive elements (see p. 11 and Chap. 8) are frequently visible just under the plasma membrane in electron micrographs of the cell surface (Fig. 4-7).

The probable functional roles of the microtubules and microfilaments associated with membranes have been worked out through the use of drugs and other treatments that specifically inhibit the two cell structures. The drug *colchicine* inhibits microtubule assembly; this drug stops motion due to microtubules and also destroys the cytoskeletal systems that depend on microtubules for their support. The drug *cytochalasin B* interferes with the assembly of microfilaments and inhibits the activity of these structures in motile and support roles. The two drugs are specific in their reactions for the two cell structures: colchicine

interferes with microtubules but has no detectable effects on microfilaments, and cytochalasin B destroys the activity of microfilaments without altering microtubule activity.

In plasma membranes, some proteins are anchored in place in the phospholipid bilayer rather than being free to move through the fluid membrane. The fixed arrangement of these proteins is destroyed in some cases by colchicine, and in others by cytochalasin B. This observation indicates that both microtubules and microfilaments form supportive networks under plasma membranes that anchor some membrane proteins in place. Among the most important proteins in this class are some of the glycoproteins that recognize and bind molecules at cell surfaces (see Chap. 5).

While both microtubules and microfilaments thus form supportive networks under the plasma membrane, active motility of the membrane seems to be associated with microfilaments. Many cell types, including the protozoan *Amoeba* and the leucocytes or white blood cells of animals, are capable of moving actively from place to place by sending out lobelike extensions of the plasma membrane. Ameboid motion of this type, involving active extensions of the plasma membrane, is generally inhibited by cytochalasin B but not colchicine, indicating that the microfilament network just under the membrane is responsible for producing this motion (ameboid movement is described in more detail in Chap. 8).

Movement of membrane subparts is also evidently produced primarily by microfilaments. Some proteins are moved actively through the phospholipid bilayer of the plasma membrane. This movement is seen in its most spectacular form in a phenomenon called *capping.* Many substances are recognized and bound to the cell surface by membrane glycoproteins. These substances initially bind at random points over the plasma membrane (Fig. 4-8a). Within minutes, however, many of the bound molecules are swept along the cell surface to one end of the cell, where they form a cap (Fig. 4-8b). Depending on the particular substance involved, the cap may remain in position, or may be absorbed into the cell interior. Like ameboid motion, capping is inhibited by cytochalasin B but not colchicine, indicating that microfilaments are also responsible for this motion associated with the plasma membrane.

Thus both microtubules and microfilaments probably act as supportive elements beneath the plasma membrane, and microfilaments in addition supply motion to the membrane or membrane parts. One of the functions of peripheral membrane proteins is evidently to link these structures to the surfaces of cell membranes. The plasma membrane diagrammed in Figure 4-9 shows these elements in place, drawn to scale in a network just under the membrane.

The Function of Membranes in Transport

Life depends on the organization of molecules inside cells. Any severe disturbance in the concentrations of substances inside the cell, or of the kinds present, will impair cell function or even cause cells to die. The internal concentrations of molecules and ions are maintained in their numbers and kinds by the plasma membrane, which regulates the passage of all substances in and out of cells. How does the plasma membrane accomplish this function?

There are two basic mechanisms by which materials are transported across plasma membranes. One, called *passive transport,* simply reflects the differences in concentration of substances inside and outside cells. If the molecules are more highly concentrated outside, the direction of movement will be from outside to inside. If the concentration is higher inside, movement will be in the opposite direction. The rate of movement increases as the difference between concentrations outside and inside increases. These differences in concentration inside and outside are called *concentration gradients.* Passive transport in response

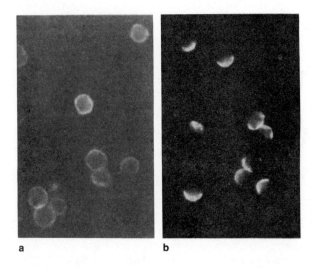

Figure 4-8 "Cap" formation in white blood cells. (**a**) Binding of proteins to the cell surface at random points. (**b**) Formation of the cap, in which the proteins are swept to one end of the cell. The proteins have been linked to a substance that fluoresces or glows in ultraviolet light to make them visible. Courtesy of G. M. Edelman, from *Proceedings of the National Academy of Sciences* 69 (1972): 608.

to concentration gradients requires no expenditure of cellular energy.

In the second mechanism transporting materials across cell membranes, *active transport,* substances are moved inside or outside *against* concentration gradients. Active transport requires cells to expend energy since substances are moved against concentration gradients; it stops if the energy-producing mechanisms of cells are experimentally inhibited.

Not all substances enter or leave cells by penetrating directly across plasma membranes. Many eukaryotic cells, in addition, are able to take up molecules by enclosing them in pockets that form in the plasma membrane. Uptake by this route, called *endocytosis,* involves active invagination of segments of the plasma membrane. The pockets formed in this way subsequently pinch off as closed, membrane-bound vesicles that sink into the cytoplasm beneath the plasma membrane. Materials can also be carried from inside the cell to the outside by mechanisms that reverse endocytosis. In this process, called *exocytosis,* molecules enclosed in membranous sacs in the cytoplasm are released to the exterior by fusion of the sacs with the plasma membrane. Cells regularly secrete large molecules, such as proteins, to the outside by this pathway.

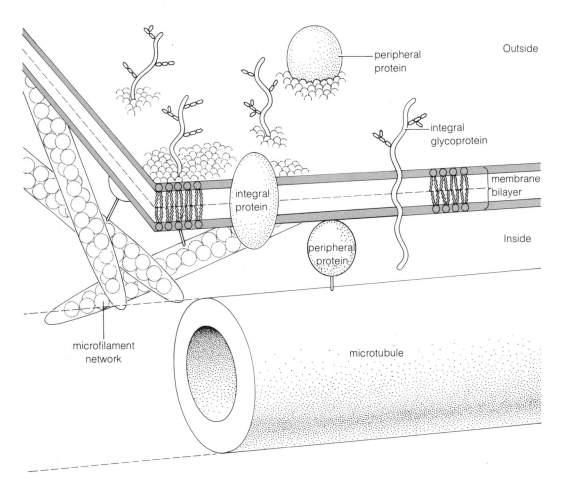

Figure 4-9 A fluid mosaic membrane with microfilaments and microtubules (drawn to scale) in position beneath the membrane.

Passive Transport

Passive transport along concentration gradients depends on the fact that the molecules in any space held at temperatures above absolute zero (−273°C) are in constant motion (see p. 53). The molecules travel in straight lines until they collide with other molecules within the space, or with molecules forming the boundaries of the space. Each molecule has a specific energy that depends on the velocity of its movement and the direction and force of its collisions with other molecules. This energy of movement is the *kinetic energy* of the collection of molecules (see also p. 53).

A net transfer of molecules from one region to another may result from this movement if the two regions have different concentrations of molecules; that is, different average numbers of molecules per unit volume. Imagine two collections of molecules in two spaces of equal volume that are initially separated by a barrier that the molecules cannot pass. The absolute temperature of the spaces is the same, but one of them contains more molecules. As a result, there is a greater amount of kinetic energy in the space containing the greater concentration of molecules because there are more moving particles with mass and energy in this space.

Suppose now that the barrier between the two compartments is removed. In the region of the boundary between the two collections of molecules, the movement and collisions on the more concentrated side will propel molecules into the less concentrated side. Molecules will also move from the less concentrated side into the more concentrated region, but over any time interval there will be more collisions and movement from the more concentrated side. As a result, there will be a *net* movement of molecules from the side of greater concentration to the side of lesser concentration. This net movement in response to concentration differences, called *diffusion,* will continue until the molecules are evenly distributed throughout the available space.

Diffusion shows how the laws of thermodynamics apply to all systems undergoing change, whether inside cells or in the inanimate world. According to the second law of thermodynamics (p. 53), systems run spontaneously toward a condition of greater disorder (or entropy). As they run to a less ordered state, they release energy to their surroundings. This energy is free energy, and it can accomplish work.

The system undergoing change in the example described above is the two collections of molecules in which one collection is initially more concentrated than the other. This arrangement is more ordered or organized than one in which the molecules are equally distributed throughout all of the available space (as they are after diffusion has taken place).[1] Once the barrier is removed, the total system will run spontaneously toward the less ordered state in which the molecules are equally distributed.

The Effects of Semipermeable Membranes on Diffusion
An artificial or natural membrane placed between two regions containing collections of molecules at different concentrations will have no effect on the final outcome of diffusion if all of the molecules or ions can pass through the membrane with equal ease. However, the net movement may be altered, sometimes in unexpected ways, if some of the molecules can pass through the membrane more readily than others, or are excluded entirely. Membranes having this effect are said to be *semipermeable.* All biological membranes have this property and thus alter the outcome of net movement in response to concentration gradients.

Osmosis One of the most surprising effects of semipermeable membranes on molecular movement can be demonstrated by a simple apparatus—a dish separated into two halves by a thin layer of cellophane (Fig. 4-10). On one side of the cellophane barrier the dish contains pure water. On the other side it contains a glucose solution. The water can pass freely in either direction, but the glucose molecules cannot penetrate through the cellophane and are retained on one side. Thus the cellophane acts as a semipermeable membrane. Within an hour enough water will move into the glucose solution to raise the level of solution noticeably on the glucose side; that is, there will be a *net* movement of water molecules from the pure water side into the glucose solution. The total system actually accomplishes work in raising the level of the glucose solution against the force of gravity. How is this work accomplished?

The free energy accomplishing this work is released by the water molecules, which follow a concentration gradient in moving toward the side of the dish containing the glucose solution. Although it is clear that a concentration gradient exists for the glucose molecules, which are obviously more concentrated on one side of the dish than the other, the gradient for water molecules is not at first so apparent. However, on the side of the membrane containing the glucose solution, much of the space available for water molecules is occupied by the dissolved glucose molecules. As a result, there are fewer water molecules per unit volume on the side containing the glucose than on the pure water side. Therefore, a concentration gradient between the two sides exists for the water. As a result, water will move across the membrane in response to the gradient.

Movement of water in response to a gradient of this type is termed *osmosis.* Osmotic movement of water in liv-

---

[1]Why this is so can be understood by considering a somewhat improbable example. Suppose that you are assigned the task of stacking bricks in a pile during an earthquake. Stacking the bricks, and keeping them stacked, requires a considerable and continuous effort on your part. At any time you stop returning bricks to the pile, they cascade downward until they assume a layer of more or less even thickness on the available surface. In falling from the pile to the surface, the bricks release the energy you expended in piling them. If harnessed in some way, the energy released as the bricks fall is free energy and could be used to do useful work. This situation is paralleled by molecules distributed unevenly on two sides of a barrier. The constant motion of molecules at all temperatures above absolute zero corresponds to the earthquake. The regions of greater concentration represent higher piles of molecular bricks on one side of the barrier. The piles tend constantly to fall until the molecules are evenly distributed throughout the available space on both sides. The amount of energy available for this movement depends on the height of the pile of molecular bricks, or in other words, on the magnitude of the concentration difference on the two sides of the barrier. Whether the molecules can run to an even distribution on both sides in response to the concentration difference depends simply on whether they can get through the barrier.

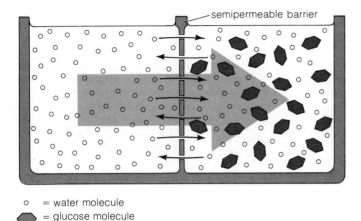

o = water molecule
⬡ = glucose molecule

Figure 4-10 Osmotic flow of water in a system in which a semipermeable barrier separates two compartments. The compartment on the left contains pure water. The compartment on the right contains a solution of glucose molecules in water. Although the water molecules can pass freely through the barrier, the glucose molecules cannot. A net movement of water molecules will occur from left to right (shaded arrow) in response to the gradient in concentration of water molecules. The gradient is set up by the presence of the glucose molecules in the right side of the dish (see text).

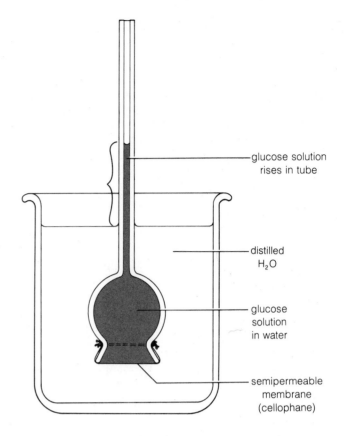

Figure 4-11 Apparatus demonstrating the ability of osmotic flow to accomplish work. Water will flow osmotically from the surroundings into the tube, raising the level of the solution in the tube. Flow continues until the weight of the column in the water (**d**) develops sufficient pressure in the solution to exactly counterbalance the tendency of water molecules to flow inward.

ing cells involves the same conditions described in the apparatus described above: (1) a semipermeable membrane must completely separate two water solutions, or a solution of molecules on one side and pure water on the other; (2) some of the nonwater molecules on one side must be unable to pass through the membrane; and (3) the concentration of substances on the two sides of the membrane must be unequal, so that there are more molecules of water per unit volume on one side of the membrane than the other. Water will then flow osmotically toward the side of the membrane containing fewer water molecules per unit volume. As the water molecules run down the gradient toward a condition of greater disorder or entropy, the system releases free energy that can accomplish work.

It is easy to show that osmotic flow of water can accomplish work. In the apparatus shown in Figure 4-11, distilled water surrounds a tube with an opening at its base. A sheet of cellophane is stretched across the opening, sealed tightly enough to prevent leakage around its sides. The tube contains a solution of glucose in water.

After a short time, the level of the solution in the tube will rise as water molecules move from the distilled water into the tube. The level of the solution will continue to rise until the pressure created by the weight of the raised so-

lution in the tube exactly balances the tendency of water molecules to move from outside to inside. At this point, the system is in balance or equilibrium, and, although water molecules will still move in both directions across the cellophane membrane, no further *net* movement of water will occur. The pressure required to exactly counterbalance the tendency of water molecules to move into the tube is a measure of the *osmotic pressure* of the solution in the tube.

Osmosis in Living Cells Cells act as osmotic devices similar to the apparatus shown in Figure 4-11 because they contain solutions of proteins and other molecules that are retained inside by a membrane impermeable to them but freely permeable to water. The resulting osmotic pressure is a force constantly operating in living cells. This force may be utilized to accomplish part of the activities of life, or it may act as a disturbance that must be counteracted for survival of the cell. The root cells of most land plants, for example, contain proteins and other large molecules in solution but are surrounded by almost pure water. As a result, net movement of water due to osmosis occurs into the root cells; the pressure developed contributes part of the force required to raise water into the stems and leaves of plants. On the other hand, protozoa and other small cells or organisms living in fresh water, to keep from bursting, must expend energy to excrete water constantly entering by osmosis. In some organisms, such as bacteria, blue-green algae, and plants, thick cell walls (see Figs. 1-3, 1-4, and 1-15) keep the cells from bursting due to the development of osmotic pressure. In such cells, the pressure developed by osmosis keeps the cell contents pressed tightly against the restraining wall. In plants, this pressure supports the softer tissues of stems and leaves against the force of gravity. Cells living in surroundings containing highly concentrated salt solutions have opposite problems and must constantly expend energy to replace water lost to the outside through osmosis.

The Effects of Membrane Lipids and Proteins on Passive Transport The cellophane film used as a semipermeable membrane in an osmosis apparatus acts essentially as a uniform molecular sieve, with no chemical or physical effects beyond stopping the movement of molecules beyond a certain size. However, the lipids and proteins of biological membranes, arranged as they are in a fluid mosaic of subregions with distinct hydrophobic and hydrophilic properties, produce a semipermeable barrier that modifies the behavior of many molecules as they diffuse in response to concentration gradients.

The effects of membrane lipids and proteins on diffusion were first given detailed study at the turn of the century by Ernst Overton, who followed the behavior of substances as they penetrated into plant and animal cells. Overton observed that the major difference between diffusion across biological membranes and artificial barriers such as cellophane is that lipid solubility modifies the rate of penetration into cells. Generally, the more soluble a substance in lipids, the more rapidly it penetrates into living cells, up to a limit determined by molecular size. Overton's work, as noted on page 69, provided the first clue that cells are surrounded by a surface layer that is lipidlike in nature.

Most nonlipid substances, in contrast, do not pass through membranes very rapidly if at all. However, there are certain exceptions to this general rule, all of them im-

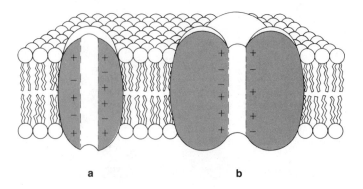

Figure 4-12 Formation of polar or charged channels through the interior of single membrane proteins (**a**) or by alignment of several membrane proteins (**b**).

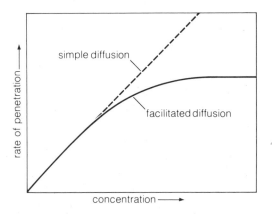

Figure 4-13 Passive transport by simple diffusion is proportional to concentration, and increases with concentration in an essentially linear fashion. Facilitated diffusion, in contrast, drops off at higher concentrations, leveling off or reaching saturation at a maximum rate that does not respond to further increases in concentration.

portant to the molecular economy of cells. Water passes through membranes rapidly even though it is a strongly polar substance. Other hydrophilic substances necessary for cellular life, including glucose, various amino acids, and ions such as Na^+ and Ca^{2+}, also pass through membranes rapidly even though they are polar or charged.

Recent research has demonstrated that the rapid penetration of these hydrophilic substances is due to the integral proteins of membranes. Their penetration stops if integral membrane proteins are denatured or removed from cellular membranes. The proteins, suspended as in-

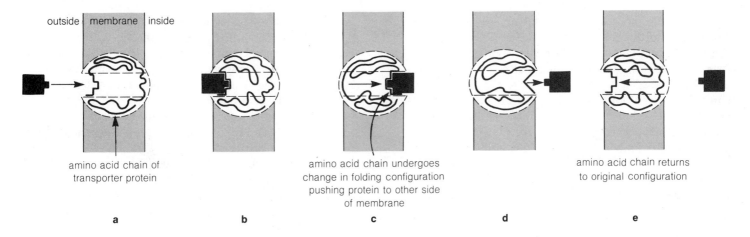

outside | membrane | inside

amino acid chain of
transporter protein

amino acid chain undergoes
change in folding configuration
pushing protein to other side
of membrane

amino acid chain returns
to original configuration

a   b   c   d   e

Figure 4-14 Facilitated diffusion by a channel forming carrier molecules (see text).

dividual particles or groups within the lipid bilayer framework of cellular membranes, evidently form polar or charged channels that extend through the membrane (Fig. 4-12). The hydrophilic nature of the channels is determined by the amino acid chains of the membrane proteins, which fold so that the channels are lined by polar or charged groups. The channels may be formed by a polar opening extending through a single protein (Fig. 4-12a) or by several proteins that become aligned to form a channel between them (Fig. 4-12b).

Although the passive penetration of the exceptional polar substances through membranes is rapid, their passage still depends on diffusion: the energy required for this transport is provided by favorable concentration gradients, and does not require an expenditure of cellular energy. Exceptional transport of polar or charged substances of this type, which follows concentration gradients but proceeds at rates significantly higher than predictions based on lipid solubility, is called *facilitated diffusion.*

Measurement of the rate of penetration of substances transported by facilitated diffusion reveals a fundamental characteristic common to all of them except water. (Water molecules are so small that they evidently pass readily through the polar membrane channels according to concentration gradients: the higher the gradient, the more rapidly water moves through biological membranes.) At successively higher concentrations of the larger polar molecules, the degree of enhancement drops off, until at some point further increases cause no further rise in the rate of penetration (Fig. 4-13). This is in sharp contrast to the behavior of nonpolar molecules that penetrate according to

their lipid solubility. For these molecules, permeability increases regularly with concentration, with no dropoff at higher levels.

The dropoff noted for facilitated diffusion at high concentrations closely resembles the behavior of enzymes in catalyzing biochemical reactions (see p. 64 and Fig. 3-10). As the concentration of reactant molecules increases, enzymes gradually become saturated and the rate of the reaction levels off. Another characteristic shared between the membrane proteins involved in facilitated diffusion and enzymes is *specificity:* only certain molecules, or groups of closely related molecules, are speeded in reaction rate by enzymes. Similarly, only certain molecules or molecular groups are speeded in transport by the membrane carrier molecules. These similarities suggest that the membrane proteins carrying out facilitated diffusion share many properties with enzymes.

These similarities to enzymes have suggested how the membrane proteins facilitating the diffusion of polar substances might work (Fig. 4-14). The membrane proteins are considered to be fixed in position, held in place by polar and nonpolar surface groups that anchor them in the lipid bilayer. The polar channel extending through the protein has an "active site" directed toward the side of the membrane facing the substance to be transported (Fig. 4-14a). Molecular collisions result in binding of the transported substance to the active site, which, like the active site of an enzyme, is tailored to fit the transported molecule (Fig. 4-14b). This combination causes a change in the folding pattern of this portion of the protein molecule, causing movement of the active site through the polar channel to the other side of the membrane (Fig. 4-14c). The change to this folding conformation alters the active site and reduces its affinity for the transported substance. As a consequence, the transported molecule is released on the other side (Fig. 4-14d). This release changes the protein to its original folding pattern, with the active site exposed in a position to combine with a second molecule (Fig. 4-14e). The energy required for movement of this type, involving limited internal regions of the carrier protein, is provided by the favorable concentration gradient.

Passive transport, involving simple and facilitated diffusion, accounts for the movement of a wide variety of substances to and from cells. By this mechanism cells are able to absorb many of the hydrophobic and hydrophilic molecules required for their biological reactions and release waste materials or secreted products to the outside. These transport processes, driven as they are by concentration gradients, take place without an expenditure of cellular energy. They greatly enhance the transport of key substances that would otherwise penetrate cells too slowly to support cellular life.

Active Transport

Transport of substances in or out of cells also takes place against concentration gradients. The cellular energy required for *active transport* of this type is supplied by linking transport to oxidation of fuel substances, usually through the medium of the ATP molecule (see p. 55). This dependence on cellular energy is the distinguishing characteristic of active transport.

Other features of active transport resemble facilitated diffusion. The process depends on membrane proteins, and stops if proteins are denatured or removed from membranes. Like facilitated diffusion, the mechanism of active transport has the property of specificity: only certain molecules or closely related groups of molecules are moved across membranes by a given active transport system. These characteristics provide the major criteria identifying active transport: (1) it goes against concentration gradients; (2) it requires cellular energy; (3) it depends on the presence and activity of membrane proteins; and (4) it is specific for certain substances or closely related groups of substances. The work of active transport may make up a major part of the total energy expenditure of cells.

Active transport mechanisms move a variety of substances in and out of cells at the expense of energy, including ions, a number of different sugars and other fuel molecules, and several amino acids. Each ionic or molec-

ular type in these categories is transported by a separate protein or group of proteins that "recognizes" and moves it across the membrane.

The active transport mechanisms moving these substances, like the facilitated diffusion mechanisms, depend on the ability of proteins to take up a variety of folding patterns (Fig. 4-15). According to current models for active transport, the end of the transport protein exposed on the inside surface of the membrane has a folded region that exactly matches the ATP molecule (Fig. 4-15a). ATP molecules are constantly colliding with the inside membrane surface; when one strikes the exposed binding site of the active transport protein, it attaches. The attachment site splits off one of the three phosphates (Fig. 4-15b), converting ATP to ADP, and releases free energy. The binding and energy release cause a change in the folding pattern of the protein. The change moves a small segment of the protein to a new position in which (1) a binding site for the substance being transported is exposed on the outside membrane surface (Fig. 4-15c), and (2) a strain is imposed on the protein chain. The effect is much like using the energy liberated by ATP breakdown to cock a spring within the protein molecule. The external site then binds a molecule of the transported substance colliding with the membrane surface (Fig. 4-15d). This binding releases the transporter "spring," pushing the transported molecule through the hydrophilic channel to the cell interior (Fig. 4-15e). As the protein segment carrying the molecule reaches this position, changes in the folding pattern rearrange the binding site so that it no longer fits the transported molecule. As a result, the transported substance is released into the cell interior (Fig. 4-15f). Release of the phosphate group also occurs at this time. These events trigger the final change in the folding pattern of the protein, returning it to the original state (Fig. 4-15a) in which it is ready to bind a fresh ATP and restart the transport cycle.

There is extensive evidence that the active transport mechanisms actually work in this way. The proteins responsible for active transport can be isolated from cell membranes. When added under the correct conditions to artificial phospholipid films, they retain at least partial activity and can continue active transport in their new location if supplied with ATP. Chemical analysis of the active transport proteins at various steps in the process show that they actually bind and split ATP, and also attach to the transported molecule. For example, the molecular system actively transporting an ion, Ca^{2+}, across membranes was first isolated and purified in the early 1970s by David H. Lennon at the University of Toronto. The complex proved to contain several polypeptides, including one ca-

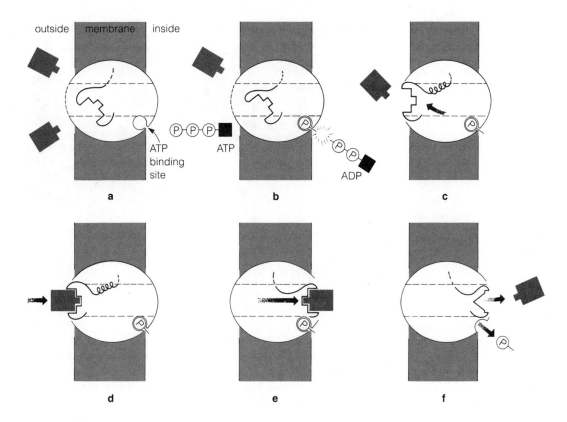

outside membrane inside

ATP
binding
site

ATP

ADP

a   b   c

d   e   f

Figure 4-15 Active transport by an ATP-dependent carrier molecule (see text).

pable of binding and splitting ATP. A second investigator, Ephraim Racker of Cornell University, showed in 1972 that introducing the purified Ca^{2+} carrier proteins into artificial phospholipid membranes gave them the capacity to move the Ca^{2+} ion against a gradient when supplied with ATP.

One interesting and highly significant finding developed through study of the Ca^{2+} transport system and other carriers actively moving ions across membranes is that the transport mechanisms can be forced to run in reverse. This was accomplished in the Ca^{2+} system, for example, by raising the Ca^{2+} concentration to abnormally high levels on the side of the membrane toward which it is transported. Under these conditions, the entire series of chemical springs and levers in the carrier system is forced to run in reverse, with the result that, as Ca^{2+} moves backwards through the transport channel, the carrier *adds phosphate groups to ADP molecules to form ATP.* These observations stand among the most important and significant experimental findings of biology, because they demonstrate that *a concentration difference can be used by membrane proteins as*

the energy source for synthesizing ATP. This principle, as Chapters 6 and 7 will show, underlies the mechanisms synthesizing ATP for cellular activities in both mitochondria and chloroplasts.

The systems that actively transport ions are among the most important transport mechanisms of both prokaryotes and eukaryotes. Almost all animal cells maintain internal Na^+ and K^+ concentrations that differ from concentrations outside by means of their active transport carriers. Normally, in these animal cells, Na^+ is excreted, so that the internal Na^+ concentration becomes lower than in the surroundings. K^+ is actively transported in the reverse direction so that it becomes more concentrated inside cells than outside.

Active transport of these ions is of special significance for animal cells because the movement of charged particles such as ions amounts to generation of an electrical current. The currents produced by active transport of Na^+ and K^+ ions across nerve cell membranes set up the conditions required for generation of electrical impulses by nerve cells. These impulses provide the basis for the complex activities of the brain and nervous system and the sensory functions coordinating the rapid and complex behavior of animals. Excretion of Na$^+$ by the same pumping systems also helps control the internal osmotic pressure of animal cells. Ca^{2+} transport by active mechanisms is important in the mechanisms regulating the activity of muscle cells in animals. Plant cells also concentrate some ions, including K$^+$, against a concentration gradient. Interestingly, most plant cells do not regulate internal Na$^+$ concentration as do animal cells; the osmotic effects of high internal concentrations of this ion, which can cause animal cells to swell and burst if not eliminated, are counteracted in plants by the cell walls.

Endocytosis and Exocytosis

The processes of passive and active transport are limited to the movement of ions and relatively small molecules between a cell's surroundings and its interior. Cells also have mechanisms that allow them to absorb or to release larger molecules such as nucleic acids or proteins, or even molecular aggregates as large as whole cells. In contrast to active and passive transport, these uptake and release mechanisms include membrane rearrangements extensive enough to be seen under the electron or even the light microscope.

Uptake of materials from the outside, called *endocytosis*, proceeds by either of two processes. One, called *pinocytosis* (meaning "cell drinking"), operates when the large molecules being taken in are in solution. The second mechanism, *phagocytosis* (meaning "cell eating"), involves the uptake of larger aggregates of particulate matter. Both pinocytosis and phagocytosis require a cellular energy input to proceed. Release of materials to the outside of the cell, called *exocytosis*, takes place by mechanisms that essentially reverse pinocytosis and phagocytosis. Apparently, cellular energy is not required for exocytosis to proceed.

Tracing the pattern of pinocytosis (Fig. 4-16) under the electron microscope shows that it occurs in three major steps (Fig. 4-17). Initially, the substance being taken up is bound to the plasma membrane (as in Fig. 4-17a). The membrane region binding the substance then invaginates (Fig. 4-17b), producing a cup-shaped depression that deepens and pinches off, forming an unattached vesicle that sinks into the underlying cytoplasm (Fig. 4-17c). Digestive enzymes may then be secreted into the vesicle,

Figure 4-16 Pinocytosis in *Amoeba*. Narrow, tubelike invaginations of the plasma membrane can be seen at several points (arrows). Beneath the plasma membrane the channels pinch off into rows of small vesicles that sink into deeper layers of the cytoplasm. Courtesy of D. M. Prescott and Academic Press, Inc., from *International Review of Cytology* 8 (1960): 484.

causing the enclosed substance to break down into small molecules that can pass across the vesicle membranes.

Pinocytosis has been detected in many types of animal cells. In mammals, ingestion of large molecules by this route occurs regularly in intestinal, liver, blood, kidney, and tumor cells among others. Whether pinocytosis takes place in plants is controversial. Surface invaginations and channels have been observed in some plant cells, particularly in roots, but whether molecules actually enter by this route has not been established.

The particles taken in by phagocytosis, which may be as large as bacteria or even other eukaryotic cells, enter by a three-step mechanism basically similar to pinocytosis, by (1) surface binding, (2) membrane invagination, and (3) vesicle formation (Fig. 4-18). The vesicles formed, however, are much larger, and often persist visibly in the cytoplasm for extended periods of time. In most cases, the material in the vesicle is broken down by enzymes released into the vesicle interior. Phagocytosis occurs commonly in protozoa such as *Amoeba* and in mammalian cells such as the white blood cells of the bloodstream, which remove foreign particles from the body by this mechanism.

In the reverse mechanisms of *exocytosis* (Fig. 4-19), material enclosed in membrane-bound vesicles is carried through the cytoplasm to the inner surface of the plasma membrane (Fig. 4-19a). The vesicle then fuses with the plasma membrane (Fig. 4-19b). This fusion releases the vesicle contents to the cell exterior (Fig. 4-19c). Proteins secreted to the outside of the cell are regularly released by

6. What experiments suppor[t]
pended in membranes as parti[cles?]

7. What holds membranes to[gether?] What are pe-]
ripheral membrane proteins?

8. What positions do the car[bohydrate groups of]
glycoproteins take in membran[es?]

9. In what ways are microtub[ules associated]
with membranes? How do th[ey associate with mem-]
branes? What evidence suppor[ts]

10. What is the difference betw[een]

11. What provides the energy [for]

12. What is a concentration gra[dient?]

13. What is diffusion? What ca[uses it?]

14. What is a semipermeable m[embrane?]

15. What is osmosis? What con[ditions cause it]
to occur? What provides the [energy for the]
movement of water?

16. What is osmotic pressure? [How is it related]
to cellular life? Compare a cell w[ith]
in Figure 4-11.

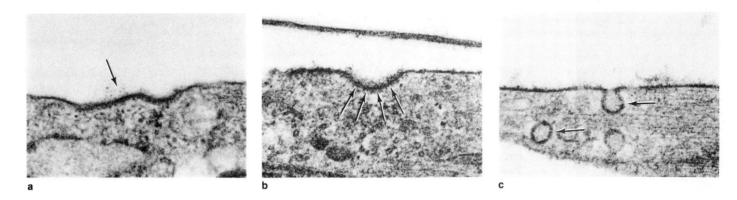

Figure 4-17 Formation of pinocytotic vesicles in cultured human cells, as seen under the electron microscope. The substance taken up, a lipoprotein, has been linked to a large, dense molecular complex to make it visible in the electron microscope. (**a**) Binding of the lipoprotein complex to the plasma membrane. (**b**) and (**c**) Invagination of membrane segments binding the complex. A free vesicle is visible in the cytoplasm beneath the plasma membrane in (**c**). × 57,000. Courtesy of R. G. W. Anderson, from *Proceedings of the National Academy of Sciences* 73 (1976): 2434.

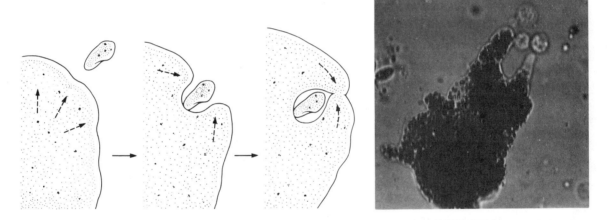

Figure 4-18 Phagocytosis. Part of the cell flows around a large particle to form a pocket in the cell membrane. The pocket pinches off as a vesicle, which sinks into the cytoplasm. The enclosed material may be hydrolyzed by enzymes secreted into the vesicle. The light micrograph shows an amoeba engulfing a food particle by phagocytosis. × 240.

exocytosis in this fashion. Residual undigested material in vesicles taken in by pinocytosis or phagocytosis may be expelled from cells by the same route (secretion of proteins by exocytosis is taken up in more detail in Chapter 10).

The basic transport mechanisms described in this section, passive transport, active transport, endocytosis, and exocytosis, account for the movement of all substances between cells and their surroundings. As we have seen, pas-

sive transport, although it may operate with some specificity as in facilitated diffusion of sugars and amino acids, depends on concentration gradients and requires no direct energy input by the cell. Active transport of substances against concentration gradients requires energy in amounts sometimes representing a significant part of the cell's total output. All of the transport mechanisms depend on the activity and properties of membrane lipids and proteins,

The Cell Surface

and their arrangement into biological membranes.

Suggestions for Further R

Bonting, S. L. and J. J. H. M. *Mammalian cell membranes.* Vol. Robinson. Butterworths: Bosto

Bretscher, M. S. and M. C. Ra branes. *Nature* 258:43–49.

Capaldi, R. A. 1974. A dynam *tific American* 230:27–33 (March

Christensen, H. N. 1975. *Biol* Reading, Mass.

Lodish, H. F. and J. E. Roth membranes. *Scientific American*

Quinn, P. J. 1976. *The molecular* Park Press: London.

Silverstein, S. C., R. M. Stein cytosis. *Annual Review of Bioch*

Singer, S. J. and G. L. Nicolso the structure of cell membran

Weissmann, G. and R. Claibor Publishing: New York.

Wolfe, S. L. 1981. *Biology of the* Calif.

Cells make contact with the outside world through the cell surface, which includes the outer face of the plasma membrane and external coats such as the walls of plant cells. Through this contact layer, cells send and receive chemical and physical signals, recognize other cells as part of the same individual or foreign, and adhere to other cells or to a nonliving substrate. The cell surface also provides the outermost line of defense against mechanical injury and attack by other cells or viruses.

The cell surface is specialized in various ways to carry out these tasks. Recognition and reception of signals from outside the cell are based on the carbohydrate groups of membrane glycolipids and glycoproteins. The carbohydrate portions of these molecules extend like antennas from the plasma membrane, forming a "sugar coating" or *glycocalyx* at the cell surface that in effect serves as the eyes and ears of the cell. Other sugar groups of the carbohydrate-containing membrane lipids and proteins function in cell adhesion. Some of this adhesive function depends on the interaction of individual glycoprotein and glycolipid molecules with other cells or with the nonliving environment. Other adhesions depend on highly specialized arrangements of carbohydrate-containing surface molecules into *cell junctions* of various kinds. These junctions, each with a distinctive structure, anchor cells to each other, seal cell boundaries, and form channels for direct transport and communication between cells.

Cells also secrete exterior coats that provide anchorage, protection, and support for the cell contents. Many animal cells secrete material to the cell exterior consisting of fibrous proteins and complex carbohydrates that serve these functions. In plants, fungi, and prokaryotes, the extracellular material forms a cell wall that functions as a supportive and protective layer. In the many-celled plants, the external walls also anchor cells in the plant and, through minute openings in the walls, provide channels of communication between cells.

The Cell Surface in Reception, Recognition, and Cell-to-Cell Adhesion

Carbohydrate-bearing lipids and proteins of the plasma membrane receive signals from outside, recognize other cells as part of the same individual or foreign, and adhere to other cells or to a substrate. The molecules active in these functions, the membrane glycolipids and glycoproteins, consist of a basal phospholipid or protein structure embedded within the membrane and a branched carbohydrate chain that extends outward, away from the cell surface (Fig. 5-1). The carbohydrate groups of both glycolipids and glycoproteins contain a group of similar sugars, including *glucose, galactose, mannose, fucose,* and *sialic acid* (Fig. 5-2). The varieties in which these sugar units may combine to form surface carbohydrates are almost endless. Glucose alone can combine with itself through different linkages by twos to form 11 different types of disaccharides and by threes to form 176 distinct trisaccharides. Other sugars combine with glucose to form complex carbohydrates containing mixtures of sugar units in both straight and branched chains.

Surface Carbohydrates and Cell Reception

One of the most exciting developments in recent years was the discovery that cells contain external receptors that can recognize and bind a wide variety of substances. Binding of many of the molecules recognized by the receptors results in transfer of a *signal* across the plasma membrane to the cell interior. The signal triggers a specific internal

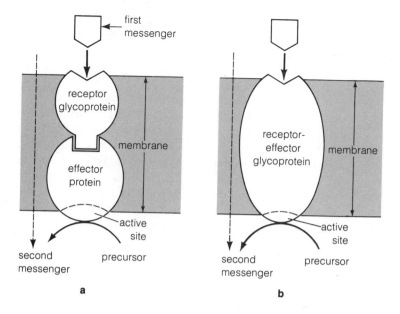

a

b

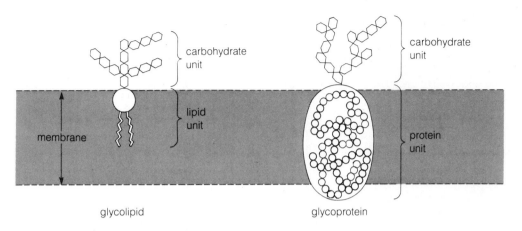

Figure 5-1 The structure of membrane glycolipids and glycoproteins.

transport of glucose across the plasma membrane, and the rapid breakdown of this fuel substance in the cytoplasm.

Persons with the disease *diabetes mellitus* are unable to produce sufficient quantities of the insulin hormone. Body cells in affected individuals, particularly in muscles, fail to take up glucose because of the low levels of the insulin acting as first messenger in the bloodstream. Uptake by liver cells, which store glucose by converting it to glycogen, is also inhibited. The condition leads to muscular weakness and an accumulation of glucose in the bloodstream. The high blood content of glucose leads directly and indirectly to a variety of other effects, including alterations in nerve and brain function that, if untreated, can produce coma and death. Fortunately for individuals with the disease, injection or oral administration of the insulin hormone restores glucose uptake by body cells and glucose metabolism returns to normal levels.

Interestingly, insulin and several other peptide hormones eventually enter their target cells through endocytosis (see p. 84). After binding to the receptors and initiating the cellular response, insulin is taken in through invaginations in the cell surface that pinch off as vesicles and sink into the underlying cytoplasm. After entrapment in the vesicles, the hormone is degraded by enzymes introduced from the cytoplasm. This process, since it takes place after initiation of the cellular response through hormone binding to surface receptors, is probably a mechanism "clearing" the cell surface for reception of further messages.

response, such as protein synthesis, cell division, or an increase in breakdown of cellular fuels to release energy.

The molecules active in surface reception have been identified in most cases as glycoproteins. Identification has been carried out by several methods. In one method a signal molecule recognized by the cell surface is marked by attaching it chemically to a radioactive label or another molecular group that can be identified biochemically. Cells are then exposed to the signal molecule; after binding, the membranes are broken down. The membrane groups binding to the marked signal molecules are then isolated and purified. Membrane glycoproteins can usually be identified as the recognition and binding molecules by this technique.

Binding and cellular response to insulin provide an excellent example of the role of these carbohydrate groups in surface reception, and of the cellular response that follows. Insulin is a peptide hormone secreted by the pancreas in mammals. Cells exposed to insulin quickly increase their rate of glucose transport and breakdown. Insulin is bound by receptor molecules at cell surfaces. Isolation of the insulin-receptor combination from cellular membranes shows that the receptor molecule is a glycoprotein.

The rate of glucose breakdown increases almost immediately after cells bind the hormone at their surfaces. The initial internal response occurs without penetration of the insulin molecule through the plasma membrane. In some way, the receptor glycoprotein transmits a signal across the membrane indicating to the cell interior that the hormone has been recognized and bound at the cell surface.

While the actual steps in transmission of the surface signal by the receptor glycoproteins are still unknown, observations made with insulin and other hormones have suggested a hypothesis for the receptor-response mechanism. Like the models developed to explain facilitated diffusion and active transport (see Chap. 4), the receptor-response hypothesis depends on the ability of protein molecules to change their folding patterns in response to combination with other molecules or ions.

In the receptor-response model (Fig. 5-3), the hormone or other molecule initiating the cellular response is called the *first messenger*. At the cell surface, the *receptor*, a glycoprotein, contains a carbohydrate unit capable of recognizing and binding the first messenger. A second unit, the *effector*, lies at the inner membrane surface. The effector may be a separate protein molecule or an inward extension of the receptor glycoprotein. Binding of the first messenger to the receptor causes a rearrangement in the folding pattern of the receptor molecule. This rearrangement, in turn, converts the effector to an active form. Activation causes the effector to generate or adjust the internal concentration of the *second messenger*, a molecule that then diffuses to the site where the biochemical response is carried out. In the case of insulin, the effector is believed to be a membrane protein regulating the penetration of Ca^{2+} across the plasma membrane. Activating the effector opens a polar channel that allows Ca^{2+} rapidly to enter the cell. Once inside, Ca^{2+} acts as the second messenger by activating a group of enzymes that depend on the presence of Ca^{2+} for their function. Ultimately, the activity of these enzymes leads to increases in the

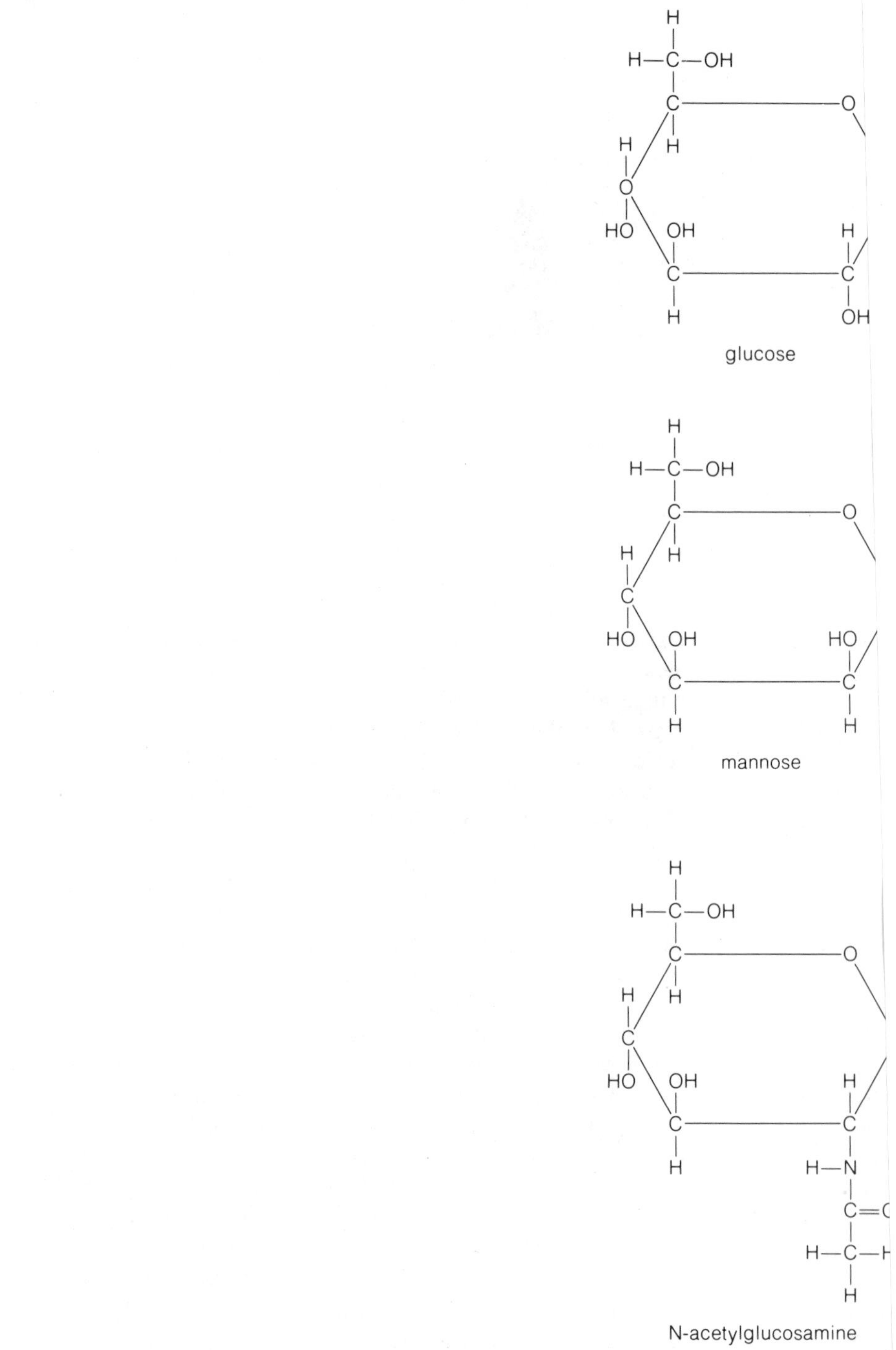

H—C—OH

HO OH H

glucose

H—C—OH

HO OH HO

mannose

H—C—OH

HO OH H

H—N
C=O
H—C—H
H

N-acetylglucosamine

Figure 5-2 The carbohydrate groups linking to lip

person, containing different HL-A markers, are transplanted into an individual, the introduced cells are recognized as foreign. The foreign cells are then destroyed by the same mechanisms that protect the body against invasion by disease organisms. Both the HL-A markers, which form the main barriers to successful tissue transplants between unrelated persons, as well as the surface receptors binding them have been identified as membrane glycoproteins. The part of the foreign HL-A glycoproteins that stimulates the body's defensive response is the carbohydrate portion of the molecule. These carbohydrate units extend from the cell surface, forming a part of the carbohydrate "antennas" covering the cell exterior. In some cell recognition systems, such as the surface markers responsible for blood groups in humans, the surface carbohydrate chains responsible are attached to membrane lipids to form glycolipids as well as glycoproteins.

Cell-to-Cell Adhesion and Aggregation

The capacity to recognize other cells as part of "self" or at least of the same species evidently extends throughout the animal kingdom, and has also been observed in protozoa and fungi such as the slime molds. This capacity has been demonstrated most convincingly by Tom Humphreys and his colleagues at the University of Hawaii, and by Max M. Burger at the University of Basel, in their research with sponges. The bodies of many sponges are so loosely organized that individuals can easily be dispersed into separate cells. Under the correct conditions the disaggregated cells can sort themselves out and reassemble into whole, functional sponges again. If dispersed cells from two different species are mixed together, reaggregation often occurs by species: the cells sort themselves out by species and adhere to produce separate, distinct individuals of the two species again. This capacity for self-recognition and adhesion has been shown to depend on surface carbohydrate groups carried by membrane glycoproteins in the sponge cells.

Cell-to-cell recognition and adhesion by means of membrane glycoproteins have also been detected in higher animal groups. Cells of the brain in mammalian embryos can reaggregate if the brain is dispersed. During reaggregation, the dispersed cells reconstruct normal brain cell associations to a remarkable degree. Other experiments have shown that kidney, heart, and other cells of animal embryos have the same capacity to reassociate. Dispersed cells from these tissues, if mixed, will sort out and form

clusters showing internal organization typical of their tissue type. Even whole embryos of many vertebrates, such as frogs and salamanders, also have the capacity to reassemble if the embryos are dispersed into single cells at early stages of development.

All of this research with membrane receptors, cell recognition, and cell-to-cell aggregation has revealed that the carbohydrate coat of animal cells acts as a sort of sensory system at the cell surface. The glycoprotein and glycolipid molecules of this system allow cells to recognize stimuli in the form of hormones and other molecules and to respond to them. They also provide the basis for recognition of the cells of an individual as self or foreign. In addition, these membrane glycoproteins and glycolipids supply a mechanism for the adhesion of cells into the tissues and organs of many-celled animals.

Cell Junctions

The initial adhesion of cells is quickly followed in animal embryonic development by the formation of cell junctions of various kinds. Once formed, the junctions usually persist throughout the life of the individual.

Cell junctions occur in three major types, each with basically different functions. *Adhesive junctions* hold cells together, acting as an intercellular "button" maintaining cells in their fixed positions in tissues. *Sealing junctions* close the space between cells to the diffusion of molecules, preventing the flow of molecules between cell layers. *Communicating junctions* provide open channels allowing direct flow of ions and small molecules from one cell to another. Junctions of these three types occur widely in the animal kingdom.

Adhesive Junctions

The primary type of adhesive junction, found in both vertebrate and invertebrate animals, is the *desmosome* (from *desmos* = bond). In the region of a desmosome (Fig. 5-4), fibrous matter forms a thick layer just under the plasma membrane in both cells forming the junction. Other, thicker fibers radiate into the surrounding cytoplasm on both sides of the junction. The thick fibers, called *tonofilaments*, evidently anchor the desmosome to the underlying cytoplasm. Tonofilaments belong to the class of cytoplasmic supportive fibers known as *intermediate filaments* (see p. 12). Thin fibers are also visible in the extracellular

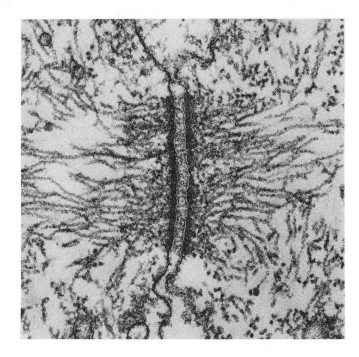

Figure 5-4 A desmosome binding adjacent cells together in the skin of a salamander. × 22,000. Courtesy of D. E. Kelly, from *Journal of Cell Biology* 28 (1966):51 by copyright permission of The Rockefeller University Press.

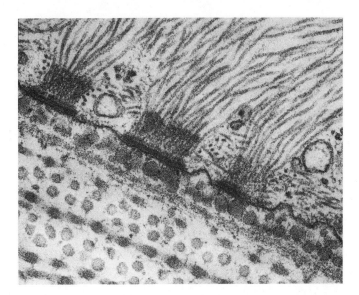

Figure 5-5 Half-desmosomes anchoring skin cells to an extracellular matrix in the salamander. × 68,000. Courtesy of D. E. Kelly, from *Journal of Cell Biology 28* (1966):51 by copyright permission of The Rockefeller University Press.

region between the plasma membranes of the two cells forming the junction. When viewed from the cell surface, in a direction 90° from the cross section shown in Figure 5-4, the entire complex looks roughly circular, much like a button holding the cells together.

Half-desmosomes (Fig. 5-5) occur where cells are anchored to extracellular material. Half-desmosomes also appear along the plasma membrane in cells grown in tissue cultures, where they anchor the cultured cells to the glass or plastic substrate on which they are grown.

The faintly fibrous material just outside the plasma membrane, in the regions between cells held together or attached to a substrate by desmosomes, has been identified as the carbohydrate portions of membrane glycoproteins. Digestion of this material with enzymes that attack either proteins or carbohydrates causes cells attached by desmosomes to come apart, indicating that membrane glycoproteins are responsible for holding the desmosome structure together.

Desmosomes are present between all cells bound together into tissues in multicellular animals. They are es-

pecially abundant in tissues subjected to physical stresses, such as the cells of the outside skin of animals, or the cells lining the surfaces of body cavities. Desmosomes appear early in embryonic development, as soon as cells take up their permanent positions in the embryo.

Other less organized adhesive junctions also occur between animal cells in addition to desmosomes (Fig. 5-6). In some of these additional junctions, the fibers anchoring the junction to the underlying cytoplasm are actin-containing microfilaments instead of intermediate filaments. The presence of microfilaments in association with these junctions suggests that they may serve a motile as well as adhesive function. Such junctions are common in the "end-plates" that link heart muscle cells together in end-to-end fashion.

Sealing Junctions

In sealing junctions, also called *tight junctions* (Fig. 5-7), the plasma membranes of the two cells forming the junction are so closely fused together that no extracellular space can be seen between them. The total width of the

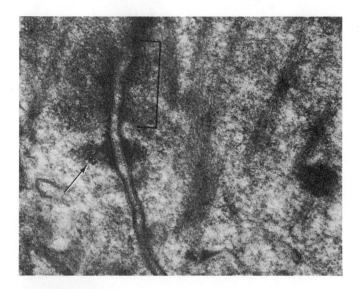

Figure 5-6 A less organized adhesive junction anchored to the underlying cytoplasm by microfilaments in intestinal cells (brackets). A desmosome is also visible in the micrograph (arrow). × 99,000. Courtesy of T. S. LeCount.

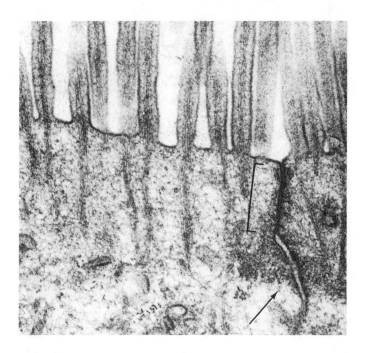

Figure 5-7 A sealing junction (brackets) between cells lining the intestinal cavity of a chick. An adhesive junction of the type anchored to the underlying cytoplasm by microfilaments is also visible between the cells (arrow). The narrow, fingerlike extensions at the top of the micrograph, called *microvilli*, greatly expand the absorptive surface of the intestinal cells. Courtesy of C. Chambers and Springer-Verlag.

membranes across the two junctions, in fact, is sometimes less than the thickness of two membranes placed side by side, suggesting that the outermost phospholipid layers join together to some degree in the region of the junction.

Sealing junctions form a complete circle around the tops of cells lining body cavities, as in the layer of cells lining the digestive tract of animals. The sealing function of these junctions was demonstrated in experiments by Daniel A. Goodenough and Jean-Paul Revel at Harvard University. In their experiments Goodenough and Revel introduced a stain into the circulatory system supplying tissue lining a body cavity. From the circulation the stain leaked into the spaces between cells, where it could be detected under the electron microscope. In these preparations, the stain penetrated into the extracellular spaces only to the margins of the cells lining the body cavity (Fig. 5-8), where it was stopped by the sealing junctions.

In addition to their function in sealing off the digestive tract in animals, sealing junctions are also found at the edges of cells in capillaries of the bloodstream in certain locations such as the brain, where a barrier is maintained against complete mixing of the blood and extracellular fluids. In some locations, as in the cell layers lining the urinary bladder, sealing junctions are suspected to prevent the flow of particles as small as ions from the bladder fluids into the spaces between cells.

Communicating Junctions

The third major type of junction, the communicating junction, also called a *gap junction* (Fig. 5-9), provides "holes" in the plasma membrane through which ions and small molecules can pass directly between cells. Communicating junctions appear as regions where the plasma membranes of adjacent cells are aligned and separated by a narrow space only 2 to 3 nanometers wide. Thin sections of communicating junctions made parallel to the cell surface reveal that the extracellular space contains closely packed hollow cylinders arranged in a regular hexagonal pattern in the region of the junctions.

The small cylinders in the junctions, which are aligned perpendicularly to the membrane surfaces, run directly between the opposed membranes and connect cells across the junction (Fig. 5-10). Each cylinder is pierced by a narrow central channel that extends entirely through both of the plasma membranes forming the communicating junc-

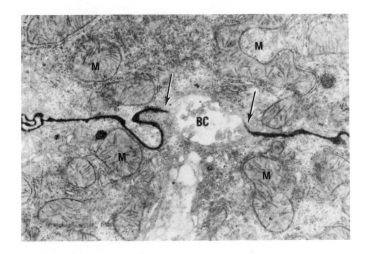

Figure 5-8 The activity of sealing junctions in mouse liver. The stain, which was introduced into the circulatory system of the liver before preparation of the tissue for electron microscopy, has penetrated through the extracellular spaces only to the region of the tight junctions (arrows) at the margins of cells lining a bile canal (*BC*). *M*, mitochondrion. Courtesy of D. A. Goodenough, from *Journal Cell Biology* 45 (1970):272 by copyright permission of The Rockefeller University Press.

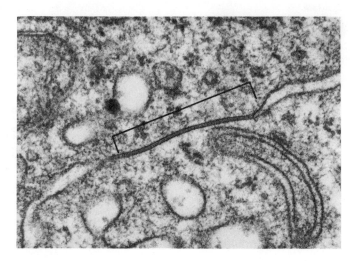

Figure 5-9 A communicating junction (brackets) between intestinal cells in a developing chick embryo. × 67,500. Courtesy of C. Chambers.

tion, opening an avenue of direct flow between the cells sharing the junction.

A variety of experiments has shown that molecules can pass freely and directly between cells through communicating junctions. In one of these experiments, Werner R. Loewenstein of the University of Miami injected a dye into one cell of a row of cells in an insect salivary gland connected by communicating junctions (Fig. 5-11). The dye moved rapidly from one cell to the next in the row, much faster than its expected rate of diffusion through plasma membranes. In other experiments Loewenstein and his colleagues placed electrodes in insect gland cells of the same type. Electrical resistance between these cells was much lower than the resistance noted across unbroken plasma membranes, indicating that ions could flow freely between the cells. The resistance noted was so low, in fact, that the rows of cells in the salivary glands could be considered electrically to contain a single, continuous unit of cytoplasm uninterrupted by membranes. The direct connections responsible for this unimpeded flow of ions and molecules between cells are believed to be the minute cylinders of communicating junctions.

Communicating junctions have been found in all animals studied to date. They occur, among other places, between some nerve and brain cells, in heart muscle, in smooth muscle, between liver cells, and between cells of the salivary glands of insect larvae. All cells in early embryos may be in communication through such junctions. Communicating junctions are subsequently lost from most embryonic cells as they differentiate, leaving only certain cells, such as nerve and heart cells, with these junctions.

Communicating junctions evidently provide avenues for the most rapid communication possible between cells. When present between nerve cells, in nerve-muscle connections, or between muscle cells, communicating junctions set up direct electrical connections that fuse the cells joined by the junctions into a single electrical unit. Responses between the cells linked together in this way are practically instantaneous.[1]

The three types of junctions described in this section, the adhesive, sealing, and communicating junctions, are adaptations characteristic of many-celled animals. The various junctions are the primary structures joining cells into

---

[1]Not all nerve cells are linked by communicating junctions; such junctions, in fact, are relatively rare between nerve cells in animals. Much more common is the *synapse,* a space between nerve cells across which communication is *chemical,* that is, by means of diffusing molecules, rather than electrical. Direct electrical connections through communicating junctions characteristically transmit nerve impulses much more rapidly than the chemical connections across synapses.

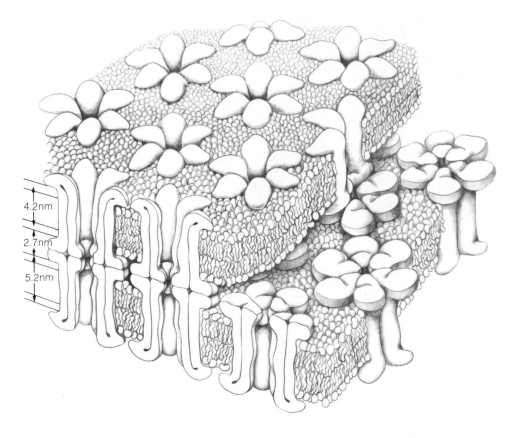

4.2nm

2.7nm

5.2nm

Figure 5-10 A model proposed by D. A. Goodenough and his coworkers for the structure of a communicating junction. Each cylinder is formed from a circle of six protein subunits. Courtesy of D. A. Goodenough, from *Journal of Cell Biology* 74 (1977):629 by copyright permission of The Rockefeller University Press.

the structural and functional groupings that make complex multicellular life possible in animals. Many of the equivalent functions in plants, as the next section will show, are supplied by cell walls.

Extracellular Surface Structures of Animal and Plant Cells

Almost all plant cells, and many types of animal cells, secrete materials into the region outside the plasma membrane. These secreted materials form extracellular structures that carry out a variety of functions. The extracellular material secreted by animal cells, primarily in supportive tissues, consists of glycoproteins of various kinds. Some of

this material occurs in very thin layers that support cells in various parts of the body. Thicker, more extensive deposits of extracellular material provide the tough, elastic structures of tendons, cartilage and bone, and supportive elements in various organs and blood vessel walls. In plants, extracellular materials consisting primarily of complex carbohydrates form the walls supporting the cells of these organisms.

In spite of their diversity, the extracellular structures of animals and plants all share two types of structural units. One type, the *fibers*, consists of long, extended molecules that resist stretching and other tensile forces. The fibers are embedded in the second element, the *matrix*, which consists of extensively branched molecules linked into a network. The network holds the fibers in place, and through retarding the flow of water molecules, resists

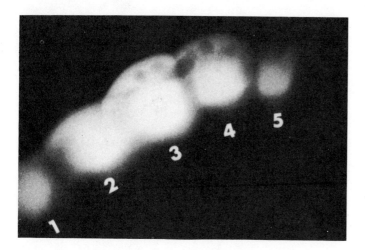

Figure 5-11 Diffusion of a fluorescent dye between adjacent cells in an insect salivary gland. The dye was injected into cell 3 and diffused rapidly to the other four. × 200. Courtesy of W. R. Loewenstein and Springer-Verlag, from *Journal of Membrane Biology* 5 (1971):51.

compression. Variations in the particular molecules present, degree of crosslinking, and amount of trapped water may produce matrix material ranging in consistency from soft, watery gels to hard, dense, almost rocklike material.

The Fibers and Matrix of Animal Extracellular Structures: Collagen and the Proteoglycans

A single type of glycoprotein, *collagen,* is the primary fiber of the extracellular supportive material secreted by animal cells. This single molecular type is produced in such quantity that it may be the most abundant body protein. In vertebrates, the collagen of supportive tissues makes up about half of the total body protein by weight.

Collagen is a unique glycoprotein characterized by a high content of a modified amino acid, *hydroxyproline* (see Fig. 2-17). Almost all of the carbohydrate portion of collagen, which makes up about 10 percent of the molecule's weight, is built up from two sugars, glucose and galactose (see Fig. 5-2). These sugars link in short two-unit chains attached to the protein component of the molecule.

Individual collagen molecules (Fig. 5-12) are linear, rigid structures about 1.5 nanometers wide and 300 nanometers long. These individual molecules line up in parallel, regularly overlapping rows to form collagen fibers. In

the electron microscope, collagen fibers, which may contain thousands of individual collagen molecules, typically show a pattern of regular cross-striations (Fig. 5-13). These cross-striations are believed to reflect, in some unknown way, the regular overlap of collagen molecules within the fiber structure.

The matrix network surrounding collagen fibers in extracellular structures is built up from a highly diversified group of glycoproteins called *proteoglycans.* Proteoglycans are distinguished from other cellular glycoproteins by their unusually high carbohydrate content, which may greatly exceed the protein part of the molecule by dry weight. In the molecule, the amino acid chain of the protein is believed to extend as a long, flexible central axis from which the carbohydrate chains extend like the bristles of a bottlebrush (Fig. 5-14). Large numbers of these bottlebrush structures link together to form the proteoglycan network.

The proteoglycan network entraps and impedes the flow of water molecules, forming the matrix that surrounds collagen fibers in the extracellular region. Depending on the relative proportions of fibers and matrix, the extracellular complex may range from watery gels to hard, incompressible masses. Cartilage, which contains matrix proteoglycans in high proportions, is relatively soft. Tendons, in contrast, contain collagen fibers in high proportion, and are tough and elastic. In bone, the matrix is impregnated with calcium phosphate crystals, producing a hard, dense, and relatively inelastic mass.

The Extracellular Fibers and Matrix Structures of Higher Plants: Cell Walls

The walls surrounding plant cells, like the extracellular material of animal cells, are constructed from fibers embedded in a matrix. However, unlike the collagen and proteoglycans of animal cells, the majority of the molecules of both the fibers and matrix in plant cell walls consists of polysaccharides of various kinds. These cell wall polysaccharides, although highly varied in type and structure, are built up from different combinations of a relatively short list of monosaccharide subunits. Proteins are also present as a minor constituent forming about 10 percent of the matrix, either as enzymes involved in wall synthesis, or as a structural part of the matrix network.

Cellulose, a long-chain molecule assembled from individual glucose units linked end to end (see Fig. 2-6a), is the predominant fiber of higher plant cell walls. In these cell walls, individual cellulose molecules combine to form

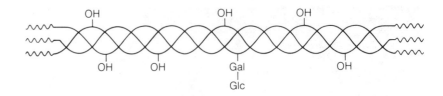

Figure 5-12 The collagen molecule, which consists of three peptide chains wound into a triple helix. Short glucose-galactose (*Glc-Gal*) carbohydrate chains are attached to points on the molecule's surface. The molecule is a fairly rigid, linear structure about 1.5 nanometers wide and 300 nanometers long. Many of these molecules combine to form a collagen fiber.

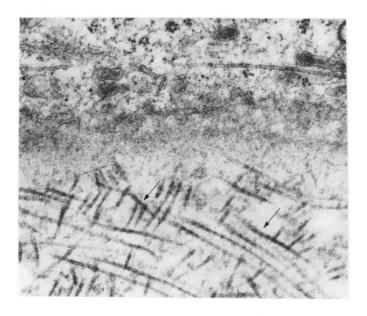

Figure 5-13 Collagen fibers (arrows) embedded in an extracellular region in a chick embryo. Note the cross-striations visible in the collagen fibers. × 41,000. Courtesy of R. L. Trelstad, from *Journal of Cell Biology* 62 (1975):815 by copyright permission of The Rockefeller University Press.

cellulose *microfibrils*, fibers about 3.5 nanometers thick and several micrometers long. With these dimensions, cellulose microfibrils are visible as fine, dense lines in thin-sectioned cell walls (Fig. 5-15). About 30 to 40 individual cellulose chains combine to form each cellulose microfibril. Within the microfibrils, individual cellulose molecules are held in parallel, unbent form by hydrogen bonds between adjacent glucose chains. The pattern of hydrogen bonds and the arrangement of the molecules within cellulose microfibrils are still unknown.

Although cellulose is the primary fiber of cell walls in higher plants, polymers of other monosaccharides or glucose in different linkages also form wall fibers, particularly in fungi and algae. Some of these additional fibers also occur in certain cells of higher plants. Long polymers formed from the sugar mannose are found in the cell walls of dates and coffee beans. Algae contain polymers of xylose, mannose, or glucose, or mixed polymers made up from all possible combinations of any two of these three.

The matrix materials of plant cell walls are made up of two major types of polysaccharide molecules, *pectins* and *hemicellulose*. Both of these molecular types consist of highly branched molecules that may contain a variety of individual sugar units linked in different ways (Fig. 5-16 shows an example of a cell wall hemicellulose). The highly branched nature of the pectins and hemicellulose enables them to build networks that trap water molecules readily. The networks, like the matrix materials of animal cells, range from soft gels to tough, highly rigid masses.

The matrix of cell walls also contains a structural protein in addition to the pectin and hemicellulose polysaccharides. The plant matrix protein, termed *extensin*, is believed to link covalently to various wall polysaccharides to form glycoproteins. In addition to glycoproteins, the matrix of the mature cell walls may contain varying amounts of *lignin*, a hard, dense substance constructed from polymers of complex alcohols. Much of the density and strength of hardwoods is due to large quantities of lignin laid down as a part of the cell wall matrix.

Working out how all these complex molecules combine in plant cell walls is one of the most difficult problems facing cell biologists. Because all the molecular classes in

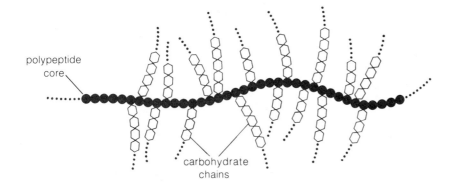

polypeptide
core

carbohydrate
chains

Figure 5-14 A structure proposed for a proteoglycan. The carbohydrate chains are believed to radiate outward from the protein axis like the bristles of a bottlebrush.

cell walls interact through both covalent linkages and hydrogen bonds, plant cell walls can be considered to contain a single "supermolecule" that surrounds individual cells and extends the length and breadth of the plant. Within this complexity, only the positions of cellulose microfibrils, because of their visibility in the electron microscope, have been worked out with any degree of certainty. These fibrils exist in the form of a mesh or network in newly synthesized cell walls (Fig. 5-17), with little or no parallel orientation. In maturing cell walls, cellulose microfibrils are laid down in a more ordered pattern, often forming nearly parallel rows (Fig. 5-18). These rows, in which the individual microfibrils may approach lengths of 2 to 5 micrometers, usually extend around the wall at an angle to the long axis of the cell. Frequently, successive layers are deposited with the angle changed so that a crossed pattern is formed in alternate layers (visible in Fig. 5-18).

Growth of Cell Walls in Higher Plants How do plant cells manage to increase in size, encased as they are in complete cell walls, and how does the nonliving wall material expand and extend during cell growth? Recent investigations have supplied at least partial answers to these questions, which have been of interest to botanists and cell biologists for more than a century.

The thin walls of embryonic plant cells are called *primary* cell walls. After the wall growth and thickening that takes place during cell elongation and maturation in plants, the walls are termed *secondary* cell walls. Primary cell walls contain cellulose microfibrils in a network (see Fig. 5-17) embedded in a relatively soft matrix. During the elongation occurring in cell maturation, which proceeds primarily in one direction, the microfibril network is first loosened and then stretched. Force for the elongation is

Figure 5-15 Section grazing the growing cell wall and underlying cytoplasm in a plant cell. Toward the left, the plane of section passes through the wall, revealing the cellulose microfibrils (arrows). Toward the right, the plane of section passes through the underlying cytoplasm, showing microtubules oriented in the same direction as the wall microfibrils. The cytoplasmic microtubules are believed to align the cellulose microfilaments as they are laid down in the cell wall. × 57,000. Courtesy of E. H. Newcomb and Annual Reviews, Inc., from *Annual Review of Plant Physiology* 20 (1969): 253.

supplied by osmotic forces built up in the cell. Throughout the period of elongation, new cellulose fibers are laid down just outside the plasma membrane, gradually thickening the wall as it extends. The newly added microfibrils of the secondary wall may be laid down in random patterns or in a more ordered arrangement similar to the one shown in Figure 5-18.

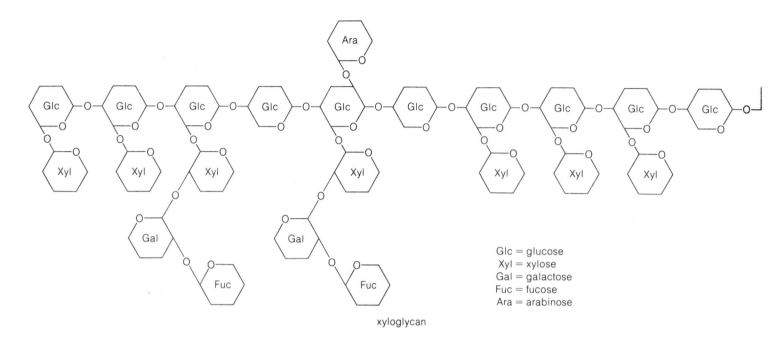

xyloglycan

Glc = glucose
Xyl = xylose
Gal = galactose
Fuc = fucose
Ara = arabinose

Figure 5-16 A structure proposed for a hemicellulose molecule of the cell wall matrix in higher plants. Courtesy of P. Albersheim, redrawn from "The walls of growing plant cells," by P. Albersheim, *Scientific American* 232 (1975):80. Copyright © 1975 by Scientific American, Inc. All rights reserved.

Figure 5-17 A newly formed cell wall in cotton, in which cellulose microfibrils are laid down in a meshlike network. × 16,000. Courtesy of A. L. Houwink, from *Acta Botanica Neelandica* 2 (1953): 218.

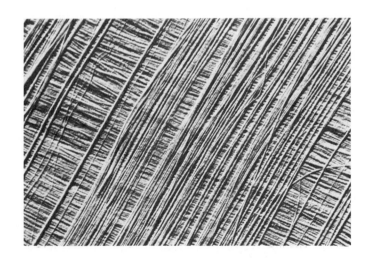

Figure 5-18 Ordered pattern of cellulose microfibril deposition in a maturing cell wall of a green alga. × 21,000. Courtesy of E. Frei and R. D. Preston, from *Proceedings of the Royal Society,* Series B, 154 (1961): 70.

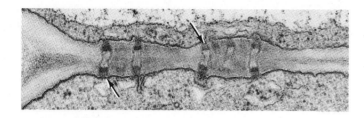

Figure 5-19 Plasmodesmata (arrows) in the wall separating two plant cells in corn. Courtesy of R. F. Evert and Springer-Verlag, from *Planta* 136 (1977): 77.

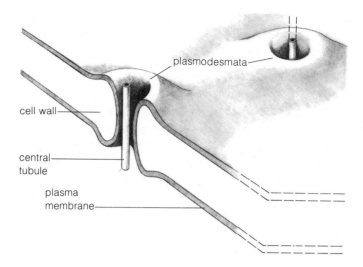

Figure 5-20 The structure of plasmodesmata. A single microtubule usually lies in the central channel of a plasmodesma.

During the first phases of elongation, sugar-digesting enzymes attack covalent linkages between the sugar residues of wall polysaccharides. These enzymes open linkages in the matrix molecules, allowing cellulose fibrils to slide over each other as the cell elongates. An interesting hypothesis about activation of these enzymes is based on the fact that experimentally lowering the pH of cell walls to pH 5 mimics the effect of some plant growth hormones in loosening cell wall structure. From this, the idea has been advanced that plant growth hormones, such as the *auxins*, activate membrane transport systems that pump H^+ ions across the plasma membrane to the cell exterior. The resultant reduction in wall pH stimulates the wall enzymes, which are proposed to be most active at pH 5.

Plasmodesmata: Communicating Junctions of Plant Cell Walls Both primary and secondary plant cell walls retain minute openings called *plasmodesmata* (singular = *plasmodesma*) through which the cytoplasm of adjacent cells communicates directly (Figs. 5-19 and 5-20). These openings originally form during deposition of the *cell plate*, the new cell wall that forms between daughter cells during cell division in plants (see Chap. 11). As the cell plate thickens into the primary wall, and during the later thickenings converting primary to secondary cell walls, the plasmodesmata persist in their original numbers, which in mature walls may vary from 1 to 140 per square micrometer. At this frequency, a mature plant cell may have as many as 1000 to 100,000 plasmodesmata forming connections to its neighbors.

Individual plasmodesmata frequently narrow down to a neck region of 30 nanometers or less. Sections of plasmodesmata usually show a tubular structure running through the center (see Fig. 5-20). Although there is some controversy over the point, this central tubule is probably a residual microtubule persisting as a remnant of the mitotic spindle (see Chap. 11). Extensions of the endoplasmic

reticulum are often seen to approach plasmodesmata closely on either side of the cell wall, opening the possibility that ER channels may also make connections between cells through the plasmodesmata.

Whether plasmodesmata actually form open channels of communication between plant cells is still controversial. Nutrients and the products of photosynthesis have been demonstrated to pass readily between plant cells; the rapidity of movement is generally correlated with the number of plasmodesmata. The cell-to-cell movement of labeled molecules too large to pass through membranes has also been detected. In addition, the electrical resistance between plant cells is approximately 50 times lower than would be expected if the cells were separated by continuous plasma membranes. The rapid transfer of ions and larger molecules demonstrated by these results is believed to occur through the plasmodesmata.

A Summary of Surface Specializations in Eukaryotic Cells

Suface specializations thus carry out a variety of functions in eukaryotic cells (the surface coats of prokaryotic cells are discussed in Supplement 5-1). In animal cells, the carbohydrate groups of glycoproteins and glycolipids allow recognition and reception of a variety of signals from outside. These signals may involve groups that identify other cells as part of the same individual or foreign, or may involve molecules such as hormones that trigger biochemical responses inside the cells receiving the signals.

Eukaryotic cells are held together by other specializations of the cell surface. The initial adhesions of animal embryos that depend on interactions of glycoprotein or glycolipid molecules at cell surfaces are converted into permanent attachments by adhesive junctions such as desmosomes. Sealing or tight junctions close off the joints between cells to the diffusion of molecules and ions. In some cell types, communicating or gap junctions open specialized channels that permit diffusion of ions and small molecules directly between the cytoplasm of adjacent cells.

Most plant cells, and many animal cell types, secrete extracellular structures that support and protect the cell contents. In animals, these molecules form cartilage, tendons, bone and other supportive elements of connective tissues. In plants, extracellular molecules form the cell walls that bind cells together, give permanent shape to cells, resist the internal forces developed by osmosis, and protect cells from mechanical injury and infection.

Suggestions for Further Reading

Albersheim, P. 1975. The walls of growing plant cells. *Scientific American* 232:80–95 (April).

Cuatrecasas, P. 1974. Membrane receptors. *Annual Review of Biochemistry* 43:169–214.

Gilula, N. B. 1974. Junctions between cells. In *Cell communication*. Ed. R. P. Cox. Wiley: New York. Pp. 1–29.

Overton, J. 1974. Cell junctions and their development. *Progress in Surface and Membrane Science* 8:401–17.

Robards, A. W. 1975. Plasmodesmata. *Annual Review of Plant Physiology* 26:13–29.

Staehelin, L. A. 1974. Structure and function of intercellular junctions. *International Review of Cytology* 39:191–283.

Wolfe, S. L. 1981. *Biology of the cell*. 2nd ed. Wadsworth: Belmont, Calif.

For Further Information

Bilayer structure, Chapters 2 and 4
Cell surface and regulation of cell division, Chapter 11
Cell wall formation during cell division, Chapter 11
Glucose oxidation, Chapter 7
Glycoproteins, insertion in membranes, Chapter 10

Microfilaments
  and the cell spindle, Chapter 11
  and furrowing in cell division, Chapter 11
Phospholipid structure, Chapter 2

Questions

1. Outline the structure of membrane glycolipids and glycoproteins. How are these molecules embedded in the fluid mosaic structure of the plasma membrane?

2. How are membrane glycoproteins believed to act in the recognition and reception of signals at the cell surface? Outline the receptor-response model for the reception and transmission of signals across the plasma membrane.

3. How is the capacity for cell-to-cell recognition related to membrane surface glycolipids and glycoproteins?

4. What are the HL-A surface markers of mammalian cells? In what way are these markers involved in the rejection of organ transplants between individuals?

5. What kinds of junctions occur between the cells of animals?

6. Outline the structure of a desmosome.

7. What evidence supports the idea that sealing junctions close intercellular spaces to the flow of molecules? How are sealing junctions structured?

8. Describe the structure of a communicating junction. What evidence indicates that communicating junctions provide open channels for the flow of molecules and ions between cells?

9. What functions do fiber and matrix molecules have in the extracellular structures of plant and animal cells?

10. What molecules provide the fibers and matrix of animal extracellular structures? Outline the structure of the collagens and proteoglycans.

11. What are the fiber and matrix molecules of plant cell walls?

12. Why are the cell wall molecules of plants considered to form a "supermolecule" that extends throughout an entire plant?

13. How do plant cell walls expand to accommodate cell growth?

14. What are plasmodesmata? In what way are plasmodesmata similar to the communicating junctions of animal cells?

Supplement 5-1:
Cell Walls in Prokaryotes

The walls of prokaryotic cells, as in plants, are extracellular structures that support and give shape to the cell contents, link cells into colonies, and protect cells from infection or attack by other cells or viruses. In addition, the exterior walls prevent prokaryotic cells from bursting due to osmotic pressure, which may reach very high levels in these organisms. Because the external walls are of such importance to the viability of prokaryotes, chemical interference with the walls or wall synthesis is an effective and much-used way to control the growth of disease-causing bacteria. Penicillin, for example, an antibiotic that revolutionized the treatment of bacterial infections and diseases, has its main effect in interfering with wall synthesis in dividing cells. The defects introduced by the drug (see below) so weaken the new wall that the cells burst from osmotic pressure.

Prokaryotic cell walls, like the walls of plant cells, are formed from complex carbohydrate molecules linked with other substances into a single "supermolecule" that surrounds the cell. Cell wall structure in bacteria falls into two major divisions that are distinguished by reaction of cells to a staining technique known as the *gram stain.* Although the chemical basis for the gram stain is unknown, the technique and results are straightforward: cells are heat-killed and heat-fixed, stained with a dye (crystal violet), and treated with iodine. After staining, they are exposed to an organic solvent such as alcohol or acetone. Gram-positive cells retain a deep blue color when treated with the solvent, and gram-negative cells are decolorized. In gram-positive cells, the color is actually retained inside the cell and not in the wall; in some unknown way, the chemical structure of the wall prevents the cell contents from becoming colorless.

Electron microscopy and chemical analysis of the two types of bacteria reveal clear differences in wall morphology and composition. Walls in gram-positive bacteria (Fig. 5-21a) consist of a single thick layer with little or nor visible differentiation into sublayers. Gram-negative bacteria, in contrast, have relatively thin walls containing several distinct layers (Fig. 5-21b). One of these layers consists of a lipid bilayer that forms an extra, membranelike layer that lies outside the plasma membrane. Thus, gram-negative cells are in effect surrounded by two membranes. One, the plasma membrane, encloses the cytoplasm and the second, called the *outer membrane,* encloses the wall.

Chemical analysis reveals similarities and differences between the molecules of gram-positive and gram-negative bacterial walls. Both types share a major structural molecule called a *peptidoglycan* (Fig. 5-22). This molecule, formed from repeating sugar and polypeptide units, provides rigidity and strength to bacterial cell walls. In gram-positive bacteria, peptidoglycans extend throughout the wall, and, by weight, make up 40 to 90 percent of the cell wall substance. In gram-negative bacteria, peptidoglycans are limited to a thin layer between the plasma membrane

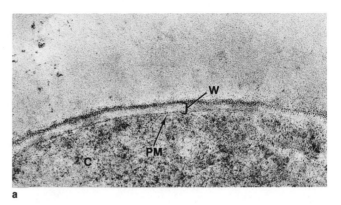

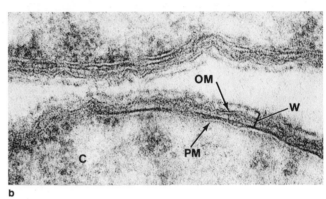

Figure 5-21 Wall structure in gram-positive and gram-negative bacteria. (**a**) The single-layered wall characteristic of gram-positive bacteria. × 103,000. (**b**) The multilayered wall of gram-negative bacteria. × 112,000. *W,* wall; *PM,* plasma membrane; *OM,* outer membrane; *C,* cytoplasm. Courtesy of J. W. Costerton, *Annual Review of Microbiology* 33 (1979):459. © 1979 by Annual Reviews, Inc.

and outer membrane. Peptidoglycans are chemically similar in all prokaryotes and differ only in the types of amino acids present in the peptide portion of the molecule, and in the way these amino acids are linked together. Proteins functioning as enzymes and recognition molecules are also present in the walls of both gram-positive and gram-negative bacteria.

The walls of the two bacterial types differ in other major molecular types. Almost all gram-positive bacteria contain long molecules called *teichoic acids* containing sugar and phosphate units linked together (Fig. 5-23a). Gram-negative bacteria lack teichoic acids but contain another group of molecules called *lipopolysaccharides* in the outer membranes of their walls (Fig. 5-23b).

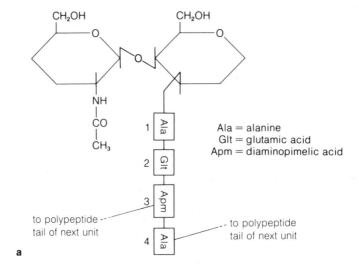

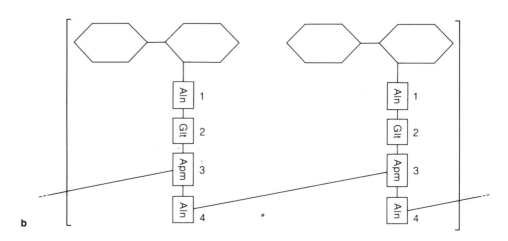

Figure 5-22 (**a**) The repeating structural unit of a wall peptidoglycan in bacteria. (**b**) How the units link together to form the wall.

The function of teichoic acids in bacterial cell walls is unknown. Mutant cells lacking these substances are apparently completely viable, with cell walls of normal strength and rigidity. However, recognition molecules developed in animals against gram-positive bacteria frequently bind specifically to teichoic acid polymers at the wall surface. Binding of bacterial viruses to the walls of gram-positive bacteria apparently also involves recognition and attachment to teichoic acids at the wall surface. Thus a part of the function of the teichoic acids may be in surface reception, in a role similar to the surface glycoproteins of animal cells.

Recognition molecules developed in animals as a defense against gram-negative bacteria bind to the lipopolysaccharides in the walls of these prokaryotes. Infection by bacterial viruses also frequently involves binding between viral particles and lipopolysaccharides at the wall surface. Thus the lipopolysaccharides may function in a recognition or reception role analogous to the teichoic acids of gram-positive bacteria. The entire outer membrane may also be a part of the defense apparatus of gram-negative bacteria because removal or interference with the outer membrane greatly increases the susceptibility of these bacteria to antibiotics.

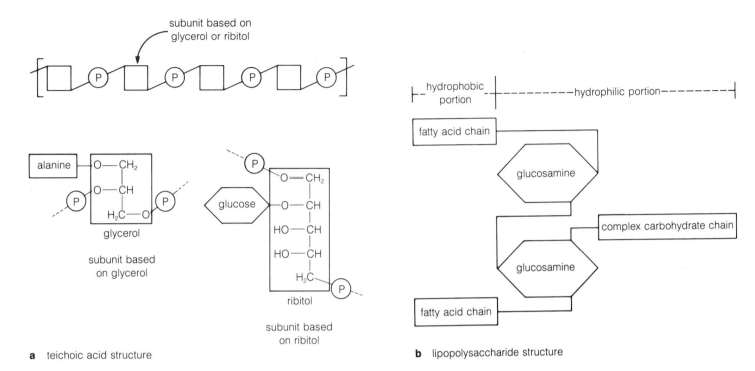

a   teichoic acid structure

b   lipopolysaccharide structure

Figure 5-23 **(a)** Structure of the teichoic acids of gram-positive bacteria. The molecules consist of a chain of subunits based on glycerol or ribitol linked by phosphate groups. **(b)** The structural unit of a lipopolysaccharide of gram-negative bacteria. These unique molecules contain a basal unit constructed from a pair of hydrophilic glucosamine subunits and hydrophobic chains derived from fatty acids. A long, highly complex carbohydrate "tail" is connected to the basal unit. The complex carbohydrate tail is believed to extend like an "antenna" from the outer membrane surface, reaching as far as 30 nanometers into the surrounding medium.

Growth of bacterial cell walls resembles the process in higher plants. Enzymes open linkages holding the peptidoglycans together, allowing the peptidoglycan subunits to slide past each other as the cell wall expands. New peptidoglycans are inserted in the wall as expansion proceeds. Penicillin interferes with wall synthesis in prokaryotes, including the blue-green algae, by either opening additional links in the peptidoglycan sheath, or preventing closure of the links opened during cell wall growth. The open links weaken the wall and allow the cells treated with penicillin to swell to the bursting point as a result of osmotic pressure.

The walls of blue-green algae resemble the gram-negative bacteria in most respects. Morphologically, cell walls in the blue-green algae and gram-negative bacteria are apparently identical (Fig. 5-24); each of the wall layers of the gram-negative bacteria has an exact counterpart in the blue-green algae. Peptidoglycans in blue-green algae occur in a thin layer just outside the plasma membrane as in the

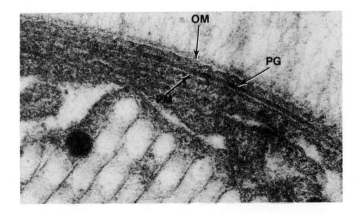

Figure 5-24 Cell wall of a blue-green alga. The layers correspond to the multilayered wall of gram-negative bacteria. *PM*, plasma membrane; *OM*, outer membrane; *PG*, peptidoglycan layer. × 96,000. Courtesy of N. J. Lang.

Table 5-1 Cell Wall Characteristics in Bacteria and Blue-Green Algae

Characteristic	Gram +	Gram −	Blue-Green Algae
Wall morphology	single layer	complex, with outer membrane	complex, with outer membrane
Peptidoglycans	+	+	+
Teichoic acid	+	−	−
Lipopolysaccharides	−	+	+
Gram stain	+	−	+

gram-negative bacteria. An outer membrane containing lipopolysaccharides is also present. Although some chemical differences are noted in these lipopolysaccharides, they closely resemble the lipopolysaccharides of the gram-negative bacteria. Curiously, in spite of the chemical and morphological similarities to gram-negative bacteria, the blue-green algae stain as gram-positive. The basis for this staining difference is unknown. (Table 5-1 summarizes the wall characteristics of gram-negative and gram-positive bacteria and the blue-green algae.)

Many prokaryotes have an external *capsule* (see Fig. 1-3) as an additional outer envelope surrounding the cell wall. This layer, which may extend outward from a few to hundreds of nanometers from the cell wall, shows great diversity in the various species of bacteria and blue-green algae. Within a single species, however, its chemical structure is relatively simple and uniform in comparison to the underlying wall. Usually, the capsule contains a single type of complex carbohydrate built up from glucose, galactose, fucose, or mannose subunits. The capsule probably serves as an outer line of defense against attack by viruses and antibiotics.

THREE

Cellular Energetics

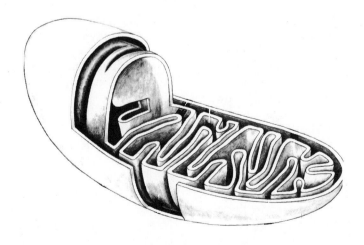

6

The Capture of Energy in Photosynthesis

The activities of life require a continuous input of energy. At the cellular level, the required energy is derived either directly from sunlight or from the breakdown of complex organic molecules such as carbohydrates and fats. If traced to its ultimate source, the energy bound into the complex organic molecules used as fuels also has its origin in the sun. Consequently, the sun is the primary energy source for life on earth.

A part of the radiant energy of the sunlight falling on the earth is absorbed by plants and transformed into chemical energy. The captured energy is then used as the energy source driving the synthesis of all the biological molecules needed by plants. Usually the assembly of these molecules proceeds from raw materials no more complex than water, carbon dioxide, and a supply of inorganic minerals. Some of the potential energy of these biological molecules is used by plants themselves as an energy source, particularly during periods of reduced light or of darkness. The complex molecules synthesized by plants also form the primary energy source for animals that live by eating plants. These plant-eating animals are eaten in turn by animals that live by ingesting other animals. These animals are eaten by other animals, and so on down the line until the last bits of chemical complexity are degraded by bacteria and fungi as any uneaten animals and plants die. Thus the energy required for life flows from the sun through plants and animals and finally to organisms of decay. Without the input of energy from the sun, all of these life forms would soon die.

Only a few organisms lie outside this path of energy flow. These organisms, including a few species of bacteria, live by using relatively high-energy inorganic molecules as an energy source. The inorganic molecules used by these *chemosynthetic* bacteria, as they are called, exist in mineral deposits in the earth and do not contain energy originating from the sun. However, even these organisms would probably cease to exist without sunlight because the conditions necessary for their survival, such as environmental temperatures within the ranges necessary for life, could not be maintained.

The basic processes in the flow of energy from the sun through living organisms take place at the cellular level. In the initial phase, *photosynthesis,* light is absorbed and converted to chemical energy. This phase takes place in the chloroplasts of eukaryotic plants and in the cytoplasm of blue-green algae and photosynthetic bacteria. The chemical energy captured in this phase is used to assemble complex organic molecules from simple inorganic raw materials in the photosynthetic cells. In the second major phase, *respiration,* the organic molecules assembled in photosynthesis are broken down to release energy for cellular activities. Respiration occurs largely in mitochondria in both plant and animal eukaryotes; in prokaryotes, the reactions of respiration are distributed throughout the cytoplasm.

This chapter describes the first phase of energy flow in living organisms, the capture and utilization of light energy in photosynthesis. In this discussion, primary emphasis is given to the way photosynthesis takes place in eukaryotic plants (photosynthesis in prokaryotes is summarized in Supplement 6-2).

Chloroplast Structure in the Eukaryotic Plants

The chloroplasts of algae and higher plant cells are easily seen under the light microscope as lens-shaped bodies about 5 to 10 micrometers in diameter. Usually, chloroplasts are considerably larger than the mitochondria in the same plant cell and occur in smaller numbers. The number

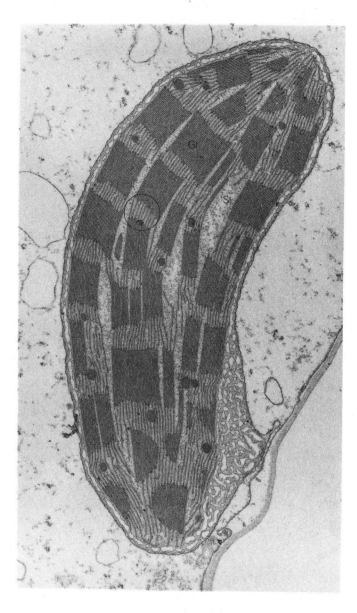

Figure 6-1 Chloroplast from a corn leaf. The chloroplast is surrounded by a smooth outer membrane and a much-folded inner membrane. Thylakoid membranes stacked into grana (G) are clearly visible inside the chloroplast. Connections called stromal lamellae (circled) run between the thylakoids of adjacent grana. The thylakoid membranes are suspended in the inner substance of the chloroplast, a solution of proteins and other molecules called the stroma (St). × 20,000. Courtesy of L. K. Shumway and T. E. Weier, from *American Journal of Botany* 54 (1967): 773.

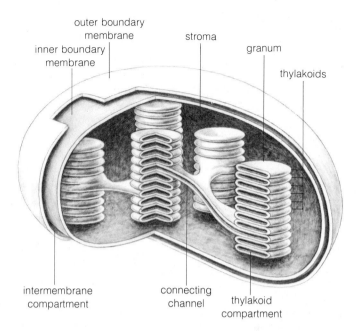

Figure 6-2 The arrangement of membranes and compartments inside a chloroplast.

may vary from a single chloroplast, as found in *Micromonas* and several other green algae, to several hundred in the cells of most higher plants.

The green pigments of chloroplasts can be seen under the light microscope to be concentrated in disclike substructures, the *grana*, in most eukaryotic plants. Grana are suspended in a colorless background substance inside chloroplasts called the *stroma*. This pigment distribution is maintained by a system of membranes and compartments visible in chloroplasts under the electron microscope (Figs. 6-1 and 6-2). Two continuous boundary membranes separate the chloroplast interior from the surrounding cytoplasm. The region between these membranes, the *intermembrane compartment*, is usually difficult to trace in electron micrographs (as in Fig. 6-1) because of extensive folding of the inner boundary membrane. The two boundary membranes enclose the fluid substance of the stroma. Within the stroma is the complex system of membranous sacs that forms the grana and their interconnections.

The individual unit of a granum is a flattened sac or vesicle, the *thylakoid* (from *thylakos* = sack or pouch), consisting of a single, continuous membrane that completely encloses an interior *thylakoid compartment*. Grana are formed from a closely fused pile of these individual sacs, set one on top of another much like a stack of coins. Chloroplasts may contain from 40 to 60 grana, each formed from stacks of 2 or 3 to as many as a hundred or more individual thylakoid sacs.

Frequent membrane connections can be seen between the thylakoids of adjacent grana (circle, Fig. 6-1). These connections, called *stromal lamellae*, enclose a channel that is continuous with the thylakoid compartments of adjacent grana (diagrammed in Fig. 6-2). The stromal lamellae may connect the thylakoid compartments into a single, continuous internal region inside chloroplasts.

The chloroplast stroma may also contain a variety of inclusions suspended in the regions surrounding the grana and stromal lamellae membranes. Most conspicuous of these are *starch granules* (Fig. 6-3), found in chloroplasts after a period of active photosynthesis in the light. The chloroplast stroma also contains DNA, ribosomes, and all the biochemical factors required for DNA replication, RNA transcription, and protein synthesis. Of the visible elements of this system, the DNA appears as faintly visible threads at scattered locations in the stroma (see Fig. 10-29). The ribosomes, conspicuously smaller than the ribosomes of the surrounding cytoplasm, are distributed throughout the stroma. The DNA and ribosomes of chloroplasts, which resemble the equivalent structures of prokaryotes much more closely than eukaryotic DNA and ribosomes, are evidently necessary for the production of some of the proteins required for photosynthesis. The similarity between the DNA and ribosomes of chloroplasts and prokaryotes suggests that chloroplasts may have evolved from prokaryotes that became established as permanent, beneficial residents in the ancestors of present-day eukaryotic plant cells (see Supplement 10-2 and Chap. 13 for details).

Photosynthesis in Chloroplasts

The Overall Reactions of Photosynthesis

The reactions of photosynthesis were pieced together in a long series of experiments that extend over a period of nearly two hundred years. In 1772 an Englishman, Joseph Priestley, showed that air in an enclosed vessel, "injured" by the presence of a burning candle, could no longer support life. Priestley found that the injured air could be "restored" if a green plant was placed in the vessel and left in the light for a time. It was soon established that air is "injured" by a burning candle, a breathing animal, or even a plant in darkness because they all produce carbon dioxide and use up oxygen. Green plants restore the injured

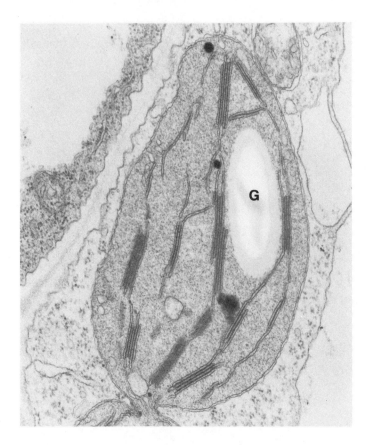

Figure 6-3 A starch granule (G) in a tobacco chloroplast. Courtesy of D. Stetler.

air in light by removing CO_2 and releasing O_2. In the 1800s water was found to be necessary for this reaction to take place. Not long after this discovery, with the basic raw materials and products known, the mechanism of photosynthesis was first written in skeletal form:

$$6CO_2 + 6H_2O \xrightarrow[\text{energy}]{\text{light}} C_6H_{12}O_6 + 6O_2 \qquad (6\text{-}1)$$

This reaction states that in photosynthesis light energy is used to drive the uphill synthesis of carbohydrates from carbon dioxide and water; oxygen is given off as a by-product of the reaction.

In the 1930s C. B. van Niel of Stanford University developed a more general statement of the basic reactions of photosynthesis through his study of photosynthetic bacteria. These bacteria do not use water as a raw material. Instead, they use other hydrogen-containing substances such as hydrogen sulfide (H_2S), alcohols, or even hydrogen itself (see Supplement 6-2). Significantly, oxygen is not given off as a by-product of photosynthesis by these bacteria. Bacteria using H_2S as a raw material, for example, release molecular sulfur instead of oxygen. Noting the similarity between H_2S and H_2O as raw materials, and the evolution of either sulfur or oxygen as by-products, van Niel proposed that the oxygen in eukaryotic photosynthesis is derived from the H_2O entering the process and not the CO_2 as was previously assumed. Van Niel rewrote the overall reaction for photosynthesis in a more general form to take his observations and hypothesis into account:

$$CO_2 + 2H_2D \rightarrow (CH_2O) + H_2O + 2D \qquad (6\text{-}2)$$

In this reaction the substance H_2D represents any raw material that can donate electrons and hydrogen to the photosynthetic mechanism. The D in H_2D represents the atom (if any) attached to the hydrogen of this raw material. The D atom is eventually released in free form as a by-product of photosynthesis. In higher plants, the H_2D in van Niel's reaction is H_2O; the O is released as molecular oxygen. The (CH_2O) in Reaction 6-2 represents a unit of carbohydrate. Six of these units are combined in plants to produce a carbohydrate such as glucose. Van Niel also found, as shown in Reaction 6-2, that water molecules are also assembled as by-products of photosynthesis.

Van Niel's general equation for photosynthesis, and his proposal that the oxygen evolved by higher plants originates from water, was confirmed when researchers were able to isolate and track the various isotopes of the elements. The heavy oxygen isotope ^{18}O was used to label oxygen atoms in CO_2 or H_2O, and was then followed through photosynthesis.[1] If an organism carrying out photosynthesis was supplied with water containing ^{18}O, the label showed up in the oxygen given off as a by-product. If carbon dioxide containing ^{18}O was supplied, the label showed up instead in the carbohydrates produced in photosynthesis and in the water molecules assembled as a by-product of the reaction. These results confirmed van Niel's hypothesis that the oxygen evolved in photosynthesis is derived from the water molecules entering the process, not the CO_2.

---

[1]Oxygen in its most common form has eight neutrons and eight protons in its atomic nucleus, giving ^{16}O; "heavy" oxygen has two additional neutrons, giving the isotope ^{18}O. The extra weight given to the substances containing ^{18}O oxygen allows them to be traced in biological reactions.

The Light and Dark Reactions of Photosynthesis

During the early part of this century biochemical investigations established that the overall reactions of photosynthesis occur in two major, interdependent parts that can be experimentally separated. One part, the *light reactions*, is directly dependent on light and stops if light energy is unavailable. The second part, the *dark reactions*, is light-independent and may continue in darkness if all of the necessary reactants are present.

The light and dark reactions were discovered in experiments conducted by Robert Emerson in the 1930s at the California Institute of Technology. Emerson studied photosynthesis in plants exposed to flashes of light lasting only a few milliseconds. He found that a brief flash of intense light could be followed by a period of darkness many times longer without slowing the production of carbohydrates in photosynthesis. In fact, a period of darkness was *required* after a bright flash to allow the most efficient use of a given quantity of light in photosynthesis. Emerson interpreted this to mean that two separate reactions occur in photosynthesis: the products of an initial rapid reaction that is dependent on light are subsequently used in a much slower reaction that can take place in the dark. A circular pathway for at least some of the intermediate compounds linking the two pathways was indicated by the fact that the light reactions were greatly inhibited if insufficient time was allowed for completion of the dark reactions between flashes of light. This suggested that the dark reactions convert products of the light reactions into a form in which they can be used again in the light reactions.

Later work concentrated on the specific reactions in the light and dark phases of photosynthesis and on identifying the chemical products linking the two reaction series together. This work showed that the rapid, light-dependent reactions synthesize two high-energy chemical products required by the dark reactions, as proposed by Emerson (Fig. 6-4). One of the products is ATP. The second is reduced *NADP* (see Fig. 6-17), a molecule that carries hydrogen and high-energy electrons. The light reactions also use H_2O as a raw material and release oxygen as a by-product. The dark reactions use the high-energy products of the light reactions as energy sources for the synthesis of carbohydrates, lipids, amino acids, and other complex organic molecules from CO_2 and other simple inorganic molecules. As a part of the dark reactions, ATP is broken down to ADP, and the hydrogen and high-energy electrons are removed from NADP. These substances,

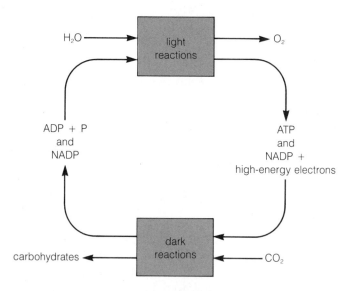

Figure 6-4 Distribution of photosynthesis between the light and dark reactions. The two reaction sequences are linked by ATP and NADP.

ADP, and NADP depleted of its hydrogen and electrons, then cycle back to the light reactions to be recharged with chemical energy derived from sunlight.

The dark reactions do not necessarily occur in darkness as their name suggests. In fact, they generally occur during the daytime, when the light reactions are active in producing the high-energy products used in the dark reactions. The name simply means that the dark reactions are light-independent and do not use light energy directly.

The Light Reactions of Photosynthesis

Light and Light Absorption

Visible light is a form of radiant energy that travels in waves through space, with wavelengths varying from about 390 nanometers (seen as blue light) to 760 nanometers (seen as red light). Although radiated in apparently continuous beams, the energy of light waves actually flows in individual packets. This is analogous to an electron beam which, although following a wave path, is formed from individual electrons. Each electron in the beam represents a separate unit of energy, following a wave path through space.

The individual units of energy forming a beam of light are called *quanta* (singular = *quantum*) or *photons*. Each quantum contains an amount of energy that is inversely proportional to its wavelength: the shorter the wavelength, the greater the energy of a quantum. For a given wavelength, the energy of a quantum of light is constant.

Certain molecules can absorb the energy of light quanta at one or more wavelengths. These molecules appear colored or pigmented because of the unabsorbed wavelengths, which are transmitted or reflected without change. The photosynthetic pigment chlorophyll, for example, absorbs red and blue light and transmits or reflects green. Absorption of light at specific wavelengths is a property of electrons occupying certain orbitals in the absorbing molecules.

Electrons in these orbitals exist at characteristic energy levels. If one of the electrons absorbs the energy of a quantum of light, it moves to a new orbital with a higher energy level, lying at a greater distance from its atomic nucleus. In the new, high-energy orbital, the electron is said to be in an *excited state*. The difference in energy level between the unexcited state (called the *ground state*) and the excited state has a fixed value for an electron in a given position in a molecule. This value is exactly equivalent to the amount contained in a quantum of light at a particular wavelength. If the energy of this quantum is absorbed, the electron jumps from the ground to the excited state orbital. All of the energy in a single activating quantum is absorbed; other quanta striking the molecule at different wavelengths have no effect and are transmitted or reflected.

The excited state is so unstable that an electron can remain in an excited orbital for only a billionth of a second or less. Release from the excited orbital may occur in one of several ways. The excited electron may simply drop back to its ground state orbital, releasing its energy as heat as it returns. Alternatively, some of the absorbed energy may be reradiated as light as the electron returns to the ground state. Light energy released in this way is called *fluorescence*. The light released in fluorescence is at a longer wavelength than the light absorbed and thus represents a quantum containing less energy than the quantum absorbed. The difference between the absorbed and fluoresced quantum is released as heat.

Another possible fate for an excited electron has greater significance for photosynthesis. If an excited molecule is situated close enough to another molecule, the unstable orbital occupied by its excited electron may extend outward far enough to overlap vacant orbitals in the second molecule. Transfer of an excited electron to a stable orbital in the second molecule may then occur. Once transferred to a suitable acceptor in this way, the energy of the

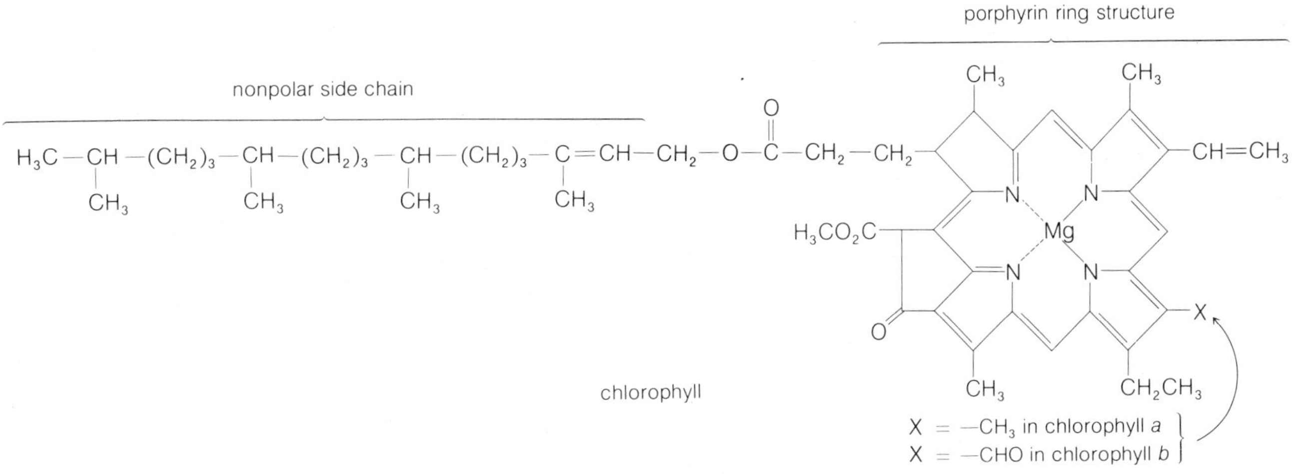

Figure 6-5 Structures of chlorophyll a and b. Chlorophyll a has a methyl (—CH₃) group at the position marked X (arrow) in the figure; chlorophyll b has an aldehyde (—CHO) group in this position. Additional chlorophylls, with minor substitutions at other points in the ring structure, are found in some algae and plants.

electron has been "trapped" as *chemical* energy. The energy absorbed by the pigmented molecules active in photosynthesis is converted to chemical energy in this way, by transfer of excited, high-energy electrons to stable orbitals in acceptor molecules.

The Molecules Absorbing Light in Photosynthesis

The molecules that absorb light energy in eukaryotic photosynthesis occur in chloroplasts. These molecules are lipidlike, and can easily be extracted from plant tissue by lipid solvents such as ether or acetone. Analysis of the extracted molecules reveals that chloroplasts contain two major classes of pigments, the green *chlorophylls* and yellow *carotenoids.*

The chlorophylls (Fig. 6-5) absorb light most strongly at red and blue wavelengths and transmit green light. The different types of chlorophyll molecules are built up from a complex central ring structure, called a *porphyrin ring*, to which is attached a long, nonpolar "tail" that gives the chlorophylls their lipidlike solubility characteristics. A single magnesium ion is bound into the center of the ring structure. The two major chlorophylls found in all higher plants, called *chlorophyll a* and *chlorophyll b,* differ only in a single substitution in one side group bound to the central ring (see Fig. 6-5). In chloroplasts, chlorophyll a and

b occur in combination with specific proteins, held in place in the amino acid chains of the proteins by hydrophobic, noncovalent bonds.

Both chlorophyll a and b have an extensive series of electrons that can absorb light energy and jump to excited orbitals. These electrons each absorb light at different wavelengths, which has the effect of greatly broadening the distribution of wavelengths absorbed from a single peak to a broad, smooth curve with several peaks (Fig. 6-6). The particular wavelengths absorbed may vary, depending on alterations in the orbitals due to the binding of chlorophyll to proteins inside chloroplasts. Purified chlorophyll a, for example, absorbs light most strongly at wavelengths of 420 and 675 nanometers when dissolved in acetone. Within membranes, chlorophyll a may show other absorption peaks at 660, 670, 678, 685, or 700 nanometers along with the shorter 420 nanometer maximum. These changes in absorption are due to interactions between the chlorophylls and their binding proteins, and not to chemical changes in the chlorophyll. The differently absorbing forms of chlorophyll a, for example, are all chemically identical.

The carotenoids (Fig. 6-7), a separate family of pigmented lipid molecules, are all built up from a single, long carbon chain containing 40 carbon atoms. Various substitutions in the side groups attached to the 40-carbon backbone give rise to the different carotenoid pigments. Two types of carotenoids occur in all green plants. The *carotenes* are entirely hydrophobic and contain no polar groups;

chief in abundance among these in higher plants is *β-car-otene* (Fig. 6-7a). The second type, the *carotenols*, contain polar hydroxyl (—OH) groups at both ends of the carbon chain. The yellow pigment *xanthophyll* (Fig. 6-7b) is predominant among the carotenols of higher plants. The carotenoids absorb light at blue wavelengths from 400 to 500 nanometers (Fig. 6-8), and transmit yellow, green, orange, and red light, producing a combination that appears predominantly yellow.

The carotenoids function as *accessory pigments* that take up light at wavelengths not directly absorbed by the chlorophylls. The light energy absorbed by these pigments, and by chlorophyll b, is eventually transferred to chlorophyll a, which is the photosynthetic pigment directly involved in transforming light into chemical energy (see Fig. 6-10). The net effect of the entire combination of photosynthetic pigments is to greatly broaden the spectrum of

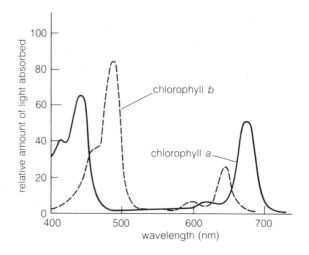

Figure 6-6 Amount of light absorbed by chlorophyll a and b at different visible wavelengths.

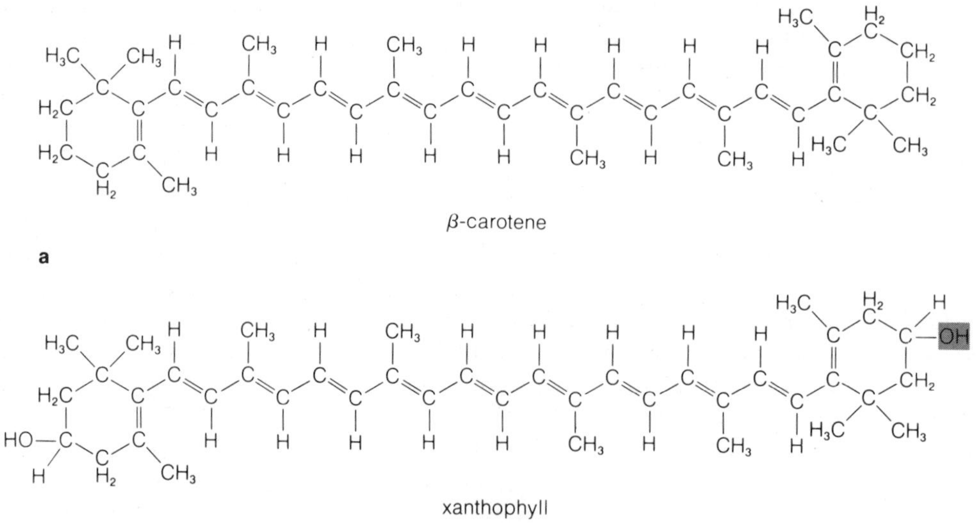

β-carotene

a

xanthophyll

b

Figure 6-7 The carotenoid pigments β-carotene (**a**) and lutein (**b**), a yellow xanthophyll pigment of the carotenol group. Carotenes are pure hydrocarbons with no oxygen atoms; carotenols contain hydroxyl (—OH) groups (shaded) at opposite ends of the molecule.

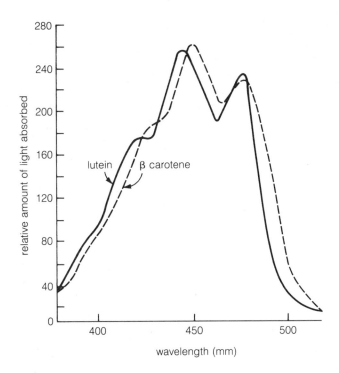

Figure 6-8 Amount of light absorbed by the cartenoids β-carotene and lutein at different visible wavelengths.

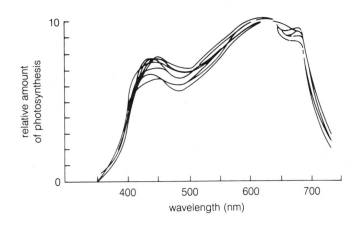

Figure 6-9 Relative amounts of photosynthesis in various crop plants driven by light wavelengths absorbed by the combined activity of the chlorophyll and carotenoid pigments. Each curve represents a different species. Modified from an original, courtesy of K. J. McCree and Elsevier/North-Holland, Inc., from *Agricultural Meteorology* 9 (1972): 191.

wavelengths usable as energy sources for photosynthesis so that light energy ranging over most of the visible wavelengths can be absorbed (Fig. 6-9).

Conversion of Light to Chemical Energy in Photosynthesis

We have noted that light is converted to chemical energy when an electron raised to an unstable excited orbital through the absorption of light energy is transferred to a stable orbital in an acceptor molecule. This transfer is the central event in the light reactions of photosynthesis, in which excited electrons are released from chlorophyll to enter stable orbitals in an electron acceptor. In accepting the high-energy electrons from excited chlorophyll molecules, the acceptor molecules are said to be *reduced* (see Information Box 6-1). The energy carried by the reduced molecules is eventually used in the synthesis of carbohydrates from H_2O and CO_2.

Much research effort has gone into tracing the pathways followed by the electrons after their transfer from chlorophyll a to the electron acceptors. This work has revealed the identity of some of the acceptor substances and has shown that in the light reactions the energy carried by the electrons is used to synthesize ATP and reduce NADP.

Generation of ATP and Reduced NADP in the Light Reactions

Photosystems and Reaction Centers It was originally believed that the energy of sunlight was absorbed and converted into chemical energy in a single center containing chlorophyll and carotenoid molecules organized with electron acceptors. Evidence that two separate centers operate in coordination in the light reactions came in the late 1950s when Robert Emerson and his coworkers discovered that the rate of photosynthesis in red light is greatly enhanced if light is supplied at two separate and distinct red wavelengths. This suggested that chlorophyll is organized into two separate systems in chloroplasts that absorb light at different wavelengths in the red range. Significantly, Emerson and his colleagues noted that the enhancement in photosynthetic rate is retained if the two wavelengths of light are flashed alternately, separated by intervals of darkness lasting up to several seconds. To Emerson and another investigator, L. N. M. Duysens, this suggested that the two light-absorbing systems operate in coordination, and that the energy absorbed by one of the systems is added to the energy absorbed by the other. Duysens called the two separate light-absorbing systems *photosystems I* and *II*.

Later work with disrupted chloroplasts revealed that the two light-absorbing photosystems proposed by Emerson and Duysens actually exist and that the pigment molecules of the photosystems are organized into groups in which chlorophyll and carotenoid molecules are linked with proteins into large particles. Analysis of the particles, and information from other sources, indicates that photosystems I and II each contain about 200 chlorophyll molecules, including both chlorophyll a and b, and about 50 carotenoid molecules, all linked with proteins into light-absorbing assemblies.

Not all of the 250 or so pigment molecules in the photosystem I and II assemblies need to absorb light energy

in order to raise electrons to excited orbitals. Instead, indications are that the energy of a light quantum absorbed by a pigment molecule anywhere in a photosystem assembly is passed along until it reaches a chlorophyll molecule that acts as a *reaction center* for the entire photosystem (Fig. 6-10). At the reaction center, the energy of the absorbed quantum is converted to chemical energy by passage of an excited electron to a stable orbital in an acceptor molecule.

While there is good evidence that the pigment molecules of the photosystems are actually organized in this way, the mechanism by which the energy of an absorbed light quantum "walks" from molecule to molecule within the assembly to reach the reaction center is not completely understood. That energy transfer can actually occur in this way is easily demonstrated by a solution containing two different pigmented molecules that absorb light at differ-

---

Information Box 6-1

Oxidation and Reduction

Many substances can accept or donate electrons. Much of the energy passed from one substance to another in photosynthesis, for example, is transferred in the form of electrons, removed from a donor molecule and accepted in the same instant by an acceptor molecule. Removal of electrons is called oxidation and acceptance is termed reduction. A substance from which electrons are removed is said to be *oxidized,* and the accepting substance is *reduced.* In general, a given substance contains more energy in its reduced state than in its oxidized state.

For any reactions involving transfer of electrons from a donor to an acceptor there is always a simultaneous reduction and oxidation; the two substances acting in the joint reduction-oxidation are often termed a redox couple. In biological reductions, a hydrogen ion (H^+) is often attached to the acceptor molecule at the instant of reduction. This is the case with NADP, the important reduced substance produced by the light reactions of photosynthesis (see equation 6-4).

The electrons removed in oxidation have a characteristic energy that depends on the orbitals they occupied in the oxidized substance. The energy associated with the removed electrons can be measured and expressed as voltage (see the figure in this box). The standard used for comparisons is the energy associated with electrons removed from hydrogen in the reaction $H_2 \rightarrow 2 H^+ + 2$ electrons. The voltage or potential of these electrons is arbitrarily designated as 0.00 millivolts, and all other voltages (often called redox potentials) are compared to this standard. Although all electrons carry a negative charge, their *relative* voltage may be greater or less than the 0.00 standard, and therefore can be called positive or negative. In general, the more negative a substance is on the redox scale, the greater is the amount of energy contained in electrons removed from the substance. Electrons are only transferred downward on the scale; that is, the donor must be higher on the scale than the acceptor. As electrons are transferred from a donor to an acceptor, an amount of free energy equivalent to the distance between the donor and acceptor on the redox scale is released.

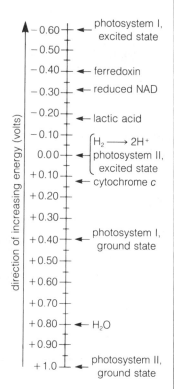

ent wavelengths. If such a solution is exposed to light at the shorter of the two wavelengths absorbed by the molecules, both pigments are converted to the excited state and quickly fluoresce, each at its characteristic wavelength. Since the light is supplied only at the wavelength absorbed by one of the molecules, the absorbed energy must be transferred in some manner to the second pigment.

The most widely accepted explanation for the transfer of energy from one molecule to another in such systems is that the absorbing molecule sets up an electromagnetic field because of the rapid vibration of its excited electrons. An adjacent pigment molecule, lying within the vibrational field created by the excited molecule, is induced to vibrate or resonate at the same frequency. In doing so, it absorbs the excitation energy of the first molecule. The mechanism works much like the transfer of energy from a broadcasting to a receiving television antenna. In the light-absorbing assemblies of photosystems I and II, the exciting quantum is considered to "walk" in this way from a carotenoid to chlorophyll, or from one chlorophyll molecule to another until it reaches the reaction center.

How is energy trapped once it reaches the reaction center? Supposedly, the center contains a type of chlorophyll that forms an energy "sink." In this chlorophyll molecule, the energy level of electrons in the excited state is slightly less than in the other chlorophyll molecules of the photosystem assembly. Although sufficient energy is supplied from the transferring molecules to raise an electron in the trapping molecule to its excited state, the reverse flow is not likely because energy released by the trapping molecule is not great enough to excite the surrounding "transfer" molecules. The energy consequently remains with the reaction center chlorophyll, to be released either as heat, long-wave fluorescence, or passed on as chemical energy to the initial electron acceptors of photosynthesis.

Analysis of the light absorbed by the chlorophyll molecules at the reaction centers has shown that the centers of both photosystems I and II are formed by specialized molecules of chlorophyll a. The binding between these chlorophyll a molecules and other molecules, probably the proteins of the photosystem assemblies, alters the orbitals of their excitable electrons so that they require slightly less energy to reach the excited state. The reaction center chlorophyll molecule of photosystem I can be excited by light at slightly longer wavelengths, about 700 nanometers, than other chlorophyll a molecules of the photosystem. Although this specialized chlorophyll, called *P700* ("P" stands for pigment), can be excited directly by light at 700 nanometers, it normally absorbs energy passed on from excited pigment molecules of the photosystem I assembly. The equivalent reaction center chlorophyll a molecule of

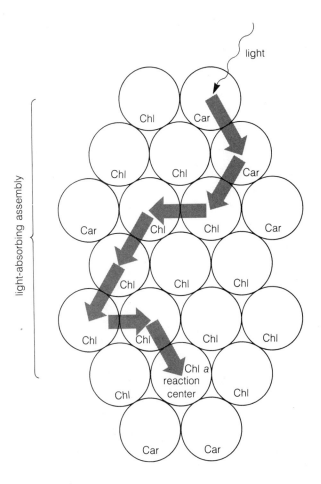

Figure 6-10 How a light quantum absorbed anywhere in the pigment molecules of a photosystem unit "walks" from molecule to molecule until it reaches the reaction center (see text).

photosystem II, which can absorb light directly at 680 nanometers, is called *P680*.

The Organization of Photosystems I and II in Chloroplasts: The Z-Pathway Emerson and Duysens's detection of the two photosystems touched off a burst of intensive research into photosynthesis in the late 1950s and early 1960s. This research led to the discovery that photosystems I and II are linked together by chains of electron acceptor molecules that act as carriers arranged in a series. In the series, high-energy electrons are passed from one member of the series to the next until they reach the final acceptor substance, NADP. Some of the energy of the electrons passing along the carrier chains is tapped off, as a part of the process, to drive the synthesis of ATP.

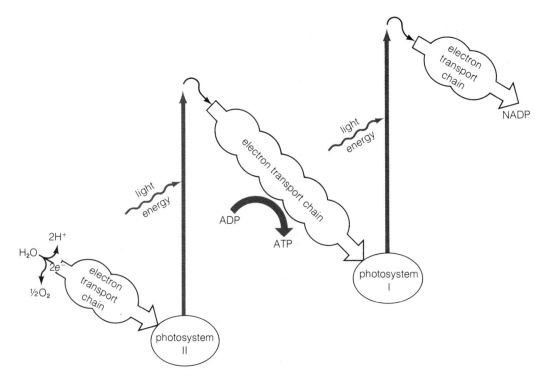

Figure 6-11 Flow of electrons through the Z-pathway (see text).

As high-energy electrons flow from one carrier to the next in the chains linking the photosystems, the carriers are alternately reduced and oxidized (see Information Box 6-1): as each carrier accepts electrons, it is reduced; as it releases electrons to the next carrier in the chain, it is oxidized.

The Arrangement of Carriers and Photosystems in the Z-Pathway
The photosystems and electron carrier chains were first pieced together with photosystems I and II into a complete system by R. Hill and Fay Bendall at the University of Cambridge in England. In the system, now known as the *Z-pathway*, electrons derived from water flow through the two photosystems, linked together by the carrier molecules, and are finally delivered to NADP.

The overall sequence of events in the Z-pathway is shown in Figure 6-11. Electrons are removed from water in the reaction:

$$H_2O \rightarrow 2H^+ + 2\ \text{electrons} + \tfrac{1}{2}O_2 \qquad (6\text{-}3)$$

The electrons, after flowing through a short transport chain, are transferred to the P680 chlorophyll at the reac-

tion center of photosystem II.[2] The electrons at this point are at a relatively low energy level, and enter ground state orbitals in the P680 molecules at the reaction center. The light-harvesting assembly of photosystem II then absorbs light and transfers energy to the reaction center, raising the electrons derived from water to excited orbitals. The excited electrons are transferred immediately to the first acceptor of photosystem II, known as the *primary acceptor* of this system. As the electrons enter stable orbitals in the primary acceptor, their energy is trapped in chemical form in the acceptor molecule. The acceptor molecule is reduced, and the P680 chlorophyll at the reaction center is oxidized as this transfer takes place. From the primary acceptor the electrons flow through an extended electron transport chain, gradually losing energy until they reach the P700 chlorophyll molecules at the reaction center of photosystem I. After excitation in photosystem I, the elec-

---

[2]The photosystems were identified and named *I* and *II* some time before they were placed in sequence in the light reactions of photosynthesis; as luck would have it, photosystem II was found to precede I in the light reactions. For better or worse, the original designations have been retained.

trons are transferred to the primary acceptor of this photosystem, and travel through a short electron transport chain to reach NADP.

The Electron Carriers and Their Sequence in the Z-Pathway With one exception all of the known carrier molecules in the chains of the Z-pathway consist of a protein molecule linked to a nonprotein group (see p. 59) that is the actual electron carrier. These protein-based carriers are the *cytochromes*, the *iron-sulfur proteins*, the *flavoproteins*, and *plastocyanin* (see Information Box 6-2). The nonprotein groups of the carriers are alternately oxidized and reduced as electrons flow through the chains. The single exceptional carrier, *plastoquinone*, is a lipidlike substance suspended in the membrane interior without direct linkage to a protein.

Figure 6-16 shows the tentative positions of the individual electron carriers of the Z-pathway as far as they are known. Following the course of electrons through the pathway shows how the two photosystems and the electron transport chains are considered to work in detail. The

---

Information Box 6-2

The Electron Carriers and Their Sequence in Chloroplasts

Four major types of electron carriers have been identified in the photosynthetic carrier chains. One type, the *cytochromes*, consists of a protein linked to a complex porphyrin ring similar to that of chlorophyll, except that the metal atom bound into the center of the ring is iron instead of magnesium (Fig. 6-12). Alternation between the oxidized and reduced forms of the cytochromes occurs by means of a single electron gained or lost by the central iron atom, which can exist as either Fe^{2+} or Fe^{3+}. The second major carrier type, the *iron-sulfur proteins*, is as yet poorly characterized and is known only to have "centers" containing iron atoms in close association with sulfhydryl (—SH) groups. In reduction and oxidation the iron atoms are believed to alternate between Fe^{2+} and Fe^{3+} states. The third carrier type, a group known as the *flavoproteins*, contains a protein linked to a nonprotein group based on a nucleotide. In the single carrier of this type occurring in chloroplasts, *FAD* (*flavin adenine dinucleotide;* see Fig. 6-13), the nonprotein group is derived from *riboflavin,* a vitamin of the B group. Since this group contains a nitrogenous base, a 5-carbon sugar, and a phosphate group, it is classified as a nucleotide. FAD accepts two electrons at a time, and also binds two hydrogens to form $FADH_2$ when it is reduced. Both the electrons and hydrogens are released when $FADH_2$ is converted back to the oxidized form. The final protein-based carrier, *plastocyanin,* has a carrier group containing copper.

The single nonprotein electron carrier, *plastoquinone* (Fig. 6-14), is built up from a ring-shaped group that is alternately oxidized and reduced during electron transport. Attached to the ring is a hydrophobic side chain. Plastoquinone, like FAD, accepts and releases both electrons and hydrogens in pairs as it is reduced and oxidized.

These electron carriers, which can all be detected in the chains linking photosystems I and II in chloroplasts, have been placed in a tentative sequence by several methods. One method is based on the fact that electrons are accepted and released by each carrier at distinct energy levels; this characteristic reflects the energy associated with the orbitals occupied by the electrons in the carriers when they are in the reduced state. The energy of the electrons released by the carriers can be measured as voltage; once measured, the carriers can be placed in order of descending energy levels, with the carriers releasing electrons at the lowest energy levels toward the end of the chains. A second method is based on the use of *inhibitors* that can bind to individual carriers and block their ability to accept and release electrons. Carriers falling before the blocked carrier in the chains tend to become reduced as electrons pile up in them; carriers falling after them tend to become oxidized as they become depleted of electrons (Fig. 6-15). By using inhibitors binding to different carriers, and noting the "crossover points" between carriers remaining in the reduced or oxidized state in each case, the carriers can be arranged in sequence.

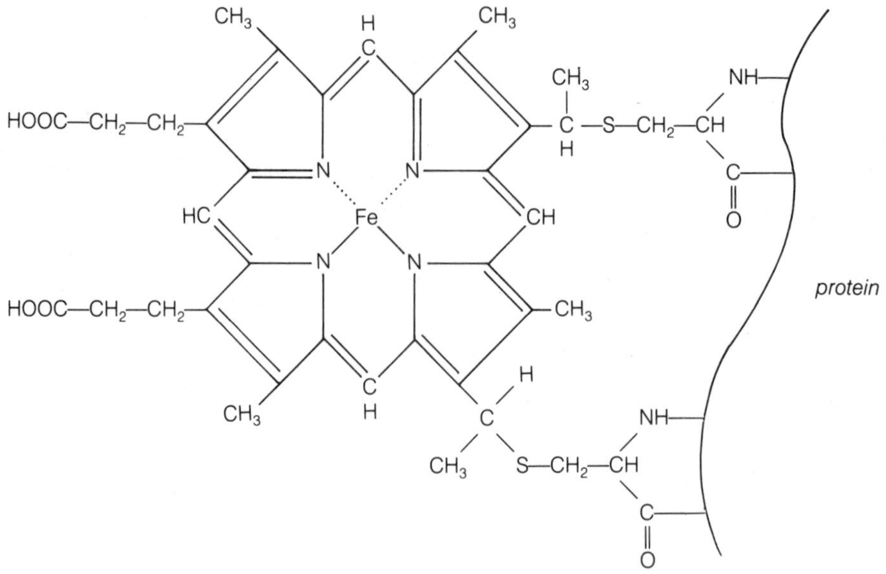

Figure 6-12 The nonprotein group of a cytochrome. The iron in the center of the ring structure, by changing alternately from the ferrous (Fe^{+2}) to the ferric (Fe^{+3}) form, acts as the electron carrier. Similar rings occur as central structures in hemoglobin, chlorophyll, and other molecules of biological importance (compare with Fig. 6-5).

first steps in the sequence remain only very incompletely characterized. Little is known about the reaction splitting water; similarly, the carriers conducting electrons from the water to the reaction center of photosystem II remain to be identified and sequenced.

After the electrons are excited at the reaction center of photosystem II through the absorption of light energy, they are passed to the primary acceptor of photosystem II. Of the various candidates proposed by different investigators for the primary acceptor of this photosystem, it now appears that the first substance to accept excited electrons from chlorophyll P680 is a specialized plastoquinone (identified as *PQ* in Fig. 6-16). From plastoquinone PQ, electrons flow to other plastoquinones and then through the remainder of the chain, consisting of cytochrome f and plastocyanin. From plastocyanin, electrons are transferred to the P700 molecule forming the reaction center of photosystem I. After excitation in photosystem I, they are transferred to the primary acceptor of this photosystem, tentatively identified as an iron-sulfur protein. From the primary acceptor the electrons then flow along the short final chain from *ferredoxin* (another iron-sulfur protein) through FAD to NADP. NADP is reduced at the final step in the Z sequence.

As NADP accepts a pair[3] of high-energy electrons at the end of the sequence, it binds a hydrogen ion from the surrounding water solution at the same time:

$$\text{NADP} + 2 \text{ electrons} + \text{H}^+ \rightarrow \text{NADPH (or reduced NADP)} \quad (6\text{-}4)$$

The electrons carried by reduced NADP are at such high energy levels that they are capable of reducing almost all other cellular electron acceptors. These electrons are therefore a source of reducing power in the cell, an energy source as vital to cellular activities as ATP. Many of the reactions synthesizing complex molecules in the cell involve a reduction, in which high-energy electrons are transferred to the substance being synthesized (as, for instance, in the dark reactions of photosynthesis building up

---

[3]Excited electrons are released one at a time by chlorophyll P700 and P680. Many of the electron carriers in photosynthesis, such as ferredoxin and the cytochromes, also carry electrons singly. Other carriers, such as FAD, plastoquinone, and the final electron acceptor, NADP, carry electrons in pairs. To balance the flow, the chlorophyll P700 and P680 molecules, and the single electron carriers may be considered to be present in twice the amount of NADP.

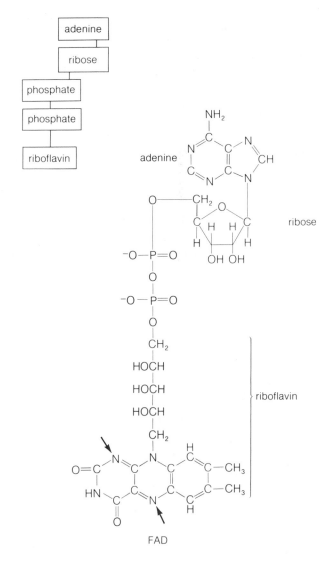

Figure 6-13 FAD (flavin adenine dinucleotide), an electron carrier linked to a protein in a flavoprotein. In going to the reduced form, a hydrogen and electron are added at the two positions marked by arrows.

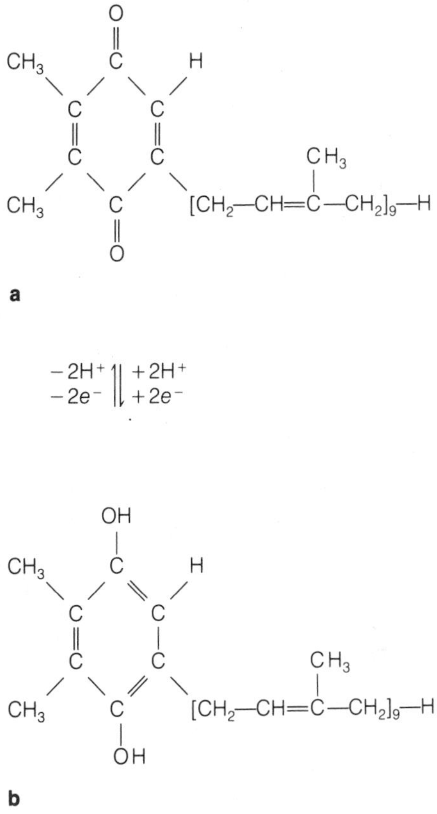

a

$$-2H^+ \quad +2H^+$$
$$-2e^- \quad +2e^-$$

b

Figure 6-14 Plastoquinone, a nonprotein electron carrier of the Z-pathway. (**a**) Oxidized form; (**b**) reduced form.

ATP Synthesis in the Z-Pathway: Photophosphorylation

Each carrier in the chains linking the two photosystems carries electrons at energy levels lower than the carrier preceding it in the series, and higher than the carrier following it. Thus the energy level of the electrons is lower in each successive member of the chains. As the electrons pass from one carrier to the next, the difference in energy level is released as free energy. Some of the free energy released as electrons traverse the carrier chains of the Z-pathway is used to drive the synthesis of ATP.

The nature of the mechanism coupling ATP synthesis to electron flow through the Z-pathway was debated for many years until its explanation by the *chemiosmotic hypothesis*, first proposed in 1961 by Peter Mitchell of the

carbohydrates from CO_2). Where reductions of this type occur, the required high-energy electrons are usually delivered to the system by reduced NADP. The NADP molecule (*nicotinamide adenine dinucleotide phosphate*, Figure 6-17) thus acts as a carrier of high-energy electrons. The hydrogen carried by NADP in the reduced state (see Reaction 6-4) is also important in the production of carbohydrates, a reaction in which CO_2 is converted to (CH_2O).

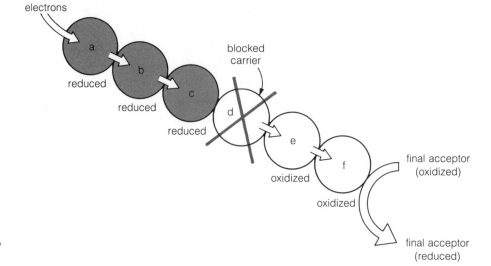

Figure 6-15 Sequencing electron carriers (**a–f**) by specific inhibitors. Carriers falling before the blocked carrier tend to remain in the reduced state (shade); carriers falling after the blocked carrier tend to remain in the oxidized state.

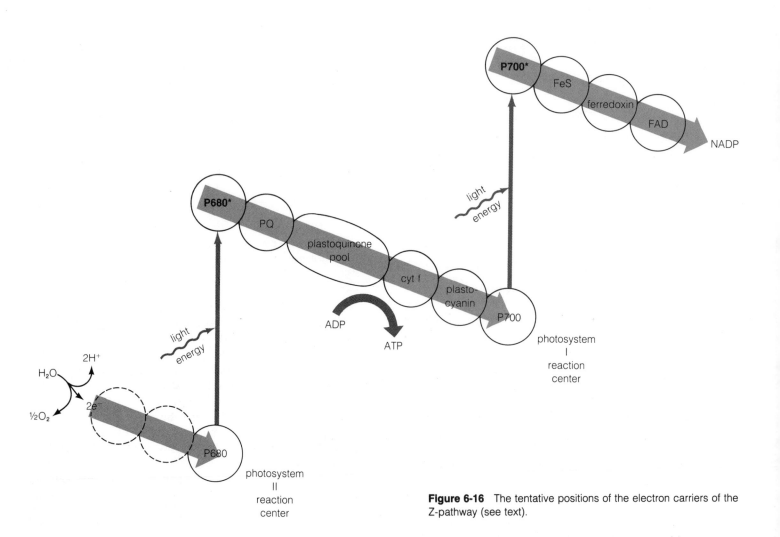

Figure 6-16 The tentative positions of the electron carriers of the Z-pathway (see text).

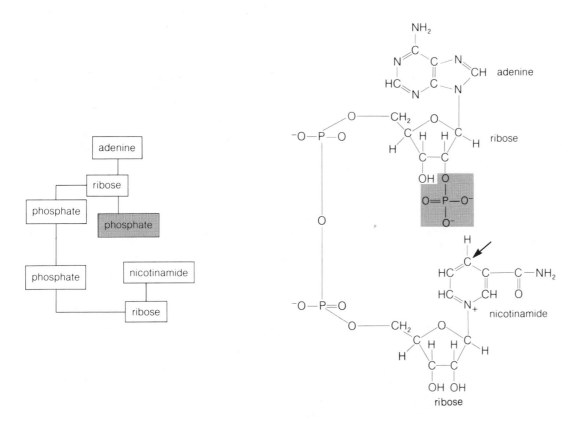

Figure 6-17 NADP (nicotinamide adenine dinucleotide phosphate), a carrier of hydrogen and high-energy electrons in cells. In going to the reduced form, the hydrogen and electrons are added at the position marked by the arrow. The shaded phosphate group is absent in the closely related electron carrier *NAD* (nicotinamide adenine dinucleotide), which is otherwise identical in structure.

Glynn Research Laboratories in England. Mitchell's hypothesis depends on two facts observed about the electron carriers of the Z-pathway: (1) some of the carrier molecules bind H^+ ions as they accept electrons in going from the oxidized to the reduced state, and some are "pure" electron carriers that do not bind H^+ during reduction, and (2) all of the carriers are tightly bound to membranes inside chloroplasts. The membranes binding the carriers inside chloroplasts form the completely closed, saclike thylakoid vesicles of the grana stacks.

According to Mitchell's hypothesis, H^+ ions are expelled across the thylakoid membranes to the inner compartment of the thylakoids as electrons flow along the Z-pathway. The resulting H^+ gradient, high inside the thylakoid sacs and low in the surrounding chloroplast stroma, acts as a source of free energy that drives the synthesis of ATP from ADP and inorganic phosphate as the gradient runs down. Synthesis of ATP in this way, through the en-

ergy released by electron transport in the light reactions of photosynthesis, is termed *photophosphorylation*.

How the H^+ Gradient is Established The mechanism producing the H^+ gradient in Mitchell's hypothesis depends on removal of hydrogens from the stroma and addition of hydrogens to the thylakoid compartment as electrons flow through the Z-pathway. According to Mitchell, each time electrons pass from a nonhydrogen carrier to a hydrogen carrier in the pathway, hydrogen is picked up from the H^+ ions in the stroma solution (Fig. 6-18a). As the electrons pass from a dual electron-hydrogen carrier to a nonhydrogen carrier, the H^+ ions are released into the thylakoid compartment (Fig. 6-18b). This builds the H^+ ion concentration inside the thylakoid compartments and reduces it in the stroma, producing a steep concentration gradient between the thylakoid compartments and the stroma.

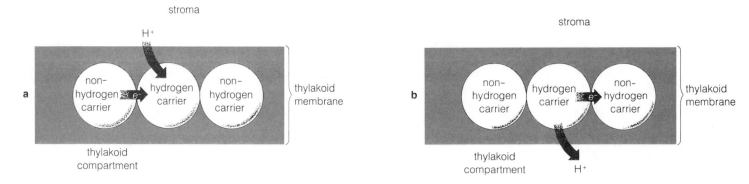

Figure 6-18 Movement of H⁺ ions from the chloroplast stroma to the thylakoid compartments as a result of electron transfer between hydrogen and nonhydrogen carriers in the Z-pathway. (**a**) As electrons pass from a nonhydrogen to a hydrogen carrier, H⁺ ions are removed from the stroma. (**b**) As electrons pass from a hydrogen carrier to a nonhydrogen carrier, electrons are expelled into the thylakoid compartment.

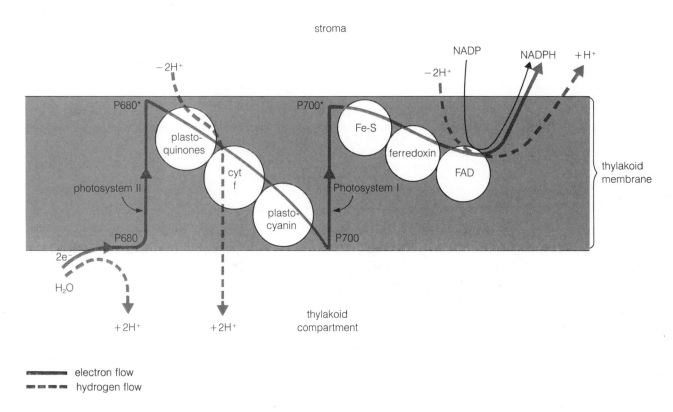

Figure 6-19 Sources of the H⁺ ions added to the gradient across the thylakoid membranes during electron flow along the Z-pathway (see text).

Mitchell's hypothesis proposes that the first H^+ ions added to the gradient established by electron flow along the Z-pathway (Fig. 6-19) are derived from the water molecules split at the first step in the pathway:

$$H_2O \rightarrow 2 \text{ electrons} + 2H^+ + \tfrac{1}{2}O_2 \qquad (6\text{-}5)$$

This reaction is considered to take place on the side of the membrane facing the thylakoid compartment. Since chlorophyll P680 is a pure electron carrier, the 2 H^+ removed from water are released to enter the thylakoid compartment as the electrons pass along the short carrier chain from water to P680. The electrons removed from water, after excitation in the P680 reaction center, pass to the primary acceptor of photosystem II, plastoquinone PQ. Plastoquinones are dual hydrogen-electron carriers; the 2 H^+ required in the reduction of plastoquinones are derived from H^+ ions in the solution on the stroma side of the membrane. These H^+ ions are released to the thylakoid compartment as the electrons pass from the plastoquinones to cytochrome f, a non-H^+ carrier. From cytochrome f, the electrons pass to plastocyanin and the P700 reaction center of photosystem I. All of the carriers of this segment are non-H^+ carriers. After excitation, the electrons pass through the short carrier chain to FAD, a dual hydrogen-electron carrier. Reduction of FAD removes a second pair of H^+ ions from the stroma, converting FAD to $FADH_2$. The electrons and one of the 2 H^+ carried by FAD are delivered to NADP at the end of the sequence. The other H^+, according to Mitchell, is released again into the stroma.

The total Z-pathway is thus considered to remove 3 H^+ from the stroma, and to expel 4 H^+ to the thylakoid compartment for each electron pair flowing from water to NADP. One hydrogen derived from the stroma remains linked to the NADP reduced in the last step. The H^+ gradient created by the mechanism, low in the stroma and high inside the thylakoid compartment, provides the energy driving ATP synthesis.

How the H^+ Gradient Drives ATP Synthesis In Chapter 4, we noted that active transport proteins use ATP as an energy source to move ions and other substances across membranes against concentration gradients. These membrane transport proteins are capable of removing a phosphate group from ATP to produce ADP + phosphate:

$$ATP + H_2O \rightleftharpoons ADP + \text{phosphate} \qquad (6\text{-}6)$$

This reaction is the direct energy source used by the membrane proteins for active transport.

We also noted that investigators working with the active transport systems discovered that by experimentally raising the concentration of ions to abnormally high levels on the side of the membrane toward which they are actively transported, the transport mechanism can be forced to run in reverse. Under these conditions, the ion moves backwards through the membrane channel, and the membrane transport proteins *add phosphate groups to ADP to form ATP.*

According to Mitchell's hypothesis, this is what happens under normal conditions in the chloroplast membranes containing the Z-pathway electron carriers. The internal chloroplast membranes, in addition to housing Z-pathway carriers, also contain a membrane transport protein that in effect is capable of actively transporting H^+ ions across the membrane through energy released by ATP breakdown. The Z-pathway electron carriers build up an H^+ gradient across the internal chloroplast membranes as electrons flow through the system. This H^+ buildup occurs in the same direction as the membrane protein would actively transport H^+ ions at the expense of ATP. As the H^+ gradient builds to high levels as a result of electron transport in the Z-pathway, the H^+ transport protein is forced to run in reverse. Under this condition, the transport protein, instead of breaking down ATP, synthesizes ATP from ADP and phosphate:

$$ADP + \text{phosphate} \rightleftharpoons ATP + H_2O \qquad (6\text{-}7)$$

The free energy required to run this reaction to the right, in the uphill direction, is supplied by the H^+ gradient.

Evidence Supporting the Chemiosmotic Hypothesis
Mitchell's chemiosmotic hypothesis for ATP synthesis in chloroplasts (and in mitochondria; see Chap. 7) is supported by an extensive and impressive list of experiments. One series of experiments has confirmed that the thylakoid membranes of chloroplasts contain a large protein complex that can run Reaction 6-7 in either direction. This complex can be seen in the electron microscope as a "lollipop" extending from the surface of the thylakoid membranes (Fig. 6-20). Ephraim Racker of Cornell University has shown that removing the lollipops from thylakoid membranes, which can be accomplished by relatively simple and reversible adjustments in the concentration of chemicals in the medium surrounding the membranes, uncouples electron transport from ATP synthesis. In this case, electrons flow through the Z-pathway without driving the synthesis of ATP. Returning the lollipops to the thylakoid membranes restores the ability of the membranes to synthesize ATP and reestablishes the close cou-

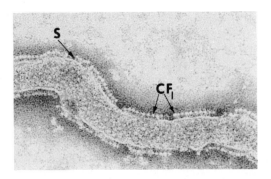

Figure 6-20 Isolated thylakoid membranes of spinach chloroplasts showing the ATPase "lollipops" (arrows). *S*, stalk connecting lollipop complex to membrane; CF_1, the enzyme complex forming the head of the lollipop that catalyzes synthesis of ATP from ADP + phosphate. × 100,000. Courtesy of M. P. Garber, from *Journal of Cell Biology* 63 (1974): 24 by copyright permission of The Rockefeller University Press.

pling between light-driven electron transport and ATP synthesis.

Other experiments confirm the dependence of ATP synthesis in chloroplasts on an H^+ gradient established by the Z-pathway. As electrons flow through the Z-pathway in response to light, H^+ ions actually become more concentrated inside the thylakoid sacs, as predicted by Mitchell's hypothesis. Experimentally breaking or disturbing the membranes of the sacs in any way, so that they become "leaky" to H^+ ions, thus destroying the H^+ gradient, stops ATP synthesis immediately. In addition, producing an artificial H^+ gradient, with H^+ ions at higher concentrations inside the thylakoid sacs, causes a burst of ATP synthesis.

The last line of evidence mentioned above, developed in experiments carried out by Andre T. Jagendorf and Ernest Uribe at the Johns Hopkins University, is among the strongest in favor of the chemiosmotic hypothesis. Jagendorf and Uribe lowered the internal pH of isolated chloroplasts by placing them in a solution containing an acid (at pH 4) that could penetrate the chloroplast membranes. This treatment raised the H^+ ion concentration inside the thylakoids. Subsequent transfer to a medium at pH 8 established an H^+ gradient, with all of the thylakoid compartments inside the chloroplasts containing H^+ ions at very high concentrations, and the stroma, now adjusted to pH 8, containing H^+ ions at relatively low concentrations. As the H^+ ions moved out of the thylakoid sacs into the stroma in response to the artificially imposed gradient, ATP was synthesized in quantity.

Thus all of the available evidence indicates that ATP synthesis in chloroplasts occurs by the chemiosmotic mechanism proposed by Mitchell: electron transport establishes an H^+ gradient which, in turn, provides the driving force for ATP synthesis. Mitchell received the Nobel Prize in 1978 for the chemiosmotic hypothesis.

The total amount of ATP synthesized as each pair of electrons follows the Z-pathway from H_2O to NADP remains somewhat in doubt. The most recent experiments indicate that movement of an electron pair through the entire pathway leads to an H^+ buildup sufficient to drive the synthesis of two molecules of ATP.

The light reactions of photosynthesis, including the Z-pathway and the system synthesizing ATP, thus use light as an energy source to reduce NADP and convert ADP to ATP. These products provide the energy (in the form of ATP) and the reducing power (in the form of reduced NADP, or NADPH) required for the production of complex organic molecules from CO_2 and H_2O in the dark reactions.

The Dark Reactions of Photosynthesis

In the dark reactions, the chemical energy produced in the light reactions is used to convert carbon dioxide into carbohydrates and a variety of other organic products. Although termed the dark reactions because they do not depend directly on light, the various interactions synthesizing carbohydrates, as noted, actually take place primarily in the daytime, when ATP and reduced NADP are readily available from the light reactions.

Tracing the Dark Reactions

Little progress was made in unravelling the dark reactions until the 1940s, when radioactive compounds first became available to biochemists. One substance in particular, CO_2 labeled with the radioactive carbon isotope ^{14}C, made possible the first real breakthroughs in research into the photosynthetic reactions that assemble carbohydrates inside chloroplasts.

Melvin Calvin, Andrew A. Benson, and their colleagues at the University of California at Berkeley used this radioactive form of CO_2 to trace the biochemical pathways of the dark reactions. In their experiments, Calvin and his colleagues allowed photosynthesis to proceed in

Chlorella, a eukaryotic green alga, in the presence of CO_2 containing the radioactive isotope. Extracts of carbohydrates and other substances were then made from the *Chlorella* cells.

If the carbohydrate extracts were made within a few seconds after exposing the cells to radioactive CO_2, most of the radioactive carbon was found in a 3-carbon sugar. Thus the CO_2 taken in by *Chlorella* was incorporated very rapidly into this 3-carbon substance, evidently one of the earliest products of photosynthesis. If the cells were allowed to photosynthesize for longer periods before extracts were made, the radioactive label showed up in more complex substances, including glucose, a 6-carbon sugar. In other experiments, Calvin and his colleagues reduced the amount of CO_2 to levels so low that photosynthesis could proceed only very slowly in the *Chlorella* cells even though adequate light was supplied. Under these conditions, a 5-carbon sugar accumulated in quantity in the cells. Accumulation of this 5-carbon sugar at low CO_2 levels suggested that it is the first substance to combine chemically with CO_2, and that it "piles up" in chloroplasts when the CO_2 supply is abnormally low. By similar methods, an extensive series of intermediate compounds between CO_2 and glucose were identified.

Using this information, Calvin, Benson, and James A. Bassham were able to piece together the dark reactions of photosynthesis. For his brilliant work with this system and his successful model for the photosynthesis of carbohydrates, now called the *Calvin cycle,* Calvin was awarded the Nobel Prize in 1961.

The Calvin Cycle

The sequence of dark reactions described by Calvin and his coworkers occurs in a repeating cycle, with CO_2, ATP, and reduced NADP as raw materials and fuel for the cycle (hydrogen is carried to the cycle by the reduced NADP), and carbohydrate units (CH_2O) as the primary product (Figure 6-21). ADP + phosphate and oxidized NADP are also given off as products of the cycle. Each step of the Calvin cycle is catalyzed by a specific enzyme.

The intermediate compounds of the Calvin cycle are complex. However, the overall progress of the cycle can easily be understood if only the carbon chains and phosphate groups are followed through the sequence (Figure 6-22; a more detailed outline of the cycle is presented in Supplement 6-1). The first steps of the cycle (reactions 1 through 3 in Fig. 6-22) take up CO_2 and combine it with a 5-carbon sugar, producing two molecules of a 3-carbon,

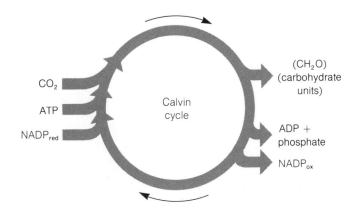

Figure 6-21 The overall reactants and products of the Calvin cycle.

one-phosphate sugar at the expense of ATP and reduced NADP. These reactions proceed in the following way. In the first reaction, CO_2 combines with the 5-carbon sugar, which contains two phosphate groups; this 5-carbon sugar is designated 5C, 2P in Figure 6-22. The reaction produces two molecules of a 3-carbon sugar, each containing one phosphate group (3C, 1P in Fig. 6-22). In the next step (reaction 2) another phosphate group is added to each of these sugars at the expense of 2 ATP derived from the light reactions of photosynthesis, yielding two molecules of a 3-carbon, two-phosphate sugar (3C, 2P in Fig. 6-22). These highly reactive products are reduced in reaction 3 by accepting electrons and hydrogen from two molecules of reduced NADP, also derived from the light reactions. As this reduction takes place, one phosphate group is removed from each of the sugars. The final product of these reactions is two high-energy molecules of *3-phosphoglyceraldehyde (3PGAL)*. This 3-carbon, one-phosphate sugar is the central product of the dark reactions.

Most of the 3PGAL produced is used to replace the 5-carbon sugar used in reaction 1 of the cycle. This sequence, which breaks down one additional molecule of ATP for each turn of the cycle, is designated reaction sequence 4 in Figure 6-22. Some 3PGAL is also released as a product of the cycle, to be converted to glucose and other substances, or to be oxidized directly by the cell as fuel. The ATP and reduced NADP used in the cycle, converted in the process to ADP + phosphate and oxidized NADP, return to the light reactions to be converted back to ATP and reduced NADP again.

The primary product of the Calvin cycle, the 3-carbon, one-phosphate sugar 3PGAL, is the starting point for synthesis of a variety of complex carbohydrates and polysaccharides. Two molecules of 3PGAL combine through a

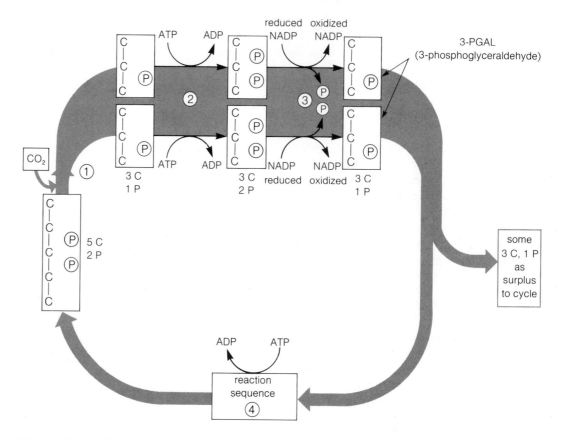

Figure 6-22 The Calvin cycle, showing only the carbon chains and phosphate groups of the intermediates of the cycle (see text).

stepwise series of reactions to produce glucose, which contains six carbons but no phosphates, and other 6-carbon sugars.[4]

Free glucose is actually formed in only very limited quantities in the chloroplasts of most plants. Instead, glucose and other 6-carbon sugars are used as the starting points for making sucrose, starch, cellulose, and a wide variety of additional organic molecules. In addition to carbohydrates, amino acids, lipids, and proteins also become labeled rapidly in illuminated chloroplasts supplied with radioactive CO_2. All of the amino acids required for protein synthesis can be synthesized by most plants, either inside chloroplasts or in the surrounding cytoplasm, by pathways starting from products of the dark reactions. Protein synthesis, occurring on ribosomes suspended in the chloroplast interior, can also be detected inside chloroplasts, along with the linkage of nucleotides into DNA and RNA (see Supplement 10-2).

The balance among carbohydrates, fats, and amino acids synthesized in chloroplasts varies widely among different species of plants. In some, nearly all of the CO_2 absorbed is incorporated into carbohydrates such as sucrose, formed from glucose and another 6-carbon sugar, fructose. In other species, such as the alga *Chlorella*, the synthesis of fats and amino acids greatly exceeds sucrose synthesis, which may account for 5 percent or less of the CO_2 absorbed. In any event, the chloroplasts of most species contain the enzymes and chemicals required to make a wide variety of substances in addition to carbohydrates, and in fact have synthetic capacity practically equivalent to entire cells.

---

[4]The reactions that convert 3PGAL to glucose are essentially the reverse of part of the sequence that breaks down glucose to release energy for cellular activities (see Chap. 7).

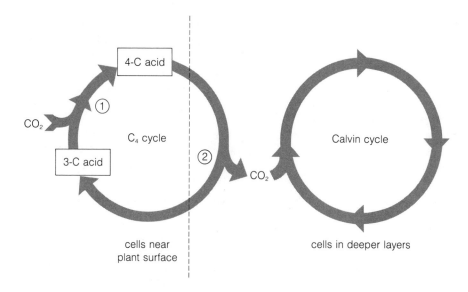

Figure 6-23 The C_4 cycle, a pathway that increases the efficiency of CO_2 utilization in some plants (see text).

The C_4 Cycle

The Calvin cycle is supplemented in some plants by an important secondary cycle that carries out a biochemical "trick" to get around a deficiency in one of the Calvin cycle enzymes. The deficient enzyme is the one that catalyzes the first step in the Calvin cycle. In this step, CO_2 combines with the 5C, 2P sugar to produce two molecules of a 3C, 1P product (reaction 1 in Fig. 6-22). The deficiency in the enzyme results from the fact that oxygen can substitute for CO_2 in this reaction, leading to the formation of products that cannot be utilized in the Calvin cycle and are lost to the dark reactions.

The supplementary C_4 cycle gets around this deficiency by combining CO_2 with a 3-carbon acid to produce the 4-carbon acid for which the cycle is named (step 1 in Fig. 6-23). This step occurs instead of the Calvin cycle in cells at the surface of the plant where oxygen is abundant enough to interfere with the deficient Calvin cycle enzyme. The 4-carbon acid then diffuses to deeper layers of the plant, where it is broken down to release CO_2 to the Calvin cycle in regions where oxygen is present in concentrations that are too low to interfere with the enzyme (step 2 in Fig. 6-23). The 3-carbon acid produced at this step diffuses back to the surface of the plant to pick up another molecule of CO_2. The C_4 cycle, which greatly increases the efficiency of photosynthesis, occurs in many plants, in-

cluding important crop grasses such as corn, sugar cane, and sorghum (the C_4 cycle is discussed in greater detail in Supplement 6-1).

Locations of the Reactions of Photosynthesis in Chloroplast Ultrastructure

Chloroplasts are relatively easy to isolate in quantity from the leaves of plants such as spinach. Once isolated and cleaned of cytoplasmic and nuclear debris, chloroplasts can be further broken down by grinding or homogenization. Two fractions are produced by this treatment: (1) a membrane fraction, including thylakoid membranes and fragments of the boundary membranes; and (2) a "soluble" fraction without membranes, originating from the chloroplast stroma. These membrane and stroma fractions retain much of their biochemical activity when isolated. Using preparations of this type, it has been possible to determine where the various light and dark reactions occur in the inner compartments of the chloroplast.

Preparations of thylakoid membranes contain all the parts and pieces of the light reactions, including the chlorophylls, carotenoids, photosystems I and II, and all of the molecules associated with electron transport, reduction of NADP, and ATP synthesis. The soluble fraction derived from the stroma contains all the enzymes and intermediates of the dark reactions. These evidently occur in solution inside chloroplasts, without attachment to the internal membrane systems.

The light and dark reactions of chloroplasts provide the raw materials needed for synthesis of the proteins, lipids, carbohydrates, and nucleic acids of the eukaryotic plants. All that most plants require in addition is a supply of inorganic minerals. The energy needed for synthesis of these molecules is derived from the light reactions in chloroplasts. The molecules of the eukaryotic plants are used in turn as energy sources for most of the remaining living organisms of the world.

Suggestions for Further Reading

Clayton, R. K. 1970. *Light and living matter*. Vols. 1 and 2. McGraw-Hill: New York.

Govindjee and R. Govindjee, 1974. The primary events of photosynthesis. *Scientific American* 231:68–82 (December).

Hinkle, P. C. and R. E. McCarty. 1978. How cells make ATP. *Scientific American* 238:104–23 (March).

Lehninger, A. L. 1971. *Bioenergetics*. 2nd ed. W. A. Benjamin: Menlo Park, Calif.

Lehninger, A. L. 1975. *Biochemistry*. 2nd ed. Worth: New York.

Miller, G. T., Jr. 1971. *Energetics, kinetics, and life: An ecological approach*. Wadsworth: Belmont, Calif.

Stryer, L. 1981. *Biochemistry*. 2nd ed. W. H. Freeman: San Francisco.

Wolfe, S. L. 1981. *Biology of the cell*. 2nd ed. Wadsworth: Belmont, Calif.

For Further Information

DNA, ribosomes, and protein synthesis in chloroplasts, Supplement 10-2
Evolutionary origins of chloroplasts, Chapter 13
Mitchell's chemiosmotic hypothesis, Chapter 7
Peroxisomes and photorespiration, Supplement 7-3

Questions

1. Trace the flow of energy from sunlight through the living organisms on the earth.

2. What structures are visible inside chloroplasts in the electron microscope? What is a thylakoid? What is the relationship between thylakoids, grana, and stromal lamellae?

3. Write the equation for photosynthesis of glucose from CO_2 and H_2O. How is this equation related to van Niel's more general equation for photosynthesis?

4. What experiment proved that the oxygen evolved in photosynthesis is derived from water, and not from CO_2?

5. How were the light and dark reactions of photosynthesis first detected?

6. What substances link the light and dark reactions together?

7. What is a quantum of light? What is the relationship between wavelength and the energy of light quanta? In what ways do the quanta of a beam of light resemble the electrons in an electron beam?

8. What makes some molecules appear pigmented or colored? What happens when light is absorbed by a pigmented molecule?

9. What molecules absorb light in chloroplasts? What wavelengths do they absorb? Why do chloroplasts appear green?

10. What is the central event in the conversion of light to chemical energy?

11. What kinds of molecules are present in photosystems I and II? How is light energy "trapped" in the reaction center of a photosystem?

12. What is chlorophyll P700 and P680?

13. What kinds of electron carriers appear in the Z-pathway? List two ways in which the carriers were sequenced.

14. According to the sequence of carriers shown in Figure 6-16, what would happen to the carriers if photosystem I is blocked by an inhibitor in the presence of light, ADP, phosphate, and oxidized NADP? If photosystem II is blocked under these conditions?

15. Trace the flow of electrons through the Z-pathway. Where does hydrogen enter and leave the pathway? What significance does this have for the mechanism synthesizing ATP?

16. How many light quanta are required to liberate $\frac{1}{2}O_2$? A molecule of oxygen? A molecule of reduced NADP?

17. How is ATP synthesis in chloroplasts related to active transport?

18. What experimental findings support Mitchell's chemiosmotic hypothesis?

19. How is ATP synthesis in chloroplasts related to thylakoid membranes?

20. How were the dark reactions worked out?

21. How are ATP and reduced NADP used in the dark reactions? How does CO_2 enter the dark reactions?

22. What major kinds of biological molecules are synthesized in chloroplasts?

23. Where are the light and dark reactions located in chloroplast ultrastructure?

24. What is the C_4 cycle? What is the significance of this cycle to the plants that possess it?

25. An investigator has detected 5 electron carriers, A, B, C, D, and E, in an electron transport system. Several inhibitors are available that prevent some of the carriers from releasing electrons. When an inhibitor for carrier E is added, A, B, and C become oxidized, and D remains reduced. When an inhibitor for A is added, carriers C, D, and E remain reduced, and B becomes oxidized. What is the probable sequence of the carriers?

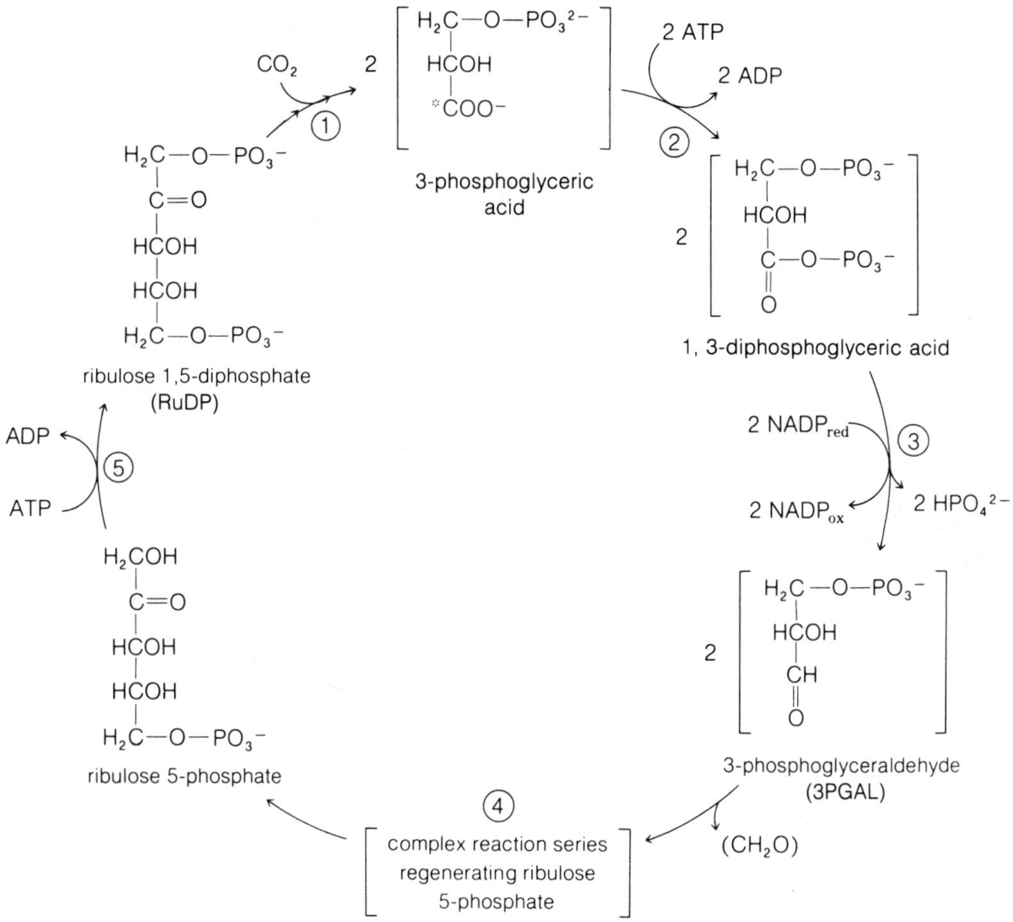

Figure 6-24 Major reactions and intermediate compounds of the Calvin cycle (see text).

Supplement 6-1: Further Light on the Dark Reactions

The Calvin Cycle

The Calvin cycle uses CO_2, ATP, and reduced NADP as raw materials. As a primary product the cycle releases a 3-carbon sugar, 3-phosphoglyceraldehyde (3PGAL), which is then used to synthesize more complex carbohydrates. Through various reactions, 3PGAL also enters the synthesis of lipids, amino acids, proteins, and nucleic acids. The cycle also releases ADP, inorganic phosphate, and oxidized NADP. As the cycle turns, all of the intermediate compounds in the sequence are continuously regenerated.

The important intermediate reactions are shown in Figure 6-24. In the first reaction of the cycle, CO_2 combines directly with the 5-carbon, 2-phosphate substance *ribulose-1,5-diphosphate (RuDP)*. The reaction, catalyzed by the enzyme *ribulose 1,5-diphosphate carboxylase (RuDP carboxylase)*, produces two molecules of *3-phosphoglyceric acid*, a 3-carbon substance. One of these contains the newly incorporated CO_2 in the position marked by an asterisk in Figure 6-24. The overall reaction requires no input of energy because the two 3-carbon products exist at a much lower energy level than RuDP, which can be considered a high-energy substance. The RuDP carboxylase enzyme catalyzing this reaction makes up as much as 25 to 50 percent of the total protein of chloroplasts. This is believed to be related to the fact that the enzyme reacts only relatively slowly with CO_2. The high content of RuDP carboxylase in chloroplasts evidently compensates effectively for the low reactivity of the enzyme.

In the next step in the cycle (reaction 2) another phosphate group is added to 3-phosphoglyceric acid to produce *1,3-diphosphoglyceric acid*. This conversion, an uphill reaction, is made possible by the simultaneous breakdown of ATP to ADP, from which the added phosphate is derived:

$$\text{3-phosphoglyceric acid + ATP} \rightarrow$$
$$\text{1,3-diphosphoglyceric acid + ADP} \quad \text{(6-8)}$$

This highly reactive 3-carbon, 2-phosphate sugar is reduced in the next step (reaction 3) by accepting two electrons and hydrogen from NADP to produce 3PGAL. At the same time, one of the phosphate groups added in reaction 2 is removed and released to the medium as inorganic phosphate. The products contain less energy and order than the reactants, and the total reaction proceeds spontaneously with the release of free energy:

$$\text{1,3-diphosphoglyceric acid + reduced NADP} \rightarrow$$
$$\text{3PGAL + oxidized NADP + HPO}_4^{2-} \quad \text{(6-9)}$$

Reactions 2 and 3 in Figure 6-24 are shown multiplied by a factor of 2 because two molecules of 3PGAL are produced for each molecule of CO_2 interacting with RuDP in reaction 1.

The remainder of the cycle replaces the RuDP used in the first reaction in the sequence and leaves one molecule of 3PGAL as a surplus to the cycle. These reactions operate in the following way. Three turns of the cycle as far as reaction 3 produce six molecules of 3PGAL. These three molecules collectively contain 18 carbon atoms. Five of the 3PGAL molecules, containing a total of 15 carbons, enter the complex series designated as reaction system 4 in Figure 6-24, yielding three molecules of ribulose-5-phosphate (15 total carbons). These are converted to RuDP (reaction 5 in Fig. 6-24) at the expense of one additional molecule of ATP for each molecule of RuDP produced. This replaces the three molecules of RuDP used in the three turns of the cycle. The surplus molecule of 3PGAL is the starting point of a wide variety of reactions yielding sugars, starch, and other complex molecules in the chloroplast. Because three turns of the cycle are required to produce one of these 3-carbon molecules as a surplus, each single turn may be considered to yield a 1-carbon carbohydrate unit, or (CH_2O). Six turns would be required to generate enough units to produce one molecule of glucose, with six carbons.

Tracing one complete turn of the cycle reveals that for each molecule of CO_2 fixed and each unit of carbohydrate generated, three molecules of ATP and two molecules of reduced NADP are converted to 3 ADP and 2 oxidized NADP:

$$CO_2 + \text{3 ATP + 2 reduced NADP} \rightarrow$$
$$(CH_2O) + \text{2 oxidized NADP + 3 ADP + 3HPO}_4^{2-} \quad \text{(6-10)}$$

The oxidized NADP, ATP, and inorganic phosphate cycle back to the light reactions of photosynthesis to be converted back to ATP and reduced NADP in the Z-pathway.

More on the C_4 Pathway

The C_4 cycle was discovered when P. Kortschak and his colleagues in Hawaii and M. D. Hatch and C. R. Slack in Australia looked for the earliest labeled intermediates produced in corn, sugar cane, and other grasses of tropical origin after exposure of the plants to labeled CO_2. Surprisingly, the earliest label appeared in a mixture of 4-carbon acids instead of the 3-carbon 3PGAL of the Calvin cycle (hence the name C_4 *cycle;* the Calvin cycle, which produces a 3-carbon acid as its end product, is sometimes called the C_3 *cycle*). Intermediates of the Calvin cycle were also found to be labeled after a delay of some seconds following the appearance of the label in the 4-carbon acids.

The appearance of the label in the 4-carbon acids was proposed by Hatch and Slack to be a part of a side cycle linked to the main Calvin cycle (Fig. 6-25). As a preparatory step in the side cycle, *pyruvic acid*, a 3-carbon substance, is pushed to a higher energy form by linkage to a phosphate group derived from ATP (step 1 in Fig. 6-25). The product, a 3C,1P acid called *phosphoenolpyruvic acid*, then reacts with CO_2 to produce *oxaloacetic acid*, a 4-carbon substance (step 2 in Fig. 6-25). The phosphate group is removed and released to the medium in this step. Oxaloacetic acid is then reduced to *malic acid* in the next step in the cycle (step 3 in Fig. 6-25). The electrons added to form malic acid are donated by NADP, which is converted to the oxidized form as a result. The CO_2 carried by malic acid is released in the next step (step 4 in Fig. 6-25), an oxidation in which electrons as well as CO_2 are removed from malic acid. NADP accepts the electrons removed in this step and is converted back to reduced NADP, replacing the molecule of reduced NADP used in step 3 of the cycle. NADP is thus converted between its oxidized and reduced forms within the cycle so that there is no net gain or loss of this electron carrier. The CO_2 released enters the

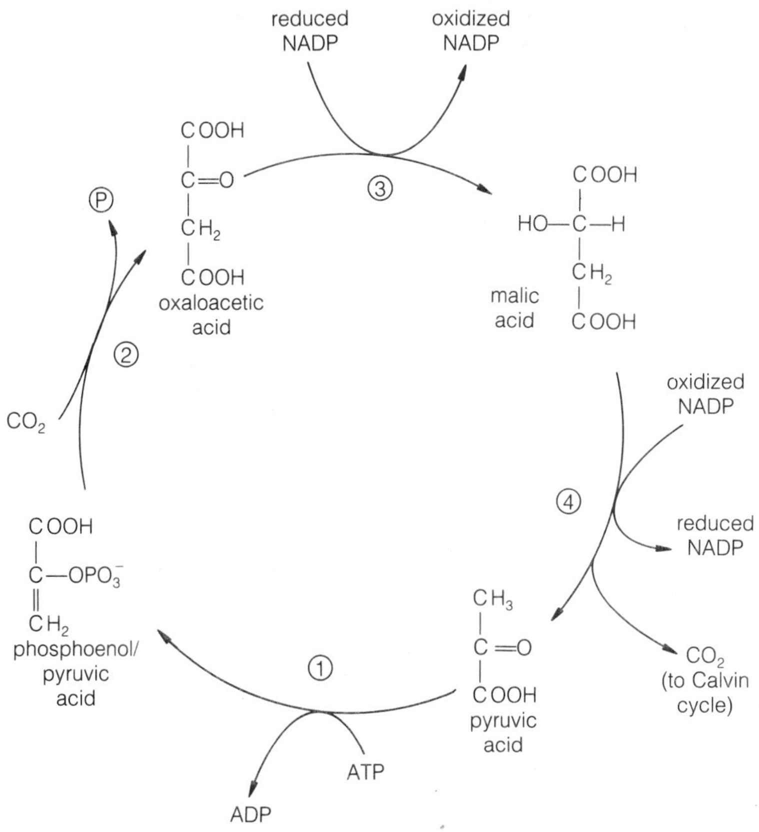

reduced NADP   oxidized NADP

COOH
|
C=O
|
CH$_2$
|
COOH
oxaloacetic acid

③

COOH
|
HO—C—H
|
CH$_2$
|
COOH
malic acid

Ⓟ

②

CO$_2$

oxidized NADP

④

reduced NADP

COOH
|
C—OPO$_3^-$
‖
CH$_2$
phosphoenol/ pyruvic acid

CH$_3$
|
C=O
|
COOH
pyruvic acid

CO$_2$ (to Calvin cycle)

①

ATP

ADP

Figure 6-25 Major reactions of the C$_4$ cycle (see text).

Calvin cycle by the regular route, by combination with RuDP, and the Calvin cycle turns as usual. The other product of reaction 4, pyruvic acid, replaces the pyruvic acid used in the first step of the C$_4$ pathway.

The C$_4$ pathway appeared at first to be a "futile" cycle that results only in net breakdown of ATP. However, investigators soon realized that it serves to increase the availability of CO$_2$ to the Calvin cycle by compensating for the imperfection in the activity of RuDP carboxylase, the enzyme catalyzing the first uptake of CO$_2$ in the Calvin cycle. As noted in Chapter 6, oxygen can compete effectively with CO$_2$ for the active site on the RuDP carboxylase enzyme, diverting it from its central role in the Calvin cycle to one in which it adds oxygen to RuDP. The products of this reaction are lost to the Calvin cycle, and the cycle fails to turn. Plants with the C$_4$ pathway, such as grasses, are able to get around this deficiency in RuDP carboxylase through the C$_4$ cycle.

In grasses, the C$_4$ cycle occurs in the chloroplasts of cells that lie close to the external surface of the plant. Because these cells are near the plant surface, oxygen is present inside them in relatively high concentration. However,

these cells lack RuDP carboxylase, effectively preventing loss of RuDP to carbohydrate production through the faulty activity of the enzyme. CO$_2$ is taken up instead in the C$_4$ pathway, leading to production of an extensive pool of malic acid. Malic acid from these cells diffuses to deeper tissues of the plant, where it is broken down to release CO$_2$ in quantity. The chloroplasts of these cells contain RuDP carboxylase in normal amounts. Since oxygen is present in reduced concentration in the deeper layers of the plant, the RuDP carboxylase carries out its Calvin cycle function of CO$_2$ fixation normally, without diversion to the use of oxygen. The end result is much greater efficiency in the activity of RuDP carboxylase in CO$_2$ fixation, and, consequently, much greater efficiency in photosynthesis. Under optimum conditions, the plants using the C$_4$ pathway carry out photosynthesis at about twice the rate of plants lacking the pathway.

Plants carrying out the C$_4$ pathway are found primarily in tropical and subtropical regions, particularly in more arid habitats. Water loss is reduced in the C$_4$ plants through a side effect on the minute openings in leaves that admit CO$_2$ and allow water vapor to escape. The greater efficiency in CO_2 uptake leads to reduction in the size of these openings and in the amounts of water lost through them. As a result, C_4 plants are about twice as economical in water use as plants lacking the pathway.

Supplement 6-2:
Photosynthesis in Prokaryotes

The two groups of prokaryotes, bacteria and blue-green algae, differ fundamentally in their photosynthetic mechanisms. The photosynthetic bacteria possess a comparatively primitive system limited essentially to the activities carried out by photosystem I in the eukaryotic plants. Because photosystem II is absent, the photosynthetic bacteria cannot use water as an electron and hydrogen donor, and do not evolve oxygen in photosyntehsis. In contrast, the blue-green algae, although typically prokaryotic in cellular organization, carry out photosynthesis by essentially the same mechanisms as higher plants.

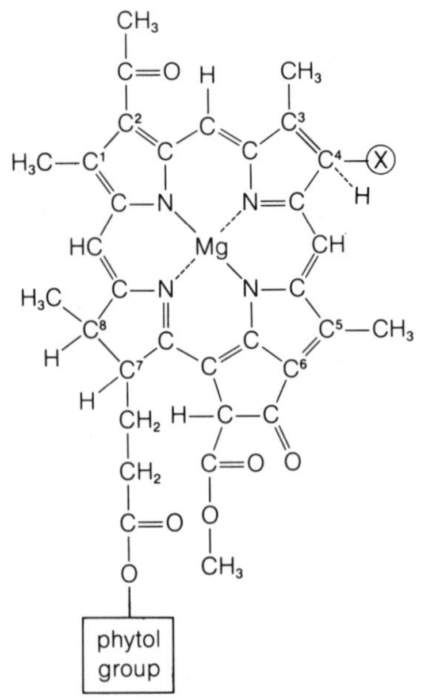

Figure 6-26 Bacteriochlorophyll. In bacteriochlorophyll a, X = C_2H_5; an additional —H occurs at C_4 in bacteriochlorophyll a (dotted line). In bacteriochlorophyll b, X = =CH—CH$_3$.

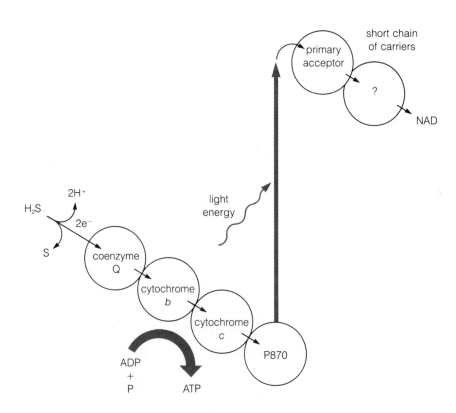

Figure 6-27 The tentative arrangement of carriers and the photosystem in a bacterial system using H_2S as electron donor.

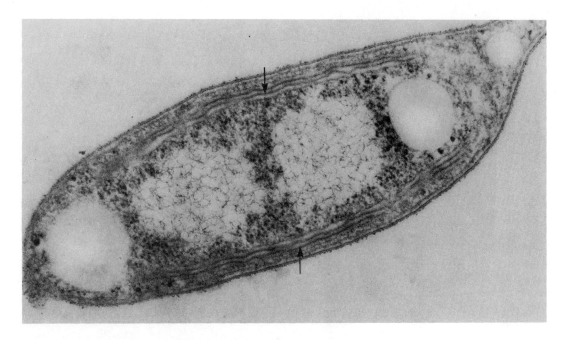

Figure 6-28 Photosynthetic membranes (arrows) in the cytoplasm of a bacterium. × 59,000. Courtesy of W. C. Trentini and the American Society for Microbiology.

The differences between the photosynthetic bacteria and the remaining photosynthetic organisms of the world are primarily in the photosystems and light reactions. All photosynthetic groups, prokaryotic and eukaryotic, use the Calvin cycle to fix CO_2 and carry out the dark reactions in essentially the same way.

The Photosynthetic Bacteria

Bacterial photosynthesis is limited to two groups named by the color of their cells, the *purple* and *green* photosynthetic bacteria. These bacteria contain a distinct family of chlorophylls, called *bacteriochlorophylls*, that differ slightly from the chlorophylls of eukaryotic plants (Fig. 6-26). The bacterial chlorophylls absorb light most strongly in the far red wavelengths, which are almost invisible to humans, and transmit almost all wavelengths of visible light. As a result, the bacteriochlorophylls contribute no distinctive color to either bacterial group. Their colors come instead from carotenoid pigments also present in the bacteria. In one group, the carotenoids present absorb most visible wavelengths but transmit green; in the other, purple light

is transmitted. Both the purple and green light in these bacteria are produced by mixtures of transmitted wavelengths.

Since the purple and green photosynthetic bacteria do not possess photosystem II, they cannot split water to provide electrons and hydrogen for photosynthesis. Thus, in the overall reaction:

$$CO_2 + 2H_2D \rightarrow (CH_2O) + H_2O + 2D \qquad (6\text{-}11)$$

the substance H_2D may be hydrogen, H_2:

$$CO_2 + 2H_2 \rightarrow (CH_2O) + H_2O \qquad (6\text{-}12)$$

or hydrogen sulfide, H_2S:

$$CO_2 + 2H_2S \rightarrow (CH_2O) + H_2O + 2S \qquad (6\text{-}13)$$

Other donors of electrons and hydrogen are also used as raw materials by different photosynthetic bacteria. All of these usable donors release electrons at relatively high energy levels when oxidized; the electrons released from water exist at energy levels too low to enter the system.

From the donor the electrons flow through a chain of carriers believed to be set up as shown in Figure 6-27 (page 134), including coenzyme Q (see Fig. 7-13), a molecule closely similar to plastoquinone in structure, and two cytochromes, b and c. From the chain, the electrons pass to the reaction center of the bacterial photosystem, which consists of a specialized bacteriochlorophyll called *P870*. From P870 the electrons, after absorbing light energy, flow to the primary acceptor and finally to the final electron acceptor, which in bacteria is *NAD* (*nicotinamide adenine dinucleotide*; see Fig. 7-6). NAD is identical to NADP except for a single phosphate group that is present in NADP but absent in NAD. (The extra phosphate in NAD is shaded in Fig. 6-17.)

The electron carriers and the single photosystem in bacteria are tightly bound to membranes, as the equivalent systems are in eukaryotes. In most bacteria, these photosynthetic membranes are suspended as closed sacs directly in the bacterial cytoplasm (Fig. 6-28, page 135); no structures equivalent to chloroplasts are present. In a few photosynthetic bacteria, the photosystem and electron carriers are concentrated in the plasma membrane instead.

Movement of electrons through the carriers produces an H$^+$ gradient across the photosynthetic membranes of bacteria, as it does in eukaryotic plants. This gradient leads to ATP synthesis, through the activity of a membrane transport protein also bound to the bacterial membranes.

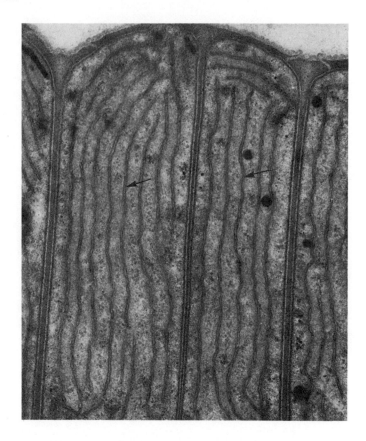

Figure 6-29 Photosynthetic membranes (arrows) suspended directly in the cytoplasm of a row of blue-green algae cells. × 44,500. Courtesy of N. J. Lang.

The Blue-Green Algae

The blue-green algae use chlorophyll a in photosynthesis, in combination with essentially the same carotenoid pigments and electron carriers as the eukaryotic plants. As in the eukaryotes, the pigments are organized into two photosystems interconnected by electron carriers into a Z-pathway. Consequently, the blue-green algae are able to use water as a source of electrons and hydrogen for photosynthesis, and evolve oxygen as a by-product. Photosynthesis in these prokaryotes thus resembles the systems of eukaryotic plants rather than bacteria, and all or most of the pathways shown in Figure 6-16 occur in the blue-green algae. However, the molecules carrying out the light reactions in the blue-green algae are bound to membranous sacs suspended directly in the cytoplasm as they are in bacteria (Fig. 6-29); no chloroplasts are present in these prokaryotic organisms.

7

Energy Release in Glycolysis and Respiration

The products of photosynthesis are used as an energy source by both plants and animals. Plants break down the photosynthetic products directly as an energy source; animals use the same products directly or indirectly by eating plants or other animals. Of the food molecules used as an energy source, carbohydrates and lipids of various kinds are most important in both plant and animal cells. However, almost all biological molecules, including proteins and nucleic acids, can be broken down by cells to release energy.

The reactions breaking down these substances are oxidations, in which high-energy electrons are removed from the fuel molecules and transferred to acceptors of various kinds. Much of the energy released in these transfers is used to synthesize ATP, the energy "dollar" of the cell.

The reactions producing ATP take place in two major stages in eukaryotic cells (Fig. 7-1). In the first stage, the food molecules—whether carbohydrates, lipids, proteins, or nucleic acids—are converted into short 2- or 3-carbon segments. Only limited quantities of ATP are produced in this first stage. The products of the first stage are the immediate fuels of the second major stage, in which the short carbon chains are oxidized completely to carbon dioxide and water. Almost all the ATP produced in cells arises from the reactions of the second stage.

The reactions of the first stage of cellular oxidations are distributed primarily in the cytoplasm outside mitochondria. All of the reactions of the second stage, and thus the major production of ATP, occur inside mitochondria.

The Structure and Occurrence of Mitochondria

In most cells, mitochondria are spherical or slightly elongated bodies, about 2 micrometers in length and 0.5 micrometers in diameter, that approximate the dimensions of a bacterial cell. Apparently mitochondria are not rigid; they may take on a variety of shapes in the cytoplasm. They can be observed under the light microscope to change slowly between spherical and longer filamentous forms, and to fuse together and divide.

The electron microscope shows that mitochondria contain two separate membrane systems (Figs. 7-2a and b). One membrane, called the *outer boundary membrane*, forms a single continuous and relatively smooth outer layer around the mitochondrion, completely separating the interior from the rest of the cytoplasm. Closely lining this membrane, inside the mitochondrion, is a second membrane, the *inner boundary membrane*. The two boundary membranes, visible in Figure 7-2a, are roughly analogous to the outer and inner boundary membranes of chloroplasts. The inner boundary membrane is thrown into folds or tubular extensions called *cristae* (singular = *crista*) that reach into the inner cavity of the mitochondrion. These folds take many forms in the mitochondria of different tissues and species. Among the most common is the arrangement shown in Figures 7-2a and b, in which the cristae consist of flattened, saclike folds extending across the interior of the mitochondrion, more or less at right angles to the boundary membranes. Tubular cristae (Fig. 7-3) also occur in many plants, the protozoa, and in some cells of higher animals.

The outer and inner boundary membranes separate the mitochondrial interior into two distinct regions (see Fig. 7-2b): the *intermembrane compartment*, between the inner and outer membranes, and the *matrix*, the innermost compartment enclosed by the inner mitochondrial membrane. Since the cristae are formed as folds of the inner boundary membrane, the space inside the cristae is continuous with the intermembrane compartment.

The mitochondrial matrix, which is analogous in location to the chloroplast stroma, may contain a variety of

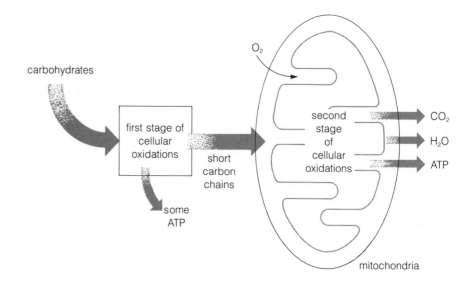

Figure 7-1 The major stages of cellular oxidation (see text).

carbohydrates

first stage of cellular oxidations

some ATP

short carbon chains

O_2

second stage of cellular oxidations

CO_2

H_2O

ATP

mitochondria

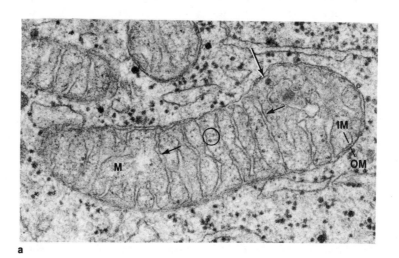

a

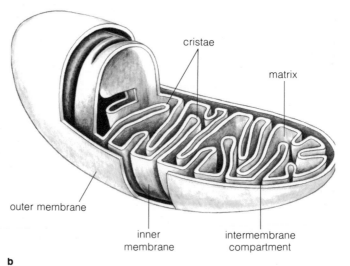

cristae

matrix

outer membrane

inner membrane

intermembrane compartment

b

Figure 7-2 (**a**) A mitochondrion from an intestinal cell of the chick. The outer boundary membrane (*OM*) is smooth and covers the entire mitochondrion; the inner boundary membrane (*IM*) folds into cristae (arrows) that extend into the mitochondrial interior. The matrix (*M*), the inner mitochondrial substance enclosed by the two membranes, contains proteins and other molecules in solution. Mitochondrial ribosomes (circle) are also present in the matrix. × 53,000. Courtesy of J. Mais. (**b**) The membranes and compartments of mitochondria.

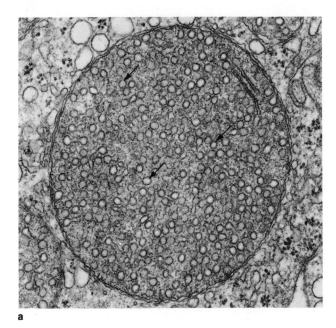

Figure 7-3 Tubular cristae (arrows) in a mitochondrion from the adrenal gland of a rat. × 35,000. Courtesy of D. S. Friend.

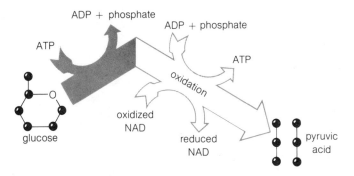

Figure 7-4 The reactions of glycolysis split glucose (6 carbons) into pyruvic acid (3 carbons) and yield ATP and reduced NAD. The reactions take place in two major parts. In the first part (shaded), 6-carbon molecules derived from glucose are raised to higher energy levels at the expense of ATP. In the second part, the high energy products of the first part are oxidized and split into pyruvic acid. The second part yields reduced NAD and a net gain in ATP.

structures, including fibrils, crystals, and dense granules. Ribosomes also occur in the matrix in large numbers (circle, Fig. 7-2a). Also embedded within the matrix are scattered deposits of DNA, closely similar in structure and properties to bacterial DNA (for details, see Supplement 10-2).

Almost all eukaryotic cells contain mitochondria. The exceptions are a few metabolically inert types such as the red blood cells of higher animals, which contain no cytoplasmic organelles of any kind when fully mature. A few algae and some protozoa contain only a single mitochondrion. Most cells, however, contain from several hundred to a thousand mitochondria. Some may contain many more. Liver cells in higher animals may contain from slightly less than a thousand to more than 2500 mitochondria. A few very large cells, such as the eggs of various animals and the giant ameba *Pelomyxa* may have hundreds of thousands of mitochondria.

The First Stage of Cellular Oxidations: Glycolysis

The initial reactions breaking cellular fuels into 3-carbon segments take place in the cytoplasm outside mitochon-

dria. These reactions all follow a similar pattern, which is best illustrated by the oxidation of glucose, a central fuel molecule in all eukaryotic plants and animals. The initial series of reactions breaking this molecule into shorter carbon chains is called *glycolysis*.

The Reactions of Glycolysis

The overall reactions of glycolysis break the 6-carbon glucose molecule into two 3-carbon fragments (Fig. 7-4). During glycolysis a single oxidation occurs, and a small amount of ATP is generated. The 3-carbon fragments produced become the fuel for the oxidations taking place inside mitochondria, which produce much larger quantities of ATP.

The reactions of glycolysis are easiest to follow if only the carbon chains and phosphate groups are considered (Fig. 7-5; a more complete biochemical description of glycolysis is given in Supplement 7-1). The first series of reactions in glycolysis (reactions 1 through 3 in Fig. 7-5) converts glucose (six carbons) into a more reactive substance by adding two phosphate groups at the expense of two molecules of ATP. At the end of this sequence, a 6-carbon, 2-phosphate sugar is produced (6C, 2P in Fig. 7-5). These initial reactions of glycolysis go uphill in terms of the energy content of the products, and proceed to completion only because they are coupled to the breakdown of 2 ATP, converted in the process to 2 ADP.

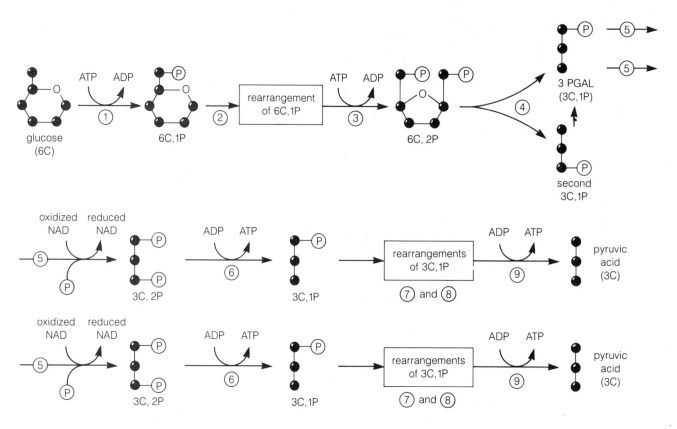

Figure 7-5 The reactions of glycolysis in terms of carbon chains and phosphate groups. The chemical structures of the intermediate molecules and the enzymes in the sequence are shown in Figure 7-21.

The ATP used in these initial steps is recovered with a net gain in the remaining reactions of glycolysis (reactions 4 through 9). These reactions break the 6-carbon, 2-phosphate product of the first series into two 3-carbon fragments, yielding reduced *NAD* (*nicotinamide adenine dinucleotide;* Fig. 7-6) in addition to a net gain in ATP. In the first of these reactions (reaction 4), the 6C, 2P sugar is broken into two 3-carbon, 1-phosphate fragments. The two 3-carbon sugars produced are different; only one of these, *3PGAL* (*3-phosphoglyceraldehyde,* already familiar as a product of the dark reactions of photosynthesis), directly enters the next step in glycolysis. However, as this 3-carbon sugar is used, a rearranging enzyme converts the second 3-carbon sugar into 3PGAL. Thus both the sugars produced by reaction 4 are used in the remainder of the sequence.

3PGAL is a high-energy substance. Some of its energy is tapped off in the next step (reaction 5), an oxidation. Remember that an oxidation involves removal of electrons and does not require reaction with an atom of oxygen (see

Information Box 6-1). In this step, two high-energy electrons are removed from 3PGAL and transferred to the acceptor molecule NAD. Except for the absence of one phosphate group, NAD is identical to the primary electron acceptor of photosynthesis, NADP (compare Figs. 7-6 and 6-17). Two hydrogen ions ($2H^+$) are removed from 3PGAL at the same time. One of the hydrogens attaches to NAD during oxidation of the 3-carbon sugar to form NADH. The second is released to the medium as an H^+ ion. The reduced NAD (or NADH) formed at this step is another high-energy substance. Most of the energy it carries is eventually used to convert ADP to ATP.

Transferring electrons from 3PGAL to NAD releases a large amount of free energy because, even though reduced NAD is a high-energy substance, the energy level of the transferred electrons is much lower in NAD than in 3PGAL. Some of the energy lost by the electrons as they are transferred to NAD is used to attach a second phosphate group to 3PGAL, yielding a highly reactive 3-carbon, 2-phosphate sugar (the 3C, 2P substance following

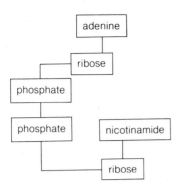

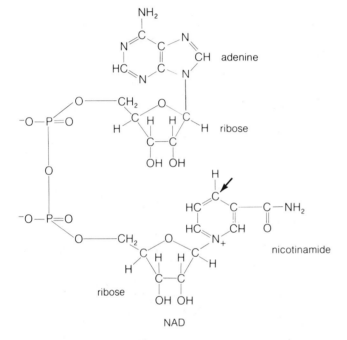

Figure 7-6 NAD, nicotinamide adenine dinucleotide, a carrier of high-energy electrons, has properties closely similar to those of NADP (compare with Fig. 6-17). In the reduced form, electrons and a hydrogen (H⁺) are added at the position marked by the arrow.

reaction 5 in Fig. 7-5). This second phosphate is derived from inorganic phosphate ions in the surrounding cytoplasm and not from ATP.

The final reactions of glycolysis (reactions 6 through 9) remove the two phosphate groups one at a time and transfer them to 2 ADP, yielding 2 ATP. In the first of these transfers (reaction 6), one phosphate is removed from the 3-carbon, 2-phosphate sugar to produce a 3-carbon, 1-phosphate sugar. Part of the free energy re-

leased at this step is captured in the transfer of the phosphate group to ADP, yielding one molecule of ATP for each 3C, 2P molecule entering the reaction. Because two molecules of the 3C, 2P sugar enter this reaction for every molecule of glucose originally entering glycolysis, there is a gain of 2 ATP at this step, replacing the two molecules used in reactions 1 through 3.

The reaction removing the first of the two phosphates is worth discussing in some detail, because it illustrates one method by which the energy of oxidation is converted into a usable chemical form in the cell. If one mole of the 3C, 2P molecule entering the reaction, a substance called *1,3-diphosphoglyceric acid*, breaks down directly to the 3C, 1P product, called *3-phosphoglyceric acid*, the reaction releases between 10,000 and 15,000 calories under standard conditions of temperature and pressure. In glycolysis, the reaction is coupled to ATP synthesis. The combined reaction still releases energy, but at the reduced level of about 4500 cal/mole:

$$1,3\text{-diphosphoglyceric acid} + ADP \rightarrow$$
$$3\text{-phosphoglyceric acid} + ATP \quad (7\text{-}1)$$

The difference in calories released between the two reactions is energy captured in the formation of ATP.

The final phosphate group is removed from the 3C, 1P product of reaction 6 after two reactions (reactions 7 and 8) that simply rearrange the atoms of the molecule. In the final reaction of glycolysis (reaction 9) the phosphate group of the rearranged 3C, 1P sugar is transferred to ADP, yielding one molecule of ATP for each molecule of 3C, 1P entering reactions 7 to 9. As in reaction 6, because two molecules of 3C, 1P enter this final series for each molecule of glucose starting down the glycolytic pathway, a total of two molecules of ATP is produced at this step. This reaction thus provides a net gain of 2 ATP for the entire glycolytic sequence. The final product of glycolysis is *pyruvic acid*, a 3-carbon substance containing no phosphate groups.[1] It is still a relatively high-energy substance, and provides the major fuel for the subsequent cellular oxidations inside mitochondria.

---

[1]*Pyruvic acid* is the name given to the uncombined form of the substance. When considered in the combined form, as in the salt produced when pyruvic acid combines with sodium, the word *acid* is dropped and the *-ic* ending is changed to *-ate*, as in *sodium pyruvate*. The *-ate* ending is also used for the ionized form of the acid; that is, when the —COOH group of the acid is ionized to —COO⁻. In this case, the substance is simply called *pyruvate*. In practice, both forms are used interchangeably for this and other organic acids. In this book, the organic acids are named as their uncombined form with the *-ic* ending.

Each of the reactions in glycolysis is increased in rate by a different enzyme (the glycolytic enzymes are listed individually in Fig. 7-21). The activity of one of the enzymes of the pathway, the enzyme adding the second phosphate in reaction 3 to form the 6C, 2P product, illustrates one of several controls regulating the rate of glycolysis in cells. This enzyme, called *phosphofructokinase*, is inhibited by high concentrations of ATP, and is stimulated by ADP and inorganic phosphate. If sufficient ATP is present in the cytoplasm, phosphofructokinase is inhibited and the subsequent reactions of glycolysis slow or stop. If energy-requiring activities take place elsewhere in the cell, resulting in the conversion of ATP to ADP and phosphate, the accumulation of ADP and inorganic phosphate stimulates phosphofructokinase, increasing the rate of glycolysis and ATP production. Regulation by phosphofructokinase is probably the most sensitive and significant of the controls of glycolysis, since it is directly keyed to the relative concentrations of ADP and ATP and thus to the rate at which cells use energy for their activities.

The various enzymes, reactants, intermediate substances, and products of glycolysis are suspended in solution in the background substance of the cytoplasm, without concentration or localization in a specific cytoplasmic organelle.

Net Products of Glycolysis

Subtracting the consumption of ATP in glycolysis from the amount produced shows that the overall reaction sequence provides a net gain of 2 ATP for each molecule of glucose entering the sequence. For every molecule of glucose entering glycolysis, a molecule of ATP is broken down to ADP in reactions 1 and 3, for a total of 2 ATP lost in the initial part of the sequence. However, two molecules of ADP are converted to ATP at reaction 6, and two more at reaction 9 for each molecule of glucose entering the pathway, giving a total of 4 ATP gained for the second, oxidative part of glycolysis. These reactions provide the net gain of 2 ATP for each molecule of glucose oxidized to pyruvic acid. Another significant product is the reduced NAD (two molecules of reduced NAD for each molecule of glucose), which is reduced by accepting electrons at reaction 5. The electrons accepted by NAD, which exist at a high energy level, can do chemical work if passed from reduced NAD to a suitable electron acceptor elsewhere in the cell. The total reactants and products of glycolysis are therefore:

$$\text{glucose} + 2\text{ ADP} + 2\text{ HPO}_4^{2-} + 2\text{ oxidized NAD}$$
$$\rightarrow 2\text{ pyruvic acid} + 2\text{ ATP} + 2\text{ reduced NAD} \quad (7\text{-}2)$$

Important Variations of the Glycolytic Pathway

Glycolysis is the central pathway for the initial breakdown of carbohydrates in both plant and animal cells. Starch in plants and glycogen in animals are both long, chainlike molecules made up from repeating glucose links (see Fig. 2-8). These molecules enter the glycolytic pathway after being broken into individual glucose molecules. Other sugars, including a wide variety of monosaccharides and disaccharides, enter glycolysis after being converted by enzymes into one of the initial molecules of the pathway.

At the opposite end of the sequence, pyruvic acid may be modified to yield other final products. In one of the most important of these modifications, pyruvic acid is converted into *lactic acid* after accepting electrons from the NAD reduced earlier in the sequence. In this form of glycolysis, lactic instead of pyruvic acid accumulates as the final product of the pathway:

$$\text{pyruvic acid} + \text{reduced NAD}$$
$$\rightarrow \text{lactic acid} + \text{oxidized NAD} \quad (7\text{-}3)$$

This modification is significant because it regenerates oxidized NAD, which is then free to cycle back to accept electrons in the oxidations of glycolysis. Because oxidized NAD is continually regenerated by this alternate pathway, glycolysis can continue to run with the net production of ATP. This pathway is vital to cells living temporarily or permanently in the absence of oxygen (reduced NAD normally transfers its electrons through a series of carriers to oxygen; see The Mitochondrial Carrier Chain, below). Lactic acid production occurs in the muscle cells of animals, including humans, for example, if intensive, sustained physical activity is carried out before increases in breathing and heart rate have a chance to meet the demand for oxygen in the muscle tissues. The lactic acid accumulating as a by-product is oxidized later, when the oxygen content of the muscle cells returns to normal levels (see Information Box 7-1).

Another important glycolytic variation occurs in organisms such as yeast. In this modification, pyruvic acid accepts electrons from reduced NAD and is converted by additional reactions into *ethyl alcohol* (a 2-carbon substance) and CO_2:

Information Box 7-1   Cell Biology and the 100-Yard Dash

The muscles of a runner poised at the start of a 100-yard dash are loaded with a high-energy substance capable of releasing the ATP required for the run. This substance, creatine phosphate, forms a storage reservoir for high-energy phosphate groups. Comparatively little ATP is stored in muscle; the reserves of creatine phosphate are about five times that of the ATP phosphate. However, phosphate groups are readily transferred from creatine phosphate to ATP, in the reaction:

$$\text{creatine phosphate} + \text{ADP} \rightarrow \text{creatine} + \text{ATP}$$

The runner's muscles at the starting line also contain reserves of glycogen (animal starch), made up from long chains of glucose units linked end to end.

At the starting gun the runner springs from rest into sudden, maximal physical exertion, straining every muscle to the limit. The immediate energy for this contraction comes from the breakdown of the small reserves of ATP. Replenishment of these reserves by transfer of phosphate groups from creatine phosphate to ATP then begins, probably within the first second after the starting gun is fired.

Besides initiating the transfer of phosphate from creatine phosphate, the ADP produced by ATP breakdown after the gun also stimulates oxidation of fuel substances within the runner's muscles to provide energy for converting ADP to ATP. Glucose units, broken off one at a time from the glycogen molecules stored in muscle, are the fuel for oxidation. Because the runner's breathing rate has not yet increased to its maximum level, the oxygen required for complete oxidation of glucose to water and CO_2 in mitochondria is quickly depleted. As a result, the main weight of ATP formation falls on glycolysis.

Continued glycolysis forms reduced NAD and depletes oxidized NAD. The reduced NAD accumulating is reoxidized by handing off its electrons to the pyruvic acid produced by glycolysis, forming lactic acid that gradually increases in concentration in the runner's muscles. By the 50-yard mark, significant quantities of lactic acid have formed; by the 75-yard mark, lactic acid has begun to reach high concentrations in the runner's muscles and is released into his bloodstream. This contributes to a feeling of fatigue that the runner now begins to notice.

As the runner crosses the finish line his breathing and heart rate, in response to the demand for oxygen in his tissues, have reached high levels. Breathing and heart rate remain high, even though the runner has now finished the race. His body begins to oxidize the lactic acid, converting it to pyruvic acid and eventually back to glycogen. (These reactions take place in the liver.) Breathing and heart rate continue above resting levels until most of the lactic acid is oxidized. The extra oxygen required by the runner during this period pays off what is called the oxygen debt. This debt was accumulated during the maximal exertion of the 100-yard dash, when oxygen in the runner's muscles was depleted below the level required for oxidation of glucose to CO_2 and water. While the runner is resting, the creatine phosphate reserves return to their normal level through ATP produced by mitochondrial oxidations:

pyruvic acid + reduced NAD →

$$\text{ethyl alcohol} + CO_2 + \text{oxidized NAD} \quad (7\text{-}4)$$

This variation is highly significant to human economy. The CO_2 released by yeast cells carrying out this reaction raises the dough used in baking breads; both the CO_2 and alcohol produced are of central importance to the brewing industry.

These variations in the glycolytic pathway, in which the electrons carried by reduced NAD are traded off to an *organic* substance such as pyruvic acid, are collectively called *fermentations*. In the alternate pathway, the electrons carried from glycolysis by reduced NAD eventually reach an *inorganic* substance, oxygen. Fermentations of various kinds, producing a wide variety of end products, are used as an ATP source by many species of bacteria. Some of these species, called *strict anaerobes*, are limited to glycolytic fermentations for ATP production and cannot use oxygen at any time as a final electron acceptor. Others can use either fermentations alone, or both glycolysis and reactions equivalent to mitochondrial oxidations (see Supplement 7-2) if oxygen is available. Bacteria in this category are called *facultative anaerobes*. A number of species, termed *strict aerobes*, are unable to live by fermentations alone. Many cells of higher organisms, including the muscle cells of humans and other vertebrates, are facultative and can switch between fermentation and complete oxidation depending on their oxygen supply. Others, such as the brain cells of vertebrates, are strict aerobes.

The glycolytic pathway is also important in photosynthesis. The 3-carbon sugar 3PGAL, which appears as an intermediate compound in glycolysis (reaction 4 in Fig. 7-5), is also the primary product of the Calvin cycle in the dark reactions of photosynthesis (see Chap. 6). This molecule is converted into glucose in chloroplasts essentially by reversing the first four reactions of glycolysis. The glucose molecules produced in these steps may then be linked end to end into starch. Alternatively, 3PGAL may serve as a fuel substance by entering the glycolytic pathway at reaction 4 and proceeding in the other direction toward pyruvic acid or other products.

Appreciation of the central role of glycolysis in cell biology and biochemistry came as a gradual development of research first taken up in the late 1800s. In Germany, Eduard and Hans Buchner discovered that alcoholic fermentation could be carried out by nonliving extracts of yeast cells. Through this work, the first enzymes were discovered and described. Research with alcoholic fermentation, involving a great many investigators in various parts of the world, continued until well into this century. By the 1930s the emphasis in these investigations had shifted from alcoholic fermentation to the more general pathway of glycolysis. Our present understanding of the glycolytic pathway is based primarily on research carried out in the 1930s by two German scientists, Gustav Embden and Otto Meyerhoff. These researchers played so central a part in tracing out the reactions and enzymes of glycolysis that the sequence is frequently called the *Embden–Meyerhoff pathway* in their honor. By 1940, glycolysis was known essentially as it is today.

The Second Stage of Cellular Oxidations: Oxidation and ATP Synthesis in Mitochondria

In the second major stage of cellular oxidations, pyruvic acid, the 3-carbon molecule produced in the final step of glycolysis, is completely oxidized to CO_2 and water. A very large quantity of free energy is released in this stage, and the ATP produced in these reactions far exceeds that obtained through glycolysis. The reactions occur in three parts, all located inside mitochondria (Fig. 7-7). In the first part, called *pyruvic acid oxidation*, pyruvic acid is oxidized and shortened into a 2-carbon segment, and CO_2 is released. In the second part, the *Krebs cycle*, this 2-carbon segment is completely oxidized to two molecules of CO_2. In the final part, *electron transport*, the electrons removed in these oxidations are delivered through a series of electron carriers to oxygen. Much of the free energy released by this electron transport is used to drive the synthesis of ATP. Because oxygen is the final electron acceptor for these reactions the oxidative activities of mitochondria are frequently termed *respiration*.

Part 1 of Mitochondrial Respiration: Oxidation of Pyruvic Acid to 2-Carbon Segments

After entering mitochondria, pyruvic acid is shortened to a 2-carbon segment in a series of reactions that remove two electrons (an oxidation), two hydrogens (as H^+), and one carbon from the pyruvic acid chain (as CO_2; see Fig. 7-8). The 2-carbon segment produced is an *acetyl* ($—CH_3CO^-$) group:

pyruvic acid →

$$\text{acetyl group} + 2 \text{ electrons} + 2H^+ + CO_2 \quad (7\text{-}5)$$

The two electrons and one of the hydrogens removed from pyruvic acid are transferred to NAD, which is re-

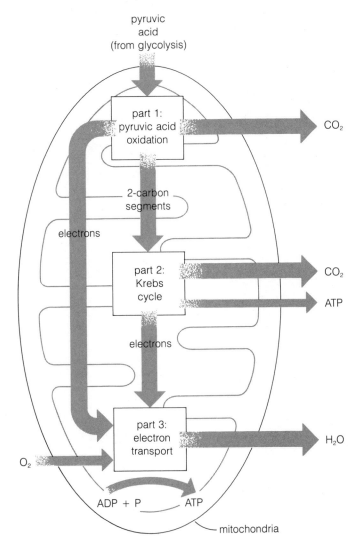

pyruvic
acid
(from glycolysis)

part 1:
pyruvic acid
oxidation

CO$_2$

2-carbon
segments

electrons

part 2:
Krebs
cycle

CO$_2$

ATP

electrons

part 3:
electron
transport

H$_2$O

O$_2$

ADP + P   ATP

mitochondria

Figure 7-7 The three major parts of the oxidative reactions of mitochondria (see text).

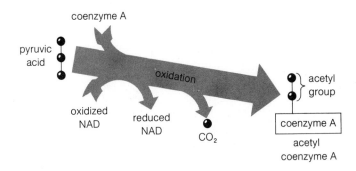

coenzyme A

pyruvic
acid

oxidation

acetyl
group

oxidized
NAD

reduced
NAD

CO$_2$

coenzyme A

acetyl
coenzyme A

Figure 7-8 The oxidation of pyruvic acid (3 carbons) produces 2-carbon acetyl groups, which provide the main fuel for mitochondrial respiration. The reaction sequence uses pyruvic acid, oxidized NAD, and coenzyme A as raw materials, and produces CO$_2$, reduced NAD, and acetyl groups attached to coenzyme A.

phosphates and the NAD/NADP molecules transport electrons.

The overall reaction sequence in pyruvic acid oxidation thus yields as net products acetyl-coenzyme A, reduced NAD, and CO$_2$:

$$2 \text{ pyruvic acid } + 2 \text{ oxidized NAD} \\ + 2 \text{ coenzyme A} \rightarrow 2 \text{ acetyl-coenzyme A} \quad (7\text{-}6) \\ + 2 \text{ reduced NAD } + 2 \text{ CO}_2$$

All of these reactants and products are multiplied by two in Reaction 7-6 because pyruvic acid oxidation is considered as a continuation of glycolysis, in which two molecules of pyruvic acid are produced for each molecule of glucose entering the pathway.

Part 2 of Mitochondrial Respiration: Oxidation of Acetyl Units to CO$_2$ in Mitochondria

The Krebs Cycle The 2-carbon acetyl groups carried by coenzyme A are oxidized to CO$_2$ in a cycle of reactions first described in 1937 by a British investigator, Hans Krebs, who received the Nobel Prize for his brilliant work with cellular oxidation. In the cycle (Fig. 7-10) which is named for Krebs, there is a continuous input of 2-carbon acetyl groups as reactants, and a continuous output of the products: CO$_2$, ATP, reduced NAD, and another electron carrier in reduced form, FAD (see Fig. 6-13 and p. 119). As in the Calvin cycle of photosynthesis, the molecules

duced in the process. The CO$_2$ and the remaining hydrogen ion are released to enter the surrounding medium. The 2-carbon acetyl group is transferred to a carrier molecule called *coenzyme A* (Fig. 7-9) to produce the high-energy substance *acetyl-coenzyme A*. Most of the free energy released by the oxidation of pyruvic acid is captured as chemical energy in this substance.

Coenzyme A is another carrier molecule based on nucleotide structure that closely resembles ATP, NAD, and NADP (compare Figs. 3-1, 6-17, 7-6, and 7-9). Coenzyme A accepts and carries a 2-carbon acetyl unit; ATP carries

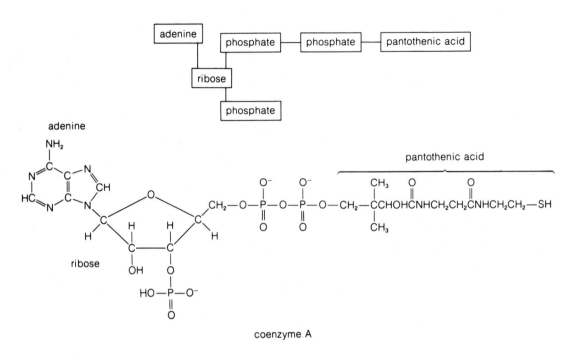

Figure 7-9 Coenzyme A, a carrier of acetyl (—CH_3CO^-) groups.

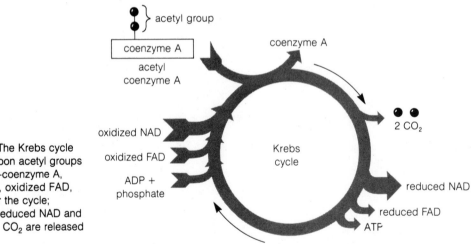

Figure 7-10 The Krebs cycle oxidizes 2-carbon acetyl groups to CO_2. Acetyl-coenzyme A, oxidized NAD, oxidized FAD, and ADP enter the cycle; coenzyme A, reduced NAD and FAD, ATP, and CO_2 are released as products.

forming intermediate parts of the Krebs cycle are continuously regenerated as the cycle turns.

Only one molecule of ATP is formed as a product of each turn of the Krebs cycle. Most of the energy released by the several oxidations of the cycle is trapped in the electrons carried from the cycle by reduced NAD and FAD. The energy carried by reduced NAD and FAD is used in the final reactions of mitochondrial oxidation to generate most of the ATP produced in cellular respiration.

The Krebs cycle (Fig. 7-11) works in the following way. In the first reaction of the cycle (reaction 1 in Fig. 7-11) the 2-carbon acetyl unit carried by coenzyme A is transferred to a 4-carbon molecule, *oxaloacetic acid*, to produce *citric acid*, a 6-carbon molecule. The reaction is cata-

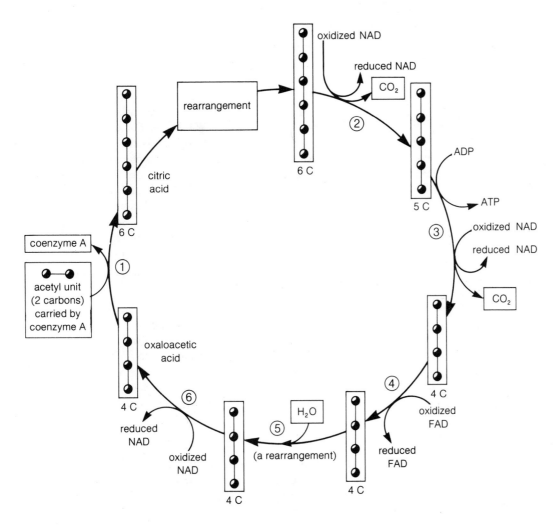

Figure 7-11 The Krebs cycle (see text). Only the carbon chains of intermediate molecules are shown in this diagram; the structures of these intermediates are shown in Figure 7-22.

lyzed by a specific enzyme, as are all the reactions of the Krebs cycle (the enzymes are listed individually in Fig. 7-22). Coenzyme A, relieved of its acetyl unit by this reaction, is free to enter another cycle of pyruvic acid oxidation:

acetyl-coenzyme A + oxaloacetic acid →
$$\text{citric acid + coenzyme A} \quad (7\text{-}7)$$

The citric acid formed in the initial reaction is converted by a rearranging enzyme into a similar 6-carbon acid, which enters the next reaction of the cycle. In this reaction (reaction 2 in Fig. 7-11), two electrons and two hydrogens are removed from the 6-carbon acid. The carbon chain is also shortened at this step, with the release

of one molecule of carbon dioxide, yielding a 5-carbon product.

Although NAD is the usual acceptor for the electrons removed in this oxidation, either NAD or NADP may be reduced, depending on the relative concentrations of ATP, ADP, and reduced NAD in the cell. Which of the acceptors is actually used illustrates one of the many control systems regulating cellular oxidation. The oxidation is catalyzed by either of two enzymes that are identical except that one uses NAD and the other NADP as acceptor for the electrons removed. When ATP concentrations are high, the enzyme using NAD is inhibited, and NADP becomes the favored acceptor. When ATP concentrations are low, inhibition of the enzyme using NAD is released, and NAD replaces NADP as the electron acceptor.

These differences are significant for ATP production because reduced NAD normally transfers its electrons to the electron transport system of mitochondria that generates ATP. Reduced NADP instead acts as electron donor for reactions of cell synthesis in which a reduction is required. As a result of this regulatory system, reaction 2 is finely tuned to the needs of the cell for ATP. Under the usual conditions, in which cellular activity demands a more or less continuous supply of ATP, the enzyme using NAD as electron acceptor predominates in Krebs cycle oxidation at this step, leading to ATP synthesis in response. For the discussion in this chapter we will consider NAD as the primary electron acceptor in this reaction.

The electrons removed in the next step in the Krebs cycle (reaction 3) are accepted by NAD, and the chain is shortened to four carbons with release of one molecule of CO_2. Enough free energy is released in the step to synthesize one molecule of ATP. Therefore, reaction 3 yields a 4-carbon product and a molecule each of CO_2, ATP, and reduced NAD. The ATP produced in this reaction is the only ATP originating from the Krebs cycle.

The electrons removed in the next step in the cycle (reaction 4) are accepted by FAD, which forms a part of the electron transport system of mitochondria (see The Mitochondrial Carrier Chain, below). In reaction 4, FAD is reduced by accepting two electrons and both of the two hydrogens removed, forming $FADH_2$ as the reduced product.

The 4-carbon product of reaction 4 is rearranged, with the addition of a molecule of water, to form another 4-carbon acid. This acid is then oxidized to oxaloacetic acid in the final reaction in the cycle (reaction 6). NAD is the electron acceptor for this final oxidation of the Krebs cycle. The molecule of oxaloacetic acid produced at this step replaces the oxaloacetic acid used in reaction 1, and the cycle is ready to turn again.

Overall Products of the Krebs Cycle We can now total up the overall products of the Krebs cycle and summarize the cellular oxidation of glucose to this point. As the Krebs cycle proceeds through one complete turn, one 2-carbon acetyl group is consumed and two molecules of CO_2 are released. High-energy electrons are removed at each of four reactions in the cycle. At three of these steps, NAD is the acceptor (assuming that NAD is used in reaction 2), producing three molecules of reduced NAD; one step reduces a molecule of FAD instead. In reaction 3 of the cycle, one molecule of ATP is generated. Oxaloacetic acid, used in the initial reaction of the cycle, is regenerated in the final reaction. Thus, as overall reactants and products, the Krebs cycle includes:

$$\text{an acetyl group (2 carbons)} + 3 \text{ oxidized NAD}$$
$$+ \text{ oxidized FAD} + \text{ADP} + HPO_4^{2-} \rightarrow 2\,CO_2$$
$$+ 3 \text{ reduced NAD} + \text{ reduced FAD} + \text{ATP} \quad (7\text{-}8)$$

With this information we can sum up the total products of the oxidation of glucose to carbon dioxide from glycolysis through the Krebs cycle. For each molecule of glucose entering the series, the Krebs cycle will turn twice. Glycolysis and acetyl-coenzyme A formation together yield 4 reduced NAD, 2 ATP, and 2 CO_2 (from Reactions 7-2 and 7-6). Adding to this the products of two turns of the Krebs cycle for every glucose entering oxidation we have:

$$\text{glucose} + 4 \text{ ADP} + 4\,HPO_4^{2-} + 10 \text{ oxidized NAD}$$
$$+ 2 \text{ oxidized FAD} \rightarrow 4 \text{ ATP} + 10 \text{ reduced NAD}$$
$$+ 2 \text{ reduced FAD} + 6\,CO_2 \quad (7\text{-}9)$$

Note at this point that little ATP has been produced. However, the electrons carried by the 10 reduced NAD and 2 reduced FAD molecules contain most of the energy obtained from oxidation of glucose to 6 CO_2. The free energy released in the transfer of these molecules to oxygen is used as the major source of energy for cellular ATP synthesis in part 3 of mitochondrial oxidations (for additional details of the Krebs cycle, see Supplement 7-1).

Part 3 of Mitochondrial Oxidations: ATP Synthesis in the Electron Transport System

The mitochondrial system transporting electrons from the Krebs cycle to oxygen is basically similar to the electron transport system of the light reactions of photosynthesis (see p. 119). As electrons flow through the system, they lose part of their energy at each step in the chain of carriers. At some of these steps, enough free energy is released to drive the synthesis of ATP from ADP. By the time the electrons reach oxygen, most of their energy has been tapped off and conserved in the form of ATP.

All but one of the known carriers in the mitochondrial electron transport system are proteins combined with a nonprotein subunit. The nonprotein subunit is the part of the total carrier molecule that actually accepts and releases electrons, and it is alternately oxidized and reduced as electrons flow along the chain. The carriers of the electron chain in mitochondria include FAD and another flavoprotein, *FMN* (*flavin mononucleotide*; see Fig. 7-12), several iron-sulfur proteins, and four cytochromes (details of these carrier types are given in Information Box 6-2). The

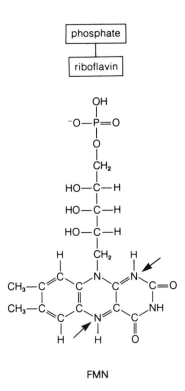

Figure 7-12 FMN (flavin mononucleotide). FMN is another nonprotein group of the flavoprotein carriers of the electron transport system. The riboflavin group in FMN binds two hydrogens at the arrows to form $FMNH_2$.

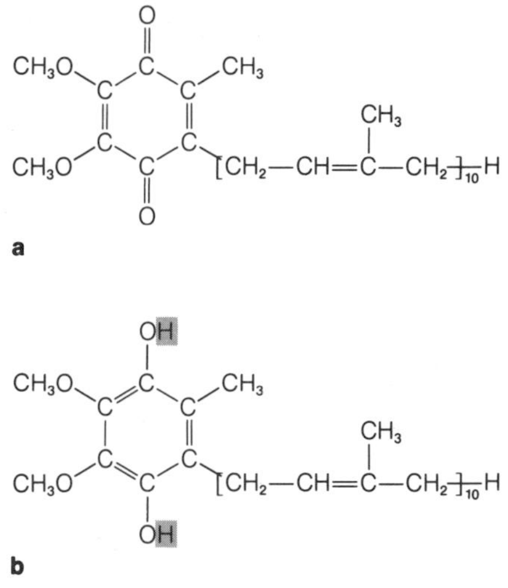

Figure 7-13 Coenzyme Q, the only carrier of the electron transport system not linked to a protein. (**a**) Oxidized form; in going to the reduced form (**b**), electrons and 2 hydrogens are added at the shaded positions.

single nonprotein carrier of the mitochondrial electron transport chain, *coenzyme Q* (Fig. 7-13), closely resembles the plastoquinone carrier of the chloroplast carrier chain. All these carriers are alternately reduced and oxidized as electrons flow through the chain to oxygen.

The Mitochondrial Carrier Chain The experimental methods used to place these carriers in sequence in the mitochondrial system are similar to the approach used in chloroplasts (described in Information Box 6-2). These methods indicate that the mitochondrial carriers transport electrons in the sequence shown in Figure 7-14. Note that electrons have two routes of entry into the chain, one at FMN and one at FAD. These routes join at coenzyme Q, which thus forms a collection point for electrons entering from the two sources. Reduced NAD, originating from cellular oxidations, transfers its electrons and hydrogen to FMN, the first carrier in this branch of the chain. In the transfer, NAD is oxidized and FMN reduced. The oxidized NAD is then free to cycle back to the reactions of pyruvic acid oxidation and the Krebs cycle. The electrons probably

then flow through one or more iron-sulfur proteins (the position of these carriers in the chain is still uncertain), and then to the coenzyme Q collection point. From coenzyme Q, electrons flow through a series of cytochromes, including cytochromes b, c_1, c, and the final carrier, a dual two-part cytochrome known as a-a_3. The a-a_3 carrier is believed to consist of two different cytochrome groups (see Fig. 6-12), a and a_3, linked to the same enzyme. At least one iron-sulfur protein is believed to be located between the cytochrome carriers in the position shown in Figure 7-14. The final carrier in the chain, cytochrome a-a_3, reduces oxygen to water in the reaction:

reduced cytochrome a-a$_3$ + ½ O$_2$ + 2 H$^+$ →

oxidized cytochrome a-a$_3$ + H$_2$O   (7-10)

The hydrogens required for this reduction are derived from the H$^+$ ions in the water solution surrounding the electron transport chain.

Electrons entering the chain at the alternate point, through FAD, originate from the single reaction of the Krebs cycle that uses FAD directly as electron acceptor (Reaction 4 in Fig. 7-11). While FAD acts as the electron acceptor for this reaction, it actually forms a part of the mitochondrial electron chain and does not cycle freely between the mitochondrial electron transport chain and the Krebs cycle reactions as does NAD. Therefore, passage of

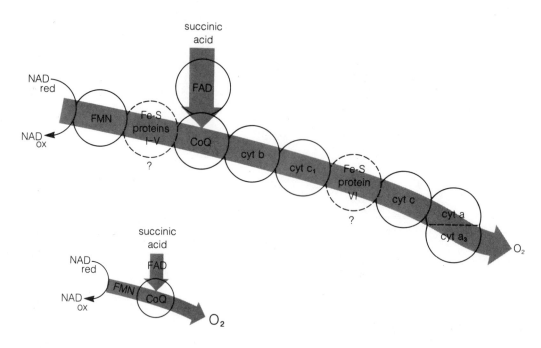

Figure 7-14 The probable sequence of carriers in the mitochondrial electron transport system (see text).

electrons from the 4-carbon acid (succinic acid; see Fig. 7-22) to FAD in reaction 4 occurs only when the acid collides with the inner mitochondrial membranes housing FAD in the electron transport system. From FAD, the electrons flow to the coenzyme Q collection point, and then through the cytochrome chain to oxygen. All of the carriers of the mitochondrial electron transport chain are embedded in the inner mitochondrial membrane, including the folds of this membrane that form the mitochondrial cristae.

ATP Synthesis in the Mitochondrial Electron Carrier Chain
The mechanism coupling electron transport to ATP synthesis in mitochondria, according to Peter Mitchell's chemiosmotic hypothesis (see Chap. 6), is considered to depend on the same two-step process that occurs in chloroplasts: (1) the free energy released as electrons flow through the carriers of the electron transport system is used to move H^+ ions across the mitochondrial membranes housing the electron transport carriers, creating an H^+ gradient, and (2) the gradient is then used as a source of free energy to synthesize ATP from ADP and phosphate.

Establishing the H^+ Gradient The mechanism establishing the H^+ gradient in mitochondria, as in chloroplasts, depends on the fact that some of the electron carriers in the mitochondrial electron transport system are nonhydrogen

carriers (the cytochromes and the iron-sulfur proteins), and some carry hydrogens as well as electrons (FMN, FAD, and coenzyme Q). As electrons pass from a nonhydrogen carrier to a dual hydrogen-electron carrier, hydrogens are picked up from the solution in the mitochondrial matrix (Fig. 7-15a); as electrons pass from dual hydrogen-electron carriers to nonhydrogen carriers, the hydrogens are released and expelled into the intermembrane compartment between the inner and outer mitochondrial boundary membranes (Fig. 7-15b; see also Fig. 6-18). Electron transport thus produces an H^+ ion gradient, with H^+ ions at high concentration in the intermembrane compartment and low in the matrix (Fig. 7-16). As in chloroplasts, this gradient forces an H^+-transport protein in the inner membrane to run backwards, driving the synthesis of ATP from ADP and phosphate (see p. 125).

Figure 7-17 shows how the mitochondrial electron transport chain is believed to work in establishing the H^+ gradient. Reduced NAD is a dual hydrogen-electron carrier, and in the reduced form, as NADH, carries one hydrogen as well as a pair of electrons. In the first step in the chain, reduced NAD passes its two electrons and one hydrogen to FMN. Since FMN carries two hydrogens in addition to an electron pair, a second hydrogen is absorbed from the H^+ ions present in the matrix solution to form $FMNH_2$. $FMNH_2$ passes its electrons to the next carrier group in the chain, the iron-sulfur proteins. Because

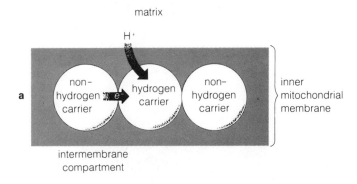

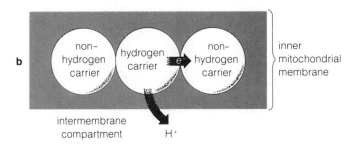

Figure 7-15 How electron transport through the mitochondrial chain sets up an H$^+$ gradient (see text). (**a**) When electrons pass from a nonhydrogen carrier to a hydrogen carrier, the H$^+$ ions required are removed from the mitochondrial matrix. (**b**) When electrons pass from a hydrogen carrier to a nonhydrogen carrier, the H$^+$ ions are released and expelled to the matrix.

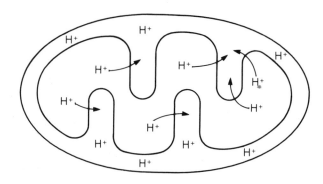

Figure 7-16 Electron transport in mitochondria increases the concentration of H$^+$ ions in the intermembrane compartment.

the iron-sulfur proteins are nonhydrogen carriers, the hydrogens carried by FMNH$_2$ are released as this substance is oxidized. Release occurs into the intermembrane compartment, adding the first H$^+$ ions on this side of the membrane. In the next step, the electrons pass to coenzyme Q, another dual hydrogen-electron carrier. In going to the reduced form (QH$_2$), coenzyme Q takes up a second pair of H$^+$ ions from the matrix solution. In the next step in the sequence, coenzyme Q passes its electrons to the next carrier in the chain, cytochrome *b*. Since the cytochromes are pure electron carriers, the coenzyme Q molecule, in going from QH$_2$ to Q, releases two more H$^+$ ions to the intermembrane compartment. The electrons carried by cytochrome *b* then move along the chain through the a-a$_3$ cytochromes to oxygen. As the electrons are accepted by an oxygen atom, an additional 2 H$^+$ is removed from

the matrix, converting ½ O$_2$ to H$_2$O. Movement of electrons along the pathway from NAD to oxygen thus expels a total of 4 H$^+$ into the intermembrane compartment and removes 5 H$^+$ from the matrix for each electron pair.

ATP Synthesis Through the H$^+$ Gradient The H$^+$ gradient, once established, is used as a source of free energy to drive the synthesis of ATP by the same mechanism operating in chloroplasts. As in chloroplasts, this ATP synthesis is carried out by membrane proteins that can be identified with lollipop-shaped structures attached to the surfaces of the cristae membranes facing the mitochondrial matrix (Fig. 7-18).

The H$^+$ gradient created as an electron pair flows from NAD to oxygen is sufficient to drive the synthesis of 3 ATP. Electrons entering the pathway from FAD (see Fig. 7-14) flow through only a part of the pathway, with the result that fewer H$^+$ ions are added to the gradient. The smaller gradient produced causes a reduction in the quantity of ATP synthesized, which amounts to 2 ATP for each pair of electrons traveling the portion of the transport chain between FAD and oxygen.

Total ATP Production from Glucose Oxidation

This information allows us to calculate the total ATP production of mitochondrial electron transport and to derive a grand total for the entire sequence of oxidations from glucose to CO$_2$ and H$_2$O. Oxidation of a glucose molecule from glycolysis through the Krebs cycle yields a total of 4 ATP, 10 reduced NAD, and 2 reduced FAD (from Reaction

matrix

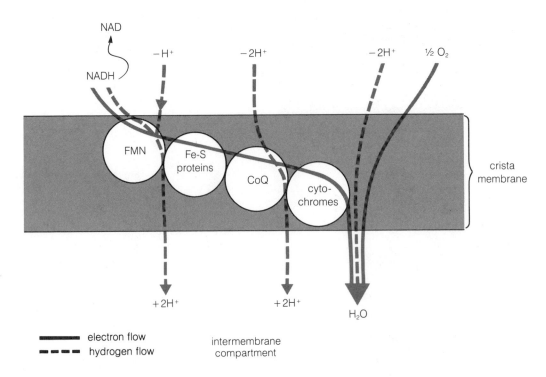

NAD

NADH

$-H^+$ $-2H^+$ $-2H^+$ ½ O_2

FMN

Fe-S proteins

CoQ

cyto-chromes

crista membrane

$+2H^+$ $+2H^+$

H_2O

———— electron flow

- - - - hydrogen flow

intermembrane compartment

Figure 7-17 The steps in mitochondrial electron transport setting up an H^+ gradient, according to the chemiosmotic hypothesis.

7-9). The electrons carried by the 10 molecules of reduced NAD will drive the synthesis of $10 \times 3 = 30$ molecules of ATP as they pass through the electron transport system. The electrons carried by the two molecules of reduced FAD drive the synthesis of only 4 molecules of ATP because they enter the sequence of carriers farther along in the chain. This gives a grand total of 34 ATP from electron transport. This 34 ATP, when added to the ATP produced in glycolysis and the Krebs cycle (Reaction 7-9) gives a grand total of 38 ATP[2] for each molecule of glucose completely oxidized to CO_2 and H_2O:

$$\text{glucose} + 38 \text{ ADP} + 38 \text{ HPO}_4^{2-} + 6 \text{ O}_2 \rightarrow$$
$$6 \text{ CO}_2 + 44 \text{ H}_2\text{O} + 38 \text{ ATP} \quad (7\text{-}11)$$

The 44 H_2O entered as products include 38 H_2O released as by-products from the formation of ATP:

$$38 \text{ ADP} + 38 \text{ HPO}_4^- \rightarrow 38 \text{ ATP} + 38 \text{ H}_2\text{O} \quad (7\text{-}12)$$

and 6 H_2O produced from oxidation of glucose to 6 CO_2 and 6 H_2O. These 6 H_2O are produced indirectly, through

[2]The total of 38 ATP given in this equation assumes that the two molecules of NAD reduced in glycolysis will each drive the synthesis of 3 ATP inside the mitochondrion. However, reduced NAD from glycolysis cannot directly enter the mitochondrion to pass electrons to the transport system because the mitochondrial membranes are impermeable to NAD. Instead, the electrons carried by reduced NAD are transferred to other substances that can shuttle back and forth

across the mitochondrial membranes. Two of these shuttle mechanisms are known. The more efficient one transfers electrons from NAD outside to NAD inside the mitochondrion. The less efficient mechanism results instead in the reduction of FAD inside the mitochondrion. If the more efficient shuttle is operating, no loss occurs and 38 ATP result from each molecule of glucose completely oxidized, as shown in Reaction 7-11. If the less efficient shuttle is operating,

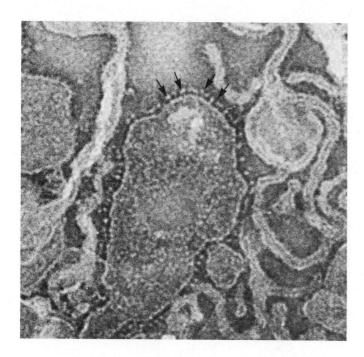

Figure 7-18 The lollipop structures (arrow) synthesizing ATP in cristae membranes isolated from beef heart mitochondria. Courtesy of D. W. Deamer.

three decades of this century. In 1913 the German scientist Otto Warburg proposed that an iron-containing enzyme must be directly involved in the use of oxygen by respiring cells. Warburg's enzyme, later identified as a cytochrome, was linked to the enzymes that remove electrons from substances oxidized in respiration by the work of Albert Szent-Gyorgyi, a Hungarian scientist who later became an American citizen. Szent-Gyorgyi proposed the basic idea that electrons flow from oxidized substances through a series of electron carriers to oxygen. Research still continues in the effort to identify all of the carriers of the series and to establish their exact sequence in electron transport in both mitochondria and chloroplasts.

The Locations of Oxidative Reactions Inside Mitochondria

Several methods are available for breaking mitochondria into fractions. In the most simple of these techniques, mitochondria are exposed briefly to a detergent, which causes breaks in the outer membrane. The outer membranes can then be easily separated from the inner membranes and purified by centrifugation. The inner membranes and the enclosed matrix remain intact and can be separately concentrated and purified. The inner membranes can then be broken by further, more prolonged treatment with detergent and isolated from the matrix by centrifugation. The three fractions, outer membrane, inner membrane, and matrix, retain much of their separate biochemical activity when separately purified in this way.

The outer membrane in such preparations proves to contain several enzymes associated with the initial breakdown of fatty acids and amino acids (see Cellular Oxidation of Fats and Proteins, below). The inner membrane fraction contains the enzymes and carriers engaged in electron transport and ATP formation. The matrix contains in solution the various enzymes and intermediates of the Krebs cycle and pyruvic acid oxidation. Only one enzyme of the Krebs cycle, the enzyme catalyzing the removal of electrons in reaction 4 of Figure 7-11 (*succinic acid dehydrogenase*) is tightly bound to the inner mitochondrial membranes. The electron acceptor for this reaction, FAD, is also tightly bound to the cristae membranes, where it forms a part of the chain of electron transport carriers.

As might be expected, the presence of enzymatic activity of this magnitude is reflected in a high content of proteins in mitochondria. The mitochondrial matrix is gel-like and about 50 percent protein; an even greater part of the inner membrane, about 70 to 80 percent, is protein. The outer boundary membrane, which is less active enzymatically than the inner membrane, is about 50 percent

hydrogens that enter the last step of the electron transport system to combine with oxygen.

Under standard conditions the hydrolysis of ATP to ADP yields 7000 cal/mole. Using this value as the energy required to synthesize ATP from ADP + HPO_4^{2-}, the total energy trapped during the oxidation of glucose, if 38 ATP are produced, amounts to $38 \times 7000 = 266,000$ cal/mole. Combustion of glucose in air yields 686,000 cal/mole. On this basis the efficiency of glucose metabolism in cells approximates $266/686 \times 100 = 39\%$. At this level, the efficiency of mitochondrial energy conversion is considerably higher than most of the energy-conversion systems designed by human engineers, which rarely perform above the 5 to 10 percent efficiency level.

The electron transport system of mitochondria was discovered in investigations carried out during the first

---

the electrons from reduced NAD are transferred to FAD inside the mitochondrion. These enter the electron transport system farther along in the chain and result in the synthesis of 2 ATP for each molecule of reduced FAD. If this less efficient shuttle operates, complete oxidation of glucose will result in a grand total of 36 ATP. The predominating shuttle seems to vary depending on the particular species involved.

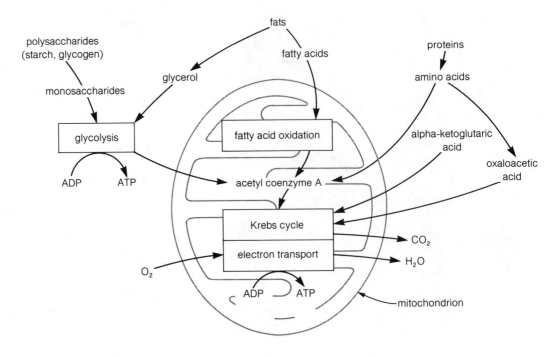

Figure 7-19 Coordination of the oxidation of carbohydrates, fats, and proteins in cells. Coenzyme A plays a central role in linking together the oxidation of these molecules.

protein. The remainder of the mitochondrial membranes is lipid.

Cellular Oxidation of Fats and Proteins

When used as an energy source, both fats and proteins are split by preliminary reactions into shorter segments that enter cellular oxidation at various levels. In many cases, the shorter segments derived from these molecules are 2-carbon acetyl groups, carried to the Krebs cycle by coenzyme A.

Fats

Initially fats are split into glycerol and fatty acids in the cytoplasm outside mitochondria. A phosphate group is then added to glycerol, at the expense of ATP, to produce the 3-carbon, 1-phosphate sugar normally produced by reaction 4 in glycolysis (see Fig. 7-5). The sugar is converted to 3PGAL and enters the glycolytic pathway directly at this point. After preliminary reactions the fatty acid chains enter mitochondria, where they are broken into 2-carbon acetyl groups linked to coenzyme A. From this point, oxidation is the same as for acetyl groups derived from glucose. Because the fatty acid chains are long, containing from 14 to 22 carbons, a large number of acetyl groups are supplied to the Krebs cycle as the chains split into 2-carbon segments. In the overall conversion of fats to CO_2 and H_2O, approximately 17 ATP are produced for each 2-carbon segment in the fatty acid residues. This ATP yield, about double that of carbohydrates by weight, explains why fats are such an excellent energy source.

Proteins

Individual amino acids split off from proteins are also oxidized through the Krebs cycle. Proteins are first split into amino acids in the cytoplasm outside the mitochondria. Some amino acids are further broken into 2-carbon acetyl groups linked to coenzyme A, which then enter the Krebs cycle. Other amino acids are converted into intermediates of the Krebs cycle, such as oxaloacetic acid, and enter the cycle directly in this form.

The cellular oxidation of carbohydrates, fats, and proteins is summarized in Figure 7-19. Inspection of this figure reveals the central position of coenzyme A in linking together the oxidation of many different substances in the cell.

Integration of Mitochondrial Oxidation with Cellular Acitivity

The rate at which pyruvic acid and acetyl groups are oxidized to CO_2 and H_2O in mitochondria is regulated by several control systems in cells. Of these controls, the most central and important is based on the concentration of ADP in the medium surrounding mitochondria. As ADP concentration increases, synthesis of ATP from ADP and phosphate begins in mitochondria. ATP synthesis in mitochondria is tightly coupled to electron transport; as ATP is synthesized, electrons flow without restriction from NAD to oxygen, and the NAD end of the transport chain becomes relatively oxidized. NAD, once oxidized, is free to recycle as an electron acceptor in pyruvic acid oxidation and the Krebs cycle. This system closely links the rate of mitochondrial oxidations to the concentration of ADP in the cell.

If cellular activity is limited, ATP concentration remains high and ADP becomes unavailable to the chemiosmotic mechanism synthesizing ATP in the cristae membranes. As a result, the reactions synthesizing ATP, and electron transport, stop. According to the Mitchell hypothesis, stoppage of electron transport is due to a buildup in the H^+ gradient, unrelieved by backflow through the membrane proteins that synthesize ATP. The H^+ gradient gradually builds until it reaches levels high enough to oppose further expulsion of H^+ across the cristae membranes by the electron transport system. As a consequence, electron transport, and, in turn, pyruvic acid oxidation and the Krebs cycle, stop. The rate of glycolysis is also linked to some extent to mitochondrial activity, since the NAD reduced in glycolysis may transfer its electrons to the mitochondrial carrier systems.

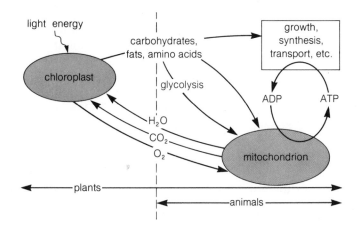

Figure 7-20 How chloroplasts and mitochondria interact in the flow of energy through cells. Plant cells contain the functions shown in the entire diagram; animal cells contain only the functions shown to the right of the dotted line. Energy flows from left to right in the diagram.

The sensitivity of mitochondrial activity to ADP concentration links oxidation with cellular activities requiring an energy input. If such cellular activities as growth, reproduction, and motility proceed at very low levels, little ATP is converted to ADP. In this case, the oxidative activity of mitochondria remains low, allowing oxidizable fuel substances to be conserved. If cellular activity increases, the demand for energy results in conversion of ATP to ADP somewhere in the cell, eventually increasing the concentration of ADP in the mitochondrial environment. Increase of ADP concentration stimulates mitochondrial oxidation, which continues until ATP concentration is restored and ADP levels are low. Mitochondrial activity is thus finely tuned to cellular activity, ready to run on demand if ATP is converted to ADP anywhere in the cell.

In plant cells with both mitochondria and chloroplasts, the combined activity of the two organelles provides all of the ATP required for cell activities (Fig. 7-20). Carbohydrates and fats are synthesized in the chloroplasts (to the left of the dotted line in Fig. 7-20) by reactions using energy derived from light; these reactions use CO_2, H_2O, and various minerals as raw materials and liberate oxygen as a by-product. These carbohydrates and fats are oxidized in mitochondria as an energy source (to the right of the dotted line in Fig. 7-20), yielding ATP, CO_2, and H_2O as products. The oxygen liberated in photosynthesis is used as final electron acceptor in mitochondria; the H_2O and CO_2 produced complete the cycle by serving as raw materials for the synthetic activities of chloroplasts. The entire cycle of activity is driven by light energy, absorbed and converted to chemical energy in the chloroplasts of green plants.

No mitochondria occur in bacteria and blue-green algae. In these prokaryotes, energy-yielding mechanisms are distributed between the cytoplasm and the plasma membrane. (The oxidative systems of bacteria and blue-green algae, which enable these organisms to obtain ATP through the breakdown of carbohydrates and other fuel molecules, are described in Supplement 7-2.)

Suggestions for Further Reading

Hinkle, P. C. and R. E. McCarty. 1978. How cells make ATP. *Scientific American* 238:104–23.

Lehninger, A. L. 1971. *Bioenergetics.* 2nd ed. W. A. Benjamin: Menlo Park, Calif.

Lehninger, A. L. 1973. *Biochemistry.* 2nd ed. Worth: New York.

Miller, G. T., Jr. 1971. *Energetics, kinetics, and life: An ecological approach.* Wadsworth: Belmont, Calif.

Munn, E. A. 1974. *The structure of mitochondria.* Academic Press: New York.

Quinn, P. J. 1976. *The molecular biology of cell membranes.* University Park Press: Baltimore.

Racker, E. 1976. *A new look at mechanisms in bioenergetics.* Academic Press: New York.

Stryer, L. 1981. *Biochemistry.* 2nd ed. W. H. Freeman: San Francisco.

Wolfe, S. L. 1981. *Biology of the cell.* 2nd ed. Wadsworth: Belmont, Calif.

For Further Information

ATP-linked transport, Chapter 5
DNA, ribosomes, and protein synthesis in mitochondria, Supplement 10-2
Electron transport and ATP synthesis in chloroplasts and prokaryotes, Chapter 6
Enzymes and enzymatic catalysis, Chapter 3
Evolutionary origins of mitochondria, Chapter 13

Questions

1. Why is ATP sometimes called the energy "dollar" of the cell?

2. Define oxidation and reduction.

3. What kinds of molecules can be used as energy sources in cells?

4. What are the major steps in the cellular oxidation of carbohydrates?

5. Outline the structure and function of the membranes and compartments of mitochondria.

6. Outline the overall sequence of events in glycolysis. What are the primary chemical inputs and outputs of glycolysis?

7. What is fermentation? What advantages does the ability to carry out fermentations have for cells? Can cells live by fermentations alone? Do fermentations occur in human cells?

8. What are aerobes, anaerobes, facultative anaerobes, and strict aerobes?

9. How are glycolysis and mitochondrial oxidations integrated in cells?

10. Outline the major chemical inputs and outputs of pyruvic acid oxidation. What overall processes occur in pyruvic acid oxidation?

11. Outline the major chemical inputs and outputs of the Krebs cycle. What overall processes occur in the Krebs cycle?

12. Is the Krebs cycle a major direct source of ATP?

13. Compare the functions and structure of ATP, NAD, NADP, FAD, and coenzyme A.

14. What is a cytochrome? A flavoprotein? A quinone? How many of these molecular groups are linked to proteins?

15. How is the mitochondrial electron transport chain believed to set up an H^+ gradient?

16. Compare the flow of electrons and hydrogens (H^+ ions) through the electron transport system of chloroplasts and mitochondria.

17. How is the H^+ gradient produced by electron transport believed to drive ATP synthesis?

18. How many molecules of ATP are synthesized through the reactions of glycolysis? Pyruvic acid oxidation? The Krebs cycle? (Consider that all electrons carried by NAD and FAD are transferred through the mitochondrial electron transport system to oxygen.)

19. The total efficiency of glucose oxidation to CO_2 and H_2O can be calculated as 39%. What part of this efficiency is contributed by pyruvic acid oxidation? By the Krebs cycle? By glycolysis? (Consider that all electrons carried by NAD and FAD flow through the mitochondrial electron transport system to oxygen.)

20. Why do electron pairs carried by NAD result in the synthesis of 3 ATP, and FAD 2 ATP, as they flow along the mitochondrial electron chain to oxygen?

21. Outline two different biochemical controls regulating the rate of cellular oxidations.

22. Where are fats oxidized in cells? Proteins?

23. Trace the cycles of chemical raw materials and products between chloroplasts and mitochondria in plant cells.

24. What effects do the relative concentrations of ATP and ADP have on the rate of cellular oxidations? What significance does this have for the efficiency of fuel molecule utilization in cells?

Supplement 7-1:
Further Information on
Glycolysis and the Krebs Cycle

Glycolysis

Glycolysis is primarily a make-ready sequence in cells that oxidize carbohydrates through the entire pathway from glucose or starch to H_2O and CO_2. It splits 6-carbon sugars and other substances into 3-carbon segments that then enter mitochondria for further oxidation. However, in cells existing either temporarily or permanently without oxygen, glycolysis becomes the central pathway for obtaining ATP.

The individual reactions in glycolysis and the enzymes catalyzing them are shown in Figure 7-21. In the first three reactions of the pathway, glucose is converted into a more reactive substance containing two phosphate groups. The energy required for this segment of the pathway is provided by the breakdown of two molecules of ATP. In the first reaction of the pathway (reaction 1), a phosphate group derived from ATP is added to glucose, producing *glucose 6-phosphate*. This substance has a higher energy content as a result of the added phosphate, and the additional energy enables it to enter the next reactions in glycolysis. The activity of the enzyme catalyzing the first reaction of the pathway, *hexokinase*, illustrates another of the many controls regulating the rate of oxidation in cells. The product of the first reaction, glucose 6-phosphate, is an inhibitor of hexokinase. If glucose 6-phosphate accumulates because the remainder of the sequence is running slowly, the hexokinase enzyme is inhibited, blocking further entry of glucose into the pathway.

Reaction 2 of the pathway simply rearranges glucose 6-phosphate into the closely related sugar *fructose 6-phosphate*. A second phosphate is added in reaction 3 to produce the highly reactive substance *fructose 1,6-diphosphate*. The second phosphate comes from another molecule of ATP, which is converted to ADP in this step. The overall sequence of reactions 1 through 3, yielding the high-energy substance fructose 1,6-diphosphate, requires energy and proceeds to completion because the reactions are coupled to the simultaneous breakdown of ATP to ADP:

$$\text{glucose} + 2 \text{ ATP} \rightarrow$$
$$\text{fructose 1,6-diphosphate} + 2 \text{ ADP} \quad (7\text{-}13)$$

The rest of the glycolytic sequence oxidizes segments derived from this highly reactive sugar and proceeds without further inputs of ATP. In the first of these reactions (reaction 4), fructose 1,6-diphosphate is split into two 3-carbon sugars. One of these, 3-phosphoglyceraldehyde (3PGAL), enters the remaining steps in glycolysis. As 3PGAL is depleted from the pool, it is replaced by conversion of the second 3-carbon product.

The next series of five reactions (reactions 5 to 9) oxidizes 3PGAL and generates ATP and reduced NAD. The oxidation occurs in reaction 5. Two electrons and two hydrogens are removed from 3PGAL; a part of the free energy released in the reaction is used to attach another phosphate group to the 3-carbon sugar, yielding *1,3-diphosphoglyceric acid*. The phosphate group added in this reaction is derived from inorganic phosphate in solution in the medium, and not from ATP. The electrons removed are accepted by NAD along with one of the two hydrogens; the second hydrogen enters the pool of H^+ ions in the cytoplasm. The overall reaction releases free energy and proceeds spontaneously to completion:

$$\text{3-phosphoglyceraldehyde} + \text{oxidized NAD} + \text{HPO}_4^{2-} \rightarrow$$
$$\text{1,3-diphosphoglyceric acid} + \text{reduced NAD} \quad (7\text{-}14)$$

Removing one of the two phosphate groups from 1,3-diphosphoglyceric acid at the next step (reaction 6) produces *3-phosphoglyceric acid* and a second large increment of free energy. Some of this free energy is captured when the phosphate group removed is attached to ADP, yielding a molecule of ATP:

$$\text{1,3-diphosphoglyceric acid} + \text{ADP} \rightarrow$$
$$\text{3-phosphoglyceric acid} + \text{ATP} \quad (7\text{-}15)$$

The 3-diphosphoglyceric acid formed in reaction 6 is rearranged in two steps (reactions 7 and 8) into *phosphoenolpyruvic acid*. The remaining phosphate group is removed in the last step in glycolysis (reaction 9), yielding pyruvic acid. Part of the free energy released in this reaction is used to attach the phosphate to ADP, forming another molecule of ATP.

Many of the intermediate reactions of glycolysis are reversible and may go in either direction in response to high concentrations of either reactants or products (the reversible steps are indicated in Fig. 7-21 by $\rightleftarrows$ arrows). Alternate enzymatic pathways are available in the cytoplasm for the irreversible steps, with the result that glycolysis can be made in effect to run in reverse. For example, the first reaction in glycolysis, the conversion of glucose to

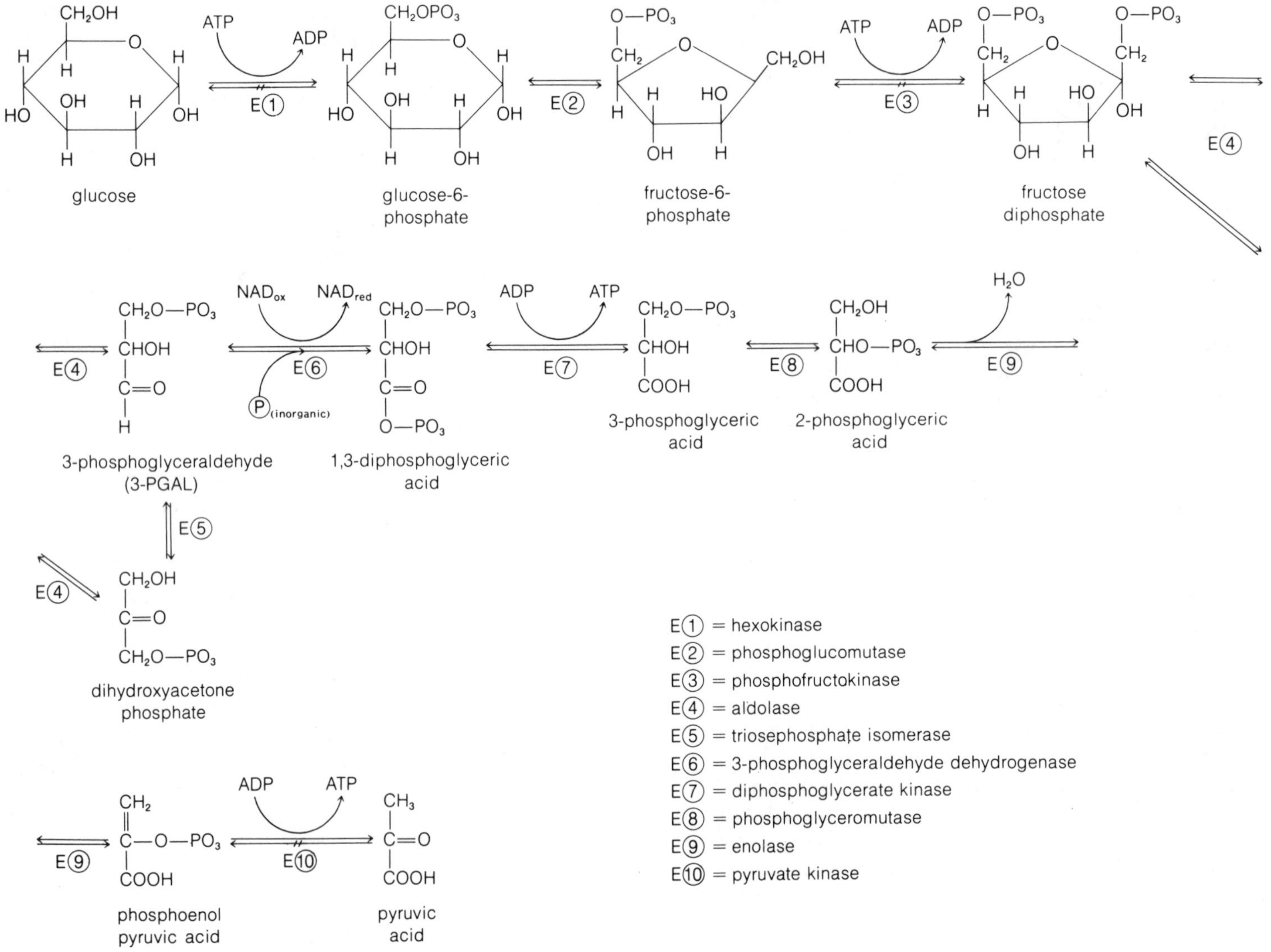

Figure 7-21 The reactions and enzymes of glycolysis (see text).

glucose 6-phosphate by hexokinase, is essentially irreversible. However, an alternate enzyme, *glucose 6-phosphatase*, can catalyze the reverse reaction.

Note that many of the intermediates and products of the final series of reactions in glycolysis are also a part of the Calvin and C_4 cycles of the dark reactions of photosynthesis (compare Fig. 7-21 with Figs. 6-22 and 6-23). Any of these intermediates of the dark reactions in plants may enter glycolysis at their respective places. The first half of the glycolytic sequence is also important in photosynthesis; glucose is synthesized from 3PGAL, the primary

product of the Calvin cycle, essentially by the reversal of reactions 1 through 4 of glycolysis.

The Krebs Cycle

The pyruvic acid (with three carbons) produced by glycolysis is subsequently broken into acetyl (2-carbon) groups in mitochondria. In this reaction, an oxidation, one mole-

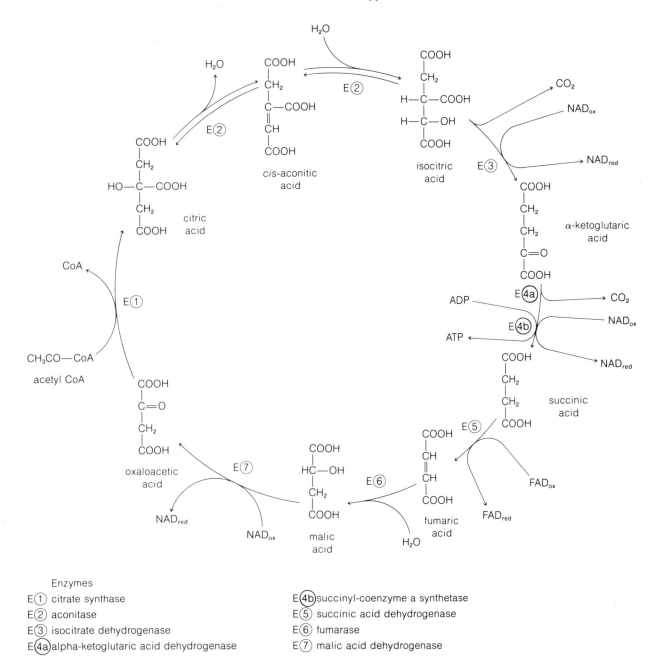

Figure 7-22 The reactions and enzymes of the Krebs cycle (see text).

Enzymes

E① citrate synthase
E② aconitase
E③ isocitrate dehydrogenase
E④a alpha-ketoglutaric acid dehydrogenase

E④b succinyl-coenzyme a synthetase
E⑤ succinic acid dehydrogenase
E⑥ fumarase
E⑦ malic acid dehydrogenase

cule of CO_2 is evolved for every molecule of pyruvic acid entering the sequence. The resulting acetyl groups, attached to coenzyme A in a high energy complex, then enter the Krebs cycle for oxidation to two molecules of CO_2.

In the first reaction of the Krebs cycle (Fig. 7-22), an acetyl group is transferred from coenzyme A to oxaloacetic acid (four carbons), yielding citric acid (with six carbons). The overall reaction is driven by the free energy released by the removal of the acetyl unit from coenzyme A. The citric acid formed in reaction 1 enters a pool of acids that are readily converted from one to the other by the enzyme *aconitase*. As *isocitric acid*, the reactant entering the next step in the cycle, is depleted, the enzyme replenishes this substance by conversion of the citric acid entering the pool.

In the third step in the cycle (reaction 3), two electrons and two hydrogens are removed from isocitric acid. At the same time, the carbon chain is reduced in length from six to five carbons, yielding *alpha-ketoglutaric acid*. The carbon removed is released as CO_2. The electron acceptor for this reaction, which also combines with one of the two hydrogens removed, may be either NAD or NADP.

The next reaction (reaction 4) is complex and is shown only in skeletal form in Figure 7-22. Alpha-ketoglutaric acid is oxidized to *succinic acid* in this step, which removes two electrons and two hydrogens (NAD acts as electron acceptor in this oxidation). The carbon chain is again shortened and CO_2 is released. As a part of the complex side cycle accomplishing this step, one molecule of ATP is synthesized.[3] As overall products, therefore, reaction 4 yields:

alpha-ketoglutaric acid + oxidized NAD + ADP + HPO_4^{2-}

$\rightarrow$ succinic acid + reduced NAD + ATP + CO_2 (7-16)

The next oxidation of the Krebs cycle (reaction 5) is catalyzed by the enzyme *succinic acid dehydrogenase*. This enzyme and its electron acceptor, FAD, are the only molecules of the Krebs cycle attached to the internal membranes of the mitochondrion. The product of reaction 5, *fumaric acid*, is then rearranged at the next step (reaction 6), with the addition of a molecule of water, to form *malic acid*. In the final reaction of the Krebs cycle (reaction 7) malic acid is oxidized to oxaloacetic acid, completing the cycle. NAD is the electron acceptor for the final oxidation.

It is important to note that the molecules of the Krebs cycle are not organized in cells in circles as shown in Figure 7-22. In the mitochondrion, all of the intermediates of the cycle are in solution and mixed intimately in concentrations that depend on the availability of the various reactants. Their interaction depends on random collisions brought about by kinetic movement.

---

[3]The complex cycle carrying out the overall reaction oxidizing alpha-ketoglutaric acid actually produces GTP instead of ATP. However, a cytoplasmic enzyme, *nucleoside diphosphate kinase*, readily converts the GTP formed to ATP. This intermediate formation of GTP has been eliminated from Reaction 7-16 for the sake of simplicity.

Supplement 7-2: Cellular Oxidations in Prokaryotes

Bacteria

Almost all bacteria are able to use carbohydrates in some form as an energy source. Organic acids and amino acids are also commonly oxidized by these organisms. Some bacteria can oxidize a wide variety of organic molecules; others are so restricted that their preferences in fuel substances can be used to identify them. However, in most bacteria carbohydrates are the primary energy source as in eukaryotes.

Most bacteria are capable of oxidizing carbohydrates by sequences closely resembling the glycolytic pathway. Because oxygen itself is not used as the final electron acceptor in glycolysis, this type of metabolism, if it is the only type used by the bacteria, occurs as a fermentation and oxidation is anaerobic. Often pyruvic acid is the final acceptor, in a reaction similar to lactic acid generation in muscle tissue:

pyruvic acid + reduced NAD $\rightarrow$

lactic acid + oxidized NAD   (7-17)

By different pathways, bacterial fermentations may yield a wide variety of final products, most commonly including acids such as lactic and acetic acid or alcohols such as ethyl and butyl alcohol. Many bacterial fermentations yield products useful to humans, such as sauerkraut, vinegar, and some types of cheese. Other bacteria are aerobic and carry out oxidations through the complete sequences of glycolysis, pyruvic acid oxidation, the Krebs cycle, and electron transport to oxygen.

The enzymes of glycolysis are suspended in solution in the cytoplasm of bacterial cells, just as they are in eukaryotes. The components of the Krebs cycle, if present, are also suspended in the cytoplasm; no membranes separate these reaction sequences from their cytoplasmic surroundings as in eukaryotic cells. The enzymes and carriers of the electron transport chain are bound tightly to the plasma membrane in bacteria and remain with the membrane fraction if bacterial cells are disrupted for biochemical analysis. Most of the enzymes of glycolysis and the

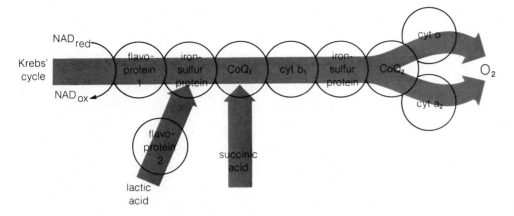

Figure 7-23 The electron transport system of the bacterium *E. coli*. The electron carriers are embedded in the plasma membrane in bacteria.

Krebs cycle in bacteria closely resemble their counterparts in eukaryotes. The bacterial electron transport chain also consists of flavoproteins, quinones, and cytochromes, as in eukaryotes. However, the electron carriers show much variation in bacteria and include flavoproteins, quinones, and cytochromes that are different from the electron transport molecules of higher organisms.

Bacterial electron transport systems are typically more branched. That is, they usually have supplemental points for entry and exit of electrons in addition to the routes typical of eukaryotes. The example shown in Figure 7-23, which diagrams the electron transport system in *Escherichia coli*, contains one such extra input, which accepts electrons removed from lactic acid (*E. coli* is the primary bacterium inhabiting the intestinal tract in humans). The final segment of the pathway in *E. coli* is also branched, offering two routes to oxygen, one through cytochrome *O* and one through cytochrome a_2. Neither of these cytochromes occurs in eukaryotes. While the electron carrier systems of aerobic bacteria use oxygen as final electron acceptor, some species are adapted to different acceptors and use substances such as sulfate or nitrate in this role instead.

The most striking relationship of bacterial systems to the oxidative mechanisms of eukaryotes is in the form taken by the ATP-synthesizing complex, which can be identified with lollipop-shaped structures extending from the plasma membrane into the cytoplasm. Peter Mitchell and his coworkers have shown that electron transport in bacteria results in expulsion of H^+ ions across the plasma membrane, from the inside of the bacterial cell to the out-side. The H^+ gradient established, as in eukaryotes, provides the driving force for ATP synthesis by the membrane-bound ATPase complex.

Blue-Green Algae

Comparatively little is known of the oxidative mechanisms of the blue-green algae. These prokaryotes appear to live primarily by photosynthesis, and evidently obtain most of the ATP they require through electron transport in the light reactions of photosynthesis. If grown in the dark, some blue-green algae are able to take up a carbohydrate, such as glucose, and slowly oxidize it to CO_2. The pathways of this oxidation are not yet clear, and it is not certain whether blue-green algae contain all of the components of glycolysis and the Krebs cycle. Although some of the enzymes of each of these oxidative sequences have been isolated from the blue-green algae, others have never been detected or identified. For example, the enzymes oxidizing an intermediate compound of the Krebs cycle, alpha-ketoglutaric acid (produced by reaction 3 in Fig. 7-22), have never been identified in blue-green algae. Thus it is not known whether these prokaryotes have a complete Krebs cycle or are able to oxidize carbohydrates only through a part of the cycle. If all of the components are present, the various enzymes and electron transport carriers are probably located in the cytoplasm as in bacteria, since blue-green algae do not have mitochondria.

Supplement 7-3:
Microbodies: Peroxisomes and Glyoxisomes

Microbodies are a class of small, relatively simple cytoplasmic organelles that appear under the electron microscope as roughly spherical structures about 0.5 to 1.5 micrometers in diameter (Fig. 7-24). Microbodies are enclosed by a single membrane; the interior is densely granular, and frequently contains a "core" particle with regular crystalline structure (visible in the plant microbody shown in Fig. 7-24).

In the mid-1960s Christian De Duve and his coworkers at the Rockefeller University established that microbodies are oxidative organelles that contain *catalase*, an enzyme capable of converting hydrogen peroxide to water. De Duve also found that microbodies from liver and kidney cells of mammals are capable of oxidizing amino acids and uric acid, a breakdown product of nucleotides. Microbodies with these oxidative capabilities were called *peroxisomes*.

Microbodies with essentially the same morphology were also described in plant tissues. Microbodies from oily plant seeds, in addition to the catalase enzyme, also contain enzymes capable of breaking down fats. In addition, the plant seed microbodies possess enzymes of the *glyoxylate cycle* (see Fig. 7-25), a series of reactions that can convert acetyl-coenzyme A produced in fatty acid oxidation into carbohydrates. Because of this capability, the plant seed microbodies were called *glyoxisomes*.

Microbodies were also found in other tissues of plants, particularly in leaf cells capable of photosynthesis. These leaf microbodies resemble animal peroxisomes in having enzymes that oxidize amino acids and uric acid. In addition, the leaf microbodies were found to contain a reaction series serving as an alternate source of sugars in photosynthesis. Somewhat confusingly, these microbodies were also termed peroxisomes. These discoveries thus identified three related types of microbodies with the following major activities:

1. animal peroxisomes: oxidize amino acids and uric acid

2. glyoxisomes: convert fats to carbohydrates through glyoxylate cycle

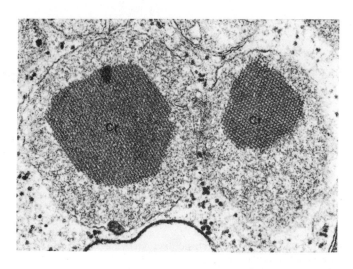

Figure 7-24   Microbodies in a castor bean cell. *Cr*, crystalline core. × 58,000. Courtesy of E. L. Vigil, from *Journal of Cell Biology* 46: (1970): 435 by copyright permission of The Rockefeller University Press.

3. plant peroxisomes: oxidize amino acids and uric acid and synthesize carbohydrates.

These microbody subclasses all share catalase activity.

The Reactions of Animal Peroxisomes   Peroxisomes in animals oxidize amino acids by removing their amino groups, yielding Krebs cycle acids and other acids as products:

$$\text{amino acid} + H_2O + \text{FMN (or FAD)} \rightarrow$$
$$\text{Krebs cycle acid} + NH_4^+ + FMNH_2 \text{ (or } FADH_2) \quad (7\text{-}18)$$

The reduced FMN or FAD produced in the reaction is converted back to the oxidized form by a reaction with oxygen:

$$FMNH_2 \text{ (or } FADH_2) + O_2 \rightarrow \text{FMN (or FAD)} + H_2O_2 \quad (7\text{-}19)$$

The hydrogen peroxide produced in this reaction is a toxic substance that is quickly broken down by the catalase enzyme of the peroxisome:

$$H_2O_2 \rightarrow H_2O + \tfrac{1}{2} O_2 \quad (7\text{-}20)$$

Peroxisomes of this type are abundant in the kidney and liver tissues of vertebrate animals. Similar bodies have

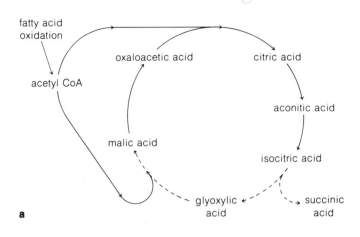

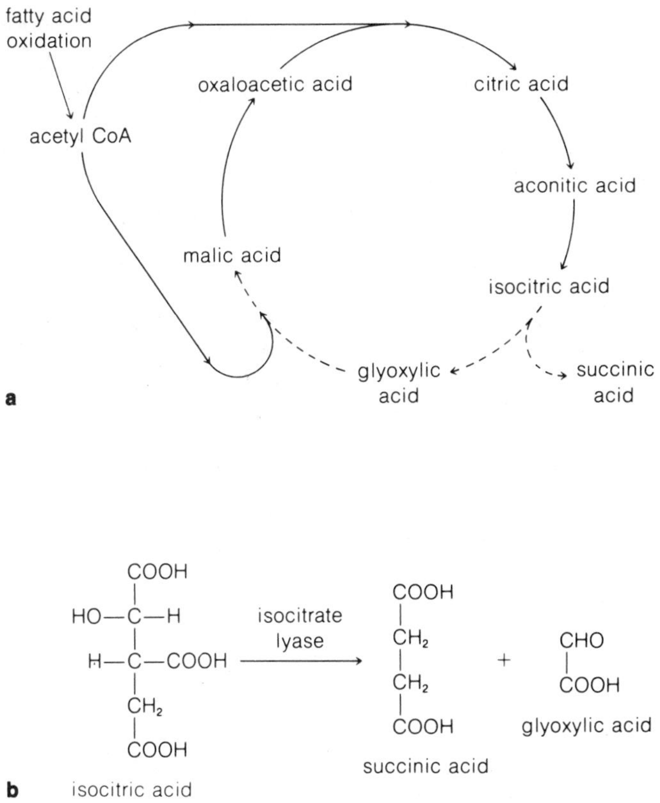

$$\underset{\text{glycolic acid}}{\begin{array}{c} CH_2OH \\ | \\ COOH \end{array}} + O_2 \longrightarrow \underset{\text{glyoxylic acid}}{\begin{array}{c} CHO \\ | \\ COOH \end{array}} + H_2O_2$$

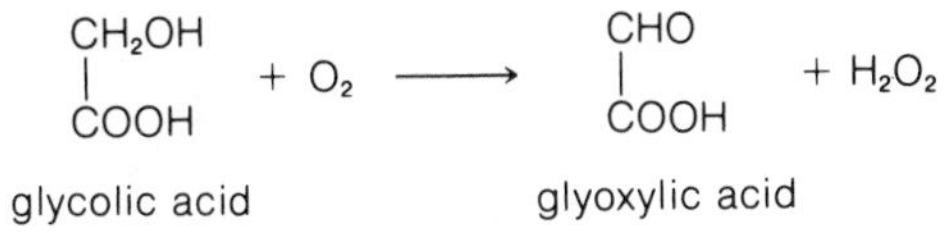

Figure 7-26 The glycolate pathway occurring in photorespiration in plant peroxisomes (see text).

$$\underset{\substack{\text{isocitric acid}\\ \textbf{b}}}{\begin{array}{c} COOH \\ | \\ HO-C-H \\ | \\ H-C-COOH \\ | \\ CH_2 \\ | \\ COOH \end{array}} \xrightarrow[\text{lyase}]{\text{isocitrate}} \underset{\text{succinic acid}}{\begin{array}{c} COOH \\ | \\ CH_2 \\ | \\ CH_2 \\ | \\ COOH \end{array}} + \underset{\text{glyoxylic acid}}{\begin{array}{c} CHO \\ | \\ COOH \end{array}}$$

Figure 7-25 (**a**) The glyoxylate cycle. The segment of the cycle in dotted lines diverges from the Krebs cycle (compare with Fig. 7-11). (**b**) The reaction producing succinic acid and glyoxylic acid in glyoxisomes.

also been described in other vertebrate tissues, including brain, small intestine, adrenal cortex, and cells of the testis. Invertebrates and protozoa, including *Tetrahymena, Paramecium,* and *Euglena* (a photosynthetic protozoan) also contain peroxisomes. The number per cell may be large; liver cells in mammals typically contain 350–400 peroxisomes.

Glyoxisomes The reactions of the glyoxylate cycle represent a partial circuit around the Krebs cycle in which some reactions are included and others bypassed (Fig. 7-25a). The glyoxylate cycle diverges from the Krebs cycle through the action of two enzymes that are unique to glyoxisomes. One splits isocitric acid into succinic acid and glyoxylic acid (Fig. 7-25b). Glyoxylic acid then enters into an interaction with a second molecule of acetyl-coenzyme A to

produce malic acid. This reaction is catalyzed by the second enzyme unique to glyoxisomes. Malic acid is then oxidized to oxaloacetic acid and the cycle is ready to turn again. The overall cycle thus uses acetyl-coenzyme A as fuel, and releases succinic acid as a product. The succinic acid released may diffuse from glyoxisomes to mitochondria, where it can enter the Krebs cycle. Alternatively, it may serve as a raw material for reactions that synthesize glucose, thus leading to the generation of a sugar from the fats originally entering oxidation in the glyoxisome. This pathway is considered to be the primary destination of the succinic acid produced in plant glyoxisomes, since these microbodies are characteristic of seeds containing high concentrations of stored fats or oils. Through this pathway, the stored fats of the seeds can contribute to products formed from glucose, such as the cellulose of new cell walls.

Plant Peroxisomes These organelles are the most complex of the various types of microbodies. In photosynthesizing tissues such as leaf cells, they contain enzymes catalyzing the *glycolate pathway* (Fig. 7-26) in addition to amino and uric acid oxidation and the glyoxylate cycle. The H_2O_2 product of glycolic acid oxidation is converted to water and oxygen by the catalase enzyme.

The glycolic acid oxidized by the sequence shown in Figure 7-26 is a product of photosynthesis in the chloroplasts of plant cells. Apparently, when plants carry out photosynthesis in conditions of adequate light, they use oxygen only in glycolic acid oxidation in peroxisomes; mitochondrial respiration is inhibited. Because oxygen is consumed by peroxisomes in photosynthesizing leaf tissue, the oxidative reactions taking place in the glycolic acid pathway are sometimes termed *photorespiration*.

The product of the reaction shown in Figure 7-26, glyoxylic acid, may enter the glyoxylate cycle to be converted to oxaloacetic acid or succinic acid. Through this route, the immediate products of photosynthesis can thus enter carbohydrate metabolism via these acids, which form a part of pathways leading to both the breakdown and synthesis of sugars. The glycolate pathway thus provides an alternate to the Calvin cycle (see Chap. 6), the usual mechanism for carbohydrate synthesis in chloroplasts. Plant peroxisomes also contain enzymes that can convert glyoxylic acid to an amino acid, glycine, thus placing products of the glycolate pathway on the route to protein synthesis.

Just why plants and animals contain glyoxisomes or peroxisomes offering these alternate biochemical pathways is unknown. There are two schools of thought on many issues in cell biology, and this one is no exception. In the case of glyoxisomes and peroxisomes, some scientists claim that these organelles are "fossils" left over from an early evolutionary period when mitochondria and chloroplasts were absent or less highly developed. According to this idea, the more rudimentary reactions of microbodies were the best available at this evolutionary time. Others maintain that the reactions carried out by these organelles are essential to the cells containing them, and that they are present because they carry out vital functions in the metabolism of fats, sugars, and amino acids.

Suggestions for Further Reading

Breidenbach, R. W. 1976. Microbodies. In *Plant Biochemistry*. Ed. J. Bonner and J. E. Varner. Academic Press: New York. Pp. 91–114.

Vigil, E. L. 1973. Structure and function of plant microbodies. *Subcellular Biochemistry* 2:237–63.

FOUR

Cell Motility

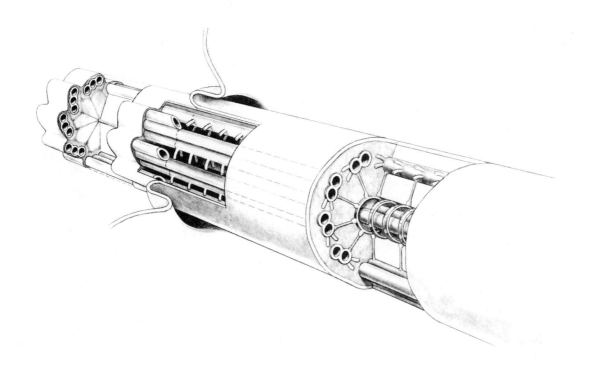

8

Microtubules, Microfilaments, and Cell Motility

One of the primary uses of the chemical energy captured in cellular oxidations is the production of motion. All cells can move; their movements range from limited rearrangements of internal structure to active motility of entire cells or groups of cells. In many-celled organisms, individual cellular movements are coordinated to produce the movements of entire individuals.

These cellular movements are accomplished by proteins that convert the chemical energy of ATP into mechanical energy. These motile activities, like enzymatic catalysis and membrane transport, depend on the ability of proteins to undergo changes in shape or conformation in response to chemical interactions.

The proteins involved in cell motility are organized into two different structures, *microtubules* and *microfilaments*. Microtubules (see Figs. 1-10 and 8-8) are unbranched, hollow cylinders about 25 nanometers in diameter that vary in length from a few nanometers to many micrometers. Microfilaments (see Figs. 1-11 and 8-17) are extremely fine, unbranched fibers of variable length and much smaller diameter than microtubules, about 5 to 7 nanometers. At these dimensions, microfilaments are not much thicker than the wall of a microtubule. Each of these motile elements is built up from distinct proteins. Microtubules are constructed from two closely related proteins collectively called *tubulin*. Microfilaments are formed from the protein *actin*. The actin molecules of microfilaments usually occur in association with a second protein called *myosin*. The movements produced by microfilaments occur through the coordinated activities of the actin and myosin proteins.

Microtubules and microfilaments act both separately and cooperatively to produce a variety of cellular movements. The whiplike motions of sperm tails and other types of flagella and cilia, for example, are based primarily or exclusively on microtubules. Microfilaments are responsible for the contractile movements of muscle cells in animals, and the streaming motions of the cytoplasm observed in both animal and plant cells. The activities of microtubules and microfilaments are coordinated in animal cell division, in which microtubules divide the chromosomes of the cell nucleus and microfilaments produce the movements dividing the cytoplasm.

Microtubule-Based Cell Motility

The 9 + 2 System of Flagella and Cilia

Much of our present information about microtubule structure and function was developed through studies of the "9 + 2" system of microtubules that forms the axis of cilia and flagella (Figs. 8-1 and 8-2). Almost all eukaryotic organisms possess cells with flagella and cilia during at least some stages of their life cycles. Among major plant and animal groups, only the flowering plants (the angiosperms) have no flagellated cells of any type. The sperm cells of animals, and plants from the algae to the primitive gymnosperms, swim by means of flagella; cells with flagella or cilia also occur in the body tissues of most animals. In humans, ciliated cells line the respiratory tract, the cavities of the brain, and parts of the female reproductive ducts. In these cells, the flagella and cilia move by beating or undulating in S-shaped waves, or by bending at the base in a stiff, oarlike motion.

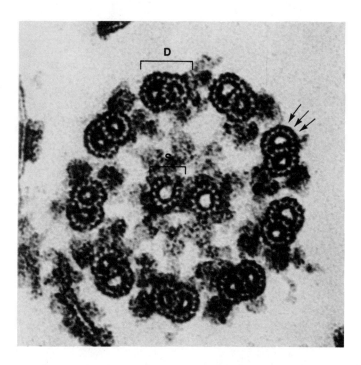

Figure 8-1 The 9 + 2 system of microtubules and connecting elements in cross section. Microtubule subunits (arrows) are clearly visible in the microtubule walls. *D*, peripheral doublet; *S*, central singlet. × 594,000. Courtesy of K. Fujiwara, from *Journal of Cell Biology* 59 (1973): 267 by copyright permission of The Rockefeller University Press.

Whether these motile organelles are called flagella or cilia is determined primarily by their number per cell. When only one or a few are present, they are called flagella. When they are present by the hundreds or thousands, they are called cilia. Usually, when present in numbers sufficient to qualify as cilia, they are relatively short, ranging around 20–25 micrometers in length. Flagella are usually 2–3 times longer than this average. In insects, sperm flagella may be as long as 1–2 millimeters or more (in some insects, such as *Drosophila*, the sperm tail is longer than the entire adult animal!).

The structural arrangement of microtubules in the axis of almost all flagella and cilia is remarkably uniform and is perhaps the most regularly repeating morphological pattern in eukaryotic organisms. This 9 + 2 microtubular complex (see Figs. 8-1 and 8-2) consists of a circle of nine peripheral double microtubules (called *doublets*), arranged around two single central microtubules (called *central singlets*). The circle of doublets has an outer diameter of about 200 nanometers or 0.2 micrometer, placing the entire structure just within the resolving power of the light microscope.

Each of the two central singlets of the 9 + 2 complex is a complete, separate microtubule. Only one of the pair making up a peripheral doublet, however, is a complete microtubule. The second, incomplete microtubule appears in cross-sections as a semicircular or elliptical "C" attached to one side of the first subtubule.

The microtubules of the 9 + 2 system are held together by connecting elements of various kinds. Among the most important of these are the *spokes,* which connect the doublets to a *sheath* surrounding the central singlets. Each doublet also bears a pair of *arms* that extend from the complete microtubule of one doublet toward the incomplete microtubule of the next doublet in the outer circle (see Figs. 8-1 and 8-2). Longitudinal sections of flagella show that the arms, spokes, and sheath elements are periodic structures that are repeated individually along the 9 + 2 system (Fig. 8-3). The entire 9 + 2 system of microtubules and connecting elements is frequently termed the *axoneme* of a flagellum. The axoneme, which extends outward from the cell surface as the axis of a flagellum, is surrounded by a thin layer of cytoplasm and covered entirely by an extension of the plasma membrane (the origin of the flagellar axoneme from cytoplasmic structures called *centrioles* is described in Supplement 8-1).

The ATPase Enzyme of the 9 + 2 System   Early biochemical work revealed that flagella contain an *ATPase* (*adenine triphosphatase*) enzyme capable of catalyzing the breakdown of ATP to ADP, with the release of free energy. In 1960 Ian R. Gibbons of the University of Hawaii found that the enzyme was a part of the arms extending from the side of the microtubule doublets of the 9 + 2 system. Gibbons noticed that when the plasma membranes of flagella were removed by exposing them to a detergent, adjustments in the salt concentration in the solution surrounding the flagella caused the arms, and the ATPase activity of flagella, to disappear. The separated arms, when isolated and purified, proved to contain a group of proteins with strong activity as an ATPase enzyme. Returning the arms to the 9 + 2 system, which could be accomplished by adjusting the salt concentration to normal values, restored ATPase activity to the flagella. Gibbons called the flagellar ATPase *dynein*. Since Gibbons's work, the arms have come to be known as the *dynein arms* of flagella.

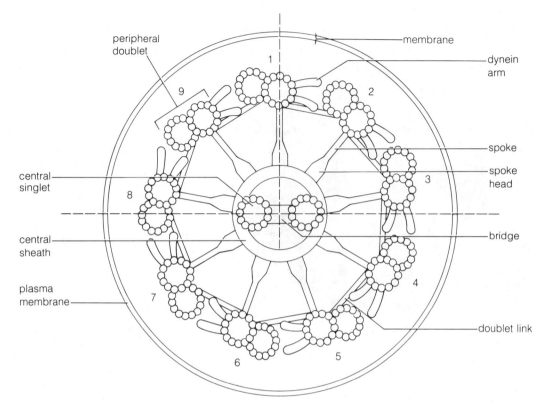

Figure 8-2 Structures of the 9 + 2 system (see text).

The Arms and Flagellar Movement These findings suggested that the arms are involved in some way, through their activity as an ATPase enzyme, in the mechanism producing motion in flagella. It seemed possible that the arms might undergo conformational changes in response to breaking down ATP and bend like small oars to push the doublets past each other. This would produce a sliding motion between the doublets that induces the flagellum to bend.

Several later experiments showed this hypothesis to be correct. One of these experiments, conducted by Peter Satir at the University of California at Berkeley, demonstrated that the doublet microtubules actually slide past each other rather than contracting during flagellar bending. Satir accomplished this by carefully analyzing sections made near the tips of straight and bent flagella. At the end of an unbent flagellum, the doublets at both sides of the flagellum extend within equal distances of the tip (Fig. 8-4a). In a flagellum bent to one side, the doublets on the side toward the direction of bending extend farther into

the tip (Fig. 8-4b). This is consistent with the idea that the doublets slide past each other rather than contract when a flagellum bends, because the radius of curvature of the doublets on the outside of the bend (side 1 in Fig. 8-4b) is greater than the radius of the doublets on the inside of the bend (side 2, Fig. 8-4b). Since the radius of curvature is smaller on the inside of the bend, the microtubules are expected to extend along a shorter distance on this side and pass farther into the tip. It is important to note that the sliding referred to is *doublet over doublet*, not sliding between the microtubules of a single doublet.

A second experiment by Gibbons and an associate, Keith E. Summers, showed that the microtubule sliding in flagella is active and not merely the passive result of some other force-generating mechanism that causes the flagellum to bend. Summers and Gibbons removed the membranes from isolated flagella by treating them with a detergent. The demembranated flagella were then treated briefly with a protein-digesting enzyme. This treatment left the dynein arms intact, but loosened or partly digested the

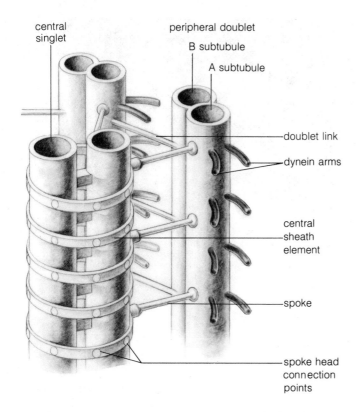

Figure 8-3 The periodic arrangement of arms and linking elements in the 9 + 2 system of flagella.

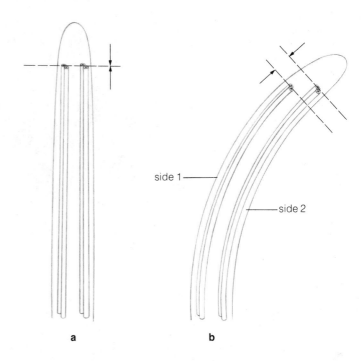

Figure 8-4 The extension of peripheral doublets in an axoneme during flagellar bending. In an unbent flagellum (**a**) the peripheral doublets on all sides of the axial complex extend an equal distance into the tip. In a flagellum bent to one side (**b**), the doublets on the inside of the bend (side 2) extend farther into the tip than those on the outside of the bend (side 1). This indicates that the doublets do not shorten in length or contract during bending of the flagellum (see text). Redrawn from an original courtesy of P. Satir, from *Journal of Cell Biology* 39 (1968): 77 by copyright permission of The Rockefeller University Press.

spokes and other linking elements of the 9 + 2 system. On adding ATP to these preparations, Summers and Gibbons noted that the microtubule doublets of the axoneme, no longer held in place by their spokes and links, slid actively and forcefully out of the 9 + 2 bundles (Fig. 8-5).

From their work Summers and Gibbons proposed, in agreement with other investigators, that the dynein arms produce the force sliding the doublets past each other in the flagellum. The spokes and links of the 9 + 2 system, according to their model, act as elastic connectors that hold the doublets together and prevent them from sliding completely out of the axoneme during movement. Instead, restriction by the spokes and links causes the flagellum to bend to accommodate the displacement of the doublets due to sliding. Thus, according to Summers and Gibbons's model, the dynein arms produce the force sliding the doublets, and the spokes and links convert the sliding into flagellar bends.

From this and other evidence, Fred D. Warner of Syracuse University and others have proposed a model for the cyclic, oarlike motion of the flagellar arms in the production of microtubule sliding. According to the model (Fig. 8-6), the cycle begins when a dynein arm binds a molecule of ATP (Fig. 8-6a). Binding causes ATP breakdown, and induces a folding change in the proteins of the arm that alters its angle with respect to the microtubule (Fig. 8-6b). The energy released by ATP hydrolysis, and the resultant movement, places the arm in a bent and "cocked" position. At the same time, the conformational change exposes a binding site at the tip of the arm, giving it strong affinity for an attachment site on an adjacent microtubule (Fig. 8-6c). Attachment to the microtubule triggers another conformational change that releases the arm from its cocked position. In response, it moves forcefully through a short arc (Fig. 8-6d), sliding the attached microtubule to a new position. The arm is now ready to bind a

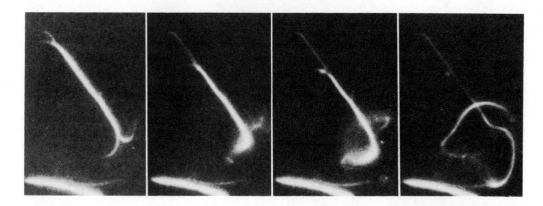

Figure 8-5 Time-lapse micrographs showing microtubule doublets sliding actively from an isolated axoneme bundle. ATP was added at frame 1. × 1,350. Courtesy of I. R. Gibbons, from *Proceedings of the National Academy of Sciences* 68 (1971): 3092.

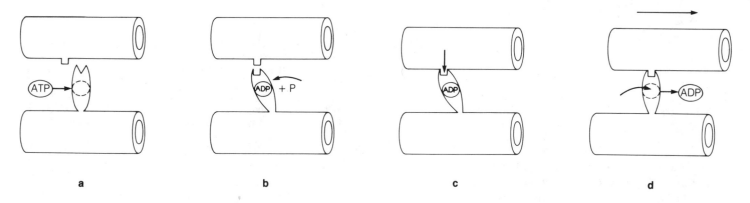

a   b   c   d

Figure 8-6 Steps in the attachment-release cycle of the dynein arms powering the sliding motion between microtubules (see text).

second molecule of ATP and repeat the cycle. Binding causes the arm to release from the microtubule and repeat steps a through d in Figure 8-6. Each cycle of a single dynein arm slides the adjacent doublets past each other by a distance of about 10 nanometers or so. The combined, rapid action of the hundreds or thousands of arms between the microtubules results in the sliding motion observed between the flagellar doublets.

Generation of Motility by Microtubules in Other Cellular Locations

Microtubules also occur in other locations in both the cytoplasm and nucleus of cells. While not organized into doublets or 9 + 2 systems, the microtubules in these lo-

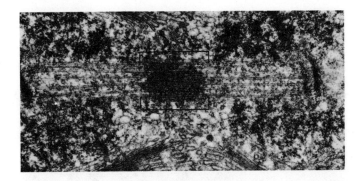

Figure 8-7 Overlap (brackets) visible between the spindle microtubules of a diatom. As the spindle elongates, the degree of overlap decreases, indicating that microtubules slide to lengthen the spindle in this species. × 28,000. Courtesy of D. H. Tippet, from *Journal of Cell Biology* 79 (1978): 737 by copyright permission of The Rockefeller University Press.

cations also show arms linking them together or to other structures in the cell. In many of these systems, microtubules produce motion by active sliding. By analogy with the dynein arms in the 9 + 2 system of flagella, the arms linking microtubules together in other cellular locations are also believed to produce the sliding motion by the same cyclic, oarlike movements dependent on ATP breakdown.

For example, during cell division the *spindle*, a structure built up from hundreds to thousands of microtubules, moves and divides the chromosomes of the cell nucleus (the details of nuclear division are presented in Chap. 11). In many organisms a part of the movement produced by the spindle has been shown to be associated with microtubule sliding. Early in division in these organisms, the microtubules running from end to end of the spindle overlap extensively at the spindle midpoint (Fig. 8-7). As division proceeds, the degree of overlap decreases, and the overall length of the spindle increases. Measurements have shown that decrease in overlap corresponds to the increase in spindle length, indicating that the elongation of the spindle in these species is produced by microtubule sliding in the region of overlap. Sections show that the microtubules in the overlapping region are linked together by arms; ATPase activity can also be detected in the same locations. Thus the elements of the sliding mechanism of flagella also appear to be present, and to produce motion by sliding in spindles as well as in flagella. The same slid-

ing mechanism, based on the cyclic motion of the connecting arms, probably operates in most locations in which microtubules produce cell movements (in some systems, microtubules also cause movements by increasing in length by the addition of tubulin subunits; see below).

Microtubule Subunits, Growth, and Assembly

In the early 1950s Shinya Inoué of the University of Pennsylvania noted that certain treatments, such as exposure to temperatures near 0°C or high pressure, cause the spindle in dividing cells to disappear. The spindle reappears quickly after removal of the disturbing treatment, sometimes within a matter of minutes. From these observations, Inoué proposed that the microtubules of the spindle exist in a *dynamic equilibrium* with a pool of unassembled microtubule subunits in the cytoplasm. Depending on cellular conditions, the equilibrium can shift to favor either assembled spindle microtubules (the *polymer*) or disassembled microtubule subunits (the *monomer pool*).

Inoué's discovery sparked a search for the microtubule monomers, the unpolymerized subunits of microtubules. The microtubule subunits were soon isolated and identified through work carried out in the laboratories of a large number of investigators.

Research with the purified microtubule subunits showed that they are formed from a protein which was subsequently called tubulin. Two closely related kinds of tubulin were discovered, called *alpha-* and *beta-tubulin;* the alpha- and beta-tubulin molecules from all species are closely similar in size and amino acid sequence.

The protein subunits are visible in the walls of microtubules as small, spherical particles about 4–5 nanometers in diameter (Fig. 8-8). The microtubule wall is made up from a circle of the tubulin subunits; in the microtubules of most species there are 13 subunits in the circle (inset, Fig. 8-8). Side views of microtubules, as in the main part of Fig. 8-8, show that the subunits line up in parallel, lengthwise rows in the walls. Since alpha- and beta-tubulin are always isolated in equal quantities from microtubules, the two forms of the tubulin are believed to alternate in the lengthwise rows. That is, one subunit in a row is an alpha-tubulin protein molecule; the next a beta-tubulin molecule, and so on along the microtubule.

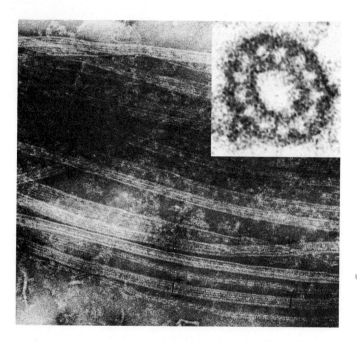

Figure 8-8 Microtubule subunits. The arrangement of the wall subunits in longitudinal rows is clearly visible in some of the microtubules (brackets). The inset shows a single microtubule in cross section. Each microtubule is constructed from a circle of 13 subunits. Main figure × 130,000; courtesy of R. Barton and Academic Press, Inc., from *Journal of Ultrastructure Research* 20 (1967): 6. Inset × 830,000; courtesy of P. R. Burton.

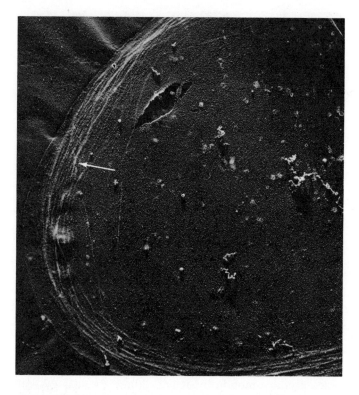

Figure 8-9 Microtubules (arrow) in the marginal band of a red blood cell of an amphibian. × 11,000. Photograph by the author, from *Journal of Ultrastructure Research* 17 (1967): 588 by permission of Academic Press, Inc.

The successful isolation of the microtubule subunits, accomplished by the early 1960s, opened the exciting prospect of assembling microtubules in the test tube. This feat was accomplished in 1972 by Richard C. Weisenberg at Princeton University. Weisenberg found that as long as the concentration of calcium ions is reduced to very low levels, isolated tubulin subunits can be induced to assemble into microtubules in the test tube by warming the solution to 37°C. The assembly is easily reversed by cooling the solution below 15°C. Assembly and disassembly can be repeated in this way indefinitely. The microtubules that polymerize when the solution is warmed to 37°C are apparently identical to microtubules in their normal locations inside cells. Weisenberg's polymerization of microtubules was among the first successful attempts to induce a cell structure to self-assemble in a test tube.

Microtubules as Supportive Elements

There are many systems in which microtubules function as supportive elements, with no direct activity in generation of force for movement. For example, the red blood cells of many higher organisms are held in a flattened shape by a band of microtubules that extends around the cell margin (Fig. 8-9). Exposure to agents causing microtubule disassembly destroys the marginal band, and in most cases, causes the blood cells or platelets to round up. Thus the microtubules in the marginal band act as cytoskeletal elements holding the cells in their flattened shape.

Microtubules in supportive roles also occur in abundance as a cytoplasmic network in many cell types in animals (see Fig. 1-13a). A similar network has also been reported in at least one plant cell type, in the carrot. In both the animal and plant systems studied, the cytoplasmic network apparently functions as a cytoskeleton maintaining cell shape.

Microfilament-Based Cell Motility

The cellular movements generated by microfilaments are of two primary types. One is contraction, involving an active shortening of cells or cell parts. The second is cytoplasmic streaming, in which elements in the cytoplasm flow directionally from one region to another. In some systems, as in ameboid movement, cytoplasmic streaming produces movements of the entire cell.

The most thoroughly studied and highly organized microfilament-based motile system occurs in the muscle cells of animals. Most of our current understanding of the structure and function of microfilaments is based on information originally developed through studies of muscle. The unique biochemical properties of the microfilament proteins, actin and myosin, which were characterized through these studies, allowed the same proteins to be identified as the motive force for contraction and streaming movements in other, nonmuscle cells.

Microfilaments in Muscle Cells

Muscle cells occur in almost all animals. In our own bodies, muscle cells provide the voluntary movements of the limbs and other consciously controlled movements such as those of the lips and eyelids. The movement produced by these muscle cells is characteristically rapid, powerful, and precisely controlled.

A single voluntary muscle, such as the biceps of the upper arm, consists of a large number of elongated, cylindrical or ribbonlike cells called *muscle fibers* (Fig. 8-10). Under the light microscope, muscle fibers show conspicuous cross-striations consisting of alternating light and dark transverse bands (Fig. 8-11). Because of the appearance given to the cells by the alternating bands, voluntary muscle is frequently called *striated* muscle.

Breaking muscle cells apart shows that the banded

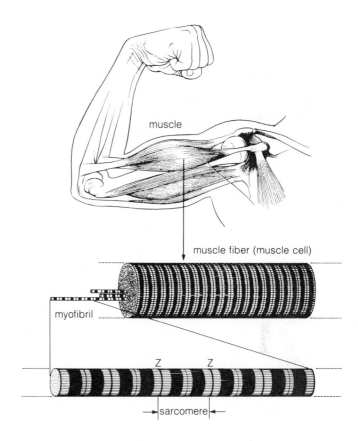

Figure 8-10 The relationship among muscles, muscle fibers, myofibrils, and sarcomeres. Redrawn from an original courtesy of D. W. Fawcett, from D. W. Fawcett and W. Bloom, *A Textbook of Histology*, 10th ed., Philadelphia: W. B. Saunders, 1975.

pattern is due to long, microscopic strands within the cells called *myofibrils* (Fig. 8-10 shows the relationship between muscle, muscle fibers, and myofibrils). Muscle cells also contain multiple nuclei and large numbers of mitochondria distributed between the myofibrils. These organelles are usually confined to a thin surface layer of nonfibrous cytoplasm surrounding the myofibrils.

The repeating cross-bands responsible for the striated appearance of myofibrils and muscle cells are clearly visible in the electron microscope (Fig. 8-12a). The boundaries of each repeated unit, called a *sarcomere*, are marked by dark, slender transverse bands called *Z-lines*. The broader

Figure 8-11 Striated muscle in the light microscope. The muscle fibers run horizontally in the micrograph; the cross striations from which striated muscle takes its name run vertically. *N*, nucleus. × 980. Courtesy of N. N. Malouf and Academic Press, Inc., from *Experimental Cell Research* 122 (1979): 233.

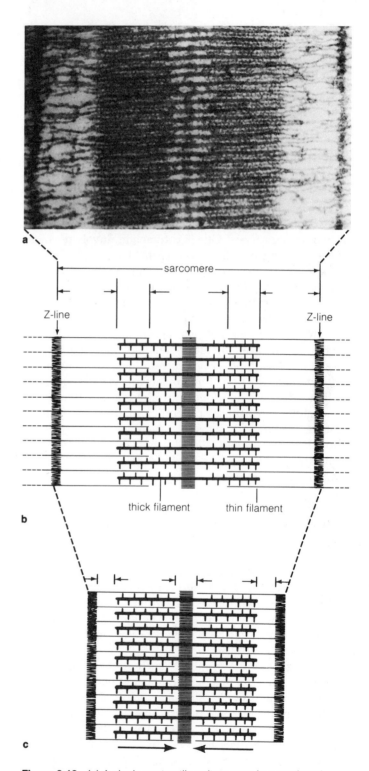

Figure 8-12 (**a**) A single contractile unit as seen in a sectioned striated muscle myofibril. × 36,000. Courtesy of H. E. Huxley. (**b**) The structures within a contractile unit in the relaxed state. (**c**) A contracted muscle myofibril unit. As muscle contracts, the thin filaments slide progressively over the thick filaments.

light and dark bands between the Z-lines are formed by overlapping filaments that run lengthwise within the myofibril. Connected to either side of the Z-lines and extending toward the center of each sarcomere are the *thin filaments*, each about 7–8 nanometers in diameter. At their ends, the thin filaments overlap with the heavier *thick filaments*, each about 14 nanometers in diameter. Faint *crossbridges* connect the thick and thin filaments where the two fiber types overlap.

In relaxed muscle, the thin filaments extend only about halfway along the thick filaments (see Fig. 8-12b). As muscle contracts the thin filaments slide progressively over the thick filaments (Fig. 8-12c). The effect is to decrease the width of each sarcomere and forcibly shorten the myofibrils, and the entire muscle, in the lengthwise direction.

The thin filaments of striated muscle cells were identified as actin microfilaments through the work of Hugh E. Huxley at the Medical Research Council Laboratory of Molecular Biology in England. In his research, Huxley took advantage of the fact that the two major proteins of muscle, actin and myosin, dissolve at different and distinct salt concentrations. Exposing striated muscle cells to salt concentrations that dissolve actin, but not myosin, caused the thin filaments to disappear. Conversely, the thick filaments disappeared if the muscle cells were treated with salt concentrations that dissolve myosin but not actin. Thus, the thin filaments were shown to contain actin, and the thick filaments myosin.

Subsequent work showed that individual actin molecules are spherical units about 5 nanometers in diameter. Under the correct conditions actin molecules can be in-

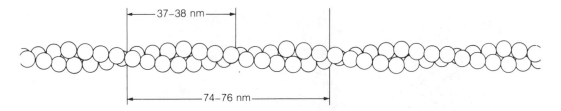

Figure 8-13 The linkage of actin molecules to form a microfilament. Each spherical unit in the double spiral is an actin molecule.

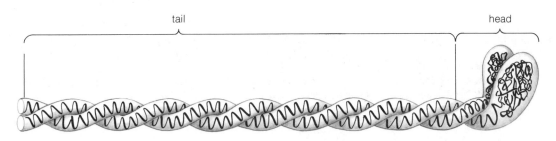

Figure 8-14 Structure of a myosin molecule. Two polypeptide chains wind together to form the double head and spiraled tail of the molecule.

duced to assemble into microfilaments in the test tube. In this assembly, actin molecules link up into two linear chains that twist around each other into a double spiral (Fig. 8-13). Sections of striated muscle and information from other sources show that a muscle thin filament consists of an actin double spiral of this type, containing a total of 300–400 individual actin molecules. Similar double spirals of actin molecules also form the microfilaments of nonmuscle cells.

Individual myosin molecules are long fibrous structures made up of two separate but identical polypeptide chains (Fig. 8-14). Each of the two peptides has a globular "head" and a long "tail." In the myosin molecule the heads align side by side and the tails twist into a double spiral, forming a structure with a double head followed by a doubly spiraled pigtail. The entire molecule is about 140 nanometers long. Biochemical work has shown that each myosin molecule can act as an ATPase enzyme, capable of splitting ATP into ADP + phosphate. This ATPase activity was found to be associated with the double head of the myosin molecule.

Evidence from a variety of physical and chemical sources was used by a number of investigators, most notably Hugh Huxley, to work out the probable arrangement of myosin molecules in the thick filaments of striated muscle. In the electron microscope, isolated thick filaments appear rough or "fuzzy" over a distance at either end, and smooth in the middle. Huxley proposed that this appearance is produced by an arrangement of individual myosin molecules in which the ATPase head units are concentrated at the two ends of a thick filament and the tails are directed toward the middle (Fig. 8-15). The head units, sticking out at the ends of a thick filament, give these regions their fuzzy appearance. In sectioned muscle, the head units are visible as the faint cross-bridges connecting the thick and thin filaments where the two fiber types overlap (see Fig. 8-12). Since only tails overlap at the center of a thick filament, this region appears smooth. Each thick filament contains hundreds of individual myosin molecules, packed together in this parallel, end-to-end array to produce a structure about 14 nanometers in diameter and more than 150 nanometers long.

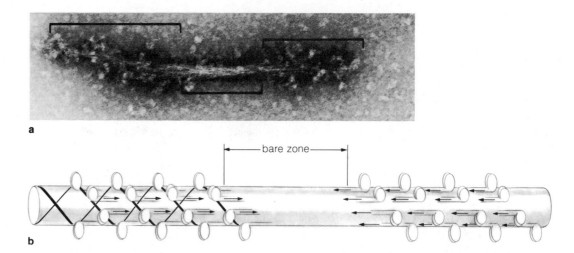

Figure 8-15 (**a**) Isolated thick filaments appear "fuzzy" at the ends (upper brackets) and bare in the middle (lower bracket). The fuzzy zones are believed to be produced by the myosin head units, which are concentrated near the ends of the thick filaments. Courtesy of H. E. Huxley, from *Science* 164 (1969): 1356, copyright 1969 by the American Association for the Advancement of Science. (**b**) The arrangement produces a bare zone at the middle of a thick filament that contains no head units (see text).

How Actin and Myosin Interact in Muscle Contraction

During muscle contraction the actin thin filaments slide progressively over the thick filaments, forcibly contracting the individual sarcomeres and the entire muscle. Huxley proposed that the force for this movement is provided by the myosin head units that stick out from the ends of the thick filaments. During contraction, according to Huxley, the myosin head units attach to the thin filaments, pull a short distance in an oarlike motion, and release. This attach, pull, and release cycle is repeated rapidly and continuously by all of the cross-bridges between actin and myosin, producing a smooth and continuous sliding between the thick and thin filaments.

In 1970 R. W. Lymn and Edwin W. Taylor of the University of Chicago proposed a model for the cross-bridge cycle that proceeds in several steps (Fig. 8-16). In the first step, a myosin head unit binds and splits a molecule of ATP (Figs. 8-16a and b). The products of ATP breakdown (ADP + phosphate) remain attached to the ATPase site of the myosin head unit. ATP breakdown induces a conformational change in the myosin, activating the actin bind-

ing site of the head unit (Fig. 8-16c). The activated myosin head unit then binds to an actin molecule in an adjacent thin filament (Fig. 8-16d). This binding releases the ADP and phosphate and induces a second conformational change in the head unit, forcibly swiveling it through an angle of about 45 degrees (Fig. 8-16e). Movement of the head through this angle pulls the attached actin filament through a distance of about 8–10 nanometers. At the end of the power stroke, the ATPase binding site of the head unit is reactivated (Fig. 8-16f). Binding a second molecule of ATP inactivates the actin-binding site, causing release of the head unit from the actin filament. At the same time, the head unit returns to the folding conformation at the start of the cycle, in which the angle of attachment of the head to its base is about 90° (Figs. 8-16g and h). The head unit now returns to its state at the beginning of the cycle, ready to break down the second ATP and attach to actin for the next power stroke.

The idea that ATP binding is necessary for release of the myosin cross-bridges at the end of their power stroke comes from observations made in muscle deprived of ATP. Muscle deprived of ATP, instead of relaxing, enters a *rigor* state in which all or most of the cross-bridges become firmly attached. This effect of ATP deprivation is the cause of *rigor mortis*, the stiffness of the limbs or body occurring

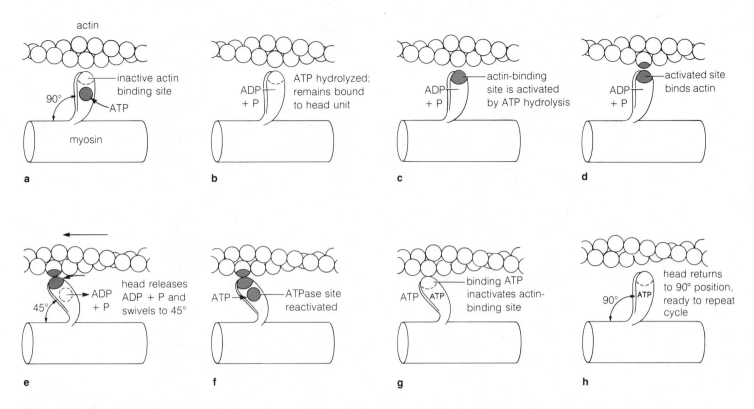

Figure 8-16 The attach-pull-release cycle of myosin cross-bridges that powers contraction in striated muscle (see text).

after death in animals. Examination of muscle cells in rigor by electron microscopy reveals that most of the myosin cross-bridges are attached to the thin filaments, locked at a 45° angle at the end of their power stroke. Addition of fresh ATP to muscles in rigor releases the cross-bridges and allows the muscle to relax.

The mechanism proposed for active filament sliding in striated muscle is therefore similar to the cycle believed to power microtubule sliding (compare Figs. 8-6 and 8-16). There are important differences in the two systems that reflect the distinct biochemical properties of myosin and dynein. Primary among these is the proposal, based on observations of dynein arms in bent, straight, and rigor flagella, that ATP binding and movement of the dynein arms of microtubules to the 45° position takes place *before* linkage to a binding site on an adjacent microtubule. Binding triggers forceful return of the arms to the 90° position, like releasing a cocked spring. As the cross-bridge swings

to the 90° angle, the attached microtubule is pushed along in a sliding motion. In muscle, the myosin cross-bridges move to the 45° position *after* linkage of myosin to actin.

Microfilament-Based Movement in Nonmuscle Cells

The highly individual biochemical properties of muscle actin and myosin provided investigators with the means to detect the two proteins in nonmuscle cells. One of the most useful tests developed from these properties is based on the ability of the myosin head unit to bind to actin microfilaments. The head units can be removed from the tails of myosin molecules by treating them with a protein-digesting enzyme that splits the molecule at a point just behind the double head. The separated head units retain

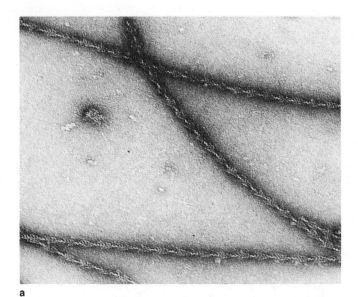

a

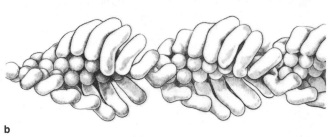

b

Figure 8-17 (a) The "arrowheads" produced when myosin head units react with actin microfilaments. × 138,000. Courtesy of J. A. Spudich, with permission from *Journal of Molecular Biology* 72 (1972): 619. Copyright by Academic Press, Inc. (London) Ltd. (b) The arrangement of myosin head units that produces the arrowheads on an actin microfilament.

their affinity for actin, and bind readily to actin microfilaments when added in purified form. In this binding, the myosin head units attach in a double spiral around the actin microfilaments, producing a pattern that appears as a row of arrowheads in the electron microscope (Fig. 8-17). Since microfilaments are the only cell structures that interact with purified myosin head units to produce the arrowhead pattern, the arrowheads are accepted as definitive evidence that actin microfilaments are present. The arrow-

head decoration technique also indicates the direction of movement powered by actin microfilaments. In both striated muscle and nonmuscle systems, active movement of the microfilaments is always in the direction pointed by the arrowheads.

Using the arrowhead decoration technique, Harunori Ishikawa and his colleagues at the University of Pennsylvania were able to detect actin microfilaments in a wide variety of nonmuscle cells in animals and protozoans. Later work by others revealed that actin microfilaments also occur in the algae, fungi, and higher plants. Biochemical analysis of the proteins isolated from these cell types confirmed that actin is present, usually in association with myosin. These discoveries have led to the concept that actin and myosin probably occur in all eukaryotic cells and produce motion in these locations by the same sliding mechanism as in striated muscle.

The actin molecules isolated from nonmuscle cells have proved to be closely similar to muscle actin. Some of these actins have been completely or partially sequenced, or at least analyzed for their content of the different amino acids. Comparisons from these studies indicate that few substitutions have occurred in the amino acids of actins in the evolutionary history of eukaryotes. For example, the actins from species as evolutionarily distant as the rabbit and *Acanthamoeba*, a protozoan, differ only in about 6 percent of their amino acids.

The nonmuscle myosins have proved to be more difficult to study by amino acid analysis or sequencing techniques. However, although the myosins from the different sources vary more widely than actins, almost all apparently possess the same three-dimensional structure as striated muscle myosin, with two globular head units set at one end of a long, rodlike double pigtail. All of these nonmuscle myosins share, with muscle myosin, the ability to bind actin molecules or microfilaments from any source and to break down ATP rapidly when bound to actin. Whether these nonmuscle myosins exist individually, or associate together into structures resembling the thick filaments of muscle cells has not yet been established.

The actin and myosin molecules of cells are responsible for a variety of cellular movements, including ameboid motion, cytoplasmic streaming in both plant and animal cells, and the movements dividing the cytoplasm in animal cell division. Like microtubules, actin microfilaments also act as supportive or cytoskeletal elements in some cells, as in the network of microfilaments supporting the cytoplasm in some mammalian cells (see Fig. 1-13b).

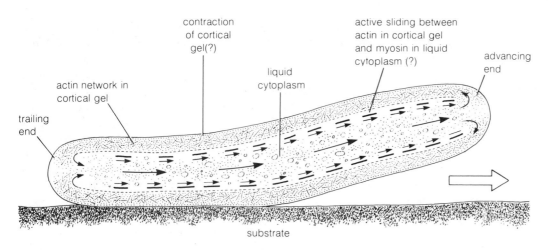

Figure 8-18 A model for the production of ameboid motion by microfilaments. At the trailing end of the cell, the microfilaments of the cortical tube continuously disassemble. At the advancing end, the tube continuously extends through the assembly of actin molecules into microfilaments. Cytoplasm moves through the cell interior by contraction of the cortical tube or interactions between microfilaments in the tube and myosin-linked structures in the liquid interior (see text).

Microfilament-Based Movement in Ameboid Cells

Of the motile systems based on microfilaments, the mechanism generating ameboid movement provides one of the best examples of the patterns in which microfilaments produce motion in nonmuscle cells. Ameboid movement, which we most often associate with protozoans like *Amoeba proteus*, actually occurs throughout the fungi and animal kingdoms. In animals, ameboid motion occurs regularly during embryonic development. Certain cells, such as the white blood cells of our bloodstream, retain the capacity for ameboid movement in adults. Other cell types, although nonmotile in their normal locations in adult tissues, may migrate by ameboid motion if grown in tissue cultures outside the body.

Ameboid movement is apparently simple enough when observed in the light microscope. At some point along the cell margin, a broad lobe forms and pushes outward. The cytoplasm of the cell streams actively into the lobe, eventually carrying with it the cytoplasmic organelles and the nucleus. Close examination of an ameboid cell shows that the cytoplasm just under the cell surface, except at the tip of an advancing lobe, is semisolid or gel-like, forming a sort of cytoplasmic tube through which the more liquid internal cytoplasm folows to reach the advancing tip.

Actin and myosin were linked to ameboid motion by a series of important discoveries in the late 1960s and early 1970s. In 1970, Thomas D. Pollard and his coworkers at Harvard University showed that the cytoplasm extracted from ameboid cells contains microfilaments that form the typical arrowheads when reacted with myosin head units. Chemical studies showed that myosin was also present in the isolated ameba cytoplasm.

Thin sections of ameboid cells reveal that most of the actin microfilaments are concentrated in a network just under the plasma membrane (diagrammed in Fig. 8-18). This layer corresponds to the cortical "tube" of gelled or semisolid cytoplasm visible just under the surface of ameboid cells. Although myosin is known to be present, thick filaments are usually not visible in ameboid cells. This may be because myosin occurs in the form of individual molecules, rather than in association as thick filaments. These structural studies were complemented by biochemical experiments demonstrating that the chemical requirements for movement in ameba are essentially the same as in striated muscle. In particular, ATP is required for ameboid motion; all movement stops if cells are treated with chemicals that interfere with the supply of ATP.

These observations, taken together, clearly indicate that actin and myosin are responsible for generating the force for ameboid motion. While there is some disagreement on details, most cell biologists agree that the semisolid cytoplasmic tube through which the cytoplasm flows in ameboid motion is formed by actin microfilaments, which assemble into a network to create the gel. At the tip of an advancing lobe actin continuously assembles into microfilaments around the margins of the tube, lengthening the tube at this end. At the trailing end of the cell, actin microfilaments continuously disassemble, adding to the liquid cytoplasm flowing into the tube at this end (see Fig. 8-18). Thus the cytoplasmic tube constantly grows at its advancing end, as liquid cytoplasm flows into the tip, and disassembles into liquid cytoplasm at its rear.

There are two major ideas about how the force is generated to make the more liquid cytoplasm in the interior of the cell flow through the tube. According to one idea, the actin microfilaments toward the rear of the tube slide over each other through the action of myosin cross-bridges, causing the tube to narrow and contract at this end. The contraction forces the more liquid cytoplasm to move passively through the tube toward the tip, in an action similar to squeezing a toothpaste tube. According to the alternate idea, actin microfilaments and myosin molecules linked to structures in the liquid part of the cytoplasm slide over microfilaments held rigidly in the tube. Thus, in this view, the sliding between actin and myosin responsible for the cytoplasmic flow occurs at the boundary between the gelled and liquid cytoplasm. There is no evidence ruling out either idea at the present time, and it is entirely possible that both mechanisms move the liquid cytoplasm through the cortical tube in ameboid motion.

The generation of motion by actin microfilaments in all nonmuscle systems is believed to operate by similar mechanisms. In all of these systems, microfilaments are considered to slide over other microfilaments anchored to the plasma membrane or to a semisolid gel created by a microfilament network. Force for the sliding is produced by myosin cross-bridges between the actin microfilaments that attach, pull, and release in a cycle powered by ATP breakdown.

Thus both microtubules and microfilaments move by a sliding mechanism produced by active cross-bridges. In both systems, ATP breakdown induces conformational changes of cross-bridges between the motile elements. The conformational changes cause the cross-bridges to move in an oarlike motion that pushes one microtubule or microfilament with respect to another. In this way, the proteins of microtubules and microfilaments act as devices that change the chemical energy released by ATP breakdown into the mechanical energy of movement.

Both microtubules and microfilaments appear single or in loosely organized groupings in the cytoplasm of all eukaryotic cells, where they move individual cell parts or regions of the cytoplasm. Both structures also occur in highly organized, complex arrangements: microtubules in the 9 + 2 system of flagella, and microfilaments in the sarcomeres of striated muscle. The organization of these complex structures is probably the most efficient arrangement of microtubules and microfilaments for the generation of rapid and powerful movement.

The parallel existence of the two motile systems, one based on microtubules and one on microfilaments, raises one of the fundamental questions of cell biology. Why have two separate but equivalent mechanisms, based on entirely different proteins, evolved and persisted in all eukaryotic cells to accomplish the same end of cell motility? The answer must lie in the distinct characteristics of the assembly and sliding reactions of microtubules and microfilaments, which make one or the other indispensible for the most efficient production of motion at certain times and places in the cell. With a very few exceptions, one or the other of the two mechanisms is responsible for all the movements of eukaryotic organisms. (The motile system of bacterial cells, which is based on a unique system for producing motion, is outlined in Supplement 8-2.)

Suggestions for Further Reading

Cohen, C. 1975. The protein switch of muscle contraction. *Scientific American* 233:36–45 (November).

Dustin, P. 1978. *Microtubules*. Springer-Verlag: New York.

Lazarides, E. and J. P. Revel. 1979. The molecular basis of cell movement. *Scientific American* 240:110–13 (May).

Murray, J. M. and A. Weber. 1974. The cooperative action of muscle proteins. *Scientific American* 230:58–71 (February).

Satir, P. 1974. How cilia move. *Scientific American* 231:44–52 (October).

Smith, D. S. 1972. *Muscles*. Academic Press: New York.

Stephens, R. E. and K. T. Edds. 1976. Microtubules: structure, chemistry and function. *Physiological Review* 56:709–75.

Wolfe, S. L. 1981. *Biology of the cell*. 2nd ed. Wadsworth: Belmont, Calif.

For Further Information

Centrioles in cell division, Chapter 11 and Supplement 11-4
Microfilaments in cytoplasmic division, Chapter 11
Microtubules and microfilaments in cell membranes, Chapter 4
Microtubules and the spindle, Chapter 11

Questions

1. Compare the structure of microtubules and microfilaments.

2. In what ways are the molecular subunits of microtubules and microfilaments similar? In what ways are they different?

3. What kinds of cellular motion are generated by microtubules? By microfilaments?

4. Outline the 9 + 2 system of microtubules and connecting elements in flagella.

5. What experiment demonstrated that the arms attached to the microtubules of the 9 + 2 system have ATPase activity?

6. What experiment demonstrated that microtubules slide past each other during flagellar motion? What experiment showed that the sliding is active?

7. How is the cross-bridge cycle producing microtubule sliding believed to work? Where and how does ATP enter the system? What property of proteins is responsible for transforming the chemical energy of ATP into the mechanical energy of movement?

8. What are alpha- and beta-tubulin? What relationships do these proteins have to the subunits seen in microtubule walls in the electron microscope?

9. How are actin microfilaments arranged in striated muscle cells? What is the structural relationship between actin and myosin in striated muscle?

10. Define myofibril, muscle fiber, and sarcomere. Why is voluntary muscle called striated muscle?

11. How were the thick and thin filaments of striated muscle identified as myosin and actin?

12. What happens to the thick and thin filaments when muscle contracts?

13. What is the basis for the "arrowhead" decoration technique? What significance has this technique had for the identification of actin in nonmuscle cells?

14. How do actin molecules assemble into microfilaments?

15. How is the cross-bridge cycle producing microfilament sliding believed to work? Where and how does ATP enter the mechanism?

16. In what ways does the cross-bridge cycle of microfilaments resemble the cycle in microtubules? In what ways is it different?

17. How are microfilaments believed to operate in ameboid motion?

Supplement 8-1: Centrioles, Basal Bodies, and the Generation of Flagella

The highly organized 9 + 2 microtubule system of flagella and cilia arises directly from a structure of equivalent complexity, the *centriole*. Centrioles are small, barrel-shaped structures, about 0.2 micrometer in diameter and 0.3 to 0.5 micrometer long, that occur in the cytoplasm of all eukaryotic cells capable of giving rise to flagella or cilia. Cross-sections show that centrioles contain nine microtubule *triplets* arranged in a circle (Fig. 8-19). Each triplet contains one complete microtubule. Attached to one side of the complete tubule are two incomplete microtubules. These microtubules join in a row to the complete subtubule to form the triplet. The microtubules are identified as the *A*, *B* and *C subtubules* of the triplet, numbered in order from the complete A subtubule (see Fig. 8-19). The triplets are cocked at an angle of about 30 to 40° to the circumference of the centriole, with the A subtubule of each set in a position closest to the center of the circle. The structure and arrangement of each centriole triplet is thus analogous to the doublets of the 9 + 2 complex, except that three, rather than two microtubular elements are present in the centriole triplets.

There are no central microtubules in centrioles equivalent to the central singlets of flagella. Centrioles appear to be almost structureless in the center except near one end, where a density pattern resembling a cartwheel is present. The cartwheel has a central hub with nine spokes that radiate from the center and connect with the A subtubule of each peripheral triplet (see Fig. 8-19). Other dense elements link the triplets around the circumference of the centrioles in a pattern that varies depending on the source of the centriole.

During generation of a flagellum, the A and B subtubules of each triplet lengthen at the end of the centriole opposite the cartwheel structure, giving rise to the two subtubules of a doublet of the 9 + 2 complex. The central singlets of the 9 + 2 complex arise in the boundary region between the centriole and the growing flagellum without any connections to microtubules of the centriole. In most flagella, no structures arise from the C subtubules of the centriole triplets.

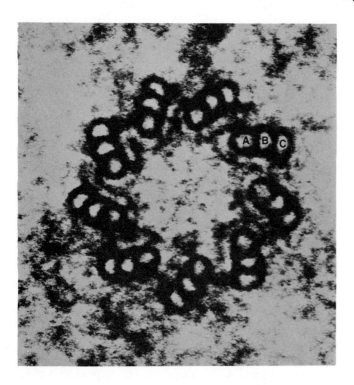

Figure 8-19 A centriole in cross section. *A, B, C,* the three subtubules of a centriole triplet. × 270,000. Courtesy of E. de Harven, from *The Nucleus,* eds. A. J. Dalton and F. Haguenau, Academic Press, Inc., 1968.

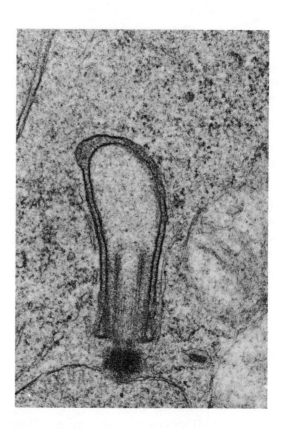

Figure 8-20 An early stage in the development of a flagellum in a mosquito. The elongating 9 + 2 complex is encased in a cuplike vesicle that will eventually fuse with the plasma membrane. × 62,000. Courtesy of D. M. Phillips, from *Journal of Cell Biology* 44 (1970): 243 by copyright permission of The Rockefeller University Press.

As the developing 9 + 2 complex elongates, one of two patterns may be followed in generation of the flagellar membrane. If the centriole is located in deeper layers of the cytoplasm, a vesicle appears over the advancing margin of the developing 9 + 2 complex. As the microtubules elongate, the vesicle spreads over the tip of the 9 + 2 complex (Fig. 8-20). The outermost vesicle membrane eventually reaches and fuses with the plasma membrane. The innermost membrane continues to extend outward and forms the flagellar membrane. Alternatively, if the centriole giving rise to the 9 + 2 complex lies just under the plasma membrane, the growing microtubules simply advance into a gradually lengthening extension of the membrane.

When development is complete, the centriole giving rise to a flagellum remains attached to the microtubules of the 9 + 2 complex at the base of the flagellum (Fig. 8-21). In this position, the centriole is frequently termed the *basal body* of the flagellum.

Centrioles apparently have little or no function in the generation of motion in cilia or flagella because isolated flagella broken from their basal bodies can be induced by addition of ATP to beat in a normal wave pattern. However, the centrioles persisting as basal bodies in mature flagella and cilia may serve to anchor the entire structure in the cytoplasm (the duplication and division of centrioles during mitosis is discussed in Chap. 11 and Supplement 11-4).

Supplement 8-2:
Bacterial Flagella

Many types of bacteria swim by means of hairlike appendages that are also called flagella. These flagella are com-

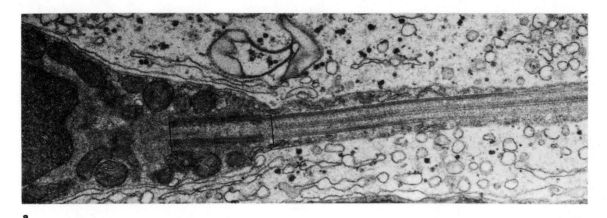

a

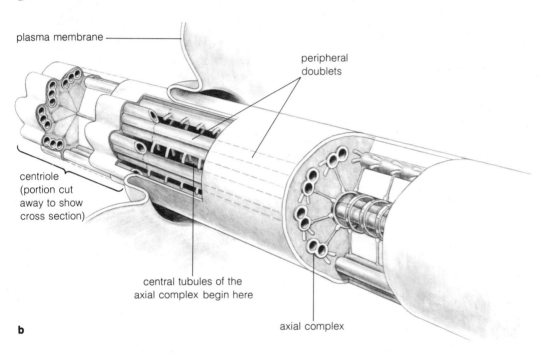

plasma membrane

peripheral
doublets

centriole
(portion cut
away to show
cross section)

central tubules of the
axial complex begin here

b

axial complex

Figure 8-21 (**a**) The centriole persisting as the basal body (bracket) in the flagellum of a frog sperm.
× 19,000. Courtesy of B. R. Zirkin. (**b**) The relationship of the microtubules of the basal body and the 9 + 2
complex in a flagellum.

pletely unrelated to eukaryotic flagella and generate motion by a mechanism completely without parallels in eukaryotes.

Bacterial flagella (see Fig. 1-5) consist of slender fibers of protein about 15 to 20 nanometers in diameter and several micrometers in length. At these dimensions, the total diameter of the structure is smaller than a single microtubule in a eukaryotic flagellum. A bacterial flagellum is built up from several intertwined, spiral chains of subunits of a protein called *flagellin*. Each bacterial species has its own variety of flagellin. In *Escherichia coli*, the best studied spe-

cies, the individual flagellin subunits are about the same size as a tubulin molecule. The flagellin proteins, however, are completely unrelated to either tubulin or actin.

Each flagellum is attached to the bacterial cell wall and plasma membrane by a basal structure consisting of a *hook* and *shaft* (Figs. 8-22 and 8-23). Depending on the bacterial type, the shaft may bear either one or two pairs of rings. The outer ring or ring pair is believed to anchor the shaft in the bacterial cell wall; the inner ring or ring pair is embedded in the underlying plasma membrane.

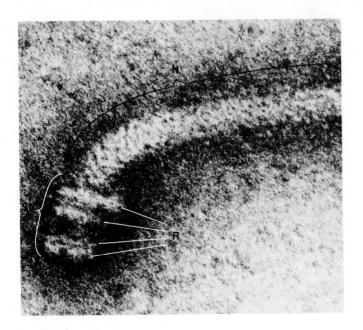

Figure 8-22 A flagellum isolated from the bacterium *E. coli* showing the basal structure. *H*, hook; *Sh*, shaft; *R*, rings. × 46,000. Courtesy of M. L. De Pamphlis and Julius Adler, from *Journal of Bacteriology* 105 (1971): 384.

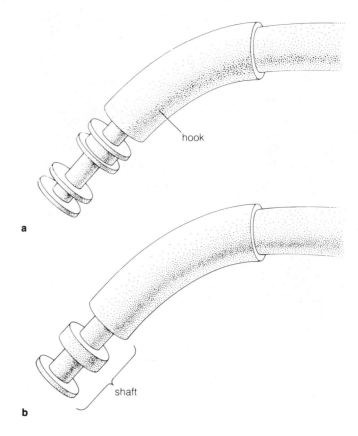

Figure 8-23 The hook, shaft, and ring structures at the base of bacterial flagella (**a**) in gram-negative bacteria and (**b**) in gram-positive bacteria.

In the natural state the long fiber of the flagellum extends from the basal structure in a fixed, open spiral resembling a corkscrew. In some bacteria, as in *E. coli*, all the flagella join together to form a corkscrew-shaped bundle that extends from one end of the cell (*E. coli* has a total of 6 flagella in its bundle). When viewed under the light microscope, the flagellar bundle in swimming cells seems to flex rapidly, pushing the cell through the medium with the bundle trailing behind.

Recent studies indicate that, instead of undulating in wavelike beats that pass successively from base to tip as in eukaryotic flagella, the fixed spiral of a bacterial flagellum actually rotates like a propellor. This motion can be duplicated by opening a paper clip and winding it into a spiral around a pencil. If one end of the wire is bent so that it lies in the axis of the spiral, twirling this end between your fingers will make it appear as if a series of waves passes from one end of the wire to the other.

To produce this screwlike motion, each flagellum must rotate at its base, just as the wire must be twirled in the paper clip model. Recent work by Michael R. Silverman and Melvin I. Simon at the University of California at San Diego demonstrated that this is actually the case. In one of their experiments, Silverman and Simon managed to tether bacterial cells to a glass slide by attaching the tips of their flagella to the glass. When the flagella were prevented from rotating in this way, the bacterial cells rotated instead.

A variety of experiments have shown that rotation of bacterial flagella is driven by an H^+ gradient across the plasma membrane rather than ATP breakdown. Flagellar motility continues at normal levels in cells in which ATP synthesis is blocked as long as electron transport is maintained in the cells. Agents that uncouple ATP synthesis from electron transport, presumably by making the plasma membrane leaky to H^+ ions (see p. 126), completely inhibit flagellar motility even though ATP reserves remain high and available. In addition, Howard C. Berg of the University of Colorado has shown that adding an acid or base to the medium around bacterial cells, so that an H^+

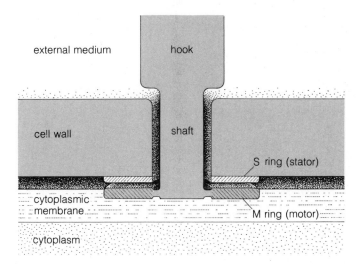

Figure 8-24 H. C. Berg's model for the biological motor powering flagellar rotation in bacteria (see text). Courtesy of H. C. Berg; redrawn from How bacteria swim, by H. C. Berg, *Scientific American* 232 (1975): 36. Copyright © 1975 by Scientific American, Inc. All rights reserved.

gradient is created between the inside and outside of the bacterial cell, can restart flagellar motion in bacterial cells that have exhausted their energy reserves and stopped moving. All of these observations indicate that an H^+ ion gradient directly powers the flagellar rotation.

Berg has used these observations as the basis for a novel hypothesis for flagellar motion in bacteria. Berg proposes that the basal structures of a bacterial flagellum, the shaft and rings, form a "biological motor" that rotates the flagellum. According to his hypothesis, one of the rings (the "S" ring in Fig. 8-24) is fixed on the membrane surface but is free to slip on the surface of the shaft. This ring acts as a "stator" in which the shaft rotates. A second ring (the "M" ring in Fig. 8-24), fixed on the shaft, is embedded in the membrane. Flow of H^+ ions across the M ring causes it to take on a charge. Through interaction between the charge on the M ring and charged groups on the S ring (either attraction between like charges, or repulsion between unlike charges), the M ring is induced to rotate. Rotation of the M ring, since it is fixed to the shaft, causes rotation of the entire flagellum. The M ring, in the model, is thus the biological equivalent of the armature in an electric motor.

While there are some skeptics, Berg's idea has attracted wide attention and, so far, has not been contradicted by experimental results. If rotation of bacterial flagella is actually driven by a biological motor of the kind Berg proposes, it is unique among the motile systems of the world's organisms.

Suggestion for Further Reading

Berg, H. C. 1975. How bacteria swim. *Scientific American* 233:36–44 (August).

FIVE

Cell Synthesis and Growth

9

Cell Synthesis I: The Role of the Nucleus

The major use of the cellular energy captured in mitochondria and chloroplasts is in the synthesis of the molecules required for cell growth. This growth occurs by the assembly of all the classes of cellular molecules—carbohydrates, lipids, proteins, and nucleic acids—powered ultimately by the ATP energy captured in chloroplasts and mitochondria.

The synthetic activity of cells follows a closely similar pattern in all living things, from the simplest bacteria to the most complex plants and animals. In all these organisms, the directions for protein synthesis are coded into DNA molecules in the cell nucleus or nucleoid (in a few viruses, the information required for synthesis is coded in RNA molecules instead). The coded information is transcribed in the nuclear region into RNA "messages" that are copied from the DNA. The RNA messages are then transferred into the cytoplasm, where they direct the assembly of proteins, including both enzymes and structural proteins. The enzymes, in turn, speed and direct the synthesis of the remaining molecules of the cell, including carbohydrates, lipids, a wide variety of smaller molecules, and even the nucleic acid information molecules themselves.

How Cell Synthesis Works: A Brief Overview

Figure 9-1 outlines the mechanism of cellular synthesis. The information required for protein synthesis in eukaryotes is coded into different sequences of the four nucleotides making up the DNA (deoxyribonucleic acid) molecules of the nucleus. Two primary kinds of information are stored in the DNA sequences. The directions for making proteins are spelled out by a code that uses the four DNA nucleotides, three at a time, in all possible combinations. Each three-nucleotide codeword stands for an amino acid (see Fig. 10-3). Reading the codewords in sequence along the DNA spells out the sequence of amino acids in a protein. These protein-encoding regions are duplicated into RNA copies called *messenger RNAs* (*mRNAs*) that carry the directions for making proteins to the cytoplasm.

Other DNA regions store the directions for making two types of accessory RNAs that act in parts of the protein synthesis mechanism. One, *ribosomal RNA* (*rRNA*), forms a part of the ribosomes, the RNA-protein structures that assemble amino acids into proteins in the cytoplasm. The second, *transfer RNA* (*tRNA*) binds directly to amino acids during protein synthesis and provides the necessary link between the nucleic acid code and the amino acid sequence of proteins.

Synthesis of the mRNA, rRNA, and tRNA copies of DNA, called *transcription*, occurs within the cell nucleus. Following transcription, the RNA copies pass through the nuclear envelope and enter the cytoplasm. Messenger RNA, once in the cytoplasm, attaches to one or more ribosomes. The ribosomes then assemble amino acids into proteins, using the information carried in the attached mRNA as a guide. In this synthesis, called *translation*, a ribosome starts at one end of an mRNA molecule and then moves along the sequence of nucleotides in the mRNA until it reaches the other end. As it moves along the mRNA, it assembles amino acids into a gradually lengthening polypeptide according to the directions coded into the mRNA. At the end of the message, both the ribosome and the completed protein molecule detach from the mRNA. At any instant, several ribosomes may be at different places on a single mRNA molecule, engaged in reading the message and assembling protein chains. Ribosomes and mRNA may recycle through this mechanism many times. In this way, each mRNA molecule may serve as a template for hundreds of identical protein molecules.

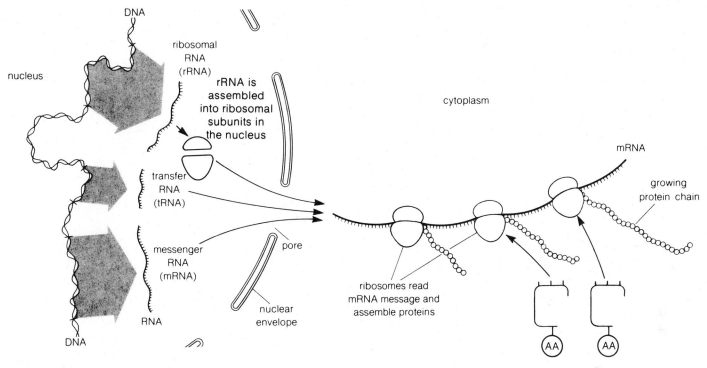

Figure 9-1 The major steps in cellular synthesis (see text).

Transfer RNA molecules function as the "dictionary" in the translation mechanism. Each kind of tRNA corresponds to one of the 20 amino acids used in protein synthesis. The different tRNAs are attached to their respective amino acids by enzymes that can recognize both a particular amino acid and the tRNA (or tRNAs) corresponding to that amino acid. As a result of the activity of these enzymes, each amino acid is linked to a specific kind of tRNA. The tRNAs, in turn, are capable of recognizing and binding to the coding triplets in mRNA specifying their attached amino acid.

Binding of tRNAs and their attached amino acids to the mRNA coding triplets takes place on ribosomes. As a ribosome encounters an mRNA coding triplet specifying a given amino acid, the tRNA carrying that amino acid binds to the ribosome. This binding places the amino acid in its correct location in the protein chain growing from the ribosome (Fig. 9-2). The ribosome then moves to the next mRNA coding triplet, causing the next tRNA–amino acid complex specified by the code to bind. As each successive amino acid arrives at the ribosome it is split from

its tRNA carrier and linked into the gradually lengthening protein chain. The process repeats until the ribosome reaches the end of the message and completes assembly of the protein.

The cell structures and mechanisms functioning in transcription and translation are taken up in the two chapters of this unit. This chapter concentrates on RNA transcription and the nuclear structures carrying out transcription. Protein synthesis, and the structure of the cytoplasmic organelles engaged in protein synthesis are covered in the following chapter.

DNA, RNA, and Transcription

DNA and RNA Structure

The two nucleic acids, DNA and RNA, are built up from nucleotides (see Information Box 9-1) linked together into

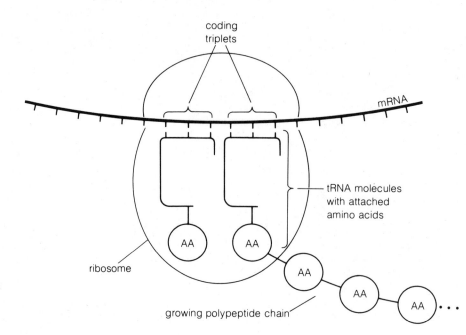

coding triplets

mRNA

tRNA molecules
with attached
amino acids

AA   AA

ribosome

AA

AA

AA . . .

growing polypeptide chain

Figure 9-2 The tRNA-mRNA pairing mechanism at the ribosome that enters successive amino acids into a growing polypeptide chain (see text).

long chains. DNA (Fig. 9-3a) is constructed from nucleotides containing the sugar *deoxyribose*, and one of the four nitrogenous bases *adenine* (A), *thymine* (T), *guanine* (G), and *cytosine* (C). RNA (Fig. 9-3b) contains nucleotides built up from the sugar *ribose* and one of the four nitrogenous bases *adenine* (A), *uracil* (U), *guanine* (G), and *cytosine* (C). The nucleotide chains of DNA and RNA thus differ in two ways: (1) DNA contains the base thymine but not uracil, and RNA contains uracil but not thymine; and (2) DNA contains the sugar deoxyribose, and RNA contains the sugar ribose (the ribose and deoxyribose sugars differ only in the presence or absence of a single oxygen atom; see Fig. 9-3). Otherwise, the nucleotide chains of the two nucleic acid molecules are identical.

The nucleotides are linked into nucleic acid chains by covalent bonds that extend between the phosphate group of one nucleotide and the sugar of the next nucleotide in the chain, as shown in Figure 9-3. These alternating sugar-phosphate-sugar-phosphate bonds form the "backbone" of a nucleic acid molecule. The nitrogenous bases, as shown in Figure 9-3, extend outward from the sugar-phosphate backbone.

DNA normally occurs in cells as a *double helix*, a molecule containing two nucleotide chains twisted together in a double spiral (Fig. 9-4). RNA, in contrast, is usually found as a single nucleotide chain; in some regions the RNA chain may fold back and twist upon itself to form a

double helix. Hybrid double helices, containing one RNA and one DNA nucleotide chain twisted together, may also form as temporary structures during transcription in the cell nucleus.

The arrangement of the two sugar-phosphate backbone chains in a spiral around the outside of the DNA molecule leaves a cylindrical space that extends through the central axis of the molecule. The nitrogenous bases attached to the sugar-phosphate backbones extend into this space, forming pairs consisting of one base from each chain. The space is just wide enough to allow a purine to pair with a pyrimidine base (purine bases, with two carbon rings, are about twice as wide as pyrimidine bases; see Information Box 9-1). Purine-purine pairs are too wide, and pyrimidine-pyrimidine pairs too narrow to fit this space.

Further pairing restrictions come from the shapes of the bases and the possibilities for hydrogen bonding between purine-pyrimidine base pairs. The shapes of the bases allow adenine and thymine to pair together like pieces of a jigsaw puzzle, and to form two hydrogen bonds that stabilize the arrangement (Fig. 9-5). Similarly, cytosine and guanine fit together perfectly, forming a pair that is stabilized by three hydrogen bonds. The other purine-pyrimidine pairing possibilities, such as adenine with cytosine, do not work: the pieces of the puzzle do not fit together, and stabilizing hydrogen bonds cannot form.

Information Box 9-1   The Nucleotides

The nucleotides of DNA and RNA consist of a nitrogenous base, a 5-carbon sugar, and from one to three phosphate groups:

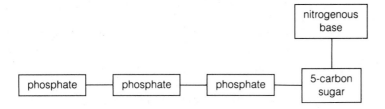

The nitrogenous bases of DNA and RNA are either purines or pyrimidines. The nucleotides of DNA contain any one of the bases adenine (A), thymine (T), guanine (G), and cytosine (C); the nucleotides of RNA contain any one of the bases adenine (A), uracil (U), guanine (G), and cytosine (C):

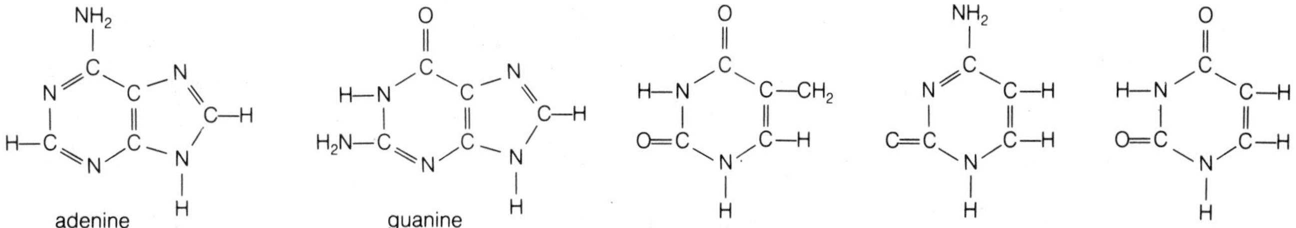

adenine   guanine   thymine   cytosine   uracil

purines   **pyrimidines**

These nitrogenous bases combine with the 5-carbon sugar deoxyribose in DNA, and ribose in RNA:

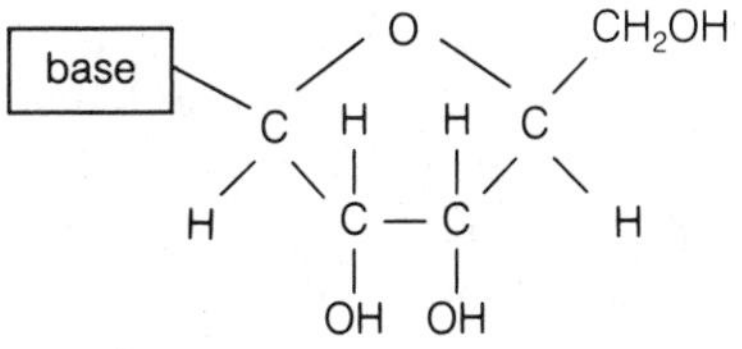

ribose

Adding one, two, or three phosphate groups to the base sugar units forms the mono-, di-, and triphosphates of the nucleotides.

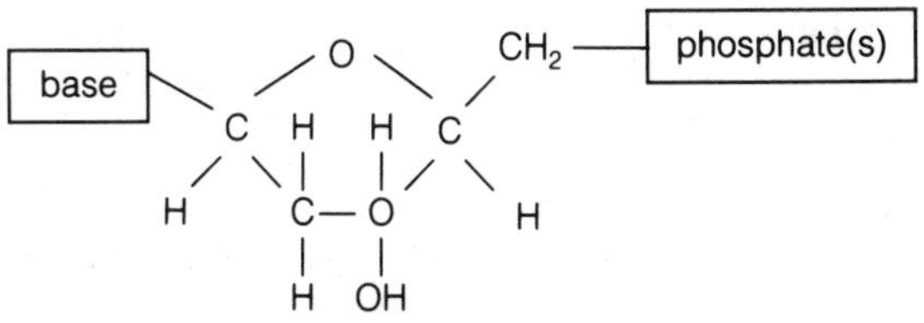

deoxyribose

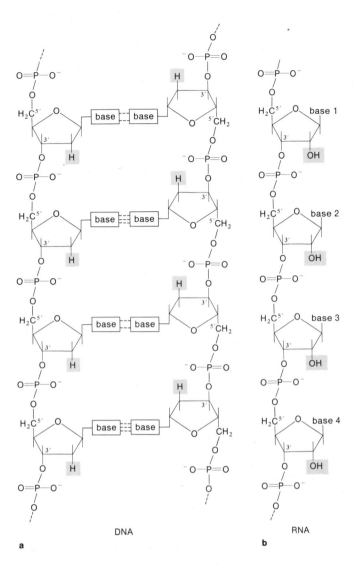

DNA   RNA

a   b

Figure 9-3 Linkage of nucleotides into the nucleotide chains of DNA (**a**) and RNA (**b**). In DNA, any one of the four nitrogenous bases adenine (A), thymine (T), cytosine (C), or guanine (G) may be bound at the positions marked as bases. In RNA, any one of the bases A, G, C, and uracil (U) occur at these sites. The deoxyribose and ribose sugars of DNA and RNA differ only in the presence or absence of a single oxygen atom at the shaded position; this oxygen is absent in DNA.

The shapes and hydrogen-bonding capabilities of the nitrogenous bases thus limit the purine-pyrimidine pairing in the center of the DNA molecule to adenine-thymine and guanine-cytosine pairs. These pairs are seen from the edge in Figure 9-4a as horizontal lines; when viewed from one end of the molecule, the pairs appear as shown in Figure 9-5. One complete turn of the double helix, which is 3.4 nanometers long, includes ten of the base pairs.

Another important feature of DNA structure is that the two nucleotide chains of the double helix run in opposite directions, or are *antiparallel* as this feature is termed. The antiparallel nature of DNA is most easily understood if the two nucleotide chains of a DNA molecule are unwound and laid out flat as in Figure 9-6. The phosphate linkages that bind the nucleotides together in the two chains extend between the 5'-carbon of one deoxyribose sugar and the 3'-carbon of the next one in line. (The carbon atoms of the sugars are numbered with primes as 1', 2', 3' . . . to distinguish them from the carbons of the bases, which are written without primes; see Fig. 2-23.) Note that if the phosphate linkages holding the nucleotide chain on the left together are traced from the bottom to the top of Fig. 9-6, they extend from the 5'-carbon of the sugar below to the 3'-carbon of the sugar above each linkage. On the other chain, the phosphate linkages run in the opposite direction: tracing from the bottom to the top of the figure shows that the linkages extend from the 3'-carbon of the sugar below to the 5'-carbon of the sugar above each linkage. Thus the chains are parallel but run in opposite directions. This feature of DNA structure has significance for the progress of both DNA duplication and RNA transcription.

Two kinds of forces hold the DNA double helix together. One is the hydrogen bonding between the base pairs in the interior of the molecule. Although individually relatively weak, the combined effect of the hydrogen bonds along the double helix forms a stable structure if the DNA molecule is more than about 10 base pairs in length. The second stabilizing force is provided by hydrophobic associations between the paired bases in the interior of the molecule. In this region, the bases, which are primarily nonpolar in character, pack tightly enough to exclude water and form a stable, nonpolar environment.

The base pairing rules are of great significance for the activities of DNA in the storage and transfer of coded information. Because of the requirement that adenine must pair with thymine, and cytosine with guanine, a sequence in one chain, once fixed, is compatible with only one sequence in the opposite chain. For example, if the sequence

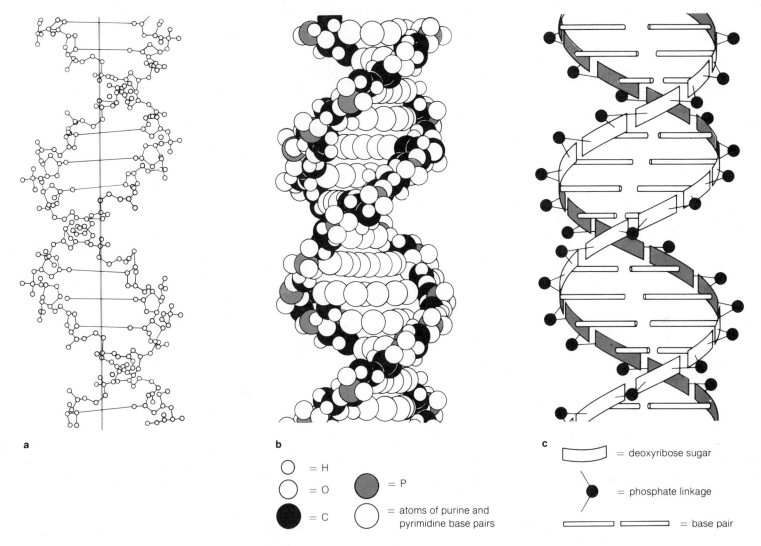

Figure 9-4 The DNA double helix. In (**a**) atoms are shown as circles and bonds as short straight lines. The bases, which lie in a flat plane, are seen on edge from this viewpoint and are shown as straight horizontal lines running between the backbone chains. Redrawn from an original courtesy of M. H. F. Wilkins, from *Science* 140 (1963): 941. Copyright 1963 by the American Association for the Advancement of Science. In (**b**) the spaces occupied by the atoms of the double helix are shown as spheres. (**c**) Diagram showing the arrangement of sugars, phosphate groups, and base pairs in the DNA double helix.

in one chain is A-T-T-G-C-G-A-C-A-T (A = adenine, T = thymine, G = guanine, C = cytosine), the opposite chain is restricted to the sequence T-A-A-C-G-C-T-G-T-A (Fig. 9-7). In the parlance of molecular biologists, the two sequences of the two chains are said to be *complementary*. Complementary base pairing provides the basis for information transfer from DNA to RNA in transcription (see below), and also for the exact duplication of DNA mole-

cules as a part of cell division (DNA duplication, called *replication*, is discussed in detail in Chap. 11).

The structure of DNA was discovered in 1953 by James D. Watson, an American, and Francis H. C. Crick, an Englishman, both working at Cambridge University, and Maurice H. F. Wilkins, another Englishman working at Kings College in London. Watson, Crick, and Wilkins were awarded the Nobel Prize for their work in 1962. Few

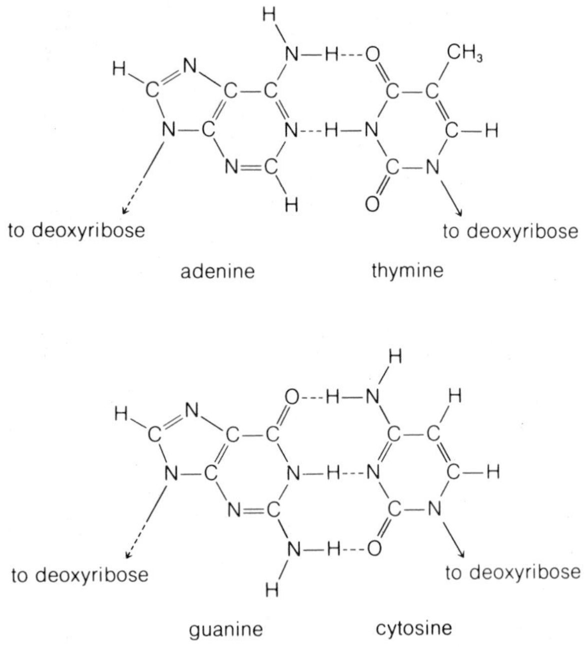

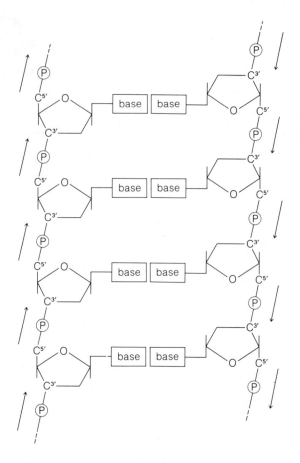

Figure 9-5 The two kinds of base pairs formed in DNA. Top: An adenine-thymine (A-T) base pair. Bottom: A guanine-cytosine (G-C) base pair. The pattern of hydrogen bonds in the two base pairs is shown by dotted lines. The total width of either base pair just fills the space between the sugar-phosphate backbone chains wound into a double helix.

Figure 9-6 The antiparallel arrangement of the two nucleotide chains in DNA. The arrows point in the 5' → 3' direction.

discoveries in the history of biology have provided as much insight into the molecular nature of life or have had as great an impact on research in all biological fields (the research leading to the discovery of DNA structure and establishing DNA as the informational and hereditary molecule of cells is described in Supplement 9-1).

Transfer of Information from DNA to RNA: Complementarity and Transcription

The fact that the two nucleotide chains of a DNA molecule are complementary—that is, that the sequence in one chain fixes the sequence in the opposite chain—provides the basis for transcription because RNA copies made from the nucleotide chains of DNA molecules must also obey the rules of complementarity. The mechanism works like this. Of the two nucleotide chains in a DNA molecule, only one carries the coded directions for synthesizing a

protein; this code, as noted, is determined by the sequence of bases in the chain. This nucleotide chain is called the *sense chain* of the two; the opposite chain of the DNA double helix, which has a complementary sequence but does not carry coded information, is called the *missense chain*. The missense chain is important in the duplication of DNA molecules in cell division but does not enter directly into transcription.

During transcription, the two nucleotide chains of a DNA molecule unwind, and the sense chain becomes the template for RNA synthesis (Fig. 9-8a). The new RNA chain is assembled on the DNA template (Fig. 9-8b); as the RNA chain is built, bases are added according to the rules for complementary pairing. Wherever a cytosine (C) occurs in the DNA template chain, a guanine (G) is placed in the growing RNA chain; wherever thymine (T) occurs in the DNA template an adenine (A) is placed in the RNA

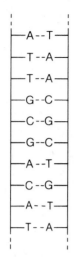

Figure 9-7 Complementarity of base sequences in the two nucleotide chains of a DNA double helix. Because A always pairs with T, and G with C, the sequence in one chain fixes the sequence of the other.

chain; and wherever guanine (G) appears in the template a cytosine (C) is placed in the RNA. Thymine does not occur in RNA. However, uracil (U) forms the same pattern of hydrogen bonds as thymine, and substitutes for thymine wherever adenine appears in the template. Thus, the new RNA molecule will contain the bases G, C, A, and U in a sequence that is exactly complementary to the C, G, T, A sequence of the template DNA nucleotide chain. When transcription is complete (Fig. 9-8c), the RNA copy unwinds from the DNA template, and the sense chain of the DNA molecule rewinds with its complementary missense chain into the double helix. RNA transcription is catalyzed by a group of enzymes called *RNA polymerases* (Information Box 9-2 outlines the mechanism of assembly of RNA by the RNA polymerase enzymes).

Each of the DNA molecules in a eukaryotic nucleus contains the codes for hundreds or even thousands of different mRNA molecules. Codes for transfer and ribosomal RNA molecules are also concentrated by the hundreds or thousands at different points on the DNA molecule. These

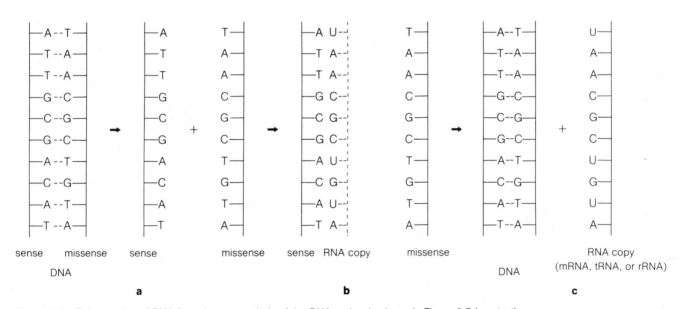

Figure 9-8 Transcription of RNA from the sense chain of the DNA molecule shown in Figure 9-7 (see text).

Information Box 9-2

The Mechanism of RNA Transcription

The mechanism of RNA transcription was worked out in experiments using *cell-free systems* in which the various enzymes and molecules involved are isolated, purified, and placed together in a test tube. In these systems, independently pioneered in the late 1950s by the American investigators Jerard Hurwitz, Samuel B. Weiss, and Audrey Stevens, RNA transcription can be detected and followed in a medium containing only a DNA template, Mg^{2+} or Mn^{2+} ion, the enzyme RNA polymerase, and the four bases adenine, guanine, cytosine, and uracil. The bases must be present as the nucleoside triphosphates ATP, GTP, CTP, and UTP (ATP = adenosine triphosphate, GTP = guanosine triphosphate, CTP = cytidine triphosphate, and UTP = uridine triphosphate). GTP, CTP, and UTP are all high-energy compounds with properties similar to ATP.

The reaction proceeds on a DNA template as shown in Figure 9-9. The RNA polymerase binds to the DNA template and recognizes the first base to be copied from the template, which in Figure he template, which in Figure 9-9a is shown as a guanine. According to the base pairing rules, the presence of a guanine at this site causes the enzyme to bind CTP from the nucleoside triphosphates available in the medium (Fig. 9-9b).

In response to binding CTP, the enzyme undergoes a conformational change, causing it to "read" the next base exposed on the DNA template, which in Figure 9-9 is shown as an adenine. The presence of adenine at this site induces the enzyme to bind UTP from the nucleotides in the medium. The UTP, once bound to the enzyme, is held opposite the thymine of the template, in a position favoring formation of the first phosphate linkage (Fig. 9-9c). The polymerase enzyme then catalyzes the linking reaction. In this reaction, the last two phosphates of the nucleotide most recently bound by the enzyme (UTP in Fig. 9-9) are split off. The remaining phosphate is linked to the 3'-carbon of the first nucleotide (Fig. 9-9d). Removal of the terminal phosphates from UTP releases free energy and greatly favors formation of the phosphate bond. Note from Figure 9-9d that a 3' —OH is always present at the newest end of the growing RNA chain, and a 5'-carbon linked to a triphosphate group is located at the beginning or oldest end. For this reason, RNA transcription is said to proceed in the 5' → 3' direction. This direction is always antiparallel to the template chain.

In response to formation of the first phosphate linkage, the enzyme undergoes a conformational change causing it to move to the next base exposed on the DNA template, shown as a thymine in Figure 9-9. The enzyme then binds an ATP nucleotide from the medium and catalyzes formation of the second phosphate linkage. This linkage, as before, is formed at the expense of two phosphate groups, split off from the nucleotide most recently bound by the polymerase enzyme. The process then repeats, binding nucleotides successively into the growing chain until the end of the coding sequence in the DNA template is reached. At the end of the sequence, the newly synthesized RNA molecule and the enzyme are released from the DNA template.

individual segments coding for either mRNA, tRNA, or rRNA molecules are the *genes* of the nucleus (each of the long DNA molecules, containing multiple mRNA, tRNA, and rRNA sequences, is a *chromosome;* see Fig. 9-27). In eukaryotes, the segments coding for mRNA molecules, that is, the segments coding for proteins, are copied individually, producing molecules that code for a single protein or polypeptide. In order to make these copies of individual segments within the long DNA molecules of the cell nucleus, the RNA polymerase enzymes must begin and complete transcription at points just preceding and following the sequences to be copied. Presumably, the RNA polymerase enzymes recognize specific control sequences in the DNA, located at the beginning and end regions of the genes, that tell the enzymes where to start and stop the transcription process.

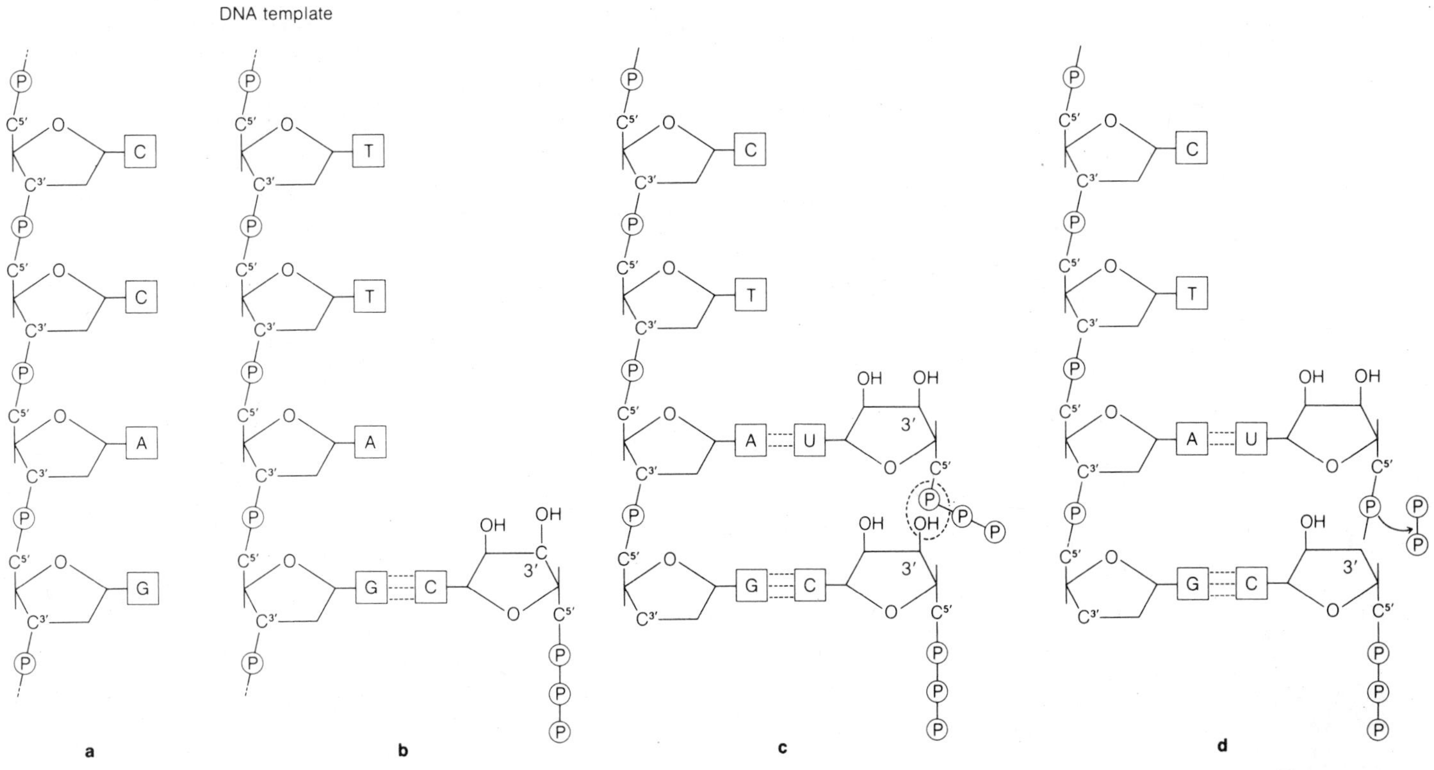

Figure 9-9 The mechanism of RNA transcription.

Most of the RNA molecules transcribed in both prokaryotic and eukaryotic cells are made in the form of *precursor* molecules that are longer than the finished RNA products active in protein synthesis in the cytoplasm. Once transcribed, these precursors are then *processed*, as it is called, to produce the finished RNA molecule. As a part of the processing, the precursor is shortened by processing enzymes that clip surplus segments from the ends or middle of the molecule. If any segments are clipped from the interior of the molecule, the free ends produced are rejoined to form a single, continuous RNA. Other enzymes may add additional RNA nucleotides to either end of the precursor once the clipping and any necessary rejoining is complete. These additional nucleotides, since they are added after transcription of the precursor, are not complementary to any sequences in the DNA template. As a final step in processing, individual G, C, A, and U bases in the precursor may be chemically modified by enzymes to other forms. Thus the finished mRNA, tRNA, and rRNA products may contain other purine and pyrimidine bases in addition to the original G, C, A, and U types

inserted during transcription. These *modified bases,* as they are called, occur in particularly high proportions in tRNA molecules (see Fig. 9-12).

Characteristics of the Major RNA Classes

Messenger RNA The completely processed mRNA molecules entering the cytoplasm to direct protein synthesis have several interesting and unusual features. All contain, as expected, a continuous sequence of nucleotides coding for the sequence of amino acids in a protein, or a polypeptide subunit of a protein. Not all of the nucleotide sequences in mRNA molecules, however, code for proteins. Studies of isolated and purified mRNA molecules from eukaryotes show that the coding sequence lies in the middle of an mRNA molecule, separated from the ends by stretches of nucleotides that are never translated into amino acid sequences (Fig. 9-10a).

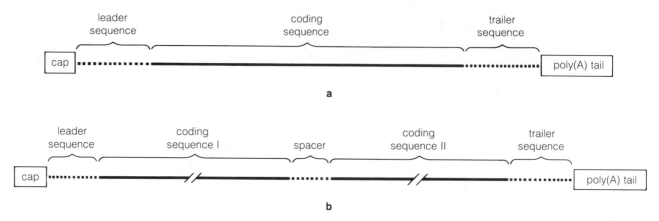

Figure 9-10 The structure of eukaryotic (**a**) and prokaryotic (**b**) messenger RNAs (see text).

The "front" end of a eukaryotic mRNA molecule is "capped" by a complex structure that contains several modified bases. The bases of this cap are added and modified as a part of the processing reactions taking place after transcription of mRNA precursor molecules. The cap is separated from the beginning of the actual coding sequence by a "leader" of 10 to 50 or more nucleotides, much like the blank leader of a roll of movie film that is used to thread the film through a projector (perhaps the mRNA leader has the same function, and threads the mRNA through a ribosome to begin synthesis). A similar stretch of 90 to 100 noncoding nucleotides follows the coding sequence as an untranslated "trailer." Both the leader and trailer sequences are copied from the DNA template during transcription, and thus correspond to complementary nucleotide sequences in the DNA. The mRNA terminates in a "tail" constructed from a sequence of adenine nucleotides that may number, in different mRNAs, from 20 or so to more than 200. The nucleotides in this *poly(A) tail*, as it is called, are added as a part of the processing reactions taking place after transcription. Thus eukaryotic mRNAs take the form: cap → leader (untranslated) → coding sequence (translated) → trailer (untranslated) → poly(A) tail.

Bacterial mRNA molecules (Fig. 9-10b) are similar in structure. However, several coding sequences, each spelling out the sequence of amino acids in different proteins or polypeptides, may be present in a single bacterial mRNA molecule. Usually, the several proteins coded together in bacterial mRNA molecules are enzymes that carry out sequential steps in the same biochemical path-

way. When several coding sequences are present in a bacterial mRNA, each is separated from the next by a short, intervening spacer sequence that is not translated. Although the coding sequences are preceded by a leader, and followed by a trailer sequence and short poly(A) tail, no cap of modified bases occurs at the front of bacterial mRNAs.

The functions of the additional, noncoding sequences in finished mRNA molecules are uncertain. The leader and trailer sequences, as noted, may serve to thread mRNAs through the ribosomes translating them, functioning in this role much like the blank leader and trailer segments of a movie film. The cap structure in eukaryotic mRNAs is believed to facilitate, in some way, the initial attachment of an mRNA to a ribosome (the initial attachment is slowed drastically if the cap is removed). The function of the poly(A) tail is completely unknown. The role of the tail is made all the more baffling since a few mRNAs lack a poly(A) tail and are nevertheless translated on ribosomes as readily as mRNAs with poly(A) tails.

Eukaryotic mRNA molecules are transcribed in the nucleus in the form of long precursor molecules that may be as much as five to ten or more times as long as their finished mRNA products. These mRNA precursors are called *heterogeneous nuclear RNA (hnRNA)* because they (1) form a group with wide variation in size and (2) occur in the nucleus of eukaryotic cells. Once transcribed, the long hnRNA precursors are processed to release mRNA. Surplus segments are clipped from the hnRNA molecules by the processing enzymes; if any segments are removed from the middle of the hnRNA precursors, the free ends

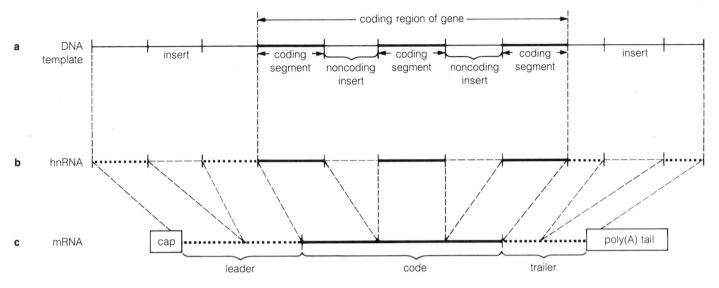

Figure 9-11 Comparisons among the DNA template (**a**), the hnRNA molecule copied from the template (**b**), and the mRNA processed from the hnRNA (**c**) show that noncoding segments occur in DNA in the coding region as well as in regions on either side of the code. These noncoding segments are copied into the hnRNA during transcription, and then are clipped out during processing to form the finished mRNA. This means that the code for the protein occurs *in pieces* in the DNA, with noncoding segments interspersed within the region of the gene that codes for the amino acid sequences of a gene. Most eukaryotic gene sequences are in pieces in the DNA and contain interspersed noncoding segments this way.

generated are rejoined. The cap and poly(A) tail are added during or following the clipping process. Why so much extra RNA is transcribed into the hnRNA precursors and then degraded almost immediately in mRNA processing is anybody's guess at the present time. The question is especially interesting when you consider that each extra nucleotide added to the hnRNA precursors (which, as noted, may be ten or more times longer than the finished mRNA products) requires a cellular energy expenditure of two high-energy phosphate groups (see Information Box 9-2).

Comparisons between hnRNA precursor molecules and the final mRNA products processed from them have led to one of the most unexpected and surprising findings of cell biology. The surplus segments clipped out of the hnRNA molecules during processing occur within the stretches that carry the code for a protein as well as in the segments that will form the leader and trailer segments of the finished mRNA molecule (Fig. 9-11b and c). Since the hnRNA precursor is an exact copy of the DNA molecule used as a template during transcription, this finding means that *the information for synthesizing proteins is coded in DNA in pieces*, with surplus, noncoding segments interspersed between the coding segments (Fig. 9-11a). The

noncoding inserts are now called *introns*, and the coding regions are called *exons*. This arrangement of sequences in eukaryotic DNA was totally unexpected. Until the discovery of the noncoding spacers or introns the sequence of nucleotides in the coding region of a gene was believed to read without interruption to spell out the amino acid sequence of a protein. Why information is coded in this manner in eukaryotes is unknown (noncoding introns do not occur within the coding regions of prokaryotic genes).

Processing of mRNAs from long hnRNA precursors is characteristic of eukaryotes and the viruses infecting eukaryotic cells. While the precursors of bacterial mRNAs may have limited surplus segments on either side of coding regions that are trimmed out during processing, extremely long precursors equivalent to the hnRNA molecule of eukaryotes evidently do not occur in bacteria.

Transfer RNA Finished tRNAs are small molecules containing from 73 to 93 nucleotides. They are distinguished from the other RNA molecules of the cell by their high content of modified bases, which may make up as much as 15 percent of the total.

Most cells contain a large variety of different tRNAs. Most of these correspond to the triplets specifying different amino acids in the nucleic acid code. There are 20 families of these tRNAs, one family for each of the 20 amino acids.

The first nucleic acid molecule of any kind to be completely sequenced was a tRNA from yeast that binds to the amino acid *alanine* in protein synthesis (Fig. 9-12). This pioneering work was accomplished by Robert W. Holley and his coworkers at Cornell University. Their research occupied seven years of effort and used a full gram of alanine tRNA, purified from more than 300 pounds of yeast cells. Holley received the Nobel Prize in 1968 for his work in nucleic acid sequencing. Since this first success, more than 350 additional tRNA molecules have been isolated, purified, and fully sequenced.

As a part of their original investigations, Holley and his colleagues noted that the sequences in many regions in the alanine tRNA molecule occur as *reverse repeats*. In these regions the sequence to the left of a point is repeated in reverse order to the right of the point in the form of complementary bases (Fig. 9-13a). The reverse repeat allows the single nucleotide chain to fold back on itself in these regions, pairing into a short double helix called a *hairpin* (Fig. 9-13b). Holley found enough of the reverse repeats to fold the alanine tRNA into the cloverleaf structure shown in Figure 9-12. Holley's cloverleaf was subsequently shown to be compatible with the internal reverse-repeat sequences of all tRNA molecules.

During the early 1970s, Alexander Rich and his colleagues, working at the Massachusetts Institute of Technology, and J. Robertus, B. F. C. Clark, and A. Klug and their colleagues at the Medical Research Council Laboratories in Cambridge, England, used x-ray diffraction (see Appendix) to deduce the three-dimensional structure of purified tRNAs. Their investigations confirm that tRNA molecules contain the cloverleaf arrangement proposed by Holley, folded further in three dimensions into an L-shaped structure (Fig. 9-14). The parts of the molecule that bind amino acids and pair with the mRNA molecules are exposed at the opposite ends of the "L." Hydrogen bonding between atoms located at different points in the arms, and hydrophobic interactions between nucleotide bases hold the molecule in its three-dimensional form.

tRNA molecules in both eukaryotes and prokaryotes are processed from precursor molecules that are somewhat longer than the finished tRNA products. Processing includes clipping out surplus segments, modifying bases at fixed points in the tRNA sequences, and adding the short

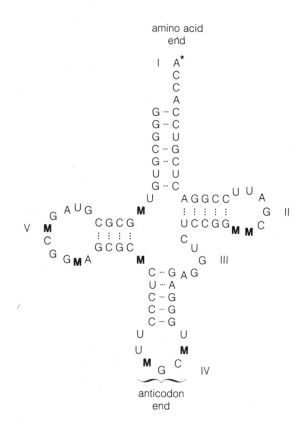

Figure 9-12 The nucleotide sequence of a eukaryotic tRNA from yeast cells that carries the amino acid alanine in protein synthesis. Modified bases occur at the positions indicated by a boldface "M"; the amino acid binds at the position marked by the asterisk. The numbered loops fold into the three-dimensional conformation shown in Figure 9-14.

C-C-A sequence at the end of the molecule that binds the amino acid (all tRNAs have the C-C-A sequence at this position). Base modifications in tRNA processing are highly varied and may include addition of a variety of chemical groups and rearrangements of the atoms within the tRNA bases.

Ribosomal RNA Ribosomal RNA is defined as the RNA that can be extracted from ribosomes, and the precursors to ribosomal RNA that occur in the cell nucleus. At least four different rRNA molecules can be detected in eukaryotic ribosomes, and three in prokaryotes (details of these molecules are given in Information Box 10-1). Sequencing studies have revealed that the rRNAs of all eukaryotes are

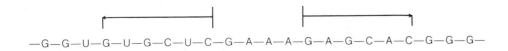

—G—G—U—G—U—G—C—U—C—G—A—A—A—G—A—G—C—A—C—G—G—G—

a

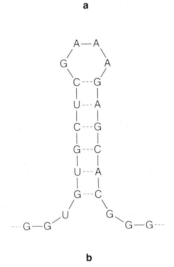

b

Figure 9-13 (**a**) A reverse repeat in an RNA molecule. The sequence under the arrow pointing to the left is repeated in reverse order in complementary bases in the sequence under the arrow pointing to the right. (**b**) Formation of foldback pairs in the reverse-repeat region to form a double helix.

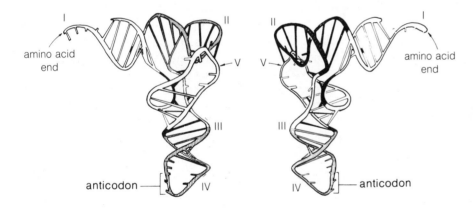

amino acid end

anticodon

amino acid end

anticodon

Figure 9-14 Opposite side views of the three-dimensional structure of a tRNA molecule. The Roman numerals show the positions of the cloverleaf loops in the folded state (see Fig. 9-12). Courtesy of A. Rich, from *Proceedings of the National Academy of Sciences* 72 (1975): 4866.

closely similar. In the same way, bacterial rRNAs are closely related in sequence among different bacteria. The rRNAs of bacteria, however, appear to be unrelated to eukaryotic rRNAs.

Three of the four rRNA molecules found in eukaryotic ribosomes are transcribed in the form of a single, large precursor molecule that is subsequently split into smaller pieces to release the three rRNAs (Fig. 9-15). As a part of the splitting reaction, surplus segments of nucleotides that do not appear in the finished rRNAs are removed from the precursor and broken down. The fourth rRNA of eukaryotes is a relatively small RNA, not much larger than a tRNA molecule, that is transcribed separately as a second rRNA precursor molecule. Processing includes removal of a small surplus segment from the precursor to this rRNA.

The rRNAs of bacterial cells are also transcribed in the form of a large precursor that contains one sequence of each of the three types. These are released by enzymes that clip the precursor into segments, as in eukaryotes.

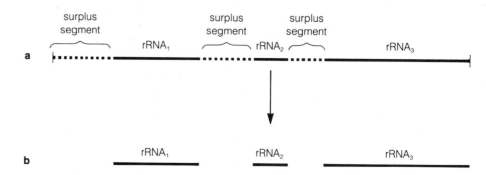

Figure 9-15 Ribosomal RNAs are also transcribed in the form of large precursors (**a**) which are subsequently clipped during processing to release the finished rRNA molecules (**b**).

As rRNA processing takes place in both eukaryotes and prokaryotes, the proteins that also form a part of ribosomes are added to the various rRNA types to form completed ribosomal subunits. In this form, the ribosomal subunits enter the cytoplasm, where they join by twos to form complete ribosomes and interact with mRNA and tRNA molecules in protein synthesis.

The Cell Nucleus and RNA Transcription

Nuclear Structure

Chromatin and the Nucleolus The most intense and prolonged RNA transcription occurs in cells during the period of the cell cycle between divisions. During this nondividing stage, called *interphase,* the organization of the nucleus is similar in all eukaryotes (Figs. 9-16 and 9-17). Two major structures can be seen in the nucleus at this time: (1) the *chromatin,* consisting of very fine fibers distributed throughout most of the nucleus, and (2) one or more roughly spherical bodies, the *nucleoli* (singular = *nucleolus*). The chromatin may be distributed evenly throughout the nuclear interior, or it may be densely packed in some regions and more loosely packed elsewhere in the nucleus. The densely packed chromatin, if present, is frequently located in a layer just inside the borders of the nucleus. In any case, the chromatin fills the entire nucleus, except for spaces taken up by the nucleoli.

The nucleus is most frequently viewed under the electron microscope in the form of very thin sections of chem-ically fixed and embedded tissue (see Appendix for details). In such preparations the fibers of nuclear chromatin appear to be about 10 nanometers in diameter, or about one-third the diameter of a ribosome.

Nucleoli are easily recognized in thin sections as dense, irregularly shaped masses of fibers and granules suspended in the chromatin (Figs. 9-17 and 9-18). The granules, which are the most conspicuous and characteristic feature of the nucleolus, are simply called the *nucleolar granules.* The *nucleolar fibrils* are indistinct and thinner than chromosome fibers. Chromatin fibers also extend into spaces within the nucleolus (as in Fig. 9-18). Investigations into nucleolar structure and function have established that the fibrils and granules are the morphological forms taken by successive steps in rRNA processing and the assembly of ribosomal subunits, both of which take place in the nucleolus.

The Nuclear Envelope The entire nucleus is surrounded by a system of two membranes. The outermost membrane covers the surface of the nucleus and faces the surrounding cytoplasm. The inner membrane lies just under the outer membrane and faces the nucleoplasm. A narrow compartment about 20 to 30 nanometers wide, the *perinuclear compartment,* separates the two membranes. The two membranes together are referred to as the *nuclear envelope.*

At frequent intervals the membranes of the nuclear envelope are perforated by *pores* that form channels of communication between the nucleus and cytoplasm. These structures are visible in cross section in Figure 9-19 and at a plane parallel to the nuclear envelope surface in Figure 9-20. At the margins of the pores, the outer nuclear membrane folds inward and becomes continuous with the

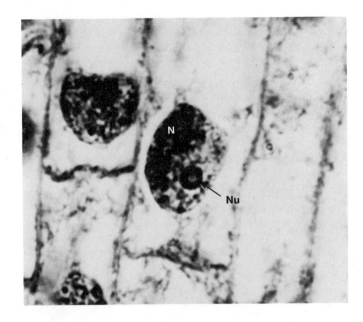

Figure 9-16 Light micrograph of a cell from an onion root tip. The nucleolus (*Nu*) is clearly visible within the nucleus (*N*).

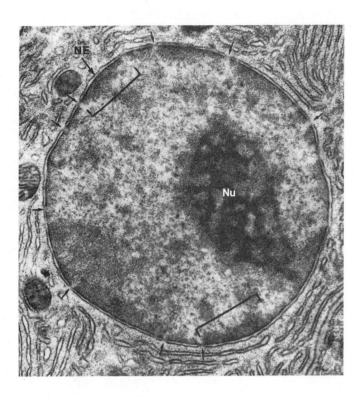

Figure 9-17 An electron micrograph of a cell nucleus from the pancreas of a bat. The nuclear envelope (*NE*) and pores (arrows) are clearly visible. Within the nucleus, the nucleolus (*Nu*) and chromatin fibers can be seen. The chromatin fibers are typically tightly packed in some regions and less densely packed in others. The tightly packed chromatin is frequently concentrated just inside the nuclear envelope (brackets). Courtesy of D. W. Fawcett, from *The Cell* © 1966 by W. B. Saunders Company.

inner membrane, forming an opening in the nuclear envelope about 70 nanometers in diameter. At this diameter, the pores are much wider than even the largest molecules or molecular complexes in the cell; they are more than twice the diameter of a ribosome. When viewed face-on, as in Figure 9-20, the pore margins often appear to trace out a regular octagon.

The pores are not completely open but appear instead to be filled in with a "plug" of electron-dense material known as the *annulus* (see Figs. 9-19 and 9-20). The annulus often appears eight-sided in its distribution around the pore margin. In many sections the annulus appears to be perforated by a narrow central channel, often occupied by a particle about the size of a ribosome (see Fig. 9-20). This particle is believed to be a ribosomal subunit, or possibly an mRNA molecule, caught in transit from nucleus to cytoplasm. The relationships of the pore and annulus, which together form the *pore complex*, are diagrammed in Figure 9-21.

Pores may take up as much as one-third of the total area of the nuclear envelope. If the pores were completely open, ions and molecules would be expected to diffuse freely between nucleus and cytoplasm. However, several experiments have shown that although ions and small molecules of the size of monosaccharides, disaccharides,

or amino acids pass freely and rapidly between the nucleus and the cytoplasm, the pores do control and limit the passage of larger molecules on the order of RNA and proteins.

There is no question that extensive traffic in RNA and proteins occurs between the nucleus and the cytoplasm. Since fully functional ribosomes are found only in the cytoplasm, all cellular proteins are probably synthesized in this region. Therefore, all of the proteins found temporarily or permanently in the nucleus must pass through the nuclear envelope to get there. And, since all of the rRNA molecules interacting in protein synthesis in the cytoplasm are transcribed in the nucleus, all of these molecules must pass in the opposite direction. Some molecules also pass in both directions. Ribosomes, for example, consist of RNA and proteins in approximately equal quantities by

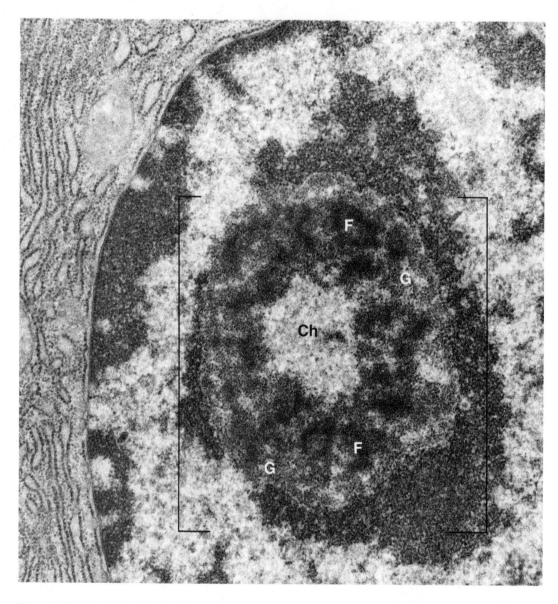

Figure 9-18 A nucleolus (brackets) from a rat pancreas cell at higher magnification. The nucleolar fibrils (*F*) and granules (*G*) are clearly visible. A part of the chromatin (*Ch*) extends into the center of the nucleolus.

weight. The proteins of ribosomes are first synthesized in the cytoplasm, from where they enter the nucleus to combine with rRNA to form the ribosomal subunits. After assembly, the ribosomal subunits pass in the opposite direction through the nuclear envelope to enter the cytoplasm.

There is ample evidence that the large RNA and pro-

tein molecules passing from nucleus to cytoplasm move through the pores. In cells that synthesize unusually large quantities of RNA, such as developing egg cells, dense material can be seen in passage through the pores (Fig. 9-22). In a few cases the material traversing the pores has been identified as RNA and protein by staining techniques.

Thus the pore complexes of the nuclear envelope are

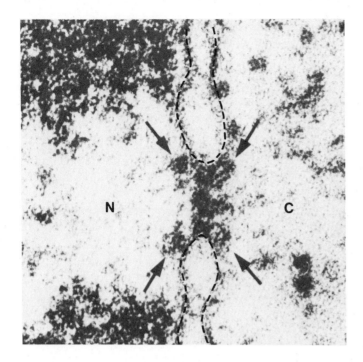

Figure 9-19 The highly magnified image of a nuclear pore complex in an onion root tip cell. The arrangement of the annular material into spherical subunits is clearly visible (arrows). *N*, nucleus; *C*, cytoplasm; dashed line, nuclear envelope membranes. × 297,000. Courtesy of W. W. Franke, from *Philosophical Transactions of the Royal Society*, London, Series B, 268 (1974): 67.

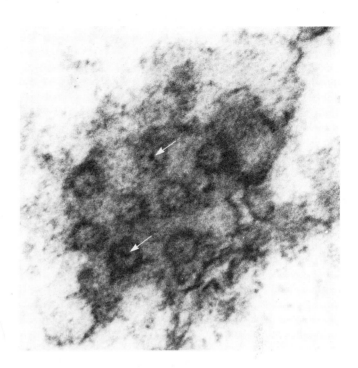

Figure 9-20 Nuclear envelope pores of an egg cell in a section that grazes the nuclear envelope surface. Dense granules (arrows) occupy the center of some of the pores. The granules may be ribosomal subunits passing from the nucleus to the cytoplasm through the pores. × 92,500. Courtesy of R. G. Kessel and Academic Press, Inc., from *Journal of Ultrastructure Research* Supplement 10, 1968.

probably specialized structures that control and facilitate the passage of large molecules and molecular complexes between the nucleus and the cytoplasm. The pore complexes probably also act as a complete barrier to some molecules, such as the DNA of the chromosomes. The membranes and pore complexes of the envelope obviously also function to separate the major organelles of the nucleus and cytoplasm, or as one investigator put it, "to keep the chromosomes in and the mitochondria out."

Prokaryotic Nucleoids Prokaryotic nucleoids, the equivalent of eukaryotic nuclei, also contain masses of dense fibers (Fig. 9-23). In both bacteria and blue-green algae, the nucleoid fibers are noticeably thinner than eukaryotic chromatin fibers. The fibers are usually more or less densely packed into a uniform mass in prokaryotic nucleoids without differentiation into densely packed and diffuse regions. Neither bacteria nor blue-green algae have

nucleoli, and no membranes equivalent to the nuclear envelope separate prokaryotic nucleoids from the surrounding cytoplasm.

When isolated by methods designed to reduce breakage, the nucleoid of bacteria proves to contain a single, large DNA molecule in the form of a closed circle. In *Escherichia coli*, the best studied bacterium, the nucleoid circle contains 1360 micrometers of DNA, equivalent to 4,000,-000 base pairs. Other bacteria have circles ranging from about 250 micrometers to a maximum not far in excess of the *E. coli* circle, about 1500 micrometers. All of this DNA is packed into the fibrous mass visible as the nucleoid, in cells that are only 1 to 2 micrometers long.

Comparatively little is known about the structure of the DNA in blue-green algae. Considerably more DNA, up to several times the amount in bacteria per cell, is present in these prokaryotes. Whether the DNA of blue-green algae exists in the form of closed circles is as yet unknown.

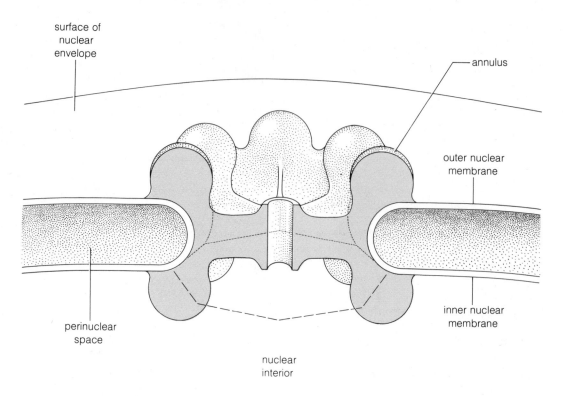

surface of
nuclear
envelope

annulus

outer nuclear
membrane

inner nuclear
membrane

perinuclear
space

nuclear
interior

Figure 9-21 The relationship between the pore and annulus in the nuclear envelope. The channel formed by the annulus can probably accommodate molecules as large as proteins and nucleic acids.

The Molecular Structure of Chromatin

Chromatin fibers can be isolated from eukaryotic cells and purified in large quantities without difficulty. Analysis of such preparations has revealed that the chromatin fibers of eukaryotes consist of DNA in association with two major classes of nuclear proteins, the *histone* and *nonhistone* proteins. Recent research with isolated chromatin has begun to reveal how these proteins, particularly the histones, are arranged with DNA in chromatin.

The Histone Proteins of Chromatin Fibers In eukaryotic nuclei DNA interacts with the histone proteins to form the basic structural units of chromatin. The histones are easily recognized against the background of proteins isolated with chromatin because they are relatively small proteins that carry a strongly positive charge at the pH ranges characteristic of living cells, a property shared by few other cellular proteins. They are easily released from isolated chromatin by increasing the salt concentration of the solution. Release by salt in this manner indicates that the forces linking the histones to DNA in chromatin are electrostatic (see p. 28) rather than covalent bonds. Five major kinds of histones, called H1, H2A, H2B, H3, and H4, are released in sequence from isolated chromatin by gradual increases in salt concentration.

The strongly basic charge of the histones is due to the presence of lysine and arginine, which make up a large part of the amino acids of these proteins. Lysine and ar-

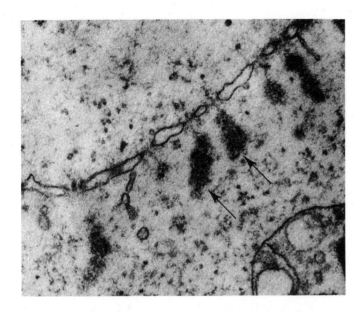

Figure 9-22 Passage of dense, RNA-containing material (arrows) from the nucleus to the cytoplasm of a developing frog egg cell. The cytoplasm is toward the bottom of the figure. × 54,000. Courtesy of H. Swift and W. Massover, from *In Vitro* 1 (1965): *The Chromosome, Structural and Functional Aspects.*

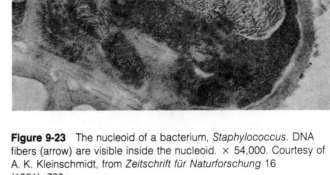

Figure 9-23 The nucleoid of a bacterium, *Staphylococcus*. DNA fibers (arrow) are visible inside the nucleoid. × 54,000. Courtesy of A. K. Kleinschmidt, from *Zeitschrift für Naturforschung* 16 (1961): 730.

ginine take on a strongly positive charge and act as bases (see Supplement 2-1) at the pH ranges characteristic of living cells. The positively charged, basic groups of the lysines and arginines of the histones are thought to be responsible for the electrostatic attraction holding DNA and the histones together in chromatin. These positively charged groups are attracted to the phosphate groups of the DNA backbones, which take on a negative charge at cellular ranges of pH.

Among the higher eukaryotes, each histone type shows relatively little variation in amino acid sequence in different species. Of the five major histones, H3 and H4 are most highly conserved in sequence. For example, histone H4 molecules from species as diverse as the calf and garden pea differ in only 2 out of a total of 102 amino acids. Although the other histones show more extensive differences among the various species of animals and plants, they are still among the most highly conserved proteins of living organisms. This close conservation of histone sequences in evolution indicates that almost all segments of the histone molecules are vital to chromatin structure and function, and that even changes as small as the substitution of a single amino acid may be lethal.

How the Histones Interact with DNA in Chromatin: Nucleosomes Until the early 1970s the histones were generally thought to be evenly distributed along the DNA helix in eukaryotic chromatin, much like the layer of insulation on an electric wire. At this time, two experiments revolutionized our view of chromatin structure. One experiment involved a new technique for isolating chromatin to be viewed under the electron microscope. In this technique, developed by Ada and Donald Olins at the Oak Ridge National Laboratory, chromatin fibers were treated briefly with a salt solution just concentrated enough to slightly loosen the association between the histones and DNA. Under these conditions, the isolated chromatin fibers appear much like a series of beads on a string, with each bead measuring about 10 nanometers in diameter (Fig. 9-24). This periodic arrangement was unexpected, and seemed incompatible with the prevailing ideas about chromatin structure.

The second experiment, carried out by Dean R. Hewish and Leigh A. Burgoyne in Australia, employed a group of enzymes called *DNA endonucleases*. These enzymes are capable of introducing breaks in the sugar-phosphate backbones of both nucleotide chains of a DNA molecule,

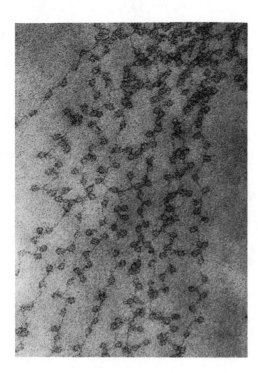

Figure 9-24 The beads-on-a-string image produced in chromatin fibers isolated in dilute salt concentrations. × 172,000. Courtesy of A. L. Olins, from *The molecular biology of the mammalian genetic apparatus.* Ed. P. O. P. Ts'o. Elsevier: North Holland, 1977.

causing the DNA to break completely at the point of attack. Hewish and Burgoyne found that if pure DNA, without any associated histone or nonhistone proteins, is broken by the endonuclease enzymes, a completely random set of DNA lengths is produced. This result shows that the endonuclease enzymes can cut "naked" DNA at any point along the sugar-phosphate backbones.

Exposure of *chromatin* to the endonuclease enzymes, however, produced some surprising results. Digestion of chromatin with the enzymes released DNA in regular lengths, each initially 200 nucleotides long, or into larger pieces that were multiples of this basic length. With more prolonged digestion of the chromatin, the DNA pieces were gradually shortened to lengths including about 140 nucleotides. These results were surprising because they also indicated that the histones join with DNA to form a periodic structure of some kind that exposes the DNA to initial digestion by the enzyme only at intervals 200 nucleotides apart.

Regular lengths of this type were obtained only if the DNA digested by the endonucleases was associated with the histone proteins in chromatin. If the histones were removed from the DNA before digestion, the endonucleases produced random lengths, with no regular periodicity evident in the preparation. Similar random lengths were obtained with bacterial DNA, which has no associated histones.

These initial observations indicating a periodic or beaded structure in chromatin were complemented by additional information developed by Roger D. Kornberg of Harvard University. Kornberg found that when the histones are removed slowly from DNA in chromatin by careful adjustments in salt concentration, the H2A, H2B, H3, and H4 histones often come out in pairs in various combinations. Under certain conditions, these histones could also be extracted from chromatin in the form of an eight-part unit or *octamer* containing two of each of the H2A, H2B, H3, and H4 types. Significantly, H1 was never found in direct association with the other histones in either the pairs or the octamers. Another significant fact discovered about H1 was that it occurs in chromatin at only half the level of any of the other histones, or in a H1:H2A:H2B:H3:H4 ratio of 1:2:2:2:2.

In 1974 Kornberg combined all of this information into a brilliant hypothesis for chromatin structure that has been widely supported by further experiments. Kornberg proposed that the histones H2A, H2B, H3, and H4 combine by twos to build an $(H2A)_2 (H2B)_2 (H3)_2 (H4)_2$ octamer that forms the *core particle* of chromatin (Fig. 9-25). Wrapped around this histone core particle is a length of DNA, 140 nucleotides long, long enough to coil through 1½ to 1¾ turns around the histone core. The entire structure, including the histone core particle and the length of DNA wrapped around it, is called a *nucleosome*. Each nucleosome is connected to the next by a short length of DNA, the *linker*, running between them. The linkers average 60 nucleotides in length, giving a total average length for each nucleosome and linker of 200 nucleotides. Kornberg proposed that H1 is associated with the linking DNA segment, in the ratio of one H1 molecule per linker. Thus, according to Kornberg's hypothesis, chromatin resembles a series of pulleys (the histone core particles) with the DNA forming a rope that winds around each pulley and connects one pulley to the next one in line.

Kornberg's hypothesis readily explains the results of chromatin breakdown by the DNA endonuclease enzymes. Initial treatment with the enzymes breaks the DNA at an exposed point on the linker between nucleosomes, giving the basic unit 200 nucleotides long. More prolonged treatment slowly chops down the linker DNA associated with H1, leaving a resistant piece 140 nucleotides long. The segment 140 nucleotides long that resists more prolonged digestion is the DNA of the nucleosome particle, protected from enzymatic attack by its association with the H2A-H2B-H3-H4 octamer. The hypothesis also explains the beads-on-a-string appearance observed in the electron microscope when chromatin is treated with dilute salt solutions. The salt evidently opens the H1-linker combination just enough to loosen the linker segments and make the linkers and nucleosomes separately visible in the electron microscope.

Nucleosomes can be induced to self-assemble by mixing DNA together with the five histones in salt solutions approximating the concentrations in living nuclei. The nonhistone proteins are unnecessary for this self-assembly, and evidently do not occur as major constituents of the nucleosome structures of chromatin.

The Nonhistone Proteins of Chromatin The nonhistone proteins are defined as the proteins, excluding the histones, that occur with DNA in chromatin. Most of the proteins in this group are negatively charged or neutral at cellular ranges of pH. A few are basic, positively charged proteins, but none of these is as basic as the histones. Typically, the nonhistone proteins occur in many more types and kinds in cell nuclei, and show much greater variability between cells and organisms than the histones.

Since nucleosomes can be reassembled from DNA and the five histones alone, the nonhistone chromosomal proteins are thought to play a functional, rather than structural, role in the nucleus. Included in this category are the enzymes catalyzing transcription and replication and proteins that regulate or control these activities. Of these nonhistone proteins, only the enzymes active in transcription and replication, such as the RNA polymerase enzymes, have been isolated and identified (the possible roles of other nonhistone proteins in the control of transcription are discussed further in Chapter 10).

The Structure of Prokaryotic DNA Proteins cannot be detected in large quantity in combination with prokaryotic DNA. Instead, in bacteria and blue-green algae, the DNA is believed to be combined with metallic ions, such as calcium (Ca^{2+}), or relatively small positively charged organic molecules. The small diameter of the nuclear fibers in these organisms probably reflects the absence of chromosomal proteins. Since no proteins resembling the eukar-

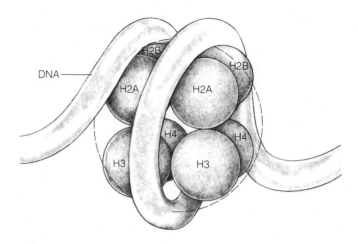

Figure 9-25 The structure of nucleosomes as proposed by R. D. Kornberg. A histone octamer forms the protein core of a nucleosome. DNA is wrapped around outside the histone core in a coil of 1½ to 1¾ turns. Histones H2A, H2B, H3, and H4 combine by twos to form the octamer; histone H1 is associated with the linker. The histones within the protein core are probably not as regularly distributed as they are diagrammed in this figure.

yotic histones are present in any quantity in prokaryotes, there are evidently no nucleosomes in these organisms.

The Function of Eukaryotic Nuclear Structures in RNA Transcription

RNA Transcription in Chromatin Fibers Several experiments have shown that RNA transcription is related to the distribution of chromatin fibers within the nucleus. One experiment demonstrating this fact uses a radioactive form of uracil, the nucleotide base that occurs in RNA but not DNA. As cells transcribe RNA in the presence of radioactive uracil, the regions of the nucleus carrying out the transcription become labeled by radioactivity and can be identified. Exposing actively transcribing cells to radioactive uracil in this way labels the loosely packed chromatin fibers, but not the tightly packed regions. This indicates that RNA transcription occurs almost entirely in the more diffusely packed regions of the nucleus. By different techniques, most of the RNA transcribed in these regions has been identified as messenger and transfer RNA. Ribosomal RNA transcription is confined almost entirely to the nucleolus.

The Nucleolus and rRNA Transcription The nucleolus was first linked directly to rRNA synthesis by experiments conducted by Donald D. Brown and J. B. Gurdon at the Carnegie Institution of Washington. For their research, Brown and Gurdon used mutants of *Xenopus*, a much studied amphibian originating in Africa. The mutants, which lacked a nucleolus, died during an early embryonic stage. Analysis of the RNAs synthesized by the mutants revealed that the absence of the nucleolus was correlated with an inability to transcribe the three larger kinds of ribosomal RNA.

Further work in 1965 by F. M. Ritossa and S. Spiegelman at the University of Illinois with the fruit fly *Drosophila* established a more direct link between the nucleolus and rRNA. Their experiments utilized mutant flies with extra nucleoli. By crossing the mutants with normal flies, they were able to produce a series of individuals with either 1, 2, 3, or 4 nucleoli in their cells. The capacity for rRNA synthesis in these individuals proved to be directly proportional to the number of nucleoli: individuals with two nucleoli made twice as much rRNA as those with one nucleolus, and so on.

These experiments also indicated that the smallest rRNA type, the one approximating tRNA in size, is transcribed in regions of the chromatin outside the nucleolus. In Brown and Gurdon's experiments with *Xenopus* mutants lacking nucleoli, the smallest rRNA molecule was found at normal levels, and in general the presence or absence of the nucleolus appeared to have little or no effect on the amount of this type of rRNA. And, in Ritossa and Spiegelman's work with *Drosophila*, the capacity of the mutants to synthesize the smallest rRNA type did not increase proportionately with the number of nucleoli, showing that the small rRNA is transcribed at sites elsewhere in the chromatin.

One of the most interesting discoveries resulting from research into ribosomal RNA transcription is that the genes for rRNA are repeated many times in chromatin. The genes for the larger rRNA types, repeated from a few hundred to thousands of times in different eukaryotes, are arranged end to end, usually at one extended site in the DNA of one chromosome or chromosome pair. All of these multiple rRNA genes are active simultaneously in transcribing rRNA precursors, which are subsequently processed and combined with the ribosomal proteins in the same location. All of this biochemical activity is visible as the fibrils and granules of the nucleolus. The fibrils correspond to early stages of rRNA processing, before the processing enzymes have clipped segments from the long rRNA precursor molecules. The granules are ribosomal subunits that are nearly ready for transport to the cytoplasm, containing finished rRNA molecules in combination with ribosomal proteins. The genes for the smallest RNA type, which may exist in as many as 20,000 copies, are distributed in blocks elsewhere in the chromatin. The many repeats of rRNA genes are reflected in the fact that cells make more rRNA than any other type; in most cells, rRNA accounts for 85 to 95 percent or more of the total RNA transcribed. The large number of rRNA genes thus increases the cell's "horsepower" for transcribing rRNA.

Figure 9-26 shows segments of nucleolar DNA caught in the act of transcription by Oscar L. Miller of the University of Virginia. The long, filamentous strands in the pictures are parts of the DNA molecule containing the repeated rRNA sequences (arrow). The brushy regions are collections of RNA molecules being transcribed on the DNA; each fiber of the brushy regions is an rRNA precursor molecule. The granule at the base of each rRNA precursor is a molecule of RNA polymerase engaged in transcribing the nucleolar DNA; many polymerase molecules are copying the DNA template at one time. At the end of the brushy region with short rRNA precursors the RNA polymerase molecules have just begun transcription. At the opposite end, the rRNA precursor molecules are longer and transcription is nearly complete. A nontranscribed DNA spacer lies on either side of the rRNA precursor gene; no RNA polymerase enzymes or rRNA molecules occupy this region. The nucleolar DNA consists of many such genes, with the nontranscribed spacers separating them.

Repetitive DNA Not all of the DNA sequences in eukaryotic chromosomes have been identified with coding mRNA, tRNA, or rRNA sequences. Eukaryotic DNA also contains noncoding regions made up of sequences that may be repeated from thousands to millions of times. The repeated sequences are usually very short; some are spelled out by only 5 to 7 bases. *Drosophila* chromosomes, for example, contain regions with millions of repeats of the short sequence AAGAG. In most eukaryotes, from 15 to 20 percent of the total DNA of the nucleus is taken up by such repeated sequences, which are collectively termed *repetitive DNA*. Some of the repeated sequences are distributed throughout the DNA of the chromosomes, where they are interspersed in small clusters between the sequences coding for mRNA, rRNA, and tRNA. Other repetitive sequences, especially the short sequences occurring

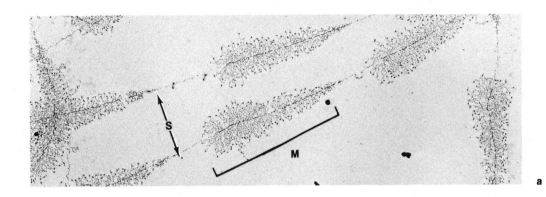

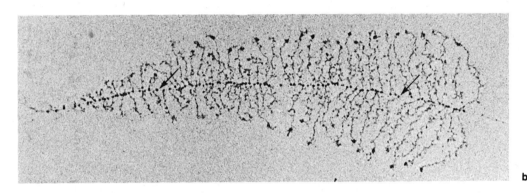

Figure 9-26 (**a**) DNA molecules (arrows) in nucleoli isolated from developing egg cells of a salamander, caught in the process of RNA transcription. The RNA molecules being transcribed appear as shorter fibers running at right angles to the DNA strands (brackets). (**b**) One of the transcribing regions at higher magnification. The spherical granules (arrows) at the base of each RNA molecule are molecules of RNA polymerase engaged in transcribing the DNA. Courtesy of O. L. Miller, Jr., and Barbara R. Beatty, Oak Ridge National Laboratory.

in millions of copies, are concentrated in a few large clusters in restricted regions of the chromosomes.

The function of repetitive DNA is unknown. Among the many ideas currently advanced is the hypothesis that the repeated sequences, particularly the ones occurring in clusters of millions of copies, are "evolutionary junk"—DNA that has been so riddled with mutations that it no longer has a function. Another idea is the opposite possibility, that repetitive DNA is evolutionary raw material: blocks of extra DNA that can mutate into functional genes. The more moderately repeated sequences, particularly those that occur in small clusters dispersed throughout the chromosomes, have been proposed as sites that may regulate and control RNA transcription or DNA duplication. None of these ideas is supported by any solid evidence at the present time, and the possible functions of the repeated sequences of eukaryotic DNA remain a mystery

(repetitive DNA does not occur in any quantity in prokaryotes).

The Structure and Function of the Cell Nucleus: A Summary

The information for synthesizing mRNA, tRNA, and rRNA is coded into the DNA of the cell nucleus. Instead of existing as one large molecule, the nuclear DNA of eukaryotes is broken into shorter lengths. Each of these DNA lengths is a *chromosome* of the cell nucleus. The nuclei of human body cells each contain 46 of these DNA lengths. The total amount of DNA per nucleus is immense, considering the size of a cell. The 46 DNA molecules of one human nucleus, for example, would measure about 1 meter

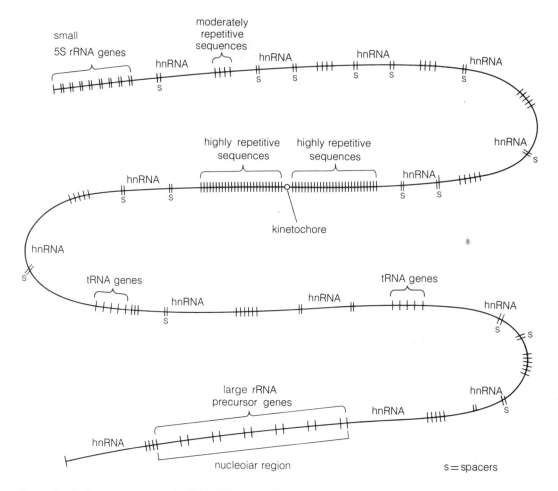

Figure 9-27 The arrangement of mRNA, tRNA, and rRNA genes in a eukaryotic chromosome (see text).

in length if laid end to end (each is about 2–3 centimeters long)!

The coded information in one of these chromosomes might be arranged as shown in Figure 9-27. Long stretches are taken up by mRNA precursor (hnRNA) coding sequences. Spacer sequences occur in the regions between the mRNA genes in most eukaryotes, and also within the tracts transcribed into hnRNA precursors. Although most of the mRNA coding sequences occur in single copies, a few are repeated (one known example of repeated mRNA genes is the gene group coding for the histones, which may occur in hundreds or thousands of copies). Clustered in one region are from hundreds to thousands of repeats of the sequence coding for the large rRNA precursor, each separated by a nontranscribed spacer sequence. This clustered region, when surrounded by transcribed rRNA pre-

cursors in various stages of processing and assembly with the ribosomal proteins, forms the *nucleolus* of the nucleus. In most organisms, only one chromosome or a pair of identical chromosomes of the total chromosome set carries the cluster of rRNA genes. Located at some distance from the nucleolar gene cluster, on the same or a different chromosome, are the repeated genes for the smallest rRNA type. These genes also consist of clustered repeats of the small rRNA coding sequence separated by nontranscribed spacers. Clustered at other points are a hundred or so repeats of the tRNA precursor genes, each also separated by spacer sequences. A cluster of highly repetitive, noncoding DNA sequences is also shown in one segment of the chromosome.

The DNA molecules of each chromosome combine with the histone and nonhistone proteins to form the *chro-*

matin fibers of the cell nucleus. Within the chromatin fibers, the DNA winds around the histone core particles to form the *nucleosomes* of the chromatin.

The RNA polymerase molecules copy segments of the DNA molecules in the nucleus, producing the precursors of mRNA, tRNA, and rRNA. These precursors are subsequently clipped into shorter pieces and modified chemically to produce the finished RNA types. Only a few of the thousands of mRNA coding segments are transcribed at any one time; these transcribed segments form the *active* chromatin of the cell nucleus, which is typically more loosely packed than inactive segments. The more tightly packed chromatin of the inactive segments is frequently concentrated in a layer just inside the nuclear envelope. The rRNA and tRNA genes of the nucleus are transcribed more or less continuously.

In bacteria, the various RNA coding sequences are distributed along a single DNA molecule that forms a closed circle. Located at points around the circle are a few repeats of the genes coding for the rRNA precursor molecule. Transfer RNA genes are spotted at other locations around the circle. Other segments are filled in with mRNA coding sequences. These sequences may be continuous, with no intervening spacers, or may be separated by small amounts of nontranscribed spacer DNA. The distribution of sequences in blue-green algae is probably similar to the bacterial pattern. Since histones cannot be detected in bacteria, the DNA circle probably exists in "naked" form, without winding into nucleosomes. The DNA molecules of blue-green algae probably take a similar form.

Suggestions for Further Reading

Brown, D. D. 1973. The isolation of genes. *Scientific American* 229:21–29 (December).

Crick, F. H. C. 1970. Split genes and RNA splicing. *Science* 204:264–71.

Elgin, S. R. C. and H. Weintraub. 1975. Chromosomal proteins and chromatin structure. *Annual Review of Biochemistry* 44:725–74.

Felsenfeld, G. 1978. Chromatin. *Nature* 271:115–22.

Franke, W. W. 1974. Structure, functions, and biochemistry of the nuclear envelope. *International Review of Cytology Supplement* 4:71–236.

Kornberg, R. D. 1977. Structure of chromatin. *Annual Review of Biochemistry* 46:931–54.

Rich, A. and S. H. Kim. 1978. The three-dimensional structure of tRNA. *Scientific American* 238:52–62 (January).

Watson, J. D. 1968. *The double helix.* Atheneum: New York.

Watson, J. D. 1976. *Molecular biology of the gene.* 3rd ed. W. A. Benjamin: Menlo Park, Calif.

Wolfe, S. L. 1981. *Biology of the cell.* 2nd ed. Wadsworth: Belmont, Calif.

For Further Information

Chromatin condensation during cell division, Chapters 11 and 12
DNA replication, Chapter 11
Genes and transmission of hereditary information, Chapter 13
Genetic code, Chapter 10
Nuclear envelope, breakdown and reformation during cell division, Chapter 11
Nucleolus, breakdown and reformation during cell division, Chapter 11
Nucleotides and nucleic acids, Chapter 2
Protein synthesis, Chapter 10
   in mitochondria and chloroplasts, Supplement 10-2
Regulation of transcription, Chapter 10
Ribosomes, structure, Chapter 10 and Information Box 10-1

Questions

1. Outline the overall pattern of protein synthesis.

2. What is mRNA? tRNA? rRNA?

3. Define transcription and translation.

4. Outline the structures of nucleotides, DNA, and RNA. How do DNA and RNA differ? How is information coded in DNA and RNA?

5. What factors determine base pairing in DNA? What holds a DNA molecule together?

6. What is complementarity? What significance does complementarity have for the process of transcription?

7. What are the sense and missense chains of a DNA molecule? What is the function of the sense chain? How do you think the missense chain might function in DNA replication?

8. Draw a sense chain containing 25 A, T, C, and G nucleotides in a random sequence. What is the sequence of the missense chain that is complementary to this sequence? What RNA sequence would be copied from the DNA sense chain?

9. Since RNA molecules contain uracil instead of thymine, how are complementary pairs formed with the DNA template during replication?

10. What is a chromosome? What is the difference between the chromatin and the chromosomes of a nucleus?

11. Outline the mechanism of RNA transcription.

12. What is RNA processing? What steps may take place in RNA processing?

13. Outline the structure of a "finished" mRNA molecule. What is hnRNA? How does mRNA differ from hnRNA? How do bacterial and eukaryotic mRNAs differ?

14. Outline the structure of a tRNA molecule. What are reverse repeats? Foldback double helices?

15. How is rRNA defined?

16. What major structures occur in eukaryotic nuclei? In prokaryotic nuclei?

17. What is the relationship between the nucleolus and the rRNA genes?

18. Outline the structure of the nuclear envelope, including pore complexes. What is the probable function of the pore complexes?

19. What are the histone and nonhistone proteins? How many major histones occur in eukaryotic nuclei?

20. What experiments showed that histones combine with DNA to form regularly spaced, periodic structures?

21. What is a nucleosome? How do DNA and the histones combine to form nucleosomes?

22. How does Kornberg's hypothesis for nucleosome structure explain the results obtained when chromatin is digested with the endonuclease enzymes?

23. Why are the nonhistone proteins not considered to be essential parts of nucleosome structure?

24. What are the possible functions of the nonhistone proteins?

25. Outline an experiment demonstrating that RNA is transcribed in the nucleus.

26. How was the nucleolus linked to rRNA transcription? What experiments support this relationship?

27. Each nucleus in an amphibian such as the leopard frog (*Rana pipiens*) contains about 15×10^{-6} micrograms of DNA. Using the fact that DNA weighs approximately 1.5×10^{-12} micrograms per micrometer, what is the total length of the DNA in a frog nucleus?

28. Draw a eukaryotic chromosome and insert the different types of sequences that are expected to be present.

Supplement 9-1:
Discovery of the Structure and Function of DNA

Early in this century cell biologists could not determine which of the major molecular types in the cell nucleus contains the stored directions for cell synthesis and heredity.

Until the decisive experiments of the 1940s and 1950s, many investigators thought that proteins were the most likely candidate for this role because they are more complex than DNA. In proteins, there are 20 amino acids in different combinations and an infinite variety of folding conformations. This structural variability seemed to offer greater opportunities for information coding than the nucleic acids, which have only four nitrogenous bases available as a basis for coding.

The first definitive evidence that DNA is actually the informational molecule of cells came with two sets of experiments conducted during the 1940s and 1950s. The first of these, by Oswald Avery and his coworkers at the Rockefeller Institute, was performed with different types of *Pneumococcus*, the bacterium causing pneumonia. The Avery experiments took advantage of two known forms of this bacterium. The cells of an infective form that can cause pneumonia are surrounded by capsules and produce smooth, gelatinous colonies when grown in pure cultures. A noninfective variety of the same bacterium has no capsule and forms colonies that appear lusterless or "rough" in texture when the bacteria are grown in culture dishes.

Genetic experiments conducted earlier had revealed that the two kinds of *Pneumococcus* bacteria are different hereditary types. When kept in pure colonies, the hereditary types bred true, that is, the cells of smooth, infective or rough, noninfective colonies normally gave rise only to cells of the same type when dividing. However, exposing rough noninfective cells to smooth, infective cells frequently caused transformation of the noninfective cells to the infective type. Transformation of this kind could be detected even if the infective bacteria were heat-killed before exposing the noninfective cells to them. Once transformed into the infective type, the bacteria bred true, indicating that the change was hereditary.

Avery and his colleagues were interested in identifying the substance in killed cells that could permanently transform rough, noninfective cells into smooth infective cells. In order to identify the substance, they treated an extract of killed cells with enzymes that can catalyze the breakdown of DNA, RNA, or proteins (Fig. 9-28). Only the enzyme capable of breaking down DNA destroyed the capacity of the extract to transform noninfective cells into the infective type. Enzymes breaking down RNA or proteins had no effect on the transforming ability of the extract. From these results, Avery proposed in 1944 that DNA is the substance responsible for transforming noninfective pneumonia bacteria to the infective kind. Since DNA can carry hereditary information in this way, it also

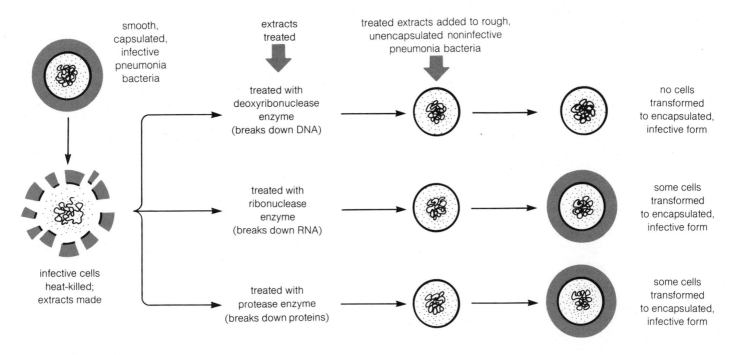

Figure 9-28 The bacterial transformation experiment carried out by Avery and his colleagues (see text).

seemed to be a likely candidate for the normal carrier of genetic information in the cell nucleus.

Avery's conclusion was directly supported and extended by a second series of experiments carried out in 1952 at the Carnegie Laboratory of Genetics by Alfred D. Hershey and Martha Chase, who studied the infection of bacteria by bacterial viruses. These viruses, called *bacteriophages* (see Fig. 1-19), cause bacterial cells to cease producing their own molecules and to make instead the DNA and protein of new virus particles (viruses consist only of a core of DNA or RNA surrounded by a surface coat of protein; see p. 17). When Hershey and Chase began their experiments, biologists assumed that, during infection of a bacterium by virus, either the DNA or the protein of the bacteriophage must enter the bacterial cell to alter its genetic and synthetic capacity.

To decide between these alternatives, Hershey and Chase used radioactive isotopes of phosphorus and sulfur, newly available at that time. Using the isotopes, they were able to label the proteins and DNA specifically, because DNA contains phosphorus but no sulfur, and the viral protein being studied contained sulfur but no phosphorus. In order to label the virus, bacterial cells were infected and allowed to make virus particles in a culture medium containing either radioactive phosphorus or sulfur. By this method, mature virus particles were obtained in which either the DNA or the protein was radioactive.

Hershey and Chase then mixed the labeled virus particles with fresh bacterial cells and allowed infection to take place (Fig. 9-29). Radioactivity was found inside the infected bacterial cells only if the virus particles contained radioactive DNA (Fig. 9-29a). When the experiment was conducted with virus particles containing labeled protein, all radioactivity remained outside the bacterial cells (Fig. 9-29b). This showed that only the DNA of the infecting virus entered the bacterial cells. Since only the DNA entered, this molecule must have carried the information required to convert the bacterial cell machinery to the synthesis of the new bacteriophage particles.

The experiments of Avery and his colleagues and Hershey and Chase established that DNA, and not protein, carries genetic information and is likely to form the basis of all information storage in the cell. These experiments touched off intensive research that eventually revealed not only the molecular structure of DNA but also how information is coded into the nucleic acids.

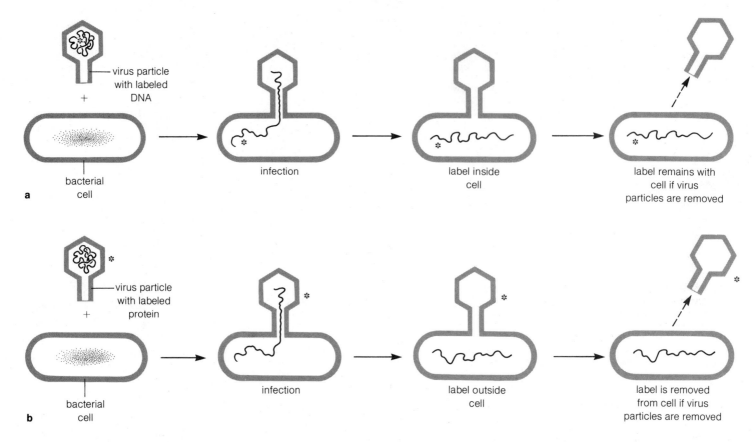

Figure 9-29 The Hershey and Chase experiment demonstrating that the DNA of viruses enters bacterial cells to alter their genetic machinery (see text).

Discovery of the Molecular Structure of DNA

When Hershey and Chase's results were announced, the complete molecular structure of DNA was unknown. Some information was available; it was known that DNA is formed from long chains of nucleotide subunits, and that these subunits are connected into chains by sugar-phosphate linkages in the molecule. The molecular structure of the individual pyrimidine and purine bases and the structure of the deoxyribose sugar had long since been worked out by organic chemists. However, the number of nucleotide chains in a DNA molecule and the manner in which the chains fold or twist to form the intact molecule were unknown.

This was the state of affairs when James D. Watson and Francis H. C. Crick took up their study of DNA structure at Cambridge University in the early 1950s. In their work, Watson and Crick relied heavily on data gathered from the x-ray diffraction (see Appendix) of DNA molecules in the laboratory of Maurice H. F. Wilkins at Kings College, London. Wilkins's evidence, developed in collaboration with his coworker Rosalind Franklin, indicated that the molecule is cylindrical, with an outside diameter of about 2 nanometers. Within the molecule, the x-ray data suggested that the nucleotide chains forming the backbone of the DNA molecule twist into a helix of regular diameter and pitch. Other evidence, from a chemical analysis of DNA carried out by Erwin Chargaff at the Rockefeller University, indicated that the amounts of adenine and guanine in DNA, while usually not equal, are always exactly paralleled by the amounts of thymine and cytosine respectively. That is, in amounts in DNA, adenine = thymine and guanine = cytosine.

From this information, Watson and Crick developed the model for DNA structure outlined in this chapter. They deduced first that the molecule consists of two nucleotide chains wound into a double helix, not three, as others had proposed. Later, they constructed scale models of the DNA nucleotides and fit the pieces together in various ways until they arrived at the correct relationship for base pairing and the three-dimensional structure of the DNA molecule. Their discovery was announced in a brief paper published in the British journal *Nature* in 1953. In the years following the publication of their work, all tests of the model supported their hypothesis and confirmed that DNA is arranged in the double helical, internally paired structure they proposed.

Suggestions for Further Reading

Watson, J. D. 1968. *The double helix*. Atheneum: New York.

Watson, J. D. and F. H. C. Crick. 1953. Molecular structure of nucleic acids: a structure for deoxyribose nucleic acid. *Nature* 171:737–38.

10

Cellular Synthesis II: Protein Synthesis in the Cytoplasm

The function of the cell nucleus, as outlined in the previous chapter, is in coding and transcribing the information required for protein synthesis. This information is coded into the sequence of nucleotide bases in DNA. It is transcribed and transferred to the cytoplasm as the directions for protein synthesis in the form of messenger RNA (mRNA) molecules. Other RNA molecules also important in protein synthesis, including ribosomal RNA (rRNA) and transfer RNA (tRNA) are transcribed at the same time. These RNAs also enter the cytoplasm after transcription and interact with mRNA in the synthesis of proteins.

Cytoplasmic protein synthesis takes place in two major steps. In a preliminary reaction sequence, the transfer RNA molecules link to their specific amino acids. This step takes place in the cytoplasmic solution. In the second step, these tRNA-amino acid complexes interact with messenger RNA in protein synthesis. All of the reactions of the second step take place on ribosomes.

The Reactions of Protein Synthesis

Step 1: The Attachment of Transfer RNA Molecules to their Amino Acids

A large number of different tRNAs occur in the cytoplasm of eukaryotes, many more than the minimum of 20 required to provide one tRNA for each of the 20 amino acids used in protein synthesis. This means that most of the 20 amino acids link to more than one type of tRNA. Because tRNAs are the molecules that match amino acids with

"words" in the messenger RNA code, most of the amino acids are therefore coded for by more than one word in the message. The amino acid serine, for example, is designated by any one of six different words in the nucleic acid code, and may link to any one of six different tRNA molecules. How this system of multiple codewords and tRNAs actually operates in protein synthesis is explained later in this chapter.

The amino acids attach to their respective tRNA molecules in a two-part reaction sequence that takes place in the cytoplasmic solution surrounding, but not directly connected to ribosomes. The final product of the sequence, an amino acid linked to tRNA, is a high-energy substance that contains energy directly derived from ATP. Because the amino acids are joined with their tRNAs in a high-energy complex, the entire two-part sequence is usually called *amino acid activation*. However, the reaction sequence is really more extensive than this name suggests because, in addition to increasing in energy level, the amino acids are also matched with their specific tRNA molecules.

In the first part of the series, an amino acid interacts directly with ATP. This enzyme-catalyzed reaction removes two phosphate groups from ATP and attaches the AMP (adenosine monophosphate) product to the amino acid:

$$\text{AA} + \text{ATP} \rightarrow \text{AA—AMP} + 2 \text{ phosphate groups}$$
$$\underset{\text{acid}}{\text{amino}} \tag{10-1}$$

The phosphate groups are released to the surrounding cytoplasmic solution. Much of the energy released as the phosphate groups are split from the ATP is trapped in the AA—AMP complex.

At this point in the process, the AA—AMP complex remains attached to the same enzyme molecule, which also speeds the next part of the reaction sequence. In this step, the amino acid is transferred from AMP to one of its specific tRNA molecules:

$$\text{AA—AMP + tRNA} \rightarrow \text{AA—tRNA + AMP} \qquad (10\text{-}2)$$
$$\text{aminoacyl-tRNA}$$

The AA—tRNA product, called *aminoacyl-tRNA*, and AMP are released from the enzyme when the reaction is complete. As in part 1 of the sequence, much of the energy released by converting ATP to AMP in Reaction 10-1 is retained in the AA—tRNA complex produced in Reaction 10-2. As a result, this complex is a high-energy substance. This energy is eventually used to drive formation of a peptide bond when amino acids link together on ribosomes.

The same enzyme, called *aminoacyl-tRNA synthetase*, catalyzes both Reactions 10-1 and 10-2. There are 20 different synthetase enzymes, each one specific in its activity for one of the 20 different amino acids. In the reaction, the enzymes recognize both a specific amino acid and one of the correct transfer RNA molecules for that amino acid, and match the two together. Thus, although tRNA molecules provide the necessary link between the nucleic acid code and the amino acid sequence of proteins, the accuracy of the system ultimately depends on the ability of the synthetase enzymes to match amino acids with their respective tRNAs.

The fact that the accuracy of protein synthesis ultimately depends on the ability of the synthetase enzymes to match amino acids and tRNAs has been amply demonstrated in experiments designed to "trick" the mRNA-tRNA pairing system on ribosomes. If an incorrect amino acid is experimentally linked to a tRNA, the ribosome system places it in a polypeptide nevertheless, with no recognition of the amino acid carried by the tRNA.

In summary, amino acid activation satisfies two requirements for protein synthesis. One of these is provision of the energy required for formation of peptide bonds. The second is matching tRNAs with their correct amino acids, which provides the basis for translation of the sequence of nucleotides in an mRNA into the sequence of amino acids in a protein. The amino acid-tRNA complexes formed as the final product of amino acid activation are the raw material as well as the primary energy source for the next phase of protein synthesis, the assembly of amino acids into polypeptide chains on ribosomes.

Step 2 in Protein Synthesis: Polypeptide Assembly on Ribosomes

In order to simplify protein synthesis we will consider the assembly of a hypothetical protein containing only two amino acids, arginine and serine, alternating in the sequence arg-ser-arg-ser . . . and so on. The information required for synthesis of this protein is coded in DNA in the sequence of the four nitrogenous bases A, T, G, and C. These are taken by threes to form the DNA codewords or *codons*. One of the DNA codewords for arginine is the sequence guanine-cytosine-guanine (GCG); one codon for serine is adenine-guanine-adenine (AGA). Therefore the sense chain of the DNA code for our hypothetical protein would carry the code GCG/AGA/GCG/AGA . . . and so on.

In the first step of the overall process, the DNA code is transcribed into an mRNA for the protein. In our example, the DNA coding sequence GCG/AGA/GCG/AGA . . . will be transcribed into the complementary mRNA sequence CGC/UCU/CGC/UCU . . . .

After transcription in the nucleus, the mRNA enters the cytoplasm and attaches to a ribosome. In the cytoplasm surrounding this mRNA-ribosome complex are all of the various kinds of transfer RNA, each attached to its specific amino acid. These elements—mRNA, the ribosomes, and the various tRNA-amino acid complexes—form the basic parts of the protein synthesis mechanism.

Among the tRNAs are several types that carry the amino acids arginine and serine. Of these, the ones of interest are the tRNAs able to recognize and bind to the CGC and UCU codons in our hypothetical messenger RNA. This recognition capacity depends on the *anticodon* of the tRNA molecule (see Figs. 10-1 and 9-12), a region able to form complementary base pairs with the codons in the message. The tRNA chain folds in such a way that the anticodon site is exposed at one end of the molecule, and the amino acid at the other end. The anticodon pairs with the codons on messenger RNA by forming the appropriate base pairs. We will consider that the tRNA carrying arginine has the anticodon GCG (Fig. 10-1a), which is able to recognize and bind to the CGC codon for arginine in the messenger RNA. The tRNA carrying serine has the anticodon AGA (Fig. 10-1b), able to recognize and bind to the UCU codon for serine in the message.

These amino acid-tRNAs interact with the messenger at the ribosome in the following way. If the mRNA sequence CGC is the first codon exposed at the ribosome

(Fig. 10-2a), a tRNA carrying arginine will attach to the ribosome and form complementary base pairs between the CGC codon on mRNA and the GCG anticodon of the tRNA (Fig. 10-2b). As the ribosome moves to the next codon on the mRNA it encounters the triplet UCU. A tRNA with the anticodon AGA, carrying the amino acid serine, pairs with the messenger at this point (Fig. 10-2c).

The two amino acids now at the ribosome are brought close together by this interaction, in a position favorable for the formation of the first peptide bond. Formation of the bond, catalyzed by an enzyme that is part of the ribosome itself, involves separating the amino acid from the first tRNA binding to the ribosome and transferring it to the amino acid carried by the most recently bound tRNA. In our example, arginine is removed from its tRNA and linked by a peptide bond to the serine carried by the second tRNA (Fig. 10-2d). The energy required to form the peptide bond comes from the free energy released when arginine separates from its tRNA. This tRNA, now free of its amino acid, is released from the ribosome.

The complex now consists of the ribosome, the messenger RNA, and a tRNA carrying the short polypeptide chain ser-arg. The ribosome now moves to the next codon on the message, which in our example is the triplet CGC. A tRNA carrying arginine will attach to the ribosome at this point (Fig. 10-2e), in a position that favors formation of the second peptide bond. In formation of this bond, the short ser-arg polypeptide is transferred to the arginine tRNA most recently attached to the ribosome (Fig. 10-2f), producing the three-amino acid chain arg-ser-arg. The entire process repeats as the ribosome moves to the next codon and continues until the end of the coding portion of the mRNA molecule is reached.

A ribosome begins translating an mRNA molecule at its 5′ end, the end synthesized first during transcription (see Information Box 9-2). Thus a ribosome reads an mRNA molecule in the 5′ → 3′ direction. The first amino acid placed into a protein under synthesis forms the NH_2-terminal end of the protein, and the last amino acid forms the —COOH terminal end (see p. 44). Thus proteins are assembled in the NH_2 → COOH direction.

Protein synthesis in the cytoplasm of living cells actually takes place by the mechanism outlined in Figure 10-2, except that the coded messages can specify any or all of the 20 amino acids. A large number of "factors," many of them enzymes, are required in addition to the basic elements (tRNA, mRNA, and the ribosome) outlined above. Some of these factors must be present for protein synthesis to begin; others are required for the process to continue; and still others are required to terminate protein

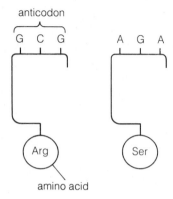

Figure 10-1 Transfer RNA molecules for arginine and serine, showing the anticodons assumed in the text.

synthesis, with release of the finished polypeptide from the ribosome. Additional energy, supplied by the nucleotide GTP, is also required for the most rapid and efficient progress of protein synthesis. The energy supplied by this nucleotide, which is broken down to GDP + phosphate in the process, is used to speed various steps in polypeptide assembly. One GTP is broken down to GDP as each successive aminoacyl-tRNA complex attaches to the ribosome, and a second GTP is converted to GDP as the ribosome advances to each new codon on the mRNA. Thus, including the ATP converted to AMP during amino acid activation (see Reaction 10-1), formation of each peptide linkage uses a total of 1ATP and 2GTP:

$$ATP + 2GTP \xrightarrow[\text{linkage}]{\text{formation of peptide}}$$

$$AMP + 2GDP + 4HPO_4^{2-} \qquad (10\text{-}3)$$

With some differences in detail, protein synthesis occurs in this general pattern in both prokaryotes and eukaryotes. The differences actually observed between the two major groups of organisms primarily involve the enzymelike factors that speed individual steps in the mechanism, and the identity of the first amino acid binding to ribosomes to begin protein synthesis. In eukaryotes, this amino acid is always methionine, coded by the codon AUG. In prokaryotes, methionine is also always used as the initial amino acid, coded by AUG in the messenger. However, in prokaryotes the initial methionine is modified by the addition of a *formyl* (—*CHO*) group. Thus all eukaryotic proteins begin with the amino acid methionine, and

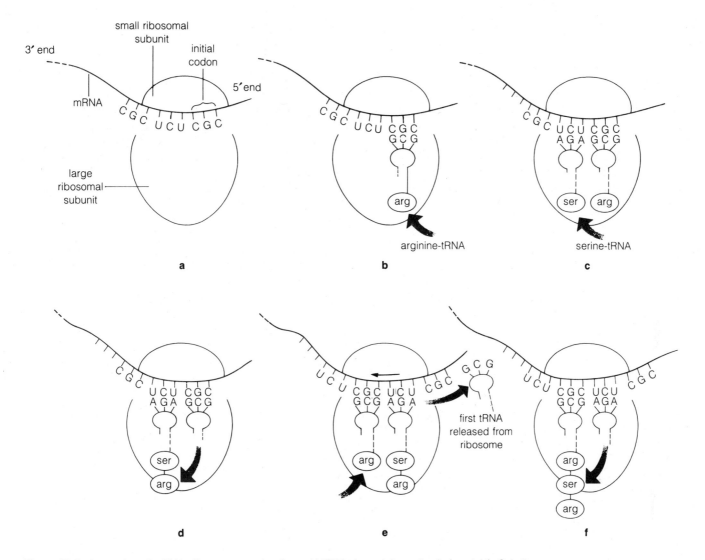

Figure 10-2 Interaction of mRNA, ribosomes, and amino acid-tRNAs in protein synthesis (see text). Only the anticodon arms of the tRNA molecules are shown in this diagram.

all prokaryotic proteins with the modified amino acid *formylmethionine.* Protein synthesis proceeds at a faster rate in prokaryotes: amino acids are added to a growing polypeptide chain at the rate of about 20 per second in these organisms. Eukaryotic protein synthesis proceeds at the rate of about one new amino acid added each second.

Once synthesis is complete, polypeptide chains may be variously modified. The methionine or formylmethionine at the "front" or NH_2 end of the new polypeptide may be removed. In prokaryotes, only the formyl group may be removed, leaving the terminal methionine in place. Amino acids in the interior of the polypeptide may be modified to other forms by the addition of acetyl, methyl, phosphate, hydroxyl, or other chemical groups. As a result of these modifications, as many as 140 different modified amino acids may occur in various natural proteins. Other modifications may involve enzymatic removal of a large part of the polypeptide chain as a part of conversion of the protein into its final, biologically active form. The digestive enzyme trypsin, for example, is converted into its active form from a larger precursor molecule, *trypsinogen,* by removal of a portion of the original polypeptide chain. Some proteins are altered by addition of metallic ions, or combination with complex lipid or carbohydrate-containing groups.

Completed proteins probably fold automatically into their final three-dimensional shape. Once the final folding pattern is taken up, it may be stabilized by formation of disulfide (S—S) linkages (see p. 46) located at different points in the polypeptide chain. The finished proteins, depending on their sequence of amino acids, become enzymes, antibodies, peptide hormones, or the structural, motile, transporter, or receptor molecules of the cell.

The Genetic Code for Protein Synthesis

Solving the Code Working out the nucleic acid codewords used in protein synthesis was one of the most exciting endeavors in the history of biology. The project attracted the interest and attention of scientists of all types: biologists, mathematicians, physicists, and even an astronomer. It stands as an unusual example of a biological conclusion contributed to by workers from practically all the major branches of science.

When the structure of DNA was discovered, it became apparent that the directions for assembling amino acids into proteins are stored in the sequence of nucleotides in DNA. The problems to be solved were how different combinations of only four different bases could designate 20 amino acids and, later, the identity of the individual codewords for the 20 amino acids.

By the early 1950s investigators realized that codewords would have to be at least three nucleotides long to specify 20 amino acids. Only four codewords could be made if the bases were taken one at a time ($4^1 = 4$); although 16 codewords could be made if all possible two-letter combinations were put together ($4^2 = 16$), 16 would still be insufficient to code for all 20 amino acids. However, three-letter codewords provide 64 combinations ($4^3 = 64$), more than enough for the total requirements of the code vocabulary. Therefore three-letter codewords seemed reasonable as a starting point for breaking the code.

Several different experimental approaches established that the genetic code actually uses three-letter codewords. One of the most definitive series of experiments was carried out by H. Gobind Khorana and his coworkers at the Massachusetts Institute of Technology, in which the coding capacity of artificial mRNAs was tested. For their experiments, Khorana and his colleagues made artificial

RNA molecules with different repeating sequences and added the mRNAs to a cell-free system containing ribosomes, amino acid-tRNA complexes and all of the other factors required for protein synthesis. Artificial mRNAs with a repeating sequence of the form ABABABAB . . . , where A and B are any two RNA nucleotides, coded for the synthesis of a polypeptide containing only *two* amino acids when added to the cell-free protein synthesis system. In the polypeptide produced by this artificial mRNA, the two amino acids alternated regularly. Assuming three-letter codewords, this is the expected result, because the code in the artificial mRNA would then be read

$$\overset{1}{ABA}-\overset{2}{BAB}-\overset{1}{ABA}-\overset{2}{BAB} \cdots;$$ only two different "words" are contained in the message. Repeating artificial mRNAs of the form ABCABCABCABC . . . coded for the synthesis of polypeptides containing only *one* type of amino acid. This is also expected because in a triplet code this message would be read $$\overset{1}{ABC}-\overset{1}{ABC}-\overset{1}{ABC}-\overset{1}{ABC} \cdots.$$ Finally, repeating artificial mRNAs of the form ABCDABCDABCDABCD . . . coded for chains of polypeptides containing alternating sequences of *four* different amino acids. This is expected only if the message ABCDABCDABCDABCDABCDABCD . . . is read in triplets as

$$\overset{1}{ABC}-\overset{2}{DAB}-\overset{3}{CDA}-\overset{4}{BCD}-\overset{1}{ABC}-\overset{2}{DAB}-\overset{3}{CDA}-\overset{4}{BCD} \cdots.$$ Khorana's results are therefore possible only if the codewords in the nucleic acid code are triplets.

The next task was to identify the codewords. Although many investigators solved parts of the puzzle, Marshall W. Nirenberg and Phillip Leder at the National Institutes of Health carried out the most definitive work. Nirenberg and Leder began their experiments by adding synthetic mRNA molecules containing only a single repeated base to cell-free protein synthesis systems. The first synthetic mRNA tested was poly(U), a nucleic acid containing only uracil in which the codeword UUU is repeated many times. Poly(U), when added to the cell-free system, was found to code for synthetic proteins containing only the amino acid phenylalanine. This established that one codeword for phenylalanine is *UUU*. Poly(A) and poly(C) were found by the same techniques to code for synthetic proteins containing only lysine and proline respectively, establishing that *AAA* stands for lysine and *CCC* for proline in the code. Thus three words of the "dictionary" were established.

In 1964 Nirenberg used a new approach to work out the remaining codewords. All 64 codewords were artificially synthesized by assembling short nucleic acid molecules containing only three nucleotides. These short mRNAs were then added one at a time to a cell-free system containing ribosomes and all the factors required for protein synthesis. The three-letter mRNAs caused single tRNA molecules, carrying a specific amino acid, to bind to ribosomes. By identifying the amino acid, the coding assignment of each triplet codon was worked out. Nirenberg and Khorana received the Nobel Prize in 1968 for their work in solving the genetic code.

This approach allowed all of the codewords to be identified (Fig. 10-3). By convention, the codewords are designated in RNA rather than DNA equivalents. The RNA codon *UUU* for phenylalanine, for example, corresponds to the DNA codon *AAA*. Three codons, *UAG*, *UAA*, and *UGA*, were found to have no coding assignments. These codons, which were at first termed "nonsense codons" because they do not code for an amino acid, were later found to be a signal for the ribosome to terminate protein synthesis. Thus the codons *UAG*, *UAA*, and *UGA*, now called *terminator codons*, represent the word "stop" in the coded message for protein synthesis. When a ribosome encounters one of these three codons, protein synthesis ceases and the mRNA and finished polypeptide are released.

Second base of codon

First base of codon	U		C		A		G		Third base of codon
U	UUU	Phe	UCU	Ser	UAU	Tyr	UGU	Cys	U
	UUC		UCC		UAC		UGC		C
	UUA	Leu	UCA		UAA *		UGA *		A
	UUG		UCG		UAG *		UGG	Trp	G
C	CUU	Leu	CCU	Pro	CAU	His	CGU	Arg	U
	CUC		CCC		CAC		CGC		C
	CUA		CCA		CAA	Gln	CGA		A
	CUG		CCG		CAG		CGG		G
A	AUU	Ile	ACU	Thr	AAU	Asn	AGU	Ser	U
	AUC		ACC		AAC		AGC		C
	AUA		ACA		AAA	Lys	AGA	Arg	A
	AUG	Met	ACG		AAG		AGG		G
G	GUU	Val	GCU	Ala	GAU	Asp	GGU	Gly	U
	GUC		GCC		GAC		GGC		C
	GUA		GCA		GAA	Glu	GGA		A
	GUG		GCG		GAG		GGG		G

Figure 10-3 The genetic code in RNA codewords. The triplets marked with an asterisk are terminator codons that signal the end of a message and cause the ribosome to release a finished protein.

Characteristics of the Code Inspection of the code reveals that all but two of the amino acids are specified by more than one codeword. This feature of the code, called *degeneracy*, means simply that many of the codewords are synonyms. The amino acid proline, for example, was found to have the four synonymous codons *CCU*, *CCC*, *CCA*, and *CCG* in the code.

The degeneracy in the genetic code probably compensates for mismatches that may occur when the anticodons of tRNA molecules pair with the codons on mRNA. According to the *wobble hypothesis* advanced by Francis H. C. Crick, one of the discoverers of DNA structure, the fact that the anticodons on tRNA molecules extend outward from the tip of a foldback double helix (see Fig. 9-10), much like the toes on the end of a foot, gives them freedom to move or "wobble." Because of this freedom, the anticodon bases can twist sufficiently to form unusual base pairs in addition to the usual A-U and G-C possibilities. For example, Crick has shown that the base U, when extended into space as the third base of an anticodon triplet, can twist enough to pair with G as well as with A. As a result, the anticodon UGU can pair with the mRNA codon ACG as well as the expected ACA codon.

This flexibility would lead to mistakes in the insertion of amino acids during protein synthesis except for the degeneracy of the code. According to Crick's analysis, the degeneracy exactly compensates for any unusual codon-anticodon pairs that may form by insuring that all of the possible "mistakes" code for the same amino acid. For example, both of the codons ACG and ACA, that might pair with the UGU anticodon, code for the same amino acid, threonine. Thus degeneracy is probably nature's way of getting around the inaccuracies in codon-anticodon pairing at the ribosome.

A second feature of the genetic code is its almost completely universal usage in living organisms. With a very few exceptions, the same codons stand for the same amino acids in all organisms from bacteria to the higher eukaryotes, and also in viruses. The exceptions, recently discovered, occur in the DNA-protein synthesis systems of mitochondria, where UGA codes for tryptophan instead of termination, AUA codes for methionine instead of isoleucine, and CUA codes for threonine instead of leucine in some polypeptides. These coding substitutions apparently depend on the unusual properties of the tRNA molecules of mitochondria (see Supplement 10-2 for details). The otherwise universal nature of the code indicates that it has very ancient origins and became established as one of the earliest events in the evolution of cellular life. It will be interesting to see, if life based on the same interdependent nucleic acid-protein system is ever discovered on another planet, if the codewords used are the same as our own—that is, if the code is really universal in the cosmic sense.

Cytoplasmic Structures in Protein Synthesis

Several cytoplasmic structures function in the synthesis and modification of proteins. The simplest of these are the ribosomes. Ribosomes occur in large numbers in the cytoplasm, either freely suspended or attached to the membranes of a system of cytoplasmic membranes, the *endoplasmic reticulum* (ER). Ribosomes assemble proteins in large quantities in either location. Evidently the existence of ribosomes in either freely suspended or attached form depends on the ultimate destinations of the proteins syn-

thesized. Freely suspended ribosomes make proteins that become a part of the soluble substance of the cytoplasm, or of structures such as microtubules that become suspended directly in the cytoplasm. The ribosomes attached to ER membranes assemble proteins that either become a part of cellular membranes or are enclosed in membranous vesicles. The contents of these vesicles are stored in the cytoplasm or released to the outside of the cell in cellular secretions. Another cytoplasmic structure, the *Golgi complex*, may be involved in modifying the proteins assembled in the ER, through chemical changes involving the attachment of lipid or carbohydrate groups to the newly synthesized proteins. Finished proteins may also be packed into storage or secretion vesicles in the Golgi complex.

Ribosome Structure

The subparts of eukaryotic ribosomes are assembled in the cell nucleus, within the nucleolus. The four kinds of ribosomal RNA molecules (see Information Box 10-1) that oc-

---

Information Box 10-1:

Ribosomes and Ribosomal RNA

Ribosomes consist of two subunits, both containing rRNA molecules in combination with proteins. Four distinct kinds of rRNA molecules can be extracted from eukaryotic ribosomes, three from the large ribosomal subunit and one from the small subunit. These rRNA molecules are usually described in terms of *Svedberg* or *S* units, which indicate the rates at which molecules descend in a test tube when spun at high speeds in a centrifuge. Generally, the larger the S number, the higher the molecular weight. In S units, the three rRNAs of the large ribosomal subunit of eukaryotes are identified as 28S, 5.8S, and 5S rRNA, with molecular weights of approximately 1,800,000, 57,000, and 50,000, respectively. The single rRNA of the smaller subunit is identified as 18S rRNA, with a molecular weight of approximately 700,000.

The rRNAs of prokaryotic ribosomes are smaller molecules with correspondingly smaller S values. Only two kinds of rRNA can be extracted from the large subunit of these organisms. These rRNAs have values of 23S and 5S and correspond to the 28S and 5S rRNAs of eukaryotes. The small ribosomal subunit of prokaryotes contains a single 16S rRNA molecule, equivalent to the 18S rRNA of eukaryotic ribosomes. The 23S, 16S, and 5S rRNAs of bacteria have molecular weights of 1,100,000, 550,000, and 42,000 respectively.

The rRNAs and proteins of the small ribosomal subunit of prokaryotes combine to form a particle centrifuging at 30S; the large subunit centrifuges at 50S. A complete prokaryotic ribosome centrifuges at 70S. The equivalent eukaryotic subunits centrifuge at 40S for the small subunit and 60S for the large subunit; complete eukaryotic ribosomes centrifuge at 80S.

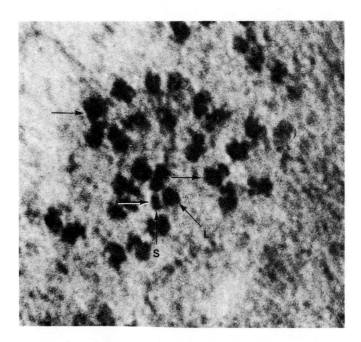

Figure 10-4 Ribosomes (arrows) in a thin-sectioned rat liver cell. The large and small subunits and the cleft marking the division between the subunits are visible in many of the ribosomes. Ribosomes occurring in rows such as these are probably reading different segments of the same messenger RNA molecule. *S*, small subunit; *L*, large subunit. Courtesy of N. T. Florendo.

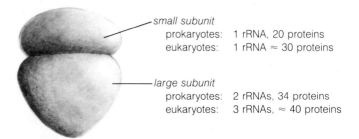

small subunit
prokaryotes: 1 rRNA, 20 proteins
eukaryotes: 1 rRNA ≈ 30 proteins

large subunit
prokaryotes: 2 rRNAs, 34 proteins
eukaryotes: 3 rRNAs, ≈ 40 proteins

Figure 10-5 Combination of a small and large ribosomal subunit to form a complete ribosome. The figures show the number of rRNAs and proteins in prokaryotic and eukaryotic ribosomes.

cur in eukaryotic ribosomes are transcribed in the nucleus, three of them within the nucleolus itself. The ribosomal proteins are made in the cytoplasm, and then enter the nucleus to be assembled with the rRNAs into ribosomal subunits in the nucleolus. The completed subunits are released from the nucleolus into the nucleoplasm as separate particles, from where they travel through the nuclear pore complexes to reach the cytoplasm. The process is similar in prokaryotes except that one fewer rRNA species is transcribed on the rRNA genes to produce three rRNA types. Transcription, processing, and assembly of these rRNAs into ribosomal subunits in prokaryotes probably takes place around the edges of the nucleoid. The completed subunits, as in eukaryotes, are released from the nucleoid as separate particles. In both prokaryotes and eukaryotes, the subunits remain separate until they join by twos with an mRNA molecule to initiate protein synthesis. The two subunits of complete ribosomes are often clearly visible in electron micrographs, in which a cleft or line marks the junction between them (arrows, Fig. 10-4).

Ribosomes can be easily isolated from cells and purified. On analysis, the ribosomes of bacteria prove to be about 60 percent rRNA and 40 percent protein by weight. Eukaryotic ribosomes contain a smaller proportion of RNA, with the ratio of RNA to protein by weight approaching 1:1.

Disassembly of isolated ribosomes is accomplished by taking advantage of the fact that the two ribosomal subunits, and their rRNA and protein molecules are held together primarily by electrostatic interactions. As a consequence, they can be induced to come apart by relatively simple procedures such as adjustments in the salt concentration of the isolating medium.

When separated in this way, the larger subunit of bacterial ribosomes proves to contain two of the three bacterial rRNAs (see Information Box 10-1) in combination with 34 proteins. The smaller subunit yields 20 proteins, in combination with the remaining rRNA type, for a total of 54 ribosomal proteins (Fig. 10-5). Eukaryotic ribosomes have a more complex assemblage of proteins. Difficulties in separating these proteins have produced conflicting results. As a consequence, the total number of proteins and the numbers per subunit are still in dispute. Most researchers report between 70 to 80 ribosomal proteins in eukaryotic ribosomes, with about 40 in the large and 30 in the small subunit. Few proteins of eukaryotic and prokaryotic ribosomes seem to be related. However, in spite of differences in the numbers and kinds of rRNAs and proteins present, eukaryotic and prokaryotic ribosomes assemble polypeptides by essentially the same mechanisms.

Some of the proteins in the ribosomal subunits are enzymes. The largest subunit, for example, contains the enzyme that catalyzes formation of peptide bonds. Other proteins undoubtedly form part of the structural framework of ribosomes and do not have enzymatic functions.

Interestingly enough, the exact function of the rRNA molecules of ribosomes is unknown. However, these RNA molecules are certain to be directly involved in the assembly of amino acids into proteins. They probably also bind the various parts of ribosomes together, and coordinate with ribosomal proteins in binding ribosomes to mRNA and tRNA.

During the late 1960s, Masayasu Nomura and his colleagues at the University of Wisconsin were successful in reassembling small ribosomal subunits from the separately purified rRNAs and ribosomal proteins of *E. coli.* Somewhat later, in 1974, Knud H. Niehaus and Ferdinand Dohme, working at the Max Planck Institute for Molecular Genetics in Germany, successfully reassembled large ribosomal subunits from the separately purified large subunit rRNAs and proteins of the same bacterium. These reassembled large and small subunits, when combined together under the appropriate conditions, formed fully functional ribosomes. The test-tube assembly of these complex RNA-protein particles is, at one and the same time, a testimony to the ingenuity of the biochemists involved and an elegant demonstration of the ability of cellular macromolecules to self-assemble. Successful reassembly of eukaryotic ribosomes in the test tube has not yet been accomplished.

Although separate large and small ribosomal subunits are easily isolated from the nucleus, complete ribosomes have never been detected anywhere in cells except in the cytoplasm. Since functional, complete ribosomes do not occur in the nucleus, all protein synthesis must therefore take place in the cytoplasm. Experiments using radioactive amino acids to detect newly synthesized proteins support this conclusion; when the amino acids are added, newly labeled proteins can be detected in the cytoplasm but not in the nucleus (see below and Fig. 10-14).

The Endoplasmic Reticulum

The membranes of the endoplasmic reticulum take two forms in eukaryotic cytoplasm. These membranes, with ribosomes attached, are termed the *rough endoplasmic reticulum* (or *rough ER*) because the ribosomes covering their surfaces give them a serrated or rough appearance in the electron microscope (see Figs. 10-6 to 10-9). Similar cytoplasmic membranes without attached ribosomes are called *smooth endoplasmic reticulum* or *smooth ER* (see Figs. 10-6 and 10-10).

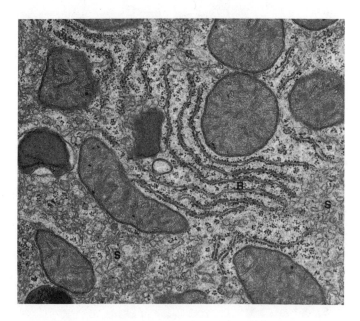

Figure 10-6 A liver cell from a hamster, showing both rough (*R*) and smooth (*S*) endoplasmic reticulum. × 21,000. Courtesy of D. W. Fawcett, from *The cell*, © 1966 by W. B. Saunders Company.

The membranes of the rough ER take the form of tubules, vesicles, and large flattened sacs, all with ribosomes attached to the surfaces facing the surrounding cytoplasm. The largest ER sacs are called *cisternae*. The sacs and tubules form an interconnected, branched network that extends through much of the cytoplasm in many cell types. In the branched networks, the boundary membrane is continuous and unbroken and completely separates the channel enclosed inside the ER from the surrounding cytoplasm.

In most cells ribosomes are also attached to the outermost membrane of the nuclear envelope, on the side facing the surrounding cytoplasm (see Fig. 10-9). Connections can sometimes be seen between the outer nuclear membrane and the rough ER (arrow, Fig. 10-9), making the compartment between the two membranes of the nuclear envelope (the perinuclear compartment) continuous with the inner ER channels at these points. These connections indicate that the outer nuclear membrane is probably closely related in structure and function to the rough ER.

The relative amounts of free and membrane-bound ribosomes vary considerably depending on the cell type. At one extreme are cells such as the pancreatic or salivary gland cells of mammals. In these cells, which actively se-

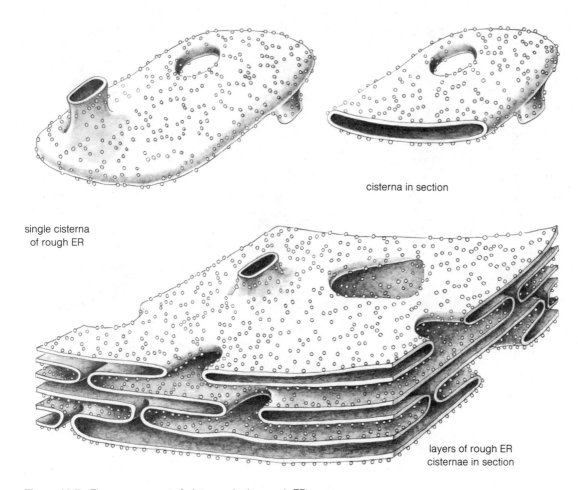

cisterna in section

single cisterna
of rough ER

layers of rough ER
cisternae in section

Figure 10-7 The arrangement of cisternae in the rough ER.

crete large quantities of protein, rough ER membranes almost completely fill the cytoplasm (as in Fig. 10-8; see also Fig. 1-6). At the other extreme are cells such as muscle and kidney cells in which rough ER is almost completely absent and most ribosomes are unattached. Other animal cell types fall at all possible points between these two extremes. Plant cells typically contain more sparsely distributed deposits of rough ER (as in Fig. 10-11).

Smooth ER membranes, which carry out a transport function in protein synthesis, lack ribosomes and form primarily tubular sacs that are generally less distinct and of smaller dimensions than the rough ER (Figs. 10-6 and 10-10). At many points, the smooth and rough ER membranes may connect together, forming a continuous inner channel enclosed by the two systems.

The Golgi Complex

The Golgi complex, named for the cell biologist who first described it in the 1800s, is formed from flattened, saclike vesicles assembled into a stack (see Figs. 1-9 and 10-12). Each flattened vesicle in the stack consists of a single, continuous membrane enclosing an inner space. The spacing in and between the stacks is very regular except at the edges, which appear swollen or dilated. The edges of the sacs fragment or bud off, producing large numbers of vesicles at the margins of the Golgi complex (see Fig. 10-12). These vesicles may fuse together to form larger storage vesicles that remain in the cytoplasm, or *secretion vesicles* (see Fig. 10-12) that are released to the outside of the cell.

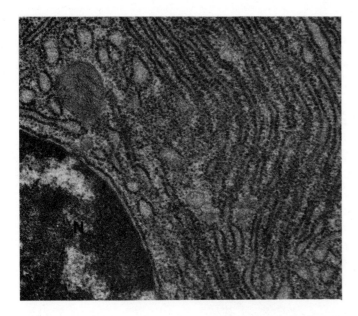

Figure 10-8 Closely packed cisternae of the rough ER in a pancreatic cell from a rat. N = nucleus. × 29,000. Photograph by the author.

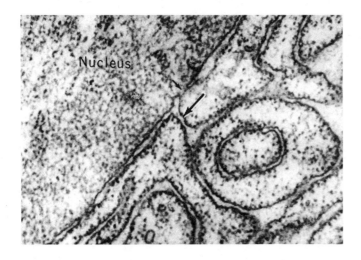

Figure 10-9 A connection between the outer membrane of the nuclear envelope and the rough ER (large arrow) in a nucleus of mouse salivary gland. The outer membrane of the nuclear envelope is typically covered with ribosomes. The small arrow points to a pore complex. × 62,000. From *The cell* by D. W. Fawcett, 1966. Courtesy of D. W. Fawcett, H. Parks, and the W. B. Saunders Company.

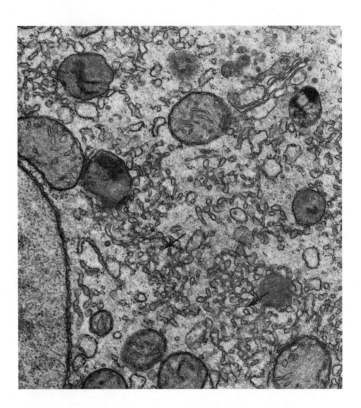

Figure 10-10 Smooth ER membranes (arrows) in a hamster liver cell. × 18,000. Courtesy of D. W. Fawcett, from *The cell*, © 1966 by W. B. Saunders Company.

No ribosomes occur between the sacs or in the region immediately surrounding the Golgi complex.

Golgi complexes are usually closely associated with the ER, separated from it only by a layer of small, smooth-walled vesicles. Some of the vesicles in this layer can be seen to arise as buds from the ER, and others to join with the nearest Golgi sac (see Fig. 10-13). From this and other evidence the vesicles of this layer, called *transition vesicles*, are believed to link the two membrane systems together, probably by budding off from the ER and fusing with the Golgi complex.

Golgi complexes are found in multiple numbers in cells. In plants, they are more or less evenly scattered through the cytoplasm (as in Fig. 10-11). In animal cells, Golgi complexes frequently occur in masses near the ER or just outside the nucleus. The total number of Golgi complexes is extremely variable in cells from different tissues or species. While the average number is about 20 per cell, there may be many more. For example, corn root tip cells contain several hundred complexes; more than 25,000

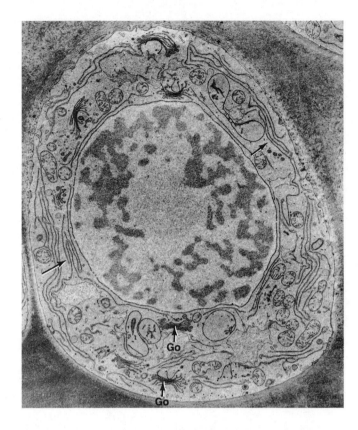

Figure 10-11 Rough ER (arrows) and Golgi complexes (*Go*) in a root cap cell of corn. The cell was fixed with potassium permanganate, a chemical that outlines membranes clearly but does not preserve ribosomes. Thus ribosomes are not visible on the ER membranes. × 7,000. Courtesy of W. G. Whaley, J. E. Kephart, H. H. Mollenhauer, and Academic Press, Inc., from *Cellular Membranes in Development*, Academic Press, 1964.

have been counted in cells of *Chara*, a green alga. Some animal tissues, such as cells of the salivary glands in insects, may also contain Golgi complexes numbering in the thousands.

Functions of the ER and Golgi Complex in Protein Synthesis

The coordination of the ER and Golgi complex in the synthesis and transport of proteins has been followed by tracing the uptake and movement of amino acids tagged by radioactivity. In studies of this type, pioneered by George E. Palade and his colleagues at the Rockefeller University, cells actively engaged in protein synthesis and secretion

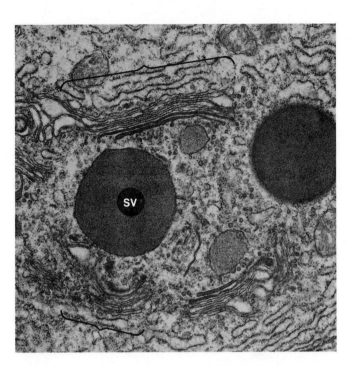

Figure 10-12 The sacs and secretion vesicles of several Golgi complexes (brackets). *SV*, secretory vesicle. Courtesy of H. W. Beams and R. G. Kessel and Academic Press, Inc., from *International Review of Cytology* 23 (1968): 209.

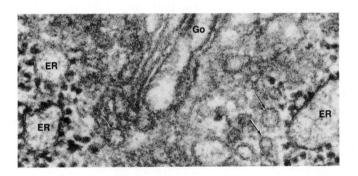

Figure 10-13 Small vesicles called *transition vesicles* (arrows) are believed to conduct proteins from the rough ER (ER) to the Golgi complex (*Go*). Cat pancreas, × 108,000. From *Cytology and cell physiology*. Ed. G. H. Bourne, 1964. Courtesy of F. S. Sjöstrand and Academic Press, Inc.

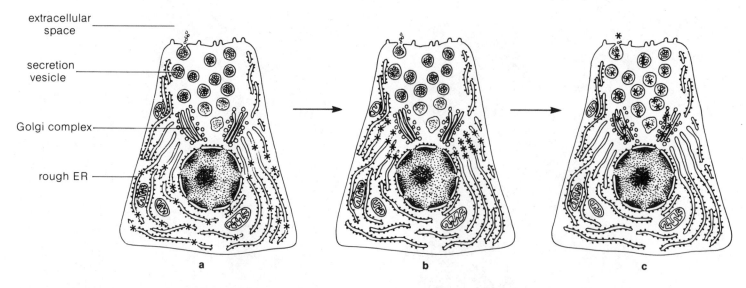

extracellular space

secretion vesicle

Golgi complex

rough ER

a   b   c

Figure 10-14 Migration of labeled proteins (asterisks) after injection of radioactive amino acids into secretory cells. (**a**) Three minutes after injection; only the rough ER is labeled. (**b**) After 10–15 minutes; label is distributed in the smooth ER near the Golgi complex. (**c**) After 30 minutes; label is located in the Golgi complex and the secretory vesicles. The autoradiography technique used in this type of experiment is described in the Appendix. Adapted; original courtesy of P. Favard.

are exposed to radioactive amino acids. If the cells are fixed and sectioned within a few minutes after exposure, radioactivity can be detected only over the rough ER, indicating that the amino acids are assembled into proteins in this region (Fig. 10-14a). If the cells are prepared about 10 to 15 minutes after exposure, radioactivity can be detected within vesicles of the smooth ER (Fig. 10-14b). This indicates that the proteins assembled on the ribosomes of the rough ER are transported across the ER membranes and concentrated inside the ER cisternae. The proteins then move within the enclosed channels to the smooth ER. If cells are fixed after 20 to 30 minutes, the radioactive label also appears over the transition vesicles between the smooth ER and the Golgi complex, and over the Golgi complex itself. Subsequently, the label appears over the larger secretion vesicles lying between the Golgi complex and the plasma membrane (Fig. 10-14c). Finally, within 1 to 4 hours after initial exposure to the labeled amino acids, radioactivity appears in the spaces just outside the secretory cells, showing that the large vesicles eventually discharge their contents to the cell exterior. These studies link the various membranous elements of the cytoplasm into a coordinated system for synthesis, transport, and secretion

of proteins, and show that proteins follow the route rough ER → smooth ER → transition vesicles → Golgi complex → secretion vesicles → cell exterior through the system.

Several variations of this basic route are observed in different cell types. In some cells, transition vesicles bud off directly from the rough ER, bypassing the smooth ER. In others, newly synthesized proteins may bypass the Golgi complex. Secretion vesicles in some cells are stored in the cytoplasm, either temporarily as a prelude to later release, or more or less permanently as specialized structures of the cytoplasm (the synthesis and functions of *lysosomes*, specialized vesicles of this type, are described in Supplement 10-1). Another much used pathway for proteins synthesized in the rough ER inserts newly synthesized proteins into cellular membranes. In this pathway, proteins synthesized on ER ribosomes penetrate into the ER membrane bilayers and are retained as the membranes successively form parts of the smooth ER, Golgi complex, secretion vesicles, and plasma membrane (see Fig. 10-15). Since ribosomes also occur on the outside of the nuclear envelope, and connections can be found between the nuclear envelope outer membrane and the ER, the nuclear envelope and the compartment enclosed between its two

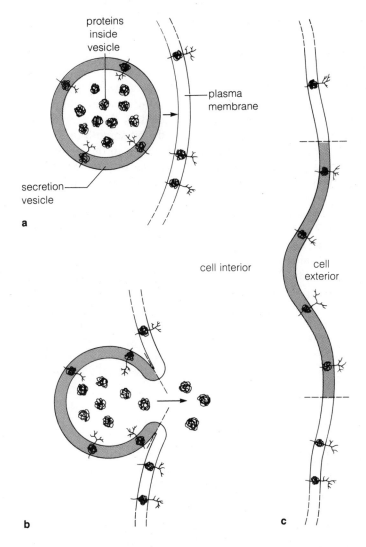

proteins
inside
vesicle

plasma
membrane

secretion
vesicle

a

cell interior

cell
exterior

b

c

Figure 10-15 Fusion of secretion vesicles with the plasma membrane. The fusion inserts the vesicle membrane, and proteins embedded in the membrane, into the plasma membrane. This is probably the major route by which new material is added to the plasma membrane.

membranes are probably also connected directly to the cytoplasmic system for synthesizing, transporting, and secreting proteins.

The role of the Golgi complex in modifying proteins has been worked out by studies using labeled monosaccharides such as glucose, or the precursors of lipid groups. These studies show that some of the labeled sugars or lipids are added to proteins to form glycoproteins and glycolipids in the Golgi complex.

A part of the information supporting these conclusions comes from experiments with labeled precursors of the carbohydrate groups of glycoproteins (see p. 88, Chap. 5). In one of these experiments, M. Neutra and C. P. Leblond of McGill University exposed mucus-secreting cells of rat intestine to labeled glucose and traced the progress of the label through the cells (mucus contains a glycoprotein as its major constituent). Within 3 to 5 minutes after exposing the cells to the label, radioactivity could be detected in the Golgi complex. After 20 minutes, the label migrated to the small vesicles at the periphery of the complex. After 40 minutes, radioactivity was found concentrated into large, mucus-containing secretion vesicles near the margin of the cell. Similar experiments also indicate that carbohydrate units are incorporated into proteins in quantity in the Golgi complex in plants. For example, root cap cells secreting a viscous glycoprotein solution first incorporate labeled glucose into the glycoprotein in the Golgi complex.

Experiments with labeled sugars suggest that the addition of carbohydrate groups to some glycoproteins is coordinated in both the ER and Golgi. In some cells, sugars such as glucosamine, mannose, and glucose are initially taken up and added to growing polypeptides in the ER. These sugars are frequently found as a part of the basal structure of the complex carbohydrate "antennae" added to glycoproteins (see Fig. 5-1). Additional glucose, and other sugars such as fucose, galactose, and sialic acid (see Fig. 5-2) that are found predominantly in the side branches of carbohydrate groups are incorporated in the Golgi complex. These observations suggest that in some cells synthesis of the core unit of the carbohydrate groups of glycoproteins is initiated in the rough ER as soon as the new polypeptide extends into the ER membrane. Addition of the complex side groups to complete the branched carbohydrate structure takes place in the Golgi complex.

Concentration of Completed Proteins in Secretion Vesicles
From the labeling studies it is clear that, once modifications are complete, newly synthesized proteins, glycoproteins, and lipoproteins are concentrated inside the secretion vesicles that bud off from the Golgi complex. Once concentrated inside the secretion vesicles the enclosed products may remain in storage in the cytoplasm or may be secreted to the exterior. Secretion is accomplished by movement of the secretion vesicle to the cell border and fusion of the vesicle membrane with the plasma membrane (Fig. 10-15a and b). This fusion releases the vesicle contents to the exterior, and introduces the former vesicle membrane as a patch in the plasma membrane (see Fig.

10-15c). As a result, any proteins or glycoproteins forming a part of the vesicle membrane are transferred to the plasma membrane, oriented so that segments formerly facing the vesicle interior now extend from the outer cell surface. This transfer has been proposed as a major source of new membrane material for the cell surface, occurring as a part of the general "membrane flow" from the ER and Golgi to the plasma membrane.

The secretory vesicles of different cell types may contain a wide variety of products. These include the digestive enzymes secreted by pancreatic cells, the mucus of intestinal goblet cells, and the slime secreted by root cap cells, already mentioned in preceding sections. Other products included in secretory vesicles are the secretions of pituitary, leucocyte, and mammary gland cells, and the yolk proteins of egg cells. This diversity in secretory product is probably reflected in wide differences in the enzymes present in the Golgi complex in different cells.

Other Functions of the ER and Golgi Complex The smooth endoplasmic reticulum carries out an extensive array of biochemical reactions in addition to serving as transport elements for proteins synthesized in the rough ER. The membranes of this system carry out the initial reactions in the oxidation of fats (see p. 154, Chap. 7). Other enzymes associated with metabolic pathways have also been detected in the smooth ER, including those synthesizing phospholipids, glycolipids, and steroids. Detoxification of drugs and poisons also takes place in the smooth ER. This diversity suggests that the membranes now classified together as smooth ER probably include several distinct types with different structures and enzymatic machinery.

In plants, the Golgi complex functions in cell wall synthesis as well as in modification of proteins assembled in the rough ER. In some algae, cellulose scales or plates that become incorporated in the cell wall are first assembled in the Golgi complex. In higher plants, the Golgi complex apparently synthesizes the noncellulose molecules that become a part of the cell wall (see p. 98).

It is possible that, in addition to their roles in cellular synthesis, segments of the ER-Golgi system may also provide a more general transport pathway in cells. The channels of the rough and smooth ER, and their possible connections with the nuclear envelope and plasma membrane, have often been proposed to act in this way, as an intracellular "waterway" that conducts substances into the cell as well as to the exterior (Fig. 10-16). There is some experimental support for this proposal. For example, L. Palay and L. J. Karlin of Harvard University followed the absorption of fats by intestinal cells in animals that had been fed after a long period of fasting. Immediately after feeding, droplets of lipid could be seen in vesicles formed from invaginations of the plasma membrane in the intestinal cells. These vesicles pinched off from the plasma membrane and appeared to merge with the smooth ER. In cells prepared 30 minutes after feeding, lipid droplets appeared within the vesicles of both the smooth and rough ER. Some lipid droplets were also seen in the perinuclear compartment. Thus in these cells, an intermittent waterway conducting materials from the outside to the cell interior seems actually to exist.

How general such waterways may be remains uncertain. Highly developed ER systems extensive enough to fill the space between the nuclear envelope and plasma membrane occur only in some cell types such as secretory cells. Many other cells, as noted above, contain only limited deposits of ER. Further, the required connections between the ER and the plasma membrane, at one end of the system, and the nuclear envelope, at the other end, are too rarely seen to be considered a regular feature of cell structure. For these reasons it is possible that the ER forms two-directional waterways only in cells highly specialized for absorption or secretion.

How Proteins Enter the ER-Golgi Pathway: The Signal Hypothesis

Following the discoveries implicating the rough ER in protein synthesis, further research centered around two major questions. How do mRNAs that code for proteins that are to be synthesized in the rough ER "know" that they are to associate with ribosomes that attach to the ER? Once the ribosomes attach, how do the proteins cross the ER membranes to get inside the membranes or ER cisternae? By 1971 work had progressed far enough for Günter Blobel and his coworkers at the Rockefeller University to advance an elegant model, the *signal hypothesis*, providing answers to both of these questions.

Blobel has proposed that the mRNAs for proteins to be assembled in the rough ER contain, usually at their "front" or 5' ends, a sequence of triplets that codes for a chain of hydrophobic amino acids. This sequence is called the *signal*. These mRNAs start protein synthesis by binding to ribosomes in the usual way. As the initial ribosome attaches to one of these mRNAs and begins to translate the coded message, the first segment of the new polypeptide to emerge is the hydrophobic signal sequence (Fig. 10-17a and b). On collision with an ER membrane, the signal,

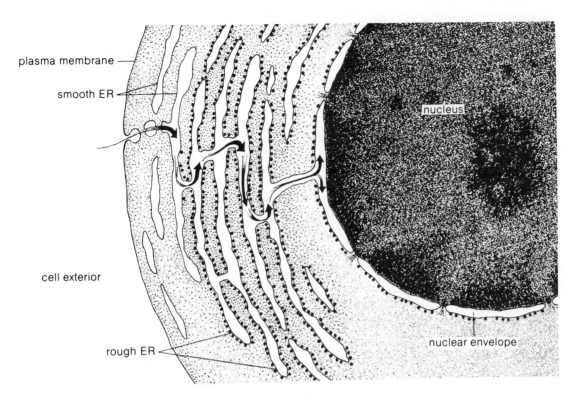

plasma membrane

smooth ER

nucleus

cell exterior

nuclear envelope

rough ER

Figure 10-16 The hypothetical "waterway" that may extend from the cell exterior through the smooth and rough ER to the perinuclear space (see text). Note that the connection between the ER channels and the cell exterior is indirect, through the medium of vesicles that invaginate from the plasma membrane.

by virtue of its hydrophobic character, attaches and penetrates into the membrane bilayer (Fig. 10-17c). As protein synthesis continues, the growing polypeptide chain follows the signal sequence into the membrane (Fig. 10-17d). If the protein is to be secreted, the entire polypeptide eventually passes through the ER membrane and into the ER cisternae. If the protein is to remain as a part of the membrane, movement of the growing polypeptide stops as it becomes fully embedded in the hydrophobic and hydrophilic membrane regions (whether the protein remains in the membrane or penetrates into the cisternae depends on the amino acid sequence of the entire protein). Soon after full extension of the growing polypeptide into or through the ER membrane, an enzyme of the membrane may clip the signal sequence from the new protein (Fig. 10-17e). Alternatively, the sequence may remain as a permanent part of the protein molecule. At completion of synthesis, the ribosomes separate from the mRNA and release from the ER membrane (Fig. 10-17f).

The signal hypothesis is supported by an extensive series of experiments carried out by the Blobel group and others. If ribosomes in a cell-free system are supplied with mRNAs for secretory proteins, the proteins synthesized are longer than the finished length actually secreted by living cells by about 16 to 23 amino acids. The extra sequence, which is usually located at the front or NH_2-terminal end of the proteins synthesized, has a high proportion of hydrophobic amino acids as proposed by the signal hypothesis. Thus newly synthesized secretory proteins actually contain a hydrophobic segment equivalent to the signal sequence, as proposed by Blobel. If isolated rough ER vesicles are added to the cell-free system, the mRNA-ribosome complexes attach to them, and the newly synthesized proteins concentrate inside the membranes or vesicles. Once inside the vesicles, the signal sequence is usually clipped off. Thus the major elements proposed in the signal hypothesis have been shown to function as Blobel's model proposes.

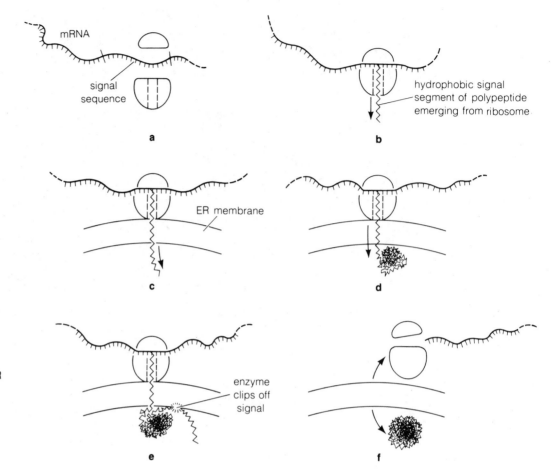

Figure 10-17 Blobel's signal hypothesis for the assembly of polypeptides on the rough ER (see text). Proteins may either penetrate entirely into the ER sacs as shown, or may remain in the ER membranes as an integral membrane protein.

A Summary of the Role of the ER and Golgi Complex in Protein Synthesis

Figure 10-18 summarizes the activity of ribosomes and cytoplasmic membrane systems in protein synthesis and secretion. Proteins are assembled on ribosomes that are either freely suspended in the cytoplasm or bound to membranes of the ER. The proteins made on freely suspended ribosomes enter the solution in the surrounding cytoplasm. The proteins synthesized on ER ribosomes penetrate into the membranes or cisternae of the endoplasmic reticulum. Within the ER the proteins are transported to the vicinity of the Golgi complex. The proteins are then transferred to the complex by transition vesicles that bud off from the ER and fuse with the Golgi complex. After modification within the Golgi complex, which may include attachment of lipid, carbohydrate, or other groups, the proteins are enclosed in vesicles that bud off from the margins of the Golgi sacs. These vesicles gradually fuse together to form large secretion vesicles that are stored in the cytoplasm or released to the cell exterior.

The Regulation of Protein Synthesis

The cells of all organisms, both prokaryotic and eukaryotic, have the capacity to make thousands of different proteins. At any given time, however, only a fraction of the total is actually synthesized. Probably no cells of any type assemble all of their encoded proteins simultaneously. For example, maturing red blood cells in vertebrate animals synthesize only hemoglobin and a few enzymatic proteins. This occurs even though these cells retain all of

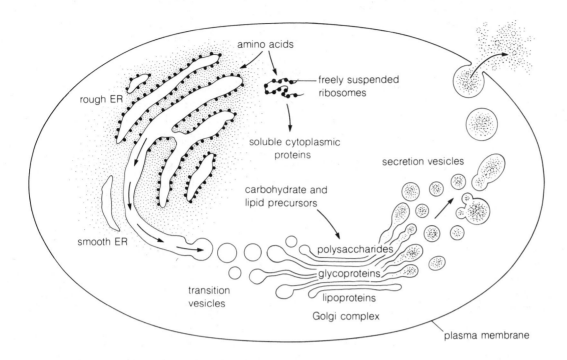

Figure 10-18 The integration of ribosomes, rough and smooth ER, Golgi complex, and secretion vesicles in protein synthesis and secretion (see text).

Labels in figure:
amino acids
freely suspended ribosomes
rough ER
soluble cytoplasmic proteins
secretion vesicles
carbohydrate and lipid precursors
smooth ER
polysaccharides
glycoproteins
lipoproteins
transition vesicles
Golgi complex
plasma membrane

the chromosomes and the entire complement of genes found in any other cell of the same organism.

Control of protein synthesis to increase the levels of some proteins and inhibit or stop the production of others is termed *regulation*. Regulation may take place at several points in the entire mechanism. It may be imposed by inhibiting messenger RNA synthesis; control at this level is termed *transcriptional regulation*. Control at the level of protein synthesis in the cytoplasm is termed *translational regulation*. Translational regulation may include any mechanism acting to control the availability of mRNA, ribosomes, or other factors in the cytoplasm, or the rate of reactions in amino acid activation or polypeptide assembly. There is good evidence that both transcriptional and translational regulation occur in both prokaryotes and eukaryotes to control the number and types of proteins synthesized in cells.

Transcriptional Regulation of Protein Synthesis

Transcriptional Regulation in Prokaryotes: The Operon Mechanism Control of RNA transcription in prokaryotes has been studied extensively and is better understood than any other type of regulatory mechanism. Bacteria can quickly regulate the synthesis of enzymes to suit the biochemical conditions of their surroundings. For example, bacteria living in a culture medium containing only inorganic salts and food substances such as glucose produce the enzymes required to make all 20 amino acids. If an amino acid is added to the medium, the enzymes synthesizing that amino acid quickly drop in quantity in the bacteria and soon reach undetectable levels. If the amino acid is then removed from the medium, the enzymes required to synthesize it reappear and quickly reach their former levels. Responses of this type in bacteria often take place within minutes after changes in the culture medium are made.

Enzymes that can be made to appear or disappear within bacterial cells by changes in the culture medium are called *inducible enzymes*. Bacteria have a variety of inducible enzymes and can respond to hundreds of organic compounds by synthesizing specific enzymes for their utilization. There are two classes of these enzymes. One is normally absent from the cell but is induced to appear if a substance acted upon by that enzyme is added to the medium. The other class, described in the amino acid system above, is reduced in quantity in the cell if the substance normally made by that enzyme is supplied in the medium.

The mechanism controlling inducible enzymes was explained in the 1950s by two investigators at the Pasteur

Institute in Paris, Francois Jacob and Jacques Monod. Jacob and Monod studied a group of enzymes that allows bacteria to use sugars called *galactosides*. If no galactosides are present in the medium surrounding the bacterial cells, few or no molecules of the enzymes catalyzing the breakdown of these sugars are present in the bacterial cytoplasm. If galactosides are added to the medium, synthesis of the required enzymes begins very quickly and the bacteria are soon able to utilize the added sugar. By genetic analysis, all of the enzymes of the galactoside pathway were found to be coded into a single stretch of DNA. These genes are transcribed as a unit into a single mRNA that directs the synthesis of the galactoside enzymes in the cytoplasm.

Several kinds of mutant bacteria were soon discovered that had interesting modifications in the galactoside enzyme system. In one of these mutants, the enzymes for galactoside breakdown were always present in the bacterial cytoplasm whether galactoside sugars were present in the surrounding medium or not. Unexpectedly, the change in the DNA information causing the mutation was found not in the region that coded for the galactoside enzymes but some distance away. This distant region, which appeared to control the activity of the galactoside part of the chromosome in normal cells, was termed the R (for *regulator*) gene; the mutant cells were designated R^-, and normal cells R^+. Jacob and Monod found that if a segment of DNA containing the R^+ gene, derived from a normal cell, was introduced into an R^- cell, the synthesis of the galactoside enzymes was quickly turned off (unless galactosides were present in the medium). This meant that a transcribed product, coded for by the normal R^+ gene, in some way controls the transcription of mRNA in the block of DNA coding for the galactoside enzymes.

Still another region of the chromosome was found to affect transcription of the block of DNA coding for the galactoside enzymes. Mutations in this region also led to an uncontrolled, maximum rate of enzyme production. This part of the DNA, which appeared to be an "on-off" switch controlling transcription of the galactoside mRNAs, was termed the *operator*. The operator was found to lie just in advance of the galactoside genes, in the region where an RNA polymerase molecule would bind to the DNA to begin transcription.

The observed activity of the regulator, operator, and galactoside regions of the chromosome formed the basis of Jacob and Monod's model, which they called the *operon hypothesis* (Fig. 10-19). According to the hypothesis, the genes coding for the enzymes breaking down galactosides are controlled as a unit; this unit is called the *operon*. Without the normal activity of the regulator and operator genes the operon would be continuously transcribed. As a result, the galactoside enzymes would be present at all times in high concentrations in the cell. According to the hypothesis, in normal (R^+) bacterial cells the regulator gene codes for synthesis of a *repressor* protein. The repressor has two sites, one that can bind to the inducing substance and one that can bind to the operator. In the galactoside system, if no galactoside sugars are present in the medium, the repressor site that binds to the galactoside operator is active, and the repressor binds to the operator next to the galactoside genes. This binding blocks access to the genes by RNA polymerase, and the genes are not transcribed. As a result, no galactoside enzymes are made in the cytoplasm (Fig. 10-19a).

If galactoside sugars are added to the medium, some of them enter the cell and bind to the inducer site on the repressor. This combination changes the folding conformation of the repressor, causing it to lose its affinity for the operator site. The operator is then vacated, and the genes controlled by the operator are turned on and transcribed (Fig. 10-19b), producing a continuous flow of galactoside mRNAs and enzymes.

The same operon mechanism also explains synthesis of enzymes that can be turned off by substances added to the culture medium, as in the amino acid enzymes that disappear when the amino acid they synthesize is added to the culture medium. To explain this effect, Jacob and Monod proposed that the regulator gene (regulator II in Fig. 10-19c and d) produces an *inactive* instead of active repressor. The inactive repressor also has two binding sites, one for the repressing substance (the amino acid in our example) and one for the operator region next to the genes that code for the enzymes catalyzing synthesis of the amino acid. The inactive repressor has no affinity for the operator, and transcription of the genes coding for the enzymes goes on continuously (Fig. 10-19c). If the amino acid is added to the surrounding medium, some of it diffuses into the cell and combines with the repressor. The combination causes a conformational change in the repressor that activates its operator binding site. The repressor then binds to the operator, turning off transcription of the DNA region coding for the enzymes that synthesize the amino acid (Fig. 10-19d).

The differences in the alternate forms of the operon hypothesis thus depend on whether the repressor, coded

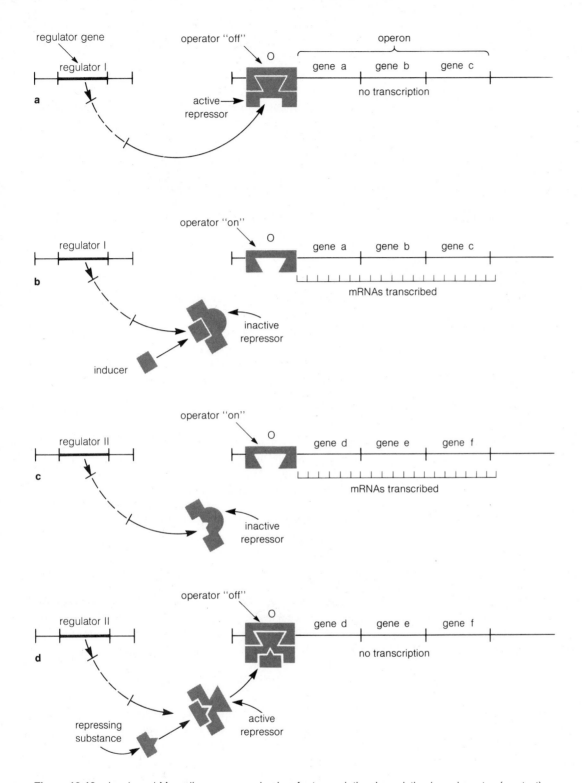

Figure 10-19 Jacob and Monod's operon mechanism for transcriptional regulation in prokaryotes (see text).

for by the R region of the DNA, is synthesized in an *active* or *inactive* state. The alternate forms can be summarized as follows:

Induction of enzyme synthesis:
active repressor + inducing substance → inactive repressor

The inactive repressor does not combine with the operator, and the genes are continuously transcribed.

Repression of enzyme synthesis:
inactive repressor + inducing substance → active repressor

The active repressor combines with the operator, and the genes are turned off.

Since Jacob and Monod first proposed the operon hypothesis, experimental support for their model has come from many sources. Some of the strongest evidence for the model, for which Jacob and Monod received the Nobel Prize in 1965, has come through isolation and identification of the repressors of various operons. For example, W. Gilbert and B. Müller-Hill of Harvard University isolated and purified a protein coded for by the galactoside regulator gene. This protein has strong affinity for bacterial DNA containing a normal operator site controlling the galactoside genes. The protein also has strong binding affinity for galactoside sugars; combination with a galactoside destroys its affinity for the galactoside operator region. Thus the protein isolated by Gilbert and Müller-Hill has all of the characteristics predicted for the repressor by the operon hypothesis.

Transcriptional Regulation in Eukaryotes   It is doubtful whether an automated system equivalent to the operon mechanism controls transcription in eukaryotes. The elements necessary for the operon mechanism, such as regulator and operator genes, have not been detected in eukaryotes. Further, clusters of genes with related function that are transcribed as a unit, such as the galactoside operon, are rare or nonexistent in eukaryotes.

The apparent absence of an operon mechanism in eukaryotes is probably a reflection of the fundamental differences in the way transcription is regulated in prokaryotes and eukaryotes. The operon mechanism provides sensitivity to changes in the environment and allows a prokaryotic cell to adjust rapidly to these changes. The resulting changes in enzyme concentrations are temporary and persist only as long as the environment remains the same. Most of the transcriptional regulation observed in eukaryotes, however, is more or less permanent. These long-

term changes are a part of the cell specialization that is a characteristic feature of embryonic development in many-celled organisms.

Although many questions remain about the mechanisms producing long-term transcriptional changes in eukaryotes, these regulatory processes are now thought to depend on the histone and nonhistone proteins occurring with the DNA in eukaryotes (see p. 206) but not in prokaryotes. These proteins are thought to cooperate in such a way that (1) the histones determine whether large blocks of genes are available for transcription, and (2) the nonhistones select which genes among the blocks made available by the histones are actually transcribed.

The Histones and Transcriptional Regulation   A variety of experiments have shown that histones can control the transcription of genes. However, the most recent evidence indicates that transcriptional control by the histones is a general effect that does not distinguish between individual genes.

"Naked" DNA, without associated histones or nonhistone proteins, serves as an excellent template for RNA transcrtiption in cell-free systems. All of the DNA, including mRNA, tRNA, and rRNA genes, is transcribed in such systems. Even the repetitive, noncoding sequences that are normally never transcribed in cells are transcribed when naked DNA is used as a template. Addition of histones from any source to DNA in such cell-free systems inhibits transcription by as much as 90 percent or more.

This inhibition is apparently random. The amount of mRNA, tRNA, and rRNA synthesized is simply reduced in overall quantity, with no significant changes in the particular genes transcribed. The greater the amount of histones added, the greater the reduction in transcription; histones from one source are as effective as another in producing this effect. Thus inhibition of transcription by the histones, as far as anyone has been able to determine, is a general effect that does not distinguish between individual genes.

This conclusion is not too surprising, since there is relatively little variability in the histones from cell to cell within the same organism, or even between different species (see p. 207). The number of different genes in eukaryote nuclei probably amounts to thousands or even tens of thousands. In order to control each gene separately, the proteins controlling them would be expected to occur in many different kinds. However, there are only five major kinds of histones in eukaryotic nuclei. Thus the histones do not seem to vary enough to control specific genes in a eukaryotic nucleus.

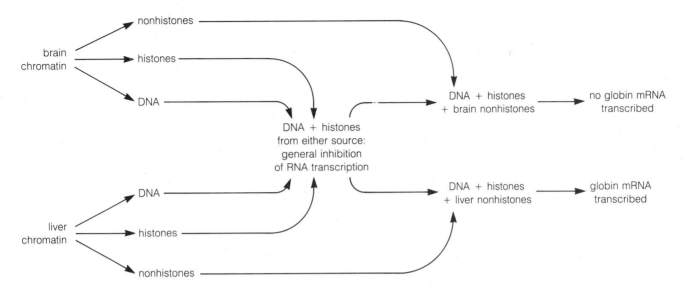

Figure 10-20 The Gilmour and Paul experiment demonstrating specific transcriptional regulation by the nonhistone chromosomal proteins (see text).

Then what is the significance of the general, nonspecific regulatory effect of the histones? An examination of one type of variability that does occur in the histones gives some clues as to what this effect might be. While no consistent differences have been observed in the number of major histone types, differences are noted in the degree of *modification* of the five major histones between chromatin regions that are active or inactive in transcription. Generally, the histones in active chromatin tend to be modified by the addition of phosphate or acetyl (CH_3CO—) groups; these groups are absent, or occur in significantly lower amounts in the histones of inactive chromatin. As expected from this difference, experimentally modifying the histones by adding acetyl groups increases the levels of RNA transcription in cell-free systems.

The main effect of modifications such as acetylation or phosphorylation is a reduction of the strongly positive charge of the histone proteins. The phosphate or acetyl groups cancel out one positive charge for each one added to a histone molecule. The effect of these changes is to reduce the strength of the electrostatic attraction between the histones and the negatively charged phosphate groups on the DNA (see p. 190). This would be expected to loosen the structure of the DNA-histone complex, separating the histones slightly from the DNA and increasing the general availability of large blocks of DNA sequences to RNA transcription.

The Nonhistones and Transcriptional Regulation In contrast to the apparently general, nonspecific control of RNA transcription by the histones, there is fairly good evidence that the nonhistone proteins can regulate the activity of single, specific genes. The histones, as noted, inhibit transcription generally when added to DNA in a cell-free system. Addition of the nonhistones to such DNA-histone complexes partially reverses the inhibition: when the nonhistones are added, some of the DNA sequences are turned on again.

Recent experiments show that the reversal of inhibition in DNA-histone preparations by the nonhistones is specific, and that particular genes are turned on when nonhistones are added. In one of these experiments (Fig. 10-20), R. S. Gilmour and J. Paul, at the Beatson Institute for Cancer Research in Scotland, tested the ability of histone and nonhistone proteins to control the transcription of an mRNA coding for a liver cell protein called a *globin*. An mRNA coding for this protein is transcribed in liver cells of a mouse, but not in mouse brain cells. Gilmour and Paul isolated the chromatin from these two mouse cell types and split the chromatin into separate DNA, histone, and nonhistone fractions. The fractions were subsequently recombined in all the possible combinations. The recombined chromatin was then added to a cell-free system containing all the factors required for RNA transcription.

Adding the histones back to the DNA in any combination (brain cell DNA + brain or liver cell histones; liver

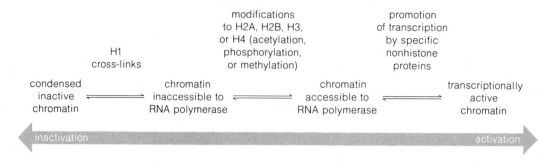

Figure 10-21 Summary of the coordination of the histone and nonhistone chromosomal proteins in transcriptional regulation (see text).

cell DNA + brain or liver cell histones) had no effect other than a general inhibition of RNA transcription. However, adding the nonhistone proteins to the DNA-histone combinations reversed the inhibition in a specific way. If nonhistone proteins from liver cells were added to any of the DNA + histone combinations, the globin was transcribed and appeared in the cell-free systems (see Fig. 10-20). The globin mRNA did not appear if brain cell nonhistones were added to any of the DNA + histone combinations. Thus, in this experiment, one or more proteins in the nonhistone fraction from the liver cells were able specifically to turn on a gene sequence coding for a typical liver protein.

These experimental findings suggest that histone and nonhistone proteins cooperate in transcriptional regulation. Modifications to the histones, such as chemical addition or removal of acetyl or phosphate groups, have a general effect and determine whether large blocks of genes are available or unavailable for RNA transcription (Fig. 10-21). Once converted to the "available" form by histone modification, specific nonhistone proteins determine which of the available genes are actually transcribed. In doing so, the nonhistones involved may recognize specific DNA sequences and combine with the DNA or histones in these regions to open the chromatin structure sufficiently to allow access to the DNA by RNA polymerase enzymes. The appearance of these specific nonhistone regulator proteins at certain times might be controlled by a variety of factors, including hormones or other chemical signals acting on the cell. The advantage of this system is that eukaryotic cells do not have to make specific nonhistone proteins to control each of the genes in the nucleus (this would re-

quire as many different nonhistone control proteins as there are genes, perhaps as many as 30,000 to 40,000 per cell). Instead, the genes are inhibited generally by the histones, which occur in only five different major kinds. Turning on specific genes by the nonhistones then requires only that a relatively few specific nonhistone proteins need be made by each cell, one for each gene that is to be turned on.

Chromatin Condensation and Transcriptional Regulation
Transcriptional inactivation is frequently reflected morphologically as a conversion of chromatin into the tightly packed form. Perhaps the best example of this conversion is the nuclear change occurring during the development of red blood cells in vertebrates. As these cells become mature, transcription gradually tapers off until the nuclei become completely inert and inactive. The reduction in transcriptional activity is accompanied by gradual condensation of the chromatin into a tightly packed mass. Condensation of one of the two X chromosomes in mammalian females into a block of tightly folded chromatin, the *Barr body* (Fig. 10-22), is also thought to reflect transcriptional inactivation. This inactivation probably makes the nuclei of mammalian males and females exactly equivalent in number of active X chromosomes. The tightly packed, or condensed, inactive form of chromatin is frequently termed *heterochromatin*.

The conversion of chromatin to the tightly packed form may depend on the properties of one of the histones, histone H1. This histone, which binds to DNA at the linkers between nucleosomes, can also form H1-H1 bonds be-

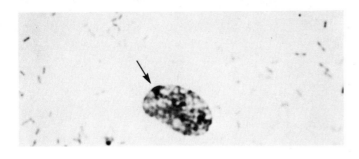

Figure 10-22 A Barr body (arrow) in the nucleus (bracket) of a human female. The Barr body is an X chromosome that condenses and becomes inactive in transcription. × 1,000.

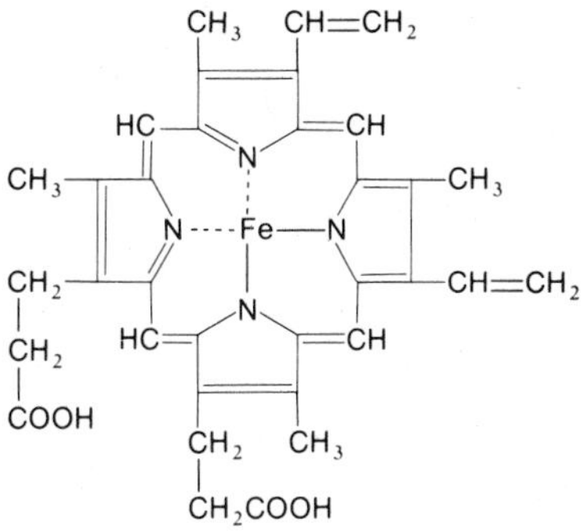

Figure 10-23 The iron-containing porphyrin ring of hemoglobin.

tween the linkers of neighboring nucleosomes. By forming such crosslinks between neighboring chromatin fibers, H1 molecules could readily fold chromatin into the condensed form.

These conclusions suggest that condensation may fit into the overall process of transcriptional regulation as shown in Figure 10-21, in which condensed chromatin is shown as the end point in the conversion of DNA to the completely inactive state. Packing chromatin into the tightly condensed form may also serve as a mechanism for the compact storage of inactive genes in the cell nucleus (polytene chromosomes, which provide direct visual evidence for transcriptional regulation and the role of chromatin condensation in the control of transcription, are described in Supplement 10-3).

Translational Regulation of Protein Synthesis The complex reactions of amino acid activation and polypeptide assembly on ribosomes include many steps that could be controlled by regulatory mechanisms. However, most of the translational controls detected in prokaryotes and eukaryotes operate at the initial steps of polypeptide assembly on ribosomes. These controls include regulation of both the activity or availability of mRNAs in the cytoplasm and the enzymelike factors required for the first steps in polypeptide assembly.

Several mechanisms are known to regulate the activity or availability of mRNAs in both eukaryotes and prokaryotes. One of these mechanisms works by controlling the rate at which messengers are degraded by RNAase enzymes. Translational control by this pathway is best known in bacteria, where differences can be detected in the rate at which different mRNAs are broken down.

Several translational controls regulating the activity of factors required for initiation of polypeptide assembly have been discovered in eukaryotes. A good example is provided by the mechanism regulating hemoglobin synthesis in developing red blood cells. In these cells, assembly of the hemoglobin protein has been shown to depend on the supply of *hemin*, the iron-containing porphyrin ring (see Fig. 10-23) incorporated into the hemoglobin molecule to complete its synthesis. Hemin affects hemoglobin synthesis through the activity of a protein called the *hemin-controlled inhibitor* (*HCI*). The HCI protein has two active sites (Fig. 10-24). One acts much like the active site of an enzyme. When active, this site adds a phosphate group to one of the factors required for an initial step in protein synthesis (binding of the first amino acid-tRNA complex to the ribosome). Addition of the phosphate group inhibits the activity of the necessary factor, and, in consequence, protein synthesis slows or comes to a stop. The activity of the phosphate-adding site is controlled by the second site on the HCI protein, which can combine with hemin. Combination with hemin causes a conformational change in the HCI protein that inactivates its phosphate-adding site. Thus, when hemin is present, it combines with the HCI protein and inactivates its ability to inhibit protein synthesis. As a result, hemoglobin synthesis proceeds only if hemin is available. The HCI system works effectively because hemoglobin is the only protein synthesized in quantity in developing red blood cells. This limitation, in turn, depends on transcriptional controls in the blood cells, which restrict transcription to the hemoglobin mRNAs.

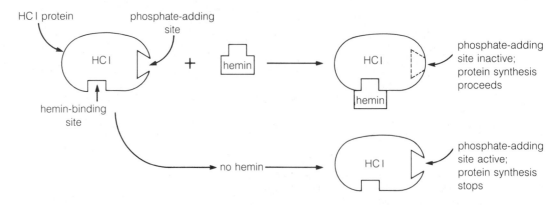

Figure 10-24 Translational regulation by the HCI protein (see text).

Translational controls are typically coordinated with transcriptional regulation along the lines illustrated by the HCI system. Transcriptional controls determine *which* proteins are to be made in a cell at a given time. Translational controls, as in the HCI system, determine *how much* of the proteins are to be made by fine-tuning synthesis to conditions in and around the cell. The end result is the highly specific and regulated activity characteristic of living cells.

Suggestions for Further Reading

Lodish, H. F. 1976. Translational control of protein synthesis. *Annual Review of Biochemistry* 45:39–72.

Palade, G. 1975. Intracellular aspects of the process of protein synthesis. *Science* 189:347–58.

Satir, Birgit. 1975. The final steps in secretion. *Scientific American* 233:28–37 (October).

Uy, R. and F. Wold. 1977. Posttranslational covalent modification of proteins. *Science* 198:890–96.

Watson, J. D. 1976. *Molecular biology of the gene.* 3rd ed. W. A. Benjamin: Menlo Park, Calif.

Wolfe, S. L. 1981. *Biology of the cell.* 2nd ed. Wadsworth: Belmont, Calif.

For Further Information

Histone and nonhistone proteins, Chapter 9
mRNA, rRNA, and tRNA, Chapter 9
Nuclear envelope in cell division, Chapter 11

Questions

1. Outline the steps in amino acid activation. Why is the reaction called an activation? Trace the flow of energy derived from ATP through the reaction sequence.

2. Compare the roles of tRNA and aminoacyl-tRNA synthetase enzymes in protein synthesis. How does the accuracy of protein synthesis depend on the function of tRNAs? Of the synthetase enzymes?

3. Outline the mechanism of polypeptide assembly on ribosomes. What is the source of energy for formation of peptide linkages as each successive amino acid is added to a growing polypeptide chain?

4. How did Khorana's experiment establish that the "words" in the genetic code are triplets?

5. How were the codons identified and assigned to amino acids? Are all 64 possibilities used? What are "nonsense" codons?

6. Define degeneracy, synonym, codon, terminator, universality, and anticodon.

7. Refer to Figure 10-3 and make a chart listing all 20 amino acids and the codons assigned to them. Can you detect any common patterns or similarities in the multiple codons for the amino acids? Which amino acids have only one codon?

8. Write out a random DNA sequence 30 nucleotides long and see what polypeptide sequence it would code for. Then introduce "mutations" by changing the sequence at five random places. Does a mutation in the DNA code necessarily cause a change in the sequence of the polypeptide coded for? Why?

9. Outline the structure of ribosomes, rough and smooth ER, and the Golgi complex. What is the function of each structure in protein synthesis?

10. Trace the pathway of a secreted protein from the rough ER to the outside of the cell. Are variations possible in this pathway? How was the pathway determined experimentally?

11. Trace the pathway followed by a membrane protein from the ER to the plasma membrane.

12. How is the nuclear envelope related to protein synthesis and secretion?

13. What factors are involved, according to the signal hypothesis, in the synthesis of proteins on ribosomes attached to the ER? How do mRNAs "know" that they are to associate with ER membranes during protein synthesis?

14. What is transcriptional regulation? Translational regulation?

15. How does transcriptional regulation differ in prokaryotes and eukaryotes?

16. Outline the operon mechanism. How does the mechanism work to induce synthesis of an enzyme? To repress synthesis of an enzyme? Define operon, operator, regulator, and repressor.

17. How are the histones believed to function in transcriptional regulation? What effect does adding phosphate or acetyl groups have on the histones?

18. Outline an experiment demonstrating that the nonhistone proteins control the transcription of individual genes.

19. Compare the effects of transcriptional and translational regulatory mechanisms.

20. In what ways does the HCI system for controlling hemoglobin synthesis resemble the operon mechanism? What are the similarities between the HCI and operon mechanisms? The differences?

21. Suppose that a regulatory mechanism controlled translation by inhibiting the termination of protein synthesis. What effect would this mechanism have on the supply of ribosomes for protein synthesis? What relationship might this have to the fact that most translational controls affect the initiation phase of polypeptide assembly?

Supplement 10-1: Lysosomes

One of the most important products of the secretory pathway in cells is a specialized class of secretion vesicles known as *lysosomes* (Figs. 10-25 to 10-27). The proteins of lysosomes are synthesized in the rough ER and subsequently packed into vesicles that remain in storage for various lengths of time in the cytoplasm. The proteins in the lysosomes consist of a variety of enzymes that, collectively, are capable of breaking down all the major biological molecules of the cell. The reactions catalyzed by the lysosome enzymes are all *hydrolytic* reactions (see Information Box 2-1) that split biological molecules into their building-block subunits by adding water:

nucleic acids + H_2O → nucleotides and nucleosides
proteins + H_2O → amino acids
polysaccharides and carbohydrate groups + H_2O →
disaccharides and monosaccharides
lipids + H_2O → fatty acids, glycerol

One of the many interesting and as yet unanswered questions about lysosomes is how, considering their impressive battery of hydrolytic enzymes, the lysosomal boundary membranes remain intact and unbroken in the cytoplasm.

Lysosomes have been identified in all animal cells examined, and in protists, several types of fungi, and lower and higher plants. Lysosomes are especially abundant in animal cells, such as the white blood cells of vertebrate animals, that ingest large quantities of extracellular material. They are also abundant in the cells of animal tissues undergoing degenerative changes or metabolic stress such as starvation, aging, or hormonal stimulation.

Lysosomes break down substances in cells by several processes. One involves digestion of material brought into the cytoplasm by endocytosis (see p. 84). In endocytosis the plasma membrane contacts the extracellular material and invaginates, creating a pocket in which the material is trapped. The pocket then pinches off from the plasma membrane as a vesicle and sinks into the cytoplasm, where it fuses with one or more lysosomes. Fusion activates the enzymes contained in the lysosomes, which proceed to break down the biological molecules inside the vesicle (Fig. 10-26 shows several lysosomes in a liver cell carrying out this process). After breakdown by the lysosomal enzymes, the products enter the cytoplasm of the cell by diffusing through the vesicle membrane. Any remaining undigested debris may be retained as a membrane-bound deposit in the cell, or may be expelled to the cell exterior.

In this function, lysosomes act as a digestive system for the cell. This activity forms a regular part of the digestive process in cells that ingest food particles, as in the protists or lower invertebrates such as the sponges and coelenterates. In higher animals, the lysosomal digestive process figures as an important part of the defense mechanisms against bacteria, virus particles, and toxic molecules. The white blood cells of vertebrates, for example, eliminate foreign cells and particles by ingesting them in vesicles that subsequently fuse with lysosomes.

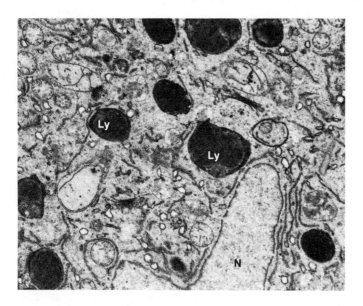

Figure 10-25 Lysosomes in the cytoplasm of a corn root cell. *Ly*, lysosome; *N*, nucleus. × 8,000. Courtesy of P. Berjak and Academic Press, Inc., from *Journal of Ultastructure Research* 23 (1968): 233.

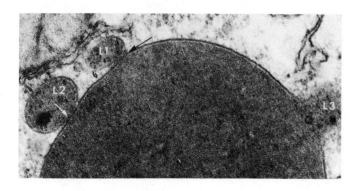

Figure 10-26 Lysosomes (L1, L2, L3) of a rat liver cell fusing with a large, membrane-bound droplet of absorbed protein. The arrow marks a point where one of the lysosomes is fusing with the protein droplet. × 45,000. Courtesy of S. Goldfischer, A. B. Novikoff, A. Albala, L. Biempica, from *Journal of Cell Biology* 44 (1970): 513, by copyright permission of The Rockefeller University Press.

In the second process, cell organelles rather than extracellular material are digested by lysosomes. In some manner, as yet unknown, a cell organelle penetrates a lysosome membrane and is digested inside (Fig. 10-27 shows the remains of two mitochondria undergoing breakdown by this process). The significance of self-digestion of

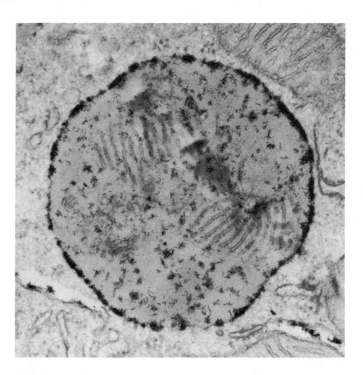

Figure 10-27 A lysosome in the cytoplasm of a mouse kidney cell. Remnants of several mitochondria are visible inside the lysosome. × 50,000. Courtesy of F. Miller.

this type, called *autolysis*, is obscure, except that it probably plays an important role in the rapid breakdown of organelles in cells undergoing differentiation or physiological stress.

The third lysosomal mechanism is a similar, but more spectacular process in which the lysosome membranes rupture and release their enzymes into the surrounding cytoplasm. This process, if large numbers of lysosomes are involved, either results in or accompanies death of the cell. Operation of lysosomes as "suicide bags" in this way forms part of normal developmental changes in both plants and animals (as, for example, in the breakdown and absorption of the tail in developing frog embryos) as well as in cell degeneration.

The effects of lysosomes in some cases extend from their microscopic and molecular world to touch the conditions and affairs of humans. The degenerative changes in bones and joints associated with arthritis are suspected by some medical scientists to be caused in part by the abnormal release of enzymes from lysosomes of bone or lymph cells into the extracellular fluids. On a somewhat different plane, lysosomal activity has also been offered as an explanation for the contraceptive action of IUDs (intra-uterine devices). According to this idea, an IUD is recognized as a foreign body by cells of the uterine lining, which release lysosomal enzymes into the uterus in response. The resultant high concentration of lysosome enzymes in the uterine fluids interferes with attachment of the embryo to the uterus, and, according to the hypothesis, thereby prevents pregnancy.

Supplement 10-2:
Transcription and Translation in Mitochondria and Chloroplasts

The DNA of Mitochondria and Chloroplasts

Both mitochondria and chloroplasts contain DNA molecules that are active in transcription. All the major RNAs, including mRNA, tRNA, and rRNA, are transcribed from the organelle DNA. These RNA species interact in protein synthesis in the organelle interior. In most features, transcription and translation in mitochondria and chloroplasts resemble the mechanisms of prokaryotes more closely than eukaryotes.

The DNA molecules of both organelles are visible as fibrous deposits suspended in the innermost cavity, the matrix in mitochondria and the stroma in chloroplasts (Figs. 10-28 and 10-29). Typically, the DNA fibers visible in the organelles are much thinner than nuclear chromatin fibers. They approach the dimensions of the fibers in bacterial nucleoids—that is, the dimensions expected for a protein-free DNA double helix.

The analysis of DNA isolated from the two organelles shows that the similarities to prokaryotic DNA are more than superficial. Both mitochondrial and chloroplast DNA take the form of closed circles, with no associated histone or nonhistone proteins. The circles in animal mitochondria are typically very small and contain only about 5 micrometers of DNA. The mitochondrial circles of fungi and plants are generally larger, ranging in size from about 20 to 30 micrometers of included DNA. Chloroplast circles, with about 40 micrometers of included DNA, are larger still. However, none of the organelle circles approach the dimensions of bacterial DNA circles, which contain about 1000 to 1500 micrometers of DNA.

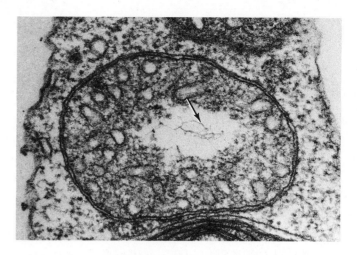

Figure 10-28 DNA deposits (arrow) within the matrix of a mitochondrion of the brown alga *Egregia*. × 58,000. Courtesy of T. Bisalputra and A. A. Bisalputra, from *Journal of Cell Biology* 33 (1967): 511, by copyright permission of The Rockefeller University Press.

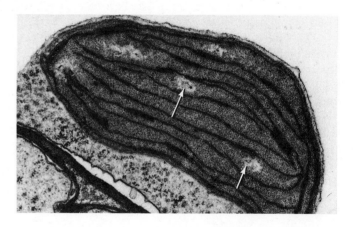

Figure 10-29 DNA deposits (arrows) in the stroma of a corn chloroplast. × 20,000. Courtesy of L. K. Shumway and T. E. Weier, from *American Journal of Botany* 54 (1967): 773.

The amount of DNA associated with mitochondria varies from as little as 0.2 percent of the total cellular DNA in mouse liver cells, to 20 percent in haploid yeast, and more than 99 percent in amphibian oocytes. Dividing the total mitochondrial DNA fraction by the number of mitochondria gives enough DNA in most cells for about 5 to 10 circles per mitochondrion. Chloroplast DNA may make up as much as 6 percent of the total DNA of a plant cell.

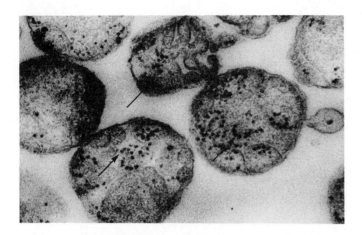

Figure 10-30 Mitochondrial ribosomes (arrows) in a yeast cell, visible in a section that just grazes the surfaces of the cristae membranes. × 46,000. Courtesy of J. André, from *Federation of European Biochemical Societies Letters* 3 (1969): 177.

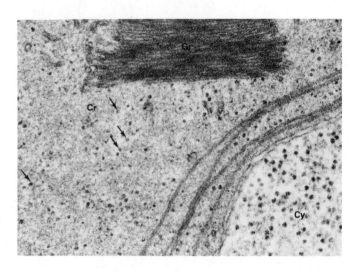

Figure 10-31 Ribosomes (arrows) inside a chloroplast of *Chlamydomonas*, a green alga. The chloroplast ribosomes are distinctly smaller than the ribosomes in the cytoplasm outside the chloroplast (*Cl*). *Gr*, granum. × 62,000. Courtesy of U. W. Goodenough and R. P. Levine, from *Journal of Cell Biology* 44 (1970): 511, by copyright permission of The Rockefeller University Press.

At these levels, there is enough DNA to form about 60 to 80 circles per chloroplast.

The rRNAs of mitochondria and chloroplasts are associated with ribosomes that, in dimensions, resemble the ribosomes of prokaryotes. As such, they are conspicuously smaller than the ribosomes in the cytoplasm surrounding the organelles. Mitochondrial ribosomes occur in the matrix, where they may be freely suspended or attached to the surfaces of cristae membranes (Fig. 10-30). Chloroplast ribosomes occupy the equivalent inner compartment, the stroma, where they may also be either freely suspended, or connected to thylakoid membranes (Fig. 10-31).

Protein Synthesis in Mitochondria and Chloroplasts

The ribosomes of mitochondria and chloroplasts carry out protein synthesis by the same overall mechanism as the cytoplasmic ribosomes of prokaryotes and eukaryotes. However, the details of the mechanism in the two organelles most clearly resemble the process of polypeptide assembly in prokaryotes.

For example, the first amino acid incorporated into the proteins synthesized inside the organelles is formylmethionine, as it is in prokaryotes, instead of the unmodified methionine used as the first amino acid in eukaryotes.

Similarly, the enzymelike factors that speed protein synthesis in mitochondria and chloroplasts resemble their equivalents in prokaryotes much more than the eukaryotic factors and, in general, are fully interchangeable with the prokaryotic factors. Another similarity is noted in the reaction of the organelle ribosomes to antibiotics that interfere with protein synthesis. For example, the antibiotic *cycloheximide*, which interferes with protein synthesis on eukaryotic but not bacterial ribosomes, has no effect on polypeptide assembly in either chloroplasts or mitochondria. Conversely, *chloramphenicol*, which inhibits prokaryotic but not eukaryotic protein synthesis, stops polypeptide assembly on chloroplast or mitochondrial ribosomes. Thus protein synthesis in mitochondria and chloroplasts, as well as DNA structure, is essentially prokaryotic rather than eukaryotic in nature.

The Polypeptides Synthesized inside Mitochondria and Chloroplasts

The responses of the cytoplasmic, mitochondrial, and chloroplast translation systems to antibiotics has provided one of the key approaches for determining which polypep-

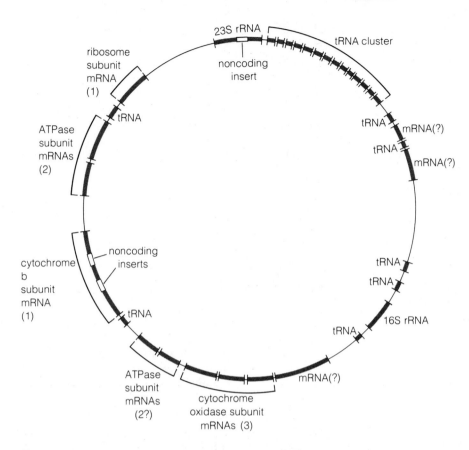

Figure 10-32 The approximate locations of mRNA, tRNA, and rRNA coding sequences in the yeast mitochondrial DNA circle. Compiled from data presented in P. Borst and L. A. Grivell, *Cell* 15 (1978): 705.

tides are synthesized inside the organelles. For example, cycloheximide, which interrupts cytoplasmic protein synthesis in eukaryotes, inhibits the production of almost all mitochondrial or chloroplast ribosomal proteins, and all the soluble factors required for protein synthesis in the organelles. The opposite inhibitor, chloramphenicol, which stops organelle protein synthesis, has no effect on synthesis of these proteins. This indicates that the organelle ribosomal proteins and soluble factors, although prokaryotic in most structural and functional characteristics, are actually made on the ribosomes in the cytoplasm outside mitochondria and chloroplasts.

Other polypeptides prove to be encoded in mitochondrial or chloroplast DNA and synthesized inside the organelles. The list is necessarily short, since the coding capacity of the organelle DNA molecules is small (Fig. 10-32 shows the known genes coded in yeast mitochon-

drial DNA). In both organelles, the polypeptides assembled inside include parts of the enzyme system synthesizing ATP in the inner organelle membranes. Other polypeptides assembled inside the organelles include parts of the carrier molecules transporting electrons in mitochondria, and a subunit of the RuDP-carboxylase enzyme of chloroplasts. (RuDP is the enzyme that fixes CO_2 in the dark reactions of photosynthesis; see p. 131.) None of these polypeptides, interestingly enough, forms a complete protein. In all cases, the polypeptides are subunits of complex proteins that also require additional polypeptides made in the surrounding cytoplasm for their completion.

This fact raises one of the most interesting questions concerning the existence of the organelle transcription-translation systems. Why, since no complete proteins are made in either mitochondria or chloroplasts, are any polypeptides encoded and synthesized inside the organelles at all? This question is especially pertinent if you consider that 100 enzymatic and ribosomal proteins, almost all encoded in the cell nucleus synthesized in the cytoplasm *outside* mitochondria and chloroplasts, must be supplied to the organelles to synthesize the relatively few and incomplete proteins encoded and made inside.

Changes in the Genetic Code in Mitochondria

Some of the polypeptides synthesized in mitochondria have been completely sequenced. Comparisons between the amino acid sequences of these polypeptides and the nucleic acid sequences coding for them in the mitochondrial DNA of yeast and human cells has revealed the unexpected fact that several substitutions have been made in the genetic code in mitochondria. Where the DNA specifies the mRNA triplet CUA, which codes for leucine elsewhere in living organisms, one of the mitochondrial polypeptides (one of the ATPase subunits) carries a threonine at the corresponding point in its amino acid sequence. Comparisons between another polypeptide and its mitochondrial DNA gene reveal two additional coding substitutions. This polypeptide (part of the electron carrier chain) carries the amino acid tryptophan at points corresponding to UGA, normally a terminator codon, in its coding sequence. At other points, methionine is inserted at points coded for by the AUA triplet, which normally specifies isoleucine.

These substitutions apparently depend on the unusual structure of the anticodon region of the tRNAs carrying threonine, tryptophan, and methionine in mitochondria. The mitochondrial tRNA carrying threonine in yeast, for example, has 8 rather than 7 nucleotides in the loop at the tip of the anticodon arm (see Fig. 9-12), a characteristic appearing in no other known tRNA. The unusual structure and properties of the anticodon loop allow the tRNA to pair with either UAA or CUA on the messenger RNA. Evidently, in entering threonine into the ATPase subunit protein, it pairs with the CUA coding triplet. The coding changes discovered in mitochondria are the first known exceptions to the triplet assignments of the genetic code, which were heretofore thought to be completely universal.

The many similarities among the transcription-translation systems of mitochondria, chloroplasts, and prokaryotes have led to the hypothesis that the two organelles originated in evolution in prokaryotic cells that were taken up as food particles by cells destined to become eukaryotes. Instead of breaking down, the prokaryotes persisted and gradually evolved into mitochondria and chloroplasts in the cytoplasm of their host cells. By this route, mitochondria may have originated from ancient bacteria, and chloroplasts from ancient blue-green algae. The transcription and translation mechanisms of the organelles, according to this hypothesis, are the last remnants of the once-independent existence of their prokaryotic ancestors (for details, see Chap. 14).

Supplement 10-3: RNA Transcription and Regulation in Polytene Chromosomes

Chromatin is typically distributed throughout the nucleus in interphase cells. This distribution makes it impossible to recognize specific genes among the chromatin fibers. Recognition of genetic sites is possible, however, in certain highly specialized interphase cells that occur in flies (the Diptera) and a few other groups in the animal and plant kingdoms. The interphase chromosomes in the nuclei of these cells are so large that they can easily be seen under the light microscope. They contain visible regions that can be identified as genes active in synthesizing RNA.

In the salivary gland cells of dipteran larvae, where they have been best studied, the giant chromosomes appear as roughly cylindrical, cross-banded bodies ranging from 5 to 15 micrometers in diameter and 200 micrometers or more in length (Fig. 10-33).

The giant chromosomes are often termed "salivary gland" chromosomes because they are commonly studied in the salivary glands of dipteran larvae. But chromosomes of this type also occur in other organs and tissues of fly larva, including cells of the esophagus, gut, gastric pouches, Malpighian tubes, body wall muscles, nerve ganglia, and tracheal walls. Some adult flies also have giant chromosomes in single large cells that occur in the footpads. At least one other insect group, the Collembola, contains cells with large, cable-like chromosomes of this type, and chromosomes with a similar appearance also occur as an intermediate stage in the development of the macronucleus in at least one ciliate protozoan. Giant chromosomes have also been observed in the embryo suspensor stalk in flowering plants. Because of their wide occurrence in a variety of species and cell types, they are now generally termed *polytene*, meaning "many-threaded," rather than salivary gland chromosomes.

Structure of Polytene Chromosomes

In *Drosophila*, polytene chromosomes appear in cells of the salivary glands early in development of the larva. Within the nuclei of these cells, the quantity of DNA increases by repeated cycles of replication. This replication takes place without division of the chromosomes or nuclei. As the replication proceeds, the chromosomes become visible in the nuclei as thick, elongated structures with the characteristic transverse bands.

Measurements of the total DNA quantity in *Drosophila* salivary gland nuclei as larval development proceeds reveal that the DNA doubles successively in ten well defined, synchronized steps. This indicates that fully mature polytene nuclei in these cells contain more than a thousand times ($2^{10} = 1024$) as much DNA as ordinary somatic nuclei. Each chromosome, therefore, is probably made up of as many molecules of DNA. The chromatin fibers containing these DNA molecules in polytene chromosomes are thought to extend from one end of the chromosome to the other and to be coiled tightly in the bands and extended in the interband regions (Fig. 10-34a).

Active Sites on Polytene Chromosomes: Bands and Puffs

Within a given species, the number and pattern of bands is the same in all the cells containing polytene chromosomes. This is true in tissues as diverse as intestinal and salivary gland cells. Thus, the differential activity of the cells in the various tissues is not reflected in variations in the number or types of bands in polytene chromosomes. However, differences can be detected in the appearance of some bands. In different tissues, or at different times in the same tissue, one or more bands swell or increase into diffuse, brushlike structures called *puffs* (see Fig. 10-34b and arrows, Fig. 10-33). The largest puffs are called *Balbiani rings*, named for the scientist who first described polytene chromosomes in the 1880s.

Under the electron microscope the puffs appear as swollen regions of the chromosomes containing large concentrations of fibers and small granules (Fig. 10-35). Some of the fibers are loops of chromatin extending outward from the chromosome axis. Other fibers and the granules

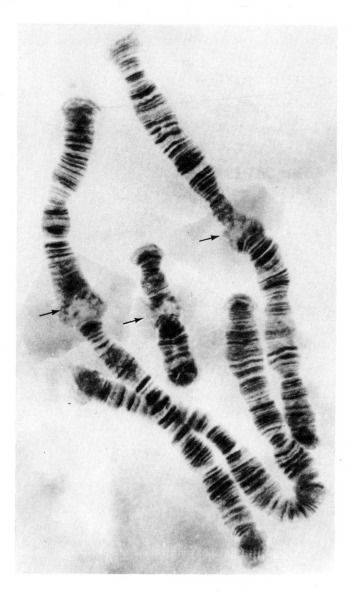

Figure 10-33 The polytene chromosomes in the salivary gland of a larva of the fly *Chironomus tentans*. The longer chromosomes in this micrograph are between 200 and 250 microns in length. The arrows point to puffs, sites of intense RNA transcription. Courtesy of W. Beerman and Springer-Verlag, from *Protoplasmatologia* 60 (1962).

are recognizably thicker than chromatin fibrils. The granules can also be seen in quantity in the nucleoplasm surrounding the puff, and in contact with the pore complexes. Similar material, extended into a rodlet, can be seen passing through the pores into the cytoplasm (Fig. 10-36). These fibers and granules are probably the mRNA prod-

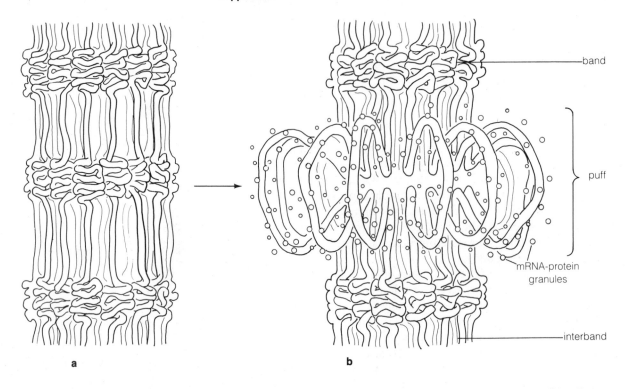

Figure 10-34 The probable structure of bands, interbands, and puffs in polytene chromosomes. (a) Unpuffed; (b) puffed.

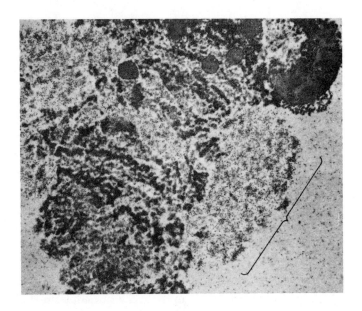

Figure 10-35 Sectioned puff (bracket) from a polytene chromosome of *Chironomus.* Many particles are visible within the puff and in the surrounding nucleoplasm. × 3,600. Courtesy of H. Swift and B. J. Stevens, from *In Vitro* 1 (1965).

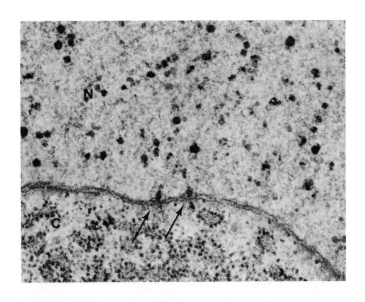

Figure 10-36 Granules in a salivary gland nucleus of *Chironomus* (*N*) apparently passing through the nuclear envelope pore complexes (arrows) to enter the cytoplasm (*C*). × 49,000. Courtesy of B. J. Stevens, H. Swift, from *Journal of Cell Biology* 31 (1966): 55, by copyright permission of The Rockefeller University Press.

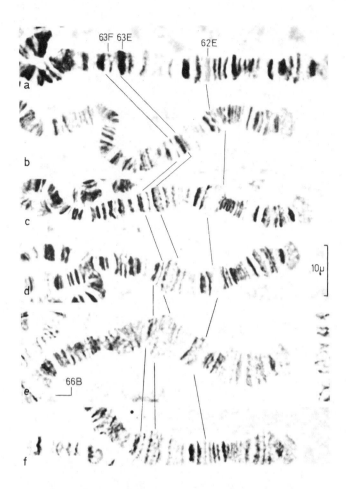

Figure 10-37 The sequence of appearance and disappearance of puffs at different points in a polytene chromosome as development proceeds in a larva of the fly *Drosophila*. Band 63F forms no puffs during the successive developmental stages in the sequence shown in (**a–f**); band 63E enlarges and forms a puff in (**c, d,** and **e**) and retracts slightly in (**f**); and band 62E puffs in (**b**) and (**c**) and retracts in (**d–f**). The lines connect the same bands in (**a–f**). × 1,300. Courtesy of M. Ashburn and Springer-Verlag, from *Chromosoma* 38 (1972): 255.

ucts of active genes, combined with proteins into particles large enough to be visible in the electron microscope.

Direct evidence that the puffs are genetic sites active in mRNA synthesis has come from the extensive experiments of B. Daneholt and his colleagues at the Karolinska Institute in Sweden. Daneholt accomplished the seemingly impossible task of removing a single chromosome from the salivary gland nuclei of another fly species, *Chironomus*, and dissecting out a single large puff. From a preparation of these dissected puffs, Daneholt isolated and purified a large RNA molecule. RNA molecules of the same size were also identified in the nuclear sap, and in the cytoplasm. In the cytoplasm, the RNA molecule attached to ribosomes and was active in protein synthesis. The product of this puff is thus an mRNA, probably coding for one of the large proteins forming a part of the salivary gland secretions in *Chironomus*. Other techniques have shown that rRNA and various tRNAs also originate from single puffed bands. Thus individual puffs can be identified as sites active in the synthesis of mRNA, rRNA, and tRNA.

Tracing polytene chromosomes under the light or electron microscope shows that as development proceeds, the puffs appear and disappear in definite patterns in a given cell type (Fig. 10-37). Since the puffs are sites of mRNA, tRNA, or rRNA synthesis, this evidence demonstrates directly that the transcription of genes is regulated, and that individual genes are turned on and off in sequence as development proceeds.

SIX

Cell Division

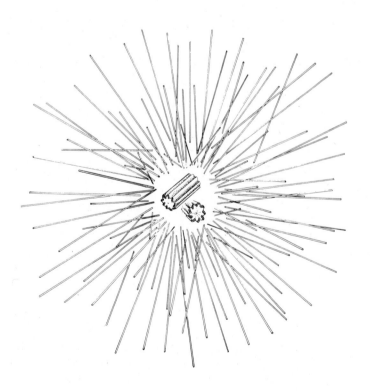

11

Mitotic Cell Division

All living organisms, from single-celled forms to the most complex animals and plants, grow and reproduce by processes taking place at the cellular level. Growth, or the increase in the mass of cells, is accomplished by the formation of new cellular molecules, including proteins, nucleic acids, lipids, and carbohydrates. Following a period of growth, cells reproduce by dividing. In the type of cell growth and division that increases the body size of higher plants and animals, the division mechanism produces daughter cells that contain exactly the same hereditary information as the parent cells. In this division mechanism, the DNA sequences are copied exactly and are passed on to daughter cells without change. All other cell parts are divided approximately equally between daughter cells, although not with the same precision as the DNA.

The cellular processes underlying cell division take place in three clearly defined phases. In the first phase, called *interphase*, a cell increases in mass by synthesizing biological molecules, among them two exact copies of its DNA. Other cellular molecules necessary for division are stockpiled. In the second phase, termed *mitosis*, the replicated DNA molecules, with their histone and nonhistone proteins, are divided and placed in two separate daughter nuclei. Following mitosis, in the final phase of cell reproduction, the cytoplasm divides and two completely separate daughter cells are produced. The cytoplasmic division taking place in the last phase, called *cytokinesis*, usually divides the cytoplasmic organelles and molecules approximately equally between the daughter cells. The entire interphase-mitosis-cytokinesis sequence makes up the *cell cycle*.

An Overview of the Cell Cycle

Cells from almost all eukaryotic organisms follow a similar pathway through the cell cycle (Fig. 11-1). A newly formed daughter cell enters a period of interphase synthesis and growth during which protein, lipid, and carbohydrate molecules are made. This stage, called G_1 of interphase, is of variable length in different cell types. In some cells, as in rapidly dividing embryonic tissue, G_1 may last for only a few minutes. Other cells, such as the muscle and nerve cells of adult mammals, remain in G_1 for the life of the animal and never divide again. In cells destined to divide again, DNA replication eventually begins, terminating G_1 and initiating the S period of interphase (S stands for "synthesis," meaning DNA synthesis). This stage, in most eukaryotes, lasts about 6 to 8 hours. After the S phase is completed, another interval, called G_2, passes before mitosis begins (the G in G_1 and G_2 stands for "gap," and indicates the periods of interphase in which there is a "gap" or pause in DNA synthesis). The G_2 phase in most cells lasts for 2 to 5 hours.

Interphase thus consists of the G_1, S, and G_2 phases. Although G_1 may last from hours to months or years, S and G_2 are usually of uniform length in cells of the same type. As a result, cells usually enter mitosis and cytokinesis within 8 to 12 hours after the beginning of the S phase. When mitosis and cytokinesis are complete, the resulting daughter cells enter the G_1 period of the next interphase. This sequence makes it obvious that cells are fixed into the pathway leading to division at the onset of

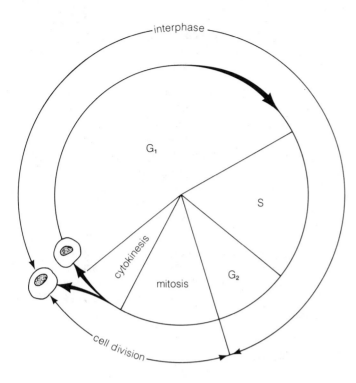

Figure 11-1 The cell cycle. The G_1 period of interphase may be of variable length, but for a given cell type the remaining S, G_2 and division segments of the cycle are usually of uniform duration.

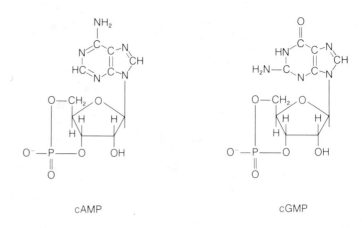

Figure 11-2 The cyclic nucleotides, cyclic AMP (cAMP) and cyclic GMP (cGMP).

S, because once S begins, G_2, mitosis, and cytokinesis usually follow without delay.

The cellular mechanisms regulating and controlling the cell cycle are presently the subject of one of the most intensive research efforts in cell biology. The outcome of this research is of fundamental importance to our understanding of the ways in which cells work, and, perhaps of greatest significance to human well-being, why cells grow and divide uncontrollably in cancer to produce malignant tumors.

Because the beginning of the S phase is the first significant event in the processes leading to cell division, most of the current research effort is directed toward identifying the mechanisms that trigger DNA replication. While many factors that modify the entry into S have been discovered, the ultimate and basic control mechanisms have yet to be established.

Many ideas have been favored for a time as the fundamental control mechanism and later discarded because they do not fit the facts. For example, it was once thought that the control of cell division is based on the size ratio

between nucleus and cytoplasm. According to this idea, once cells reach a certain size through cytoplasmic growth, division is triggered. However, although cell division usually parallels increases in cell size, there are many important exceptions to this pattern. During embryonic development in many animals, cell division takes place with no net increase in size of the embryo. This is the case in bird embryos, which begin with a large egg cell (the yolk of a hen's egg is a single, large cell), and divide to complete embryonic growth with no increase in overall size or weight. Cells in these embryos, therefore, actually become progressively smaller as the divisions proceed.

Research in this field is now concentrating on a variety of other factors that may possibly regulate the cell cycle. One of these is the relative concentrations of two nucleotides, AMP and GMP, changed by an enzyme into a form in which both ends of the single phosphate connect to the sugar of the nucleotide (Fig. 11-2). These forms, because of the circular connections made by the phosphate group, are called *cyclic nucleotides*, or *cyclic AMP (cAMP)* and *cyclic GMP (cGMP)*. Both of these nucleotides vary regularly in concentration as cells progress through the cell cycle. In many cells cAMP concentration is at relatively high levels during G_1 and drops by about two-thirds as cells enter S and mitosis. At the following G_1, cAMP concentration rises again. Frequently, cGMP concentrations vary in the reverse pattern: cGMP is low during G_1, peaks at the beginning of S, and then falls again to low levels by the time division is complete. Adding or removing either of the two cyclic nucleotides in many of these cells can stop or start entry into S and the subsequent cell division.

These findings created great excitement among cell biologists until a gradually growing list of exceptions appeared. Many cell types were discovered in which the cyclic nucleotides worked in the reverse of the patterns described above, or remained more or less constant in concentration throughout the cell cycle. And, in plants, neither the cyclic nucleotides, nor the enzymes producing them from the regular forms of AMP and GMP, appear to be present at all.

Because of these exceptions current opinion is swinging toward the idea that in some cell types the cyclic nucleotides modify, but do not control the cell cycle. In other cell types, the cyclic nucleotides have no effect or are not even present. The same may be said for other conditions that vary regularly with the cell cycle in some species, such as changes in internal calcium ion concentration or pH. Where changes in these conditions do modify the cell cycle, they probably act as elements that make adjustments in other, more fundamental regulatory mechanisms that as yet remain unknown.

The pattern of another series of changes that occur in step with progress through the cell cycle suggests what these fundamental controls may be. This is a variation noted in the number of phosphate groups added to the histone chromosomal proteins (see p. 206). The number of phosphate groups added changes regularly as the cell cycle turns, particularly to histone H1. Typically, phosphate groups are added to H1 as cells approach and enter the S phase; more are added until the following mitotic division is well underway. Then, as mitosis is complete, the phosphate groups are removed so that few or none remain on the H1 molecule by the time the following G_1 phase begins. Addition and removal of phosphate groups to the nonhistone proteins has also been detected in coordination with the cell cycle. These changes in the histone and nonhistone proteins occur in organisms as diverse as animals, plants, and the fungi.

Changes in the histones and nonhistones in coordination with the cell cycle is especially interesting because these chromosomal proteins have otherwise been implicated in regulating the activity of genes in RNA transcription during interphase (see p. 238). This suggests that the fundamental controls of the cell cycle may actually be a series of genes that is turned on and off in sequence to trigger entry into S and passage through G_2, mitosis, and cytoplasmic division. The effect of other factors, such as changes in pH or the concentrations of Ca^{2+} ion or the cyclic nucleotides, may simply be to modify the activity of one or more genes in this sequence in some animals or plants.

Chromosome Duplication During S

The primary activity of the S period of interphase is duplication of the chromosomes. Replication of chromosomal DNA produces two copies that, with very rare exceptions, are exact, sequence-by-sequence duplicates of their parent molecules. The duplication of chromosomal proteins produces essentially the same result: the histone and nonhistone chromosomal proteins are duplicated and combined with the replicated DNA molecules to produce two daughter chromosomes that are exact duplicates of each other and their G_1 chromosomal "parent." Of the mechanisms accomplishing chromosome duplication, only DNA replication is understood in any detail. Many uncertainties still surround the synthesis of chromosomal proteins and the pattern in which they combine with replicated DNA to form daughter chromosomes.

DNA Replication

When Watson and Crick discovered the molecular structure of DNA (see p. 216), they pointed out that the complementarity of the two nucleotide chains of the DNA molecule provides a mechanism for DNA duplication (Fig. 11-3). Complementarity (see also p. 194) refers to the fact that a given base in one of the two nucleotide chains of a DNA molecule will pair only with one kind of base in the opposite chain: thymine will pair only with adenine, and guanine only with cytosine (Fig. 11-3a). As a result of complementarity, the sequence of bases in one chain fixes the sequence of bases in the opposite chain. From this, as Watson and Crick pointed out, the two nucleotide chains of a DNA molecule, if unwound and separated (Fig. 11-3b), can act as patterns or *templates* for the synthesis of their missing halves (Fig. 11-3c and d). The two molecules of DNA produced from these templates are each identical in sequence to the original molecule.

This mode of replication, as first outlined by Watson and Crick, set the stage for subsequent work. It was soon established that in replication the two original nucleotide chains, after serving as templates, remain paired with their newly synthesized copies (Fig. 11-4). As a result, each of the two molecules formed contains one old and one new nucleotide chain. This pattern of DNA replication is called *semiconservative* replication because one-half of the original parent DNA molecule is conserved in each of the

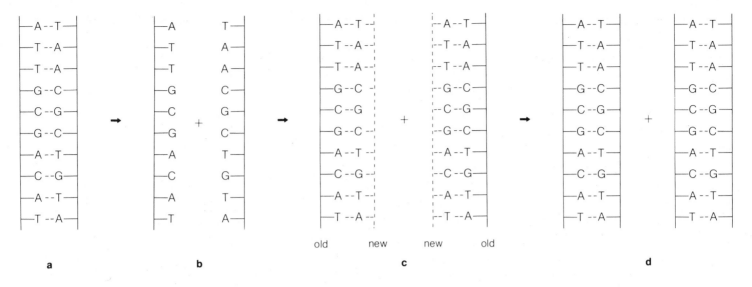

Figure 11-3 DNA replication. (**a**) An intact DNA double helix before replication. (**b**) The two nucleotide chains unwind. (**c**) Each half of the "old" molecule acts as a template for the synthesis of a complementary, "new" nucleotide chain. (**d**) The two molecules produced are exact duplicates of the molecule entering replication; each consists of one old and one new chain.

daughter molecules (one of the key experiments demonstrating that replication is semiconservative is described in Supplement 11-1).

Semiconservative DNA replication proceeds by a mechanism that closely resembles RNA transcription (see Information Box 9-2), at least as far as the addition of nucleotides to the growing nucleotide chains is concerned. The reaction is catalyzed by *DNA polymerase*, an enzyme similar in its activity to the RNA polymerase enzymes catalyzing transcription. In the reaction (see Information Box 11-1), the DNA polymerase enzyme moves along the template chain, adding nucleotides to the growing new chain by matching up complementary base pairs. The energy required for the DNA replication is carried to the reaction by the DNA nucleoside triphosphates (dATP, dGTP, dCTP, and dTTP; see p. 196) that are linked into the new chain; each of these is a high-energy molecule with properties similar to ATP. The new nucleotide chain, once synthesized, remains wound into a double helix with its template chain.

While the addition of nucleotides to a growing DNA chain resembles RNA transcription, the process of replication differs fundamentally in the fact that DNA is a double helix. Because it is double helical in structure and because replication is semiconservative, the parental DNA

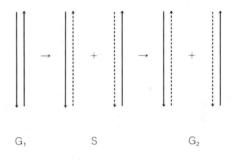

Figure 11-4 The pattern of semiconservative replication (see text). All living organisms replicate their DNA by the semiconservative pathway.

molecule must unwind completely for replication to occur, at a rate estimated to approach 13,000 revolutions per minute in bacteria and about 1000 to 2000 revolutions per minute in eukaryotes. Unwinding DNA involves an additional group of enzymes that have no parallels in RNA transcription (how these enzymes operate to unwind the template DNA chains and the details of other steps in DNA replication are outlined in Supplement 11-2).

Duplication of the Chromosomal Proteins

The histone and nonhistone chromosomal proteins of eukaryotes are also duplicated during interphase. Histones are synthesized during the S stage of interphase at the same time as DNA replication. The total quantity of histones doubles during S, as would be expected if duplication of these chromosomal proteins is closely coupled with DNA replication. The duplicated chromosomes at the completion of S, as a result, contain old and new histones in approximately equal quantities as well as old and new DNA nucleotide chains. The nonhistone chromosomal proteins are also duplicated during interphase, some of them during G_1 and some during S, so that by the time S is complete the duplicated chromosomes also have a full complement of these proteins.

Mitosis

Once chromosome duplication is complete and the brief G_2 stage of interphase is over, the cell is ready to enter mitosis. In contrast to the molecular synthesis that occurs

---

Information Box 11-1

The Mechanism of DNA Replication

DNA replication requires, among other factors, a DNA template chain, the DNA polymerase enzyme, and the four nucleotides occurring in DNA, *deoxyadenosine triphosphate* (*dATP*), *deoxyguanosine triphosphate* (*dGTP*), *deoxycytidine triphosphate* (*dCTP*), and *deoxythymidine triphosphate* (*dTTP;* see Information Box 9-1). Each of these nucleotides differs from its counterpart in RNA synthesis by the presence of deoxyribose, rather than ribose, as the 5-carbon sugar of the molecule. Since the DNA polymerase enzyme can add nucleotides only to the end of an existing nucleotide chain, a short segment called the *primer* must already be in place opposite the template chain. Although both short DNA and RNA chains can serve as primers for DNA synthesis, almost all DNA molecules in nature are assembled on RNA primers (the short RNA primers are assembled by specialized RNA polymerase molecules that act in this role in replication; see Supplement 11-2).

Replication proceeds as shown in Figure 11-5. In the first step in the reaction sequence, DNA polymerase binds to the template chain at the end of the short length of primer. In the diagram, the first base exposed at the end of the primer is cytosine; base pairing and the activity of the enzyme will match the complementary nucleoside triphosphate dGTP to the template at this point (Fig. 11-5b). dGTP binds to the enzyme in a position close to the end of primer. The total complex now consists of the enzyme, the template DNA with primer, and the molecule of dGTP. The enzyme now catalyzes removal of the two end phosphates from dGTP; at the same time, the remaining phosphate is linked to the terminal primer nucleotide (Fig. 11-5c). Energy for the formation of this linkage comes from the removal of the phosphates from dGTP. The enzyme now moves one step along the template and binds at the end of the nucleotide just added (Fig. 11-5d). Adenine is exposed on the template DNA at this point; the complementary nucleoside triphosphate dTTP is bound from the medium and the cycle repeats. The enzyme moves along the template in this way, catalyzing the assembly of the complementary strand in stepwise fashion until it reaches the end of the template.

Each successive nucleotide is added to the —OH group bound to the 3' carbon at the end of the growing chain. As each nucleotide is added, it provides the 3'—OH group (shaded in Fig. 11-5d) for the next base to be added. Thus a 3'—OH group is always present at the "newest" end of a growing nucleotide chain, and DNA synthesis is said to proceed in a 5' → 3' direction. All the known DNA polymerase enzymes add successive nucleotides only in this direction. RNA transcription, as noted in Information Box 9-2, also proceeds only in this direction.

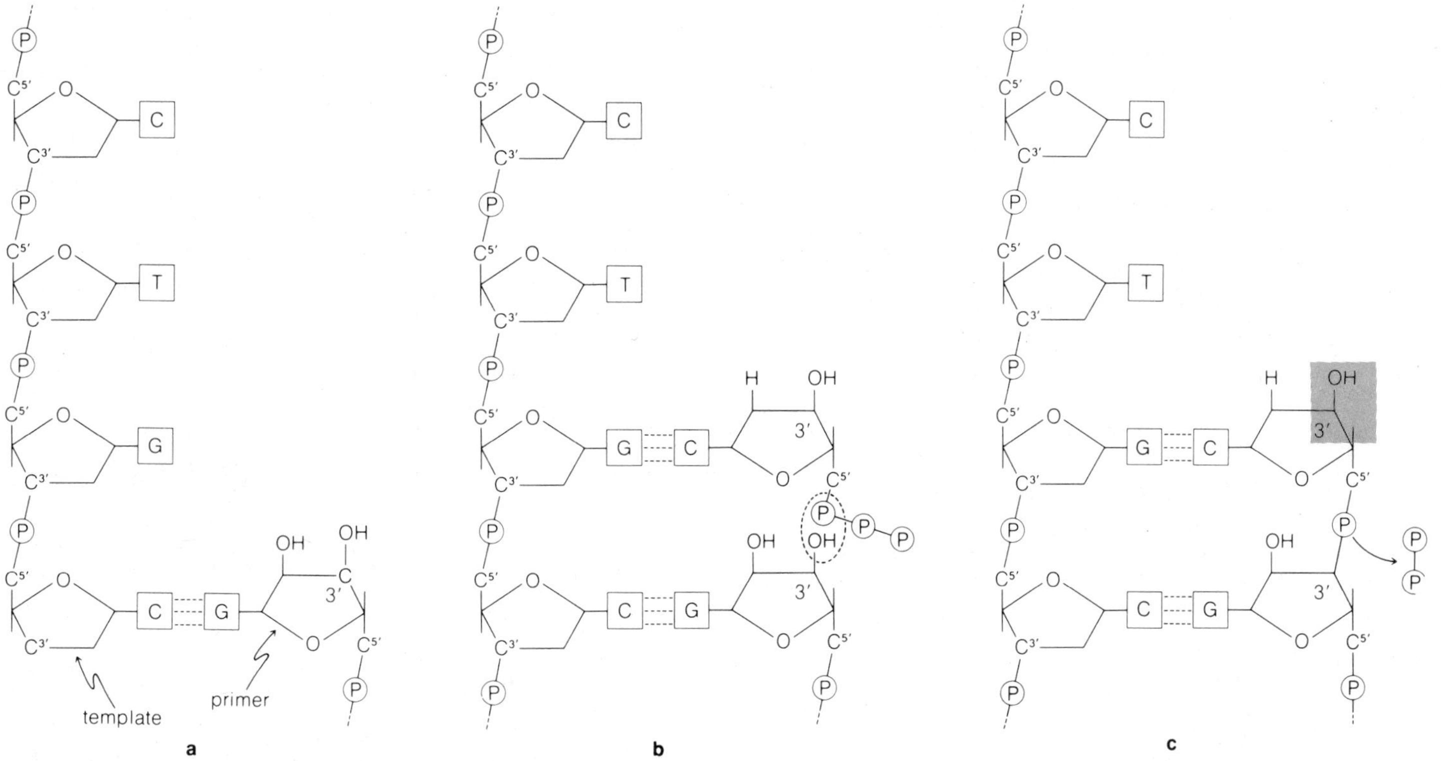

Figure 11-5 The assembly of nucleotides into a DNA chain (see Information Box 11-1).

during S and G_2, which produce no visible rearrangements of cell organelles, mitosis involves extensive structural alterations. These changes are easily observed under the light microscope, and were correctly interpreted as early as 1880 by many investigators. Most prominent of these early workers were the European scientists Eduard Strasberger and Walter Flemming, who are credited with the first description and interpretation of mitosis.

Mitosis is a continuous process and takes place with no significant pauses or interruptions. However, for convenience in study, the process is usually broken down into four stages: *prophase* (from *pro* = before), *metaphase* (*meta* = between), *anaphase* (*ana* = back), and *telophase* (*telo* = end). These stages are shown in light micrographs in Figure 11-6 and in diagrammatic form in Figure 11-7.

Prophase

Eukaryotic nuclei contain a collection of long DNA molecules combined with histone and nonhistone proteins. The individual DNA molecules, with their associated proteins, are the *chromosomes* of the nucleus (see Information Box 11-2). During interphase, the chromosomes, taken together, form the *chromatin* of the nucleus, which is so generally distributed at this stage that individual chromosomes cannot be distinguished (as in Fig. 9-17). Other than the chromatin fibers, the interphase nucleus shows little internal differentiation except for the nucleolus.

The Chromosomes at Prophase The beginning of prophase (Figs. 11-6a and 11-7a) is marked by the first appearance of chromosomes as recognizable threads in the nucleus (the word mitosis, from *mitos* = thread, is derived from the threadlike appearance of the chromosomes as they begin to pack into thicker structures as prophase begins). This progressive folding and packing of the chromatin fibers into thicker structures is called chromosome *condensation*.

As soon as the chromosomes appear as distinct threads, it is clear that each is split lengthwise into two subunits (double arrows, Fig. 11-6b and c). The two subunits, called

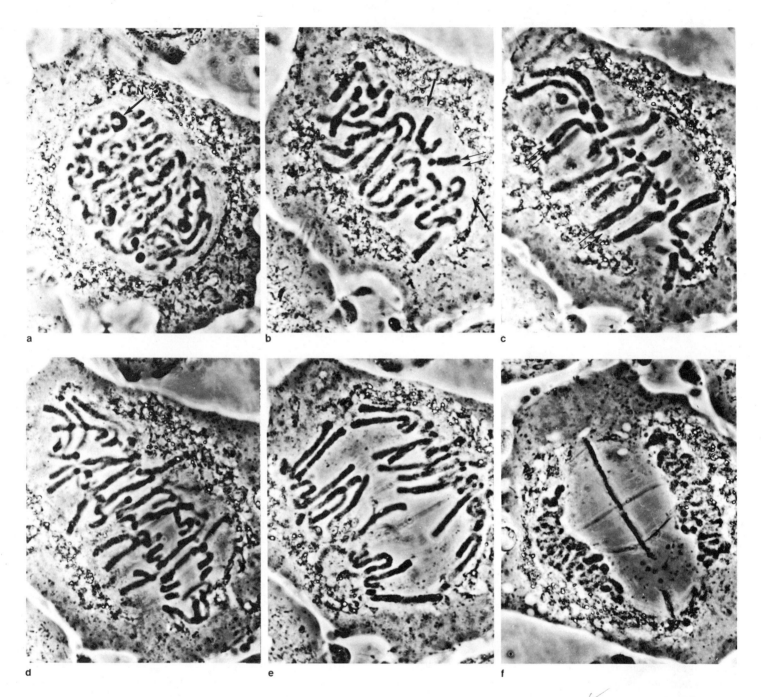

Figure 11-6 Mitosis in a living cell of the plant *Haemanthus*, observed in the light microscope. (**a**) Prophase. The chromosomes have condensed into distinct structures. The nucleolus (single arrow) is still visible at these early stages. The "clear zone" around the nucleus is the first indication of spindle formation in plants. Electron micrographs at this stage show the clear zone to be occupied by large numbers of spindle microtubules. (**b**) Beginning of metaphase. The nuclear envelope has broken down but fragments (single arrows) are still visible. (**c**) Metaphase. Although the kinetochores lie in a plane at the midpoint of the spindle, the arms of each chromosome on either side of the kinetochores extend in random fashion toward either pole. (**d**) and (**e**) Anaphase. (**f**) Telophase. The chromosomes at the poles are no longer individually distinguishable. The double arrows in (**b**) and (**c**) mark points where both chromatids of a chromosome are visible. × 700. From *Cinematography in cell biology*, 1963. Courtesy of A. Bajer and Academic Press, Inc.

a   b   c   d   e   f

Figure 11-7 Diagram of mitosis; compare with Figure 11-6.

chromatids, are the result of the duplication of DNA and chromosomal proteins during interphase (see Fig. 11-14). As such, the two chromatids of each chromosome are exact duplicates of each other, and contain exactly the same genetic information. The longitudinal split marking the division between chromatids becomes more distinct as the chromosomes progressively thicken during prophase. By the end of prophase, the chromosomes have condensed down to relátively short, thick double rods (Figs. 11-6b and 11-7b).

Just how the chromosomes fold from the extended interphase state to the tightly packed condition of late prophase remains one of the unsolved problems of mitosis. The most widely accepted hypothesis maintains that one of the chromosomal proteins, possibly histone H1 (see p. 206), forms crosslinks between chromatin fibers that fold and hold them together. As more of the crosslinks form, the long chromatin fibers gradually condense into the compact, tightly folded form characteristic of late prophase.

By late prophase, when condensation is nearly complete, it is apparent that many of the chromosomes have distinctive shapes and sizes (see Information Box 11-3). This distinctive structure depends on differences in length and width of the chromosomes and the locations of narrow regions called *constrictions.* Each chromosome usually has at least one prominent narrow region called the *primary constriction,* which marks the point at which the spindle microtubules attach when prophase changes to metaphase (see Metaphase, below). Usually, one or more chromosomes also have other narrow regions known as *secondary constrictions.* The pattern and location of primary and secondary constrictions, along with the chromosome number and variations in the length and thickness of the chromosomes, provide a collective morphology known as the *karyotype* of an organism. Frequently, the karyotype is so distinctive that a species can be identified from this information alone (Fig. 11-8 shows the distinctive structure of the human karyotype).

Other Nuclear Changes During Prophase Nucleoli usually disappear after the early stages of prophase. As condensation proceeds in early prophase, it becomes clear that the nucleolar material is attached to one or more of the chromosomes at prominent secondary constrictions (other chromosomes may have constrictions with no obvious connections to nucleoli). Toward the close of prophase the nucleoli grow smaller and usually disintegrate by the onset of metaphase. This happens because rRNA transcription, as well as synthesis of other RNA types, drops to very low levels or stops entirely during prophase and metaphase of mitosis (the breakdown and reformation of the nucleolus during mitosis is described in Supplement 11-3).

As prophase draws to a close the nuclear envelope breaks down, releasing the chromosomes and other contents of the nucleus into the surrounding cytoplasm. By this time, the chromosomes have packed into short, dou-

---

Information Box 11-2   An Overview of Mitosis

If all the DNA in a eukaryotic cell, containing all of the genetic information of the cell nucleus, existed in a single piece, it could be represented as a single line of definite length:

No eukaryotic cells contain DNA in a single piece like this, however. Instead, the DNA is broken into shorter subunits:

These DNA subunits, with their attached histone and nonhistone proteins, are the chromosomes of the cell nucleus. During interphase, the chromosomes are exactly duplicated:

In mitosis, the duplicated chromosomes first condense into shorter, thicker units:

They then attach to spindle microtubules in such a way that the two parts of each chromosome connect to microtubules leading to opposite ends of the cell:

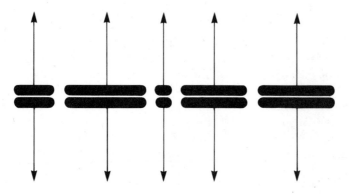

---

ble rodlets that are more or less randomly distributed in the region formerly occupied by the nucleus. The nucleolar material, if still visible at this time, disperses and becomes indistinguishable from the surrounding cytoplasm.

Spindle Formation During Prophase As these nuclear events follow their course, changes also take place in the cytoplasm, some beginning as early as the S stage of in-

terphase. Almost all animal cells, and the cells of many lower plants, contain a pair of small, barrel-shaped structures called *centrioles* in the cytoplasm just outside the nuclear envelope (see Fig. 8-19; Supplement 8-1 describes centriole structure in detail). The centrioles at G_1 are surrounded by short lengths of microtubules that radiate outward in all directions from the centriole pair. As the S period begins, the centrioles separate slightly and duplicate; this duplication produces two pairs of centrioles still

The spindle microtubules then develop tension, separating the duplicate parts of each chromosome and moving them to the opposite ends of the cell:

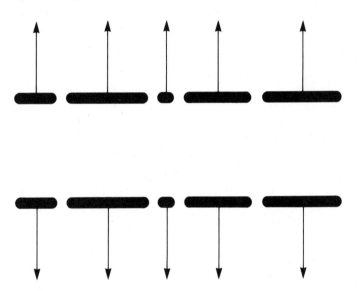

The chromosomes then decondense and become enclosed in separate cell nuclei:

When the cytoplasm divides, the two nuclei are separated into two daughter cells, the products of mitotic divison.

Information Box 11-3

DNA, Nucleosomes, and the Structure of Metaphase Chromosomes

Each of the two chromatids of a chromosome contains a single long DNA molecule. This molecule, on the order of several centimeters long in most species, is combined with the histones into nucleosomes. The nucleosomes, with the nonhistone proteins, form the chromatin fibers of the nucleus. Spaced along the fibers are reactive groups capable of forming interfiber crosslinks. These groups, probably H1 histone molecules, act as "folders" that pack the chromatin fibers from the interphase state into metaphase chromosomes. The final folding pattern of the chromatin fibers in chromosomes, although apparently irregular, is probably highly ordered and produces the structural features recognizable at metaphase: a primary constriction, and arms of characteristic length and thickness. Secondary constrictions may also be present at one or more locations within the arms. Two kinetochores, one for each chromatid of the chromosome, form the attachment sites for spindle microtubules within the primary constriction. (The primary constriction containing the kinetochores is sometimes termed the *centromere* or *centromere region* of a metaphase chromosome.) The results of electron microscopy have made it clear that the entire structure is fully replicated and double at all points along its length, including the primary constriction and kinetochores.

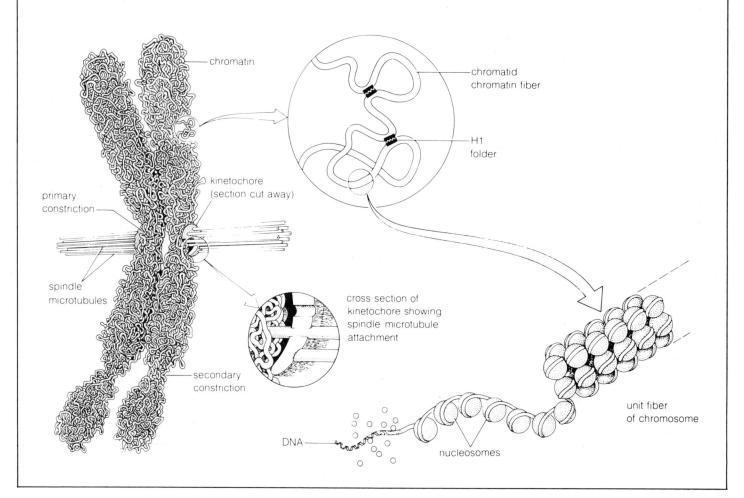

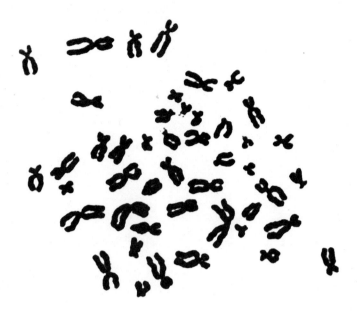

Figure 11-8 Human chromosomes isolated from a metaphase cell. × 1,500. Courtesy of S. Brecher.

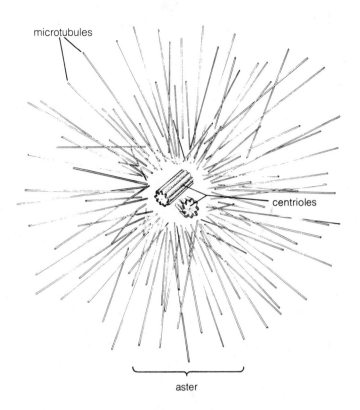

Figure 11-9 The relationship of the centriole and aster. Note that none of the astral microtubules touch the centrioles.

contained within the radiating microtubules. During centriole duplication, these surrounding microtubules increase in length and number, shaping the surrounding granules and vesicles of the cytoplasm into a starlike array called the *aster* (Fig. 11-9).

At the initiation of prophase, the two pairs of centrioles begin to separate. By late prophase the pairs have moved to opposite ends of the cell just outside the nucleus (Fig. 11-10). As this shift takes place, bundles of microtubules lengthen between the separating centrioles, stretching in the direction of movement. By late prophase, when the centrioles have reached the opposite ends of the cell, the microtubules form a mass extending completely around one side of the nucleus. This mass is the *primary spindle* (Fig. 11-10c).

The cells of some higher plant species, particularly the flowering plants, do not contain centrioles (in plants, centrioles are limited to species that produce motile reproductive cells). Some animal cells, such as certain protozoa and the developing eggs of many higher animals, also lack centrioles. In these cells, spindle formation follows a different pathway. No conspicuous changes take place until prophase, when under the light microscope a clear space becomes visible in a narrow zone surrounding the nucleus (Fig. 11-11a). This zone increases in size until a spindle-

shaped formation develops (Fig. 11-11b). Electron microscopy at this time reveals that the clear zone is packed with microtubules extending around the nucleus from pole to pole. No asters are present at the poles in this type of spindle (see Fig. 11-12b). The absence of centrioles or asters has no apparent effect on spindle function (the replication of centrioles, and the significance of the movements of centrioles and asters in animals and other organisms with asters are discussed further in Supplement 11-4).

Metaphase

The transition from prophase to metaphase (Figs. 11-6c and 11-7c) is gradual but can be conveniently marked by the fragmentation and breakdown of the nuclear envelope. Three major rearrangements take place after the nuclear envelope breaks down. First, the spindle moves into the region formerly occupied by the nucleus. At the same

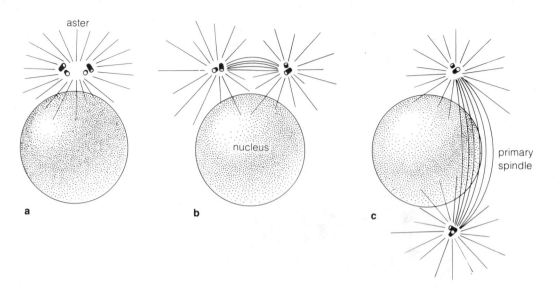

Figure 11-10 Formation of the primary spindle in animals (see text).

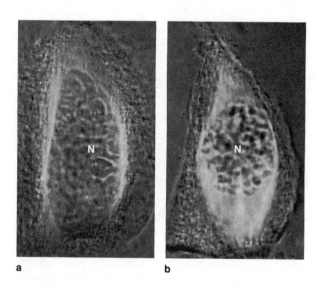

Figure 11-11 Spindle formation in plants. (**a**) A clear zone forms around the nucleus (*N*). Under an electron microscope, this clear zone proves to be packed with microtubules. (**b**) The clear zone develops into the spindle by continued formation of microtubules. × 700. Courtesy of S. Inoué and A. Bajer and Springer-Verlag, from *Chromosoma* 12 (1961):48.

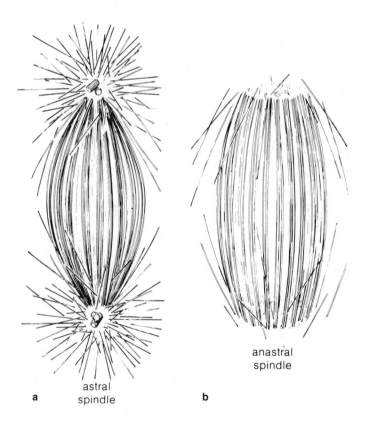

Figure 11-12 The two types of spindles at maturity. (**a**) With centrioles and asters. (**b**) With no centrioles or asters. Either spindle type functions normally in mitosis.

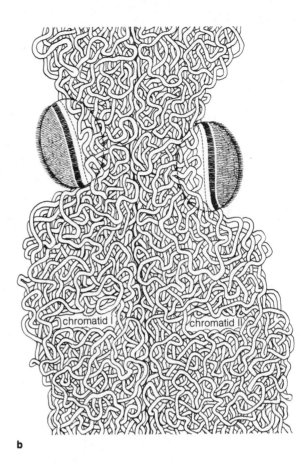

chromatid I

chromatid II

Figure 11-13 (**a**) Kinetochores (brackets) of a Chinese hamster. Both kinetochores of this chromosome have been caught in the plane of section. × 45,000. Courtesy of B. R. Brinkley and Springer-Verlag, from *Chromosoma* 19 (1966):28. (**b**) Diagram outlining the kinetochores visible in the electron micrograph.

time, the chromosomes move to the midpoint of the spindle. Finally, during movement to the midpoint of the spindle, each chromosome attaches to bundles of spindle microtubules. These three events establish metaphase: the spindle is completely formed and the chromosomes, each attached to spindle microtubules, are aligned at the midpoint of the spindle.

By this time, the spindle in higher eukaryotes contains from 500 to 1000 or more microtubules, all stretching from one end of the cell to the other (Fig. 11-12). The microtubules are more closely spaced near the ends or *poles* of the spindle, producing a pronounced narrowing of the structure at the tips.

One of the most significant events of mitosis is the way in which the chromosomes attach to the spindle at metaphase. Much of the precision of mitosis depends on this attachment. Remember that each chromosome at this

stage consists of two replicated subparts, the chromatids, which are exact duplicates of each other. At the primary constriction of the chromosome, each of the two chromatids of the metaphase chromosomes has a *kinetochore* (brackets, Fig. 11-13), a disclike structure that forms the point of attachment for spindle microtubules. The two kinetochores of each chromosome attach to the spindle in such a way that they face and connect to microtubules leading to *opposite* poles of the spindle (Fig. 11-14). This pattern ensures that the two chromatids of each chromosome separate and move to opposite poles when the spindle microtubules develop their pulling force at anaphase. As a result, the chromatids are divided equally and precisely between the two poles when separation and movement of all the chromatids is complete.

By the time the kinetochore connections are made, growth of the spindle has progressed to such an extent

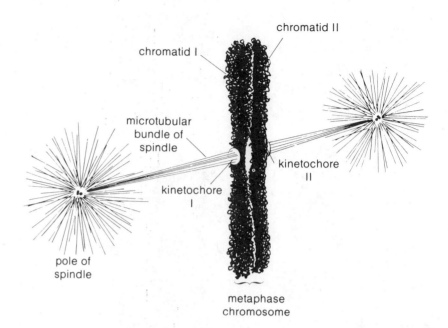

Figure 11-14 Kinetochore connections at metaphase. Two kinetochores are present on each metaphase chromosome, one for each of the two chromatids. The kinetochores are directed toward and make microtubule connections leading to opposite poles of the spindle.

that most of the cytoplasm is filled by spindle microtubules. This development, and the connection of chromosomes to spindle microtubules at the spindle midpoint, completes the events of metaphase.

Anaphase

Tension is developed by the spindle microtubules as the chromosomes align at the midpoint of the spindle. This tension is sufficient to pull the two chromatids apart slightly at the primary constriction, so that the attachment points, the kinetochores, stretch toward the opposite spindle poles. Tension on the kinetochores continues to develop, and after a brief pause in this condition, the two chromatids of each chromosome separate completely and begin moving to opposite poles of the spindle. The separation of the chromatids and the initiation of movement to the spindle poles marks the transition from metaphase to anaphase (Figs. 11-6d and 11-7d).

The movement of the chromatids to opposite ends of the spindle, the most spectacular feature of mitosis, has intrigued scientists for nearly a hundred years. The basis for this rapid movement is still not completely understood. Most investigators now believe that the spindle microtubules generate the force for poleward movement by a combination of controlled growth and active sliding movements (see also p. 171).

The anaphase movement has two parts that contribute to the total distance moved by the chromosomes. The first part of the movement depends on the microtubules that extend from the kinetochores of the chromosomes to the spindle poles. As the anaphase movement progresses, these microtubules become shorter (Figs. 11-15a and b); as they shorten, the chromatids are separated and pulled toward the poles. This part of the anaphase movement may be produced by active sliding between the kinetochore microtubules and the microtubules that extend between the spindle poles, or by a process in which the kinetochore microtubules are reduced in length through the disassembly of microtubule subunits at their tips.

The second part of the movement is produced by an increase in length of the entire spindle during the anaphase movement. This movement depends on microtubules that extend between the spindle poles without connecting directly to chromosomes. These microtubules overlap in the center of the spindle (Fig. 11-15b; see also Fig. 8-7). As the anaphase movement progresses, the zone of overlap decreases (Fig. 11-15c). This movement, produced by active sliding between the microtubules in the zone of overlap, increases the overall length of the spindle by an amount equivalent to the decrease in overlap. In some species, lengthening of the entire spindle may also

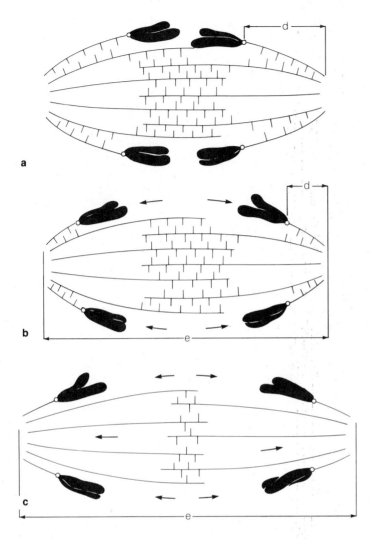

Figure 11-15 The two parts of the anaphase movement. One part of the movement is produced by the microtubules that extend between the kinetochores and the poles; these microtubules get shorter as anaphase proceeds (compare distance *d* in **a** and **b**). The second part of the movement is produced by an overall lengthening of the spindle (compare distance *e* in **b** and **c**).

be produced by the addition of microtubule subunits to the tips of the microtubules that extend between the poles; that is, by growth of these microtubules. As they grow longer, they also push the poles farther apart.

The anaphase movement continues until the separated chromatids are collected at the opposite spindle poles (Figs. 11-6e and 11-7e). Since each end of the spindle receives one chromatid from every metaphase chromosome, the two collections of chromatids have exactly the

same hereditary information. Once the chromatids have completed their movement to the spindle poles, the cell enters telophase, the final stage of mitosis.

Telophase

During telophase (Figs. 11-6f and 11-7f) the chromatids at the poles unfold and become indistinct. Segments of a new nuclear envelope appear at the borders of the unfolding chromatids. These segments gradually extend around the mass until the chromatids, by now closely resembling interphase chromatin, are completely separated from the surrounding cytoplasm by a new, continuous nuclear envelope.

Soon after the beginning of telophase, nucleolar material begins to appear at one or more secondary constrictions in the chromatids. If nucleoli form at more than one secondary constriction, the several nucleoli produced may fuse together or remain separate, depending on the cell type and the species. By the time nucleoli are completely formed, the polar masses of chromatin are again surrounded by nuclear envelopes and are indistinguishable from G_1 chromosomes. This transition is accompanied by a gradual return to full transcription of all RNA types.

The distinction between chromatids and chromosomes at this stage is somewhat arbitrary. For most purposes, the transition from chromatids to chromosomes at telophase can be considered complete when the chromatids of each chromosome entering mitosis have been separated and enclosed in separate daughter nuclei.

The spindle starts to disappear as soon as the change from anaphase to telophase begins. The asters, if present, become smaller until the centriole pair near each daughter nucleus is surrounded by a limited number of relatively short microtubules. The centrioles then take up their characteristic interphase position at one side of the nucleus, just outside the nuclear envelope. After telophase, all that remains of the spindle is a layer of short microtubules at the former spindle midpoint. This persistent layer marks the region where the cytoplasm will divide to separate the daughter cells at the close of mitosis.

The entire mitotic sequence, from the onset of prophase to the close of telophase, may require as little as 5 to 10 minutes in rapidly developing animal embryos, or as much as 3 hours in various plant and animal tissues. Of the different stages, prophase is usually the most extended, with metaphase and telophase requiring less time. Anaphase proceeds most rapidly, rarely taking more than a few minutes in most species.

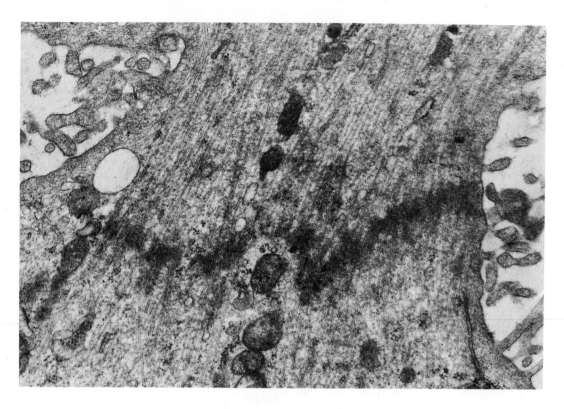

Figure 11-16 Midbody of a human cell at late telophase. × 35,000. Courtesy of A. Krishan and R. C. Buck, from *Journal of Cell Biology* 24 (1965):443, by copyright permission of The Rockefeller University Press.

The Significance of Replication and Mitosis

The result of replication and mitosis is the production of two daughter nuclei, each with genetic capacity equivalent to the original parent nucleus. The accuracy of mitotic division depends on two basic features of the mechanism: (1) the arrangement of spindle microtubules, which forms two distinct ends or poles in the cell, and (2) the opposite spindle pole connections made by the two chromatids of each chromosome. These connections result in the separation and delivery of the two chromatids of each chromosome to the opposite poles.

Cytokinesis

Cytokinesis, the division of the cytoplasm, completes the process of cell division by enclosing the daughter nuclei produced in mitosis in separate cells. Although the details of cytokinesis differ in plants and animals, spindle remnants function similarly in both processes.

Cytokinesis in Animals: Furrowing

In animals, short lengths of microtubules persist at the spindle midpoint after mitosis is complete. The rest of the spindle breaks down and disappears after anaphase except for the short lengths of microtubules left around the centrioles. The microtubules at the spindle midpoint become surrounded with patches of dense, apparently structureless material. This material forms a layer called the *midbody*, which soon extends completely across the cell (Fig. 11-16).

After the midbody develops, a depression or *furrow* appears in the plasma membrane around the outside of the cell at the level of the midbody (Fig. 11-17). This furrow gradually deepens, following the plane of the mid-

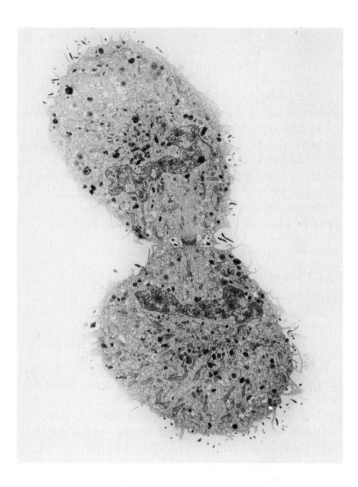

Figure 11-17 Furrow formation in a dividing human cell in culture. The furrow gradually deepens until the cytoplasm is divided into two parts. × 4,000. Courtesy of G. G. Maul and Academic Press, Inc., from *Journal of Ultrastructure Research* 31 (1970):375.

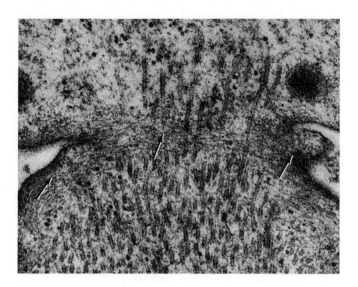

Figure 11-18 Microfilaments (arrows) at the edge of the advancing furrow in a dividing egg cell of a rat. × 35,000. Courtesy of D. Szollosi, from *Journal of Cell Biology* 44 (1970):192, by copyright permission of The Rockefeller University Press.

body, until the two daughter cells are completely separated by continuous plasma membranes. As the deepening furrow penetrates into the cell, the midbody is compressed and becomes smaller, usually disappearing as the furrow cuts off the daughter cells. Mitochondria, endoplasmic reticulum, Golgi complex, vesicles, and other cytoplasmic organelles are roughly divided between the daughter cells as the furrow deepens.

The manner in which the furrow develops makes it obvious that the original position of the spindle determines the plane of cytoplasmic division. Ordinarily, the spindle is situated with its midpoint at the cell equator, so that the daughter cells formed by cytokinesis are of equal size. In some cases, as in the unequal division of cytoplasm during egg development in animals, the spindle is positioned at one side of the dividing cell. The furrow forms opposite the spindle midpoint as usual, cutting the cytoplasm into two unequal parts. The factors governing the alignment and position of the spindle in animals, which determine the later plane of furrowing, remain unknown.

Cells that are unattached to their neighbors, such as some developing eggs or cells in tissue culture, look as if a drawstring is being tightened around them during furrowing. This impression is directly supported by electron micrographs of the advancing furrow in dividing cells, which reveal large numbers of microfilaments at the furrow edge (see Fig. 11-18). These microfilaments, which are also found in many motile systems in cells (for details, see Chap. 8) are believed to produce contractile force for furrowing by actively sliding over each other, progressively tightening the furrow until the dividing cell is separated into two parts.

Cytokinesis in Plants: Phragmoplast and Cell Plate Formation

The initial stages in plant cytokinesis resemble midbody formation in animals. During telophase, portions of the

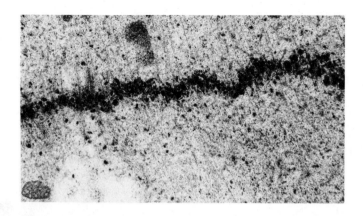

Figure 11-19 Phragmoplast formation in a plant cell of *Haemanthus.* Microtubules are embedded in the phragmoplast layer. × 12,000. Courtesy of A. Bajer and Springer-Verlag, from *Chromosoma* 24 (1968):383.

spindle microtubules persist at the spindle midpoint. These microtubules become surrounded by a layer of dense material (Fig. 11-19), as in midbody formation in animals. However, in plants much of this dense material is enclosed in membrane-bound vesicles that originate from the Golgi complex. These vesicles gradually increase in number until a continuous layer extends across the cell at the former spindle midpoint. This layer of microtubules and vesicles, called the *phragmoplast,* determines the position of the new wall that will separate the daughter cells.

Cell wall formation begins in the central area of the phragmoplast and gradually extends outward toward the plasma membrane, in a direction opposite to furrowing in animals. This takes place as the vesicles fuse together, forming two layers of continuous plasma membranes (Figs. 11-20a and b and 11-21a and b). As the vesicles fuse, their contents are released into the developing extracellular space between the daughter cells. When the fusion process reaches the original cell walls, the daughter cells are completely separated by two continuous plasma membranes with an enclosed space between them (Figs. 11-20c and 11-21c). This space is filled with the dense material from the fused vesicles. When fully formed, the layer of wall material separating the daughter cells is called the *cell plate.*

Once formed, the cell plate is progressively thickened and strengthened by the deposition of new cell wall material between the membranes separating the daughter cells. At points, this new wall is perforated by cytoplasmic connections that remain intact between the daughter cells

(see Fig. 11-21c). These narrow connections, termed *plasmodesmata* (singular = *plasmodesma*), evidently serve as channels of communication between the cells of plant tissue (see also p. 101).

The plane of division in plant cells, as in animal cells, is determined by the position of the spindle at metaphase. Usually, the spindle occupies the center of the cell, so that the subsequent cytokinesis separates the dividing cell into two equal parts. In some plant tissues, however, the spindle takes a position at one side of the cell, so that the following cytoplasmic division is unequal. As in animals, the factors determining the position of the spindle and the resultant plane of cytoplasmic division are unknown.

The Overall Effects of Replication, Mitosis, and Cytokinesis: A Review

In order to review the overall effects of mitotic cell division we will consider a diploid organism (see Information Box 11-4) with only two pairs of chromosomes, one long pair and one short pair. At interphase (Fig. 11-22a), the chromosomes extend throughout the nucleus. As a result of replication during the S period (Fig. 11-22b), each chromosome becomes double at all points and now consists of two chromatids. During prophase of mitosis, the chromosomes condense into short rodlets (Fig. 11-22c). At metaphase (Fig. 11-22d), they become aligned at the midpoint of the spindle. Spindle attachments are made so that the two kinetochores of each chromosome connect to microtubules leading to opposite poles of the spindle. At anaphase (Fig. 11-22e), the two chromatids of each chromosome separate and move to the opposite spindle poles. At the poles, the chromatids unfold and return to the interphase state during telophase (Fig. 11-22f). The formation of new nuclear envelopes during telophase, and the subsequent cytoplasmic division (Fig. 11-22g) completes the separation of the two daughter cells. As a result of replication and mitosis, the nucleus in each of these daughter cells contains exactly the same number and types of chromosomes as the original parent cell.

The processes of cell division usually occur in sequence as outlined in this chapter. However, replication, mitosis, and cytokinesis are potentially separable and in some organisms may proceed independently. In the salivary glands of *Drosophila* larvae, for example, DNA replication takes place without mitosis or cytokinesis, producing

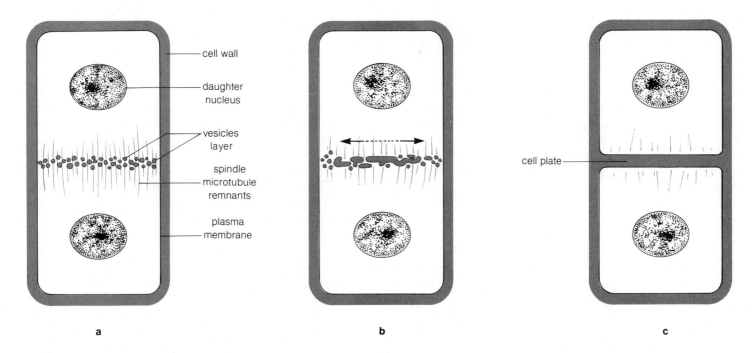

Figure 11-20 Phragmoplast and cell plate formation in plants (see text).

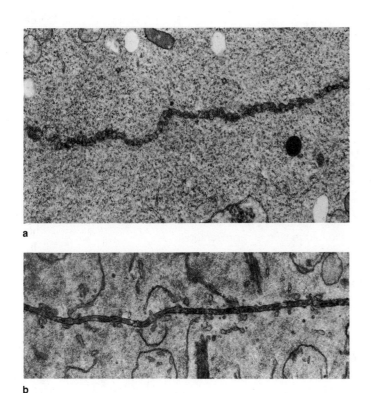

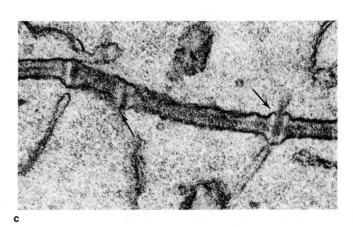

Figure 11-21 Successive stages in the fusion of phragmoplast vesicles to form the cell plate. Protoplasmic connections called *plasmodesmata* (arrows in **c**) remain open between the daughter cells. (**a**) *Haemanthus*. × 10,000. Courtesy of W. G. Whaley, M. Dauwalder, J. E. Kephart, and Academic Press, Inc., from *Journal of Ultrastructure Research* 15 (1966):169. (**b**) and (**c**) *Phalaris*. (**b**) × 19,000; (**c**) × 70,000. Courtesy of A. Frey-Wyssling, J. R. Lopez-Saez, K. Muhlethaler, and Academic Press, Inc., from *Journal of Ultrastructure Research* 10 (1964):422.

large nuclei containing hundreds or thousands of copies of each DNA molecule (see Supplement 10-3). Or, as happens regularly in some groups of fungi, replication and mitosis proceed without cytokinesis. This sequence produces cells with many nuclei enclosed in a common cytoplasm. Another variation eventually produces cells containing single nuclei, but does so in a pattern in which mitosis and cytokinesis are widely separated in time. This occurs, for example, in developing insect eggs, in which replication and mitosis proceed rapidly for a time without cytokinesis, producing an early embryo with several hundred nuclei suspended in a common cytoplasm. Eventually, rapid cytokinesis occurs in these embryos, enclosing the many daughter nuclei in separate cells.

Mitotic cell division may serve as a method of reproduction for whole organisms if one or more daughter cells are released from the parent and grow separately into complete individuals. Reproduction of this type, which occurs in many kinds of plants, animals, and single-celled organisms, is called *vegetative* or *asexual reproduction*. Because the cell products in asexual reproduction result from mitosis, all of the offspring are genetically identical.

Cell Division in Prokaryotes

Cell division in the bacteria and blue-green algae, although as precise as division in eukaryotes, does not proceed by mitosis. Instead, in bacteria, where replication and cell division have been most intensively studied, replicated DNA is probably separated by the activity of the plasma membrane.

Current hypotheses of bacterial division are based on the observation that bacterial DNA, when isolated from the cell, frequently has one or more attachments to the plasma membrane. These attachments, which are sometimes visible in sectioned bacteria, are believed to divide replicated bacterial DNA molecules between daughter cells according to the mechanism shown in Figure 11-23.

In bacteria all of the hereditary information is coded into a single DNA molecule. This DNA molecule, the single "chromosome" of a bacterial cell, exists as a closed circle with no free ends. At at least one point, the circle is considered to be attached to the plasma membrane. Rep-

---

Information Box 11-4

Haploidy and Diploidy

Almost all body cells of the individuals in a particular species have the same number of chromosomes. Depending on the life history of the organism, this number is usually the *haploid* or *diploid* number for the species, although it may be some higher multiple of these numbers. In a haploid nucleus, only one copy of each chromosome is present (haploid nuclei are also sometimes called *monoploid* nuclei). Diploid nuclei contain two copies of each chromosome, so all the genetic information is represented twice.

The two copies of each chromosome in diploids make up a pair. The pairs are described as *homologs* because they contain the same genes in the same sequence (slightly different forms of the genes, called *alleles*, may be present on either chromosome of a pair). Because higher plants and animals are diploid throughout most of their life cycle, mitosis is described in this chapter as it would occur in a diploid organism. However, the mechanism and outcome of mitosis are independent of the chromosome number and follow the same pattern in both haploids and diploids. During mitotic cell division, the two members of the homologous pairs remain separate and proceed independently through all the stages of mitosis.

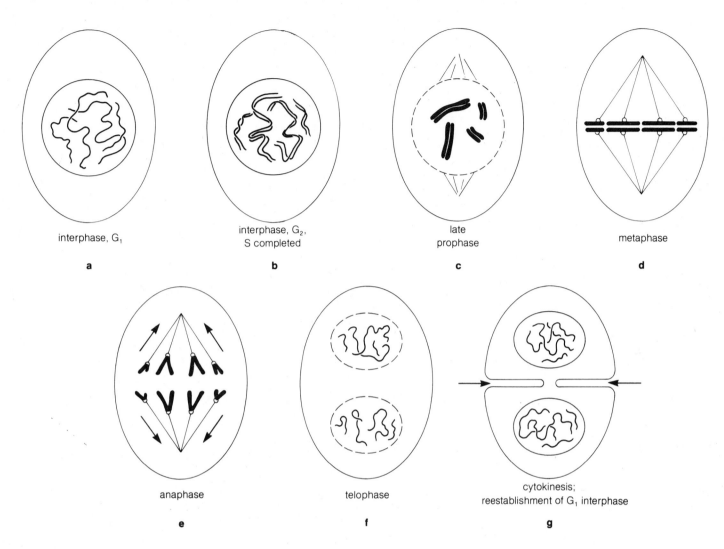

interphase, G₁

interphase, G₂, S completed

late prophase

metaphase

a

b

c

d

anaphase

telophase

cytokinesis; reestablishment of G₁ interphase

e

f

g

Figure 11-22 A review of mitosis (see text).

lication starts at one point on the molecule and proceeds in both directions from this point, producing two replication "forks" that gradually advance around the circle (Fig. 11-23a and b). As the forks complete their circuit around the DNA molecule, the duplicated circles become completely separate (Fig. 11-23c). Presumably, the membrane attachment point is also duplicated at the same time, with the result that both circles are now attached to the plasma membrane at two closely spaced points. The membrane then begins to grow between the attachment points (Fig. 11-23d), separating and pushing the two DNA circles to opposite ends of the cell (Fig. 11-23e). Thus the plasma membrane rather than a spindle divides the replicated DNA molecules.

The cytoplasm then divides by a mechanism resembling furrowing in animal eukaryotes. An indentation in the plasma membrane appears around the midpoint of the cell between the separated DNA circles (Fig. 11-23f). This indentation or furrow deepens, eventually pinching the cytoplasm into two completely separate halves. The cell wall follows the inward path of the furrow; as the membranes separate, the growing wall becomes complete and extends as a partition between the daughter cells (Fig. 11-23g). Inward growth of the plasma membrane and cell wall of a bacterium can be seen in progress in Figure 11-24. The entire process of replication and cell division takes no more than about 20 minutes in rapidly growing bacteria.

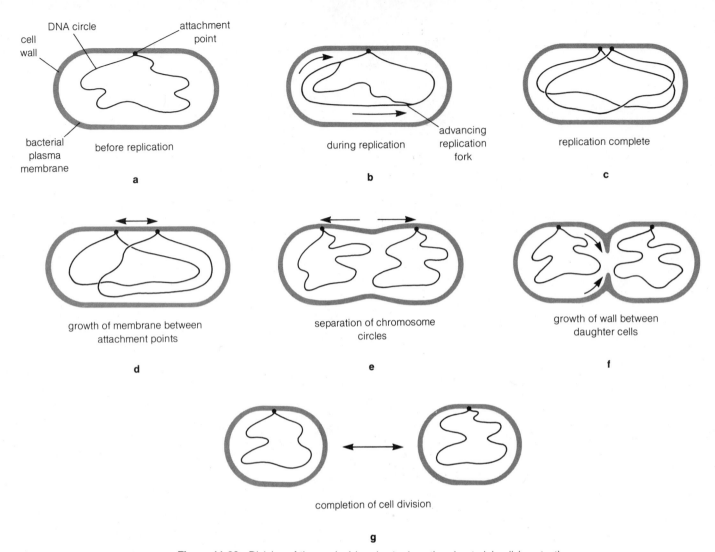

Figure 11-23 Division of the nucleoid and cytoplasm in a bacterial cell (see text).

The prokaryotic mechanism works effectively because there is only one chromosome per cell. The greater number of chromosomes in eukaryotic cells would no doubt cause many mistakes in distribution if division proceeded as it does in the bacteria. Mitosis probably arose in evolution as an adaptation in response to the increasing length and complexity of the genetic message as eukaryotes first appeared. As the length of the chromosomal DNA increased, restrictions on the total time required for replication, the danger of breakage, and the mechanical difficulties of dividing long chromosomes favored subdivision of the genetic information into the shorter subunits that become the chromosomes of eukaryotes. As the chromosomes appeared, a mechanism was required to divide their replicated products precisely and equally between daughter cells.

The required precision, as noted in this chapter, is supplied by the mitotic spindle and the attachments made by the chromosomes to the spindle at metaphase. The parallel arrangement of microtubules in the spindle, running from end to end of the cell, sets up two distinct ends or poles to receive the divided chromosomes. The chromatids are separated into two equal groups and delivered to these poles as a result of the connections made by the kinetochores of the chromosomes at metaphase, which always lead to opposite spindle poles.

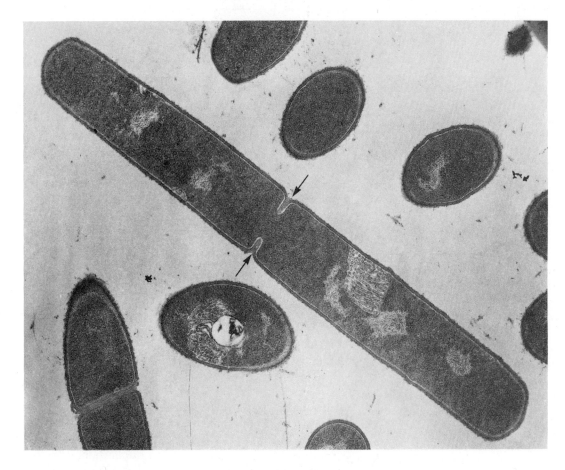

Figure 11-24 Division of the cytoplasm in a bacterium, *Bacillus cereus*. Inward growth of a new wall (arrows) is separating the daughter cells. × 23,000. Courtesy of L. Santo.

The entire cycle of replication, mitosis, and cytokinesis, in all of its elegant complexity, occurs countless billions of times in the growth, development, and maintenance of structure in many-celled eukaryotes. In the maintenance of red blood cells alone in humans, mitotic divisions occur in each individual at the rate of more than 2 million per second. The perfection of the mechanism is such that these repeated cycles of division occur almost without error throughout the lifetime of the organism.

Suggestions for Further Reading

Alberts, B. and R. Steinglanz. 1977. Recent excitement in the DNA replication problem. *Nature* 269:655–61.

Baserga, R. 1976. *Multiplication and division in mammalian cells.* Marcel Dekker: New York.

Kornberg, A. 1980. *DNA replication.* Freeman: San Francisco.

Mazia, D. 1974. The cell cycle. *Scientific American* 230:54–65 (January).

Pardee, A. B., R. Dubrow, J. L. Hamlin, and R. F. Kletzein. 1978. Animal cell cycle. *Annual Review of Biochemistry* 47:715–80.

Prescott, D. M. 1976. *Reproduction of eukaryotic cells.* Academic Press: New York.

Wolfe, S. L. 1981. *Biology of the cell.* 2nd ed. Wadsworth: Belmont, Calif.

Yeoman, M. M. 1976. *Cell division in higher plants.* Academic Press: New York.

For Further Information

Cell wall formation and growth in plants and prokaryotes, Chapter 5
Centrioles, Supplement 8-1
Genes and alleles, Chapter 13
Haploidy and diploidy, Chapters 12 and 13
Kinetochore activity in meiosis, Chapter 12
Nuclear structure and function, Chapter 9
Microfilament structure and function, Chapter 8
Microtubule structure and function, Chapter 8

Questions

1. List the major stages in the cell cycle. What happens in each stage?

2. What happens in G_1, S, and G_2 of interphase?

3. What does the complementarity of the two nucleotide chains of a DNA molecule have to do with replication?

4. Define semiconservative replication. What happens to the "old" and "new" nucleotide chains in semiconservative replication?

5. Outline the steps that occur in the addition of nucleotides to a growing DNA chain during replication.

6. In what ways are DNA replication and RNA transcription alike? In what ways do they differ?

7. What happens during mitotic prophase? What events mark the beginning and end of this stage?

8. What structures are visible on the chromosomes during prophase and metaphase? Define chromosome and chromatid. At what times in the life cycles of cells are chromatids present? What is a karyotype? A kinetochore?

9. Trace the development of the spindle in organisms with and without centrioles.

10. What happens during metaphase? What events mark the beginning and end of this stage?

11. What happens during anaphase? What events mark the beginning and end of this stage? How are the spindle microtubules believed to move the chromosomes during anaphase?

12. What happens during telophase? What events mark the beginning and end of this stage? What events are considered to convert the separated chromatids into chromosomes?

13. What is the outcome and significance of mitosis? What features of mitosis ensure that the chromatids are equally divided between daughter nuclei?

14. Trace the patterns of cytoplasmic division in plants and animals. In what ways is cytoplasmic division similar in plants and animals? In what ways is it different?

15. How is the plane of cytoplasmic division related to the position taken by the spindle in mitosis?

16. How are the activities of microtubules and microfilaments coordinated in animal cell division?

17. Define midbody, furrow, phragmoplast, cell plate, and plasmodesma.

18. Trace mitotic cell division in an organism with three pairs of chromosomes (a diploid). Do the same for an organism that has three chromosomes that exist singly rather than in pairs (a haploid or monoploid; see Information Box 11-4). Does the existence of chromosomes singly or in pairs make any fundamental difference in the way mitosis proceeds?

19. How is cell division believed to occur in prokaryotes? Compare prokaryotic and eukaryotic cell division.

20. What happens if replication occurs without mitosis and cytokinesis? If replication and mitosis occur without cytokinesis?

21. What is vegetative or asexual reproduction?

Supplement 11-1:
The Experimental Demonstration that Replication is Semiconservative

At the time DNA structure and the mechanism of DNA synthesis were determined, semiconservative replication (see Figs. 11-3 and 11-4) was only one of several pathways considered as possibilities for DNA duplication. Semiconservative replication was established as the pathway followed in nature by Matthew Meselson and Franklin W. Stahl at the California Institute of Technology, who investigated replication in a prokaryote, and by J. Herbert Taylor at Columbia University, who worked with a eukaryote.

Meselson and Stahl's experiment used the bacterium *Escherichia coli*. As a first step, they grew *E. coli* for several generations in a medium containing the heavy nitrogen isotope ^{15}N (the most common form of nitrogen is the less dense isotope ^{14}N). This period of growth was continued long enough to ensure that all of the *E. coli* DNA contained the ^{15}N isotope. Meselson and Stahl then removed the bacteria from the ^{15}N source and placed them in a medium containing only ^{14}N nitrogen. After the transfer to the unlabeled medium, cells were removed at intervals and the DNA was isolated and purified.

The DNA was then compared to pure ^{15}N and ^{14}N DNA standards. To carry out these comparisons, Meselson and Stahl used a technique known as *buoyant density centrifugation* (Fig. 11-25). In this technique, a centrifuge tube is filled with a solution of cesium chloride (CsCl), made up to a concentration approximating the density of

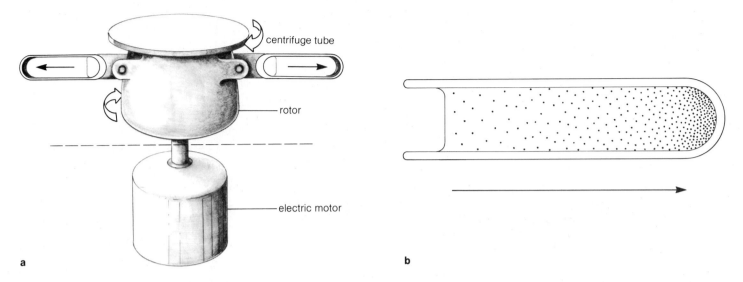

Figure 11-25 Buoyant density centrifugation. (**a**) The arrangement of tubes and rotor. The tubes are hinged so that they spin at right angles to the axis of rotation. (**b**) Production of a density gradient in a centrifuge tube by centrifugal force (see text). The dots represent cesium chloride molecules.

DNA. As the tube spins at high speeds in the centrifuge, the cesium chloride molecules become more concentrated toward the bottom of the tube, producing an even gradient of density from the top to the bottom. At the top of the tube, the CsCl, because it is less concentrated, is lower in density than the DNA sample. At the bottom of the tube, centrifugal force packs the CsCl molecules more closely together, producing a region higher in density than the DNA sample. In response, the DNA descends or ascends in the tube until it reaches the level at which it matches the density of the surrounding CsCl solution. If the DNA is of uniform density, it will form a sharply defined band at this point. Because it contains the heavier nitrogen isotope, ^{15}N DNA is denser and will form a distinct band in the centrifuge tube at a lower level than ^{14}N DNA. This process enables the two kinds of DNA to be separated and identified.

The results of Meselson and Stahl's experiments are summarized in Figure 11-26. DNA extracted from bacterial cells at the instant of transfer from ^{15}N to ^{14}N growth medium, before any DNA replication in ^{14}N medium could occur, centrifuged down to a single band at the level expected for ^{15}N DNA (Fig. 11-26a). After about 20 minutes following transfer from ^{15}N to ^{14}N medium (the time required for one complete bacterial cell cycle in ^{14}N medium, including DNA replication and cell division), all of the DNA isolated from the cells spun down to a level inter-

mediate in density between ^{15}N and ^{14}N DNA (Fig. 11-26b). DNA removed from the bacteria after two generations of growth in the unlabeled medium spun down in two bands, one at the intermediate level and one at a level characteristic of pure ^{14}N DNA (Fig. 11-26c).

This finding effectively eliminated a possible pathway for DNA replication called *conservative replication*. In conservative replication (Fig. 11-27), the nucleotide chains of a parent DNA molecule, after serving as templates, would unwind from their copies and rewind together again. The two newly synthesized chains would likewise wind together into a double helix. Thus one of the DNA molecules produced by conservative replication would consist entirely of "old" DNA, and one entirely of "new" DNA. If replication were conservative, two bands would therefore be expected to appear in the centrifuge tube after one replication in the ^{14}N medium, one band of pure ^{15}N DNA and one band of pure ^{14}N DNA. This result would be expected because the two old ^{15}N nucleotide chains of the parent DNA molecule, after serving as templates for replication, would reassociate into an all old ^{15}N–^{15}N DNA molecule. The two newly synthesized chains, assembled entirely from nucleotides containing ^{14}N nitrogen (because the bacteria were placed in a medium containing only ^{14}N nitrogen as a growth source), would wind together into an all new ^{14}N–^{14}N DNA molecule. The actual appearance of a single band of intermediate density at this point in the

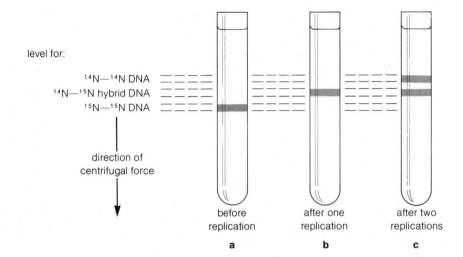

level for:

^{14}N—^{14}N DNA

^{14}N—^{15}N hybrid DNA

^{15}N—^{15}N DNA

direction of
centrifugal force

before
replication

after one
replication

after two
replications

a   **b**   **c**

Figure 11-26 The Meselson and Stahl experiment demonstrating semiconservative DNA replication (see text).

experiment, rather than two distinct bands at the ^{14}N and ^{15}N levels, therefore ruled out the possibility of conservative replication.

The single intermediate band observed at this time is fully explained if replication is semiconservative. In semiconservative DNA replication, the parent DNA molecule, consisting of two ^{15}N nucleotide chains, unwinds and serves as template. After the synthesis of its complementary chain from nucleotides containing ^{14}N nitrogen, each template remains wound into a double helix with the new ^{14}N copy, producing a hybrid molecule consisting of the old ^{15}N nucleotide chain wound into a double helix with a new ^{14}N nucleotide chain. This hybrid DNA molecule is of intermediate density, between pure ^{15}N and ^{14}N DNA, and spins down to an intermediate level in the centrifuge. After one semiconservative replication, all the bacterial DNA consists of ^{15}N–^{14}N hybrids of this type, and produces only one intermediate band in the centrifuge. As we have seen, this was the result actually obtained by Meselson and Stahl.

DNA removed from bacteria allowed to grow for two full generations after transfer from a ^{15}N to a ^{14}N medium produced two bands after centrifugation, one at the ^{15}N–^{14}N intermediate density and one at the level of pure ^{14}N DNA. This observation is also exactly as expected if replication is semiconservative. Following one of the ^{15}N–^{14}N hybrid DNA molecules produced in the first generation through another cycle of semiconservative replication in ^{14}N medium will explain these results (Fig. 11-28). During the second replication, the two nucleotide chains will separate and serve as templates for DNA syn-

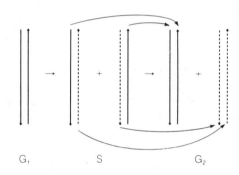

G_1 S G_2

Figure 11-27 Conservative DNA replication (see text).

thesis. The ^{15}N chain will remain with its newly synthesized copy, producing another ^{15}N–^{14}N hybrid of intermediate density. The ^{14}N template chain will also remain with its ^{14}N complementary copy, producing a pure ^{14}N–^{14}N DNA molecule, with both component chains synthesized from nucleotides containing only ^{14}N nitrogen. This DNA will centrifuge to a position characteristic of ^{14}N DNA. Thus, after two generations, the DNA molecules would be expected to centrifuge into two bands, one characteristic of hybrid ^{15}N–^{14}N DNA and one equivalent to pure ^{14}N DNA. These were the results actually obtained by Meselson and Stahl; therefore the results of the experiment are completely explained by semiconservative replication.

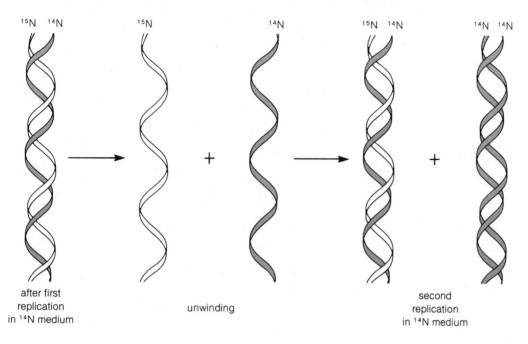

after first
replication
in ^{14}N medium   unwinding

second
replication
in ^{14}N medium

Figure 11-28 Explanation of Meselson and Stahl's results (see text). The ^{14}N nucleotide chain is shaded.

Meselson and Stahl's results with *E. coli* were reported in 1958. At about the same time Taylor, working independently with cells of a higher plant, *Vicia faba* (the broad bean), showed by equivalent experiments that replication in eukaryotes also proceeds by the semiconservative pathway.

Supplement 11-2:
Unwinding, Primer Synthesis, and Fork Movement in DNA Replication

While the basic mechanism assembling nucleotides into nucleic acid chains is similar in DNA and RNA synthesis (see Information Boxes 9-2 and 11-1), the double helical structure of DNA, and the fact that the DNA polymerase enzymes need a primer (noted in Information Box 11-1) to begin synthesis introduce complexities in DNA replication that require a battery of enzymes with no counterparts in RNA transcription.

Unwinding the DNA Double Helix

Two major enzymes cooperate to unwind the nucleotide chains of a DNA double helix during replication. The first of these, called the *unwinding enzyme,* uses the energy of ATP to open the DNA helix and unwind it into single nucleotide chains. The enzyme works in such a way that one molecule of ATP is broken down to ADP for each turn of the helix unwound (Fig. 11-29a). Unwinding is promoted by a group of nonenzymatic *binding proteins* (see Fig. 11-29a). These bind strongly to each single nucleotide chain exposed by the unwinding enzymes. By stabilizing the chains in single-stranded form, the binding proteins greatly reduce the energy required to unwind the DNA from its double helical state. Unwinding creates a "fork" in the DNA, with two single nucleotide chains trailing behind the fork and an intact double helix in front of it.

DNA molecules are so long, and the cellular medium containing them so viscous that neither the single nucleotide chains behind the replication fork nor the intact helix in front of it is free to rotate in response to the unwinding. As a result, *supercoils* are created in advance of the fork at the rate of one supercoil for each turn of the helix unwound. To understand why this happens, take two lengths

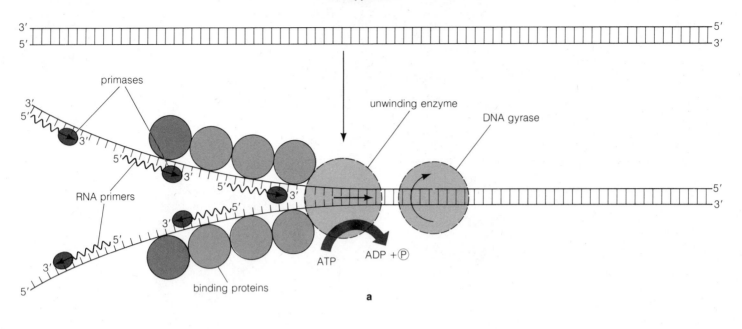

primases

unwinding enzyme

DNA gyrase

RNA primers

ATP

ADP + ℗

binding proteins

a

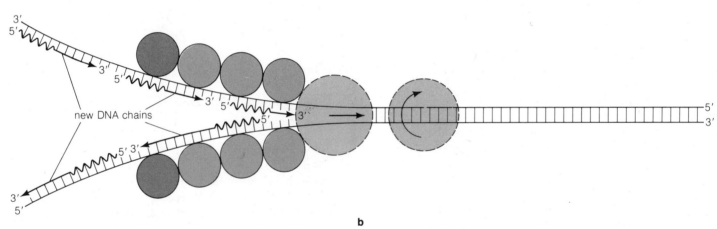

new DNA chains

b

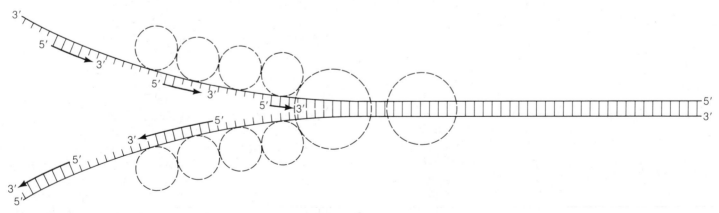

c

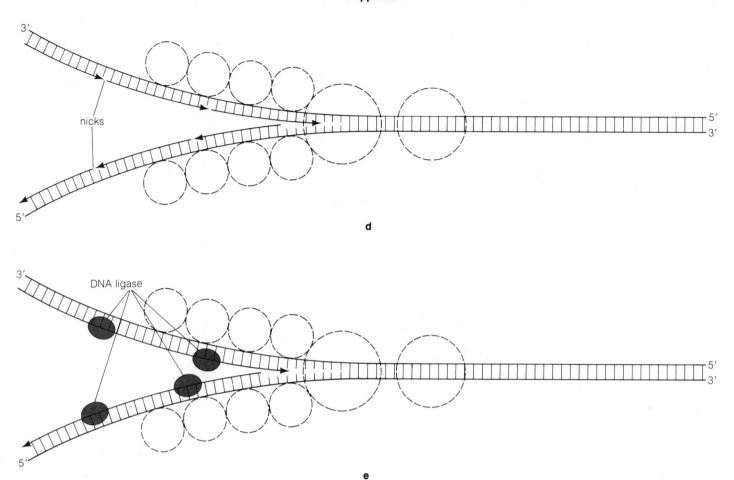

Figure 11-29 Details of the mechanism of DNA replication (see text). (**a**) Unwinding by the unwinding enzyme, DNA gyrase, and the binding proteins, and primer synthesis. (**b**) Synthesis of new DNA chains. (**c**) Removal of primers. (**d**) Completion of DNA chains to remove gaps left by primer removal. Nicks still separate the short DNA chains. (**e**) Closure of nicks to join the short DNA lengths into continuous nucleotide chains.

of string and twist them together into a double helix. Now place a weight on one end of the twisted string to keep it from rotating. Take one of the two strings at the opposite end of the helix in each hand, and create a replication fork by pulling them apart. You will notice that pulling the ends apart throws the helix in front of the fork into extra turns or supercoils at the rate of one supercoil for each turn of the helix pulled apart. Eventually, you will reach a point at which the strain imposed by the supercoils prevents the strings from unwinding further. The strain created by unwinding at the fork is even more intense in

DNA than in the string example because the DNA double helix cannot wind into a tighter helix (more turns per unit length) as the string can to compensate for the turns pulled apart. Instead, the helix in front of the fork is thrown into loops (see Fig. 11-30) at the rate of one loop for each turn unwound.

These loops, and the strain imposed on the helix in front of the fork are relaxed by *DNA gyrase*, the second major enzyme involved in unwinding DNA (see Fig. 11-29a). This enzyme acts by opening a "nick" in one of the two template chains in front of the replication fork.

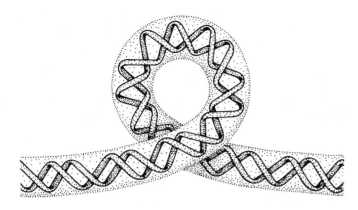

Figure 11-30 A supercoil loop thrown into DNA by the activity of the unwinding enzyme (see text).

Table 11-1 Major Enzymes Active in DNA Replication

Enzyme	Activity
Unwinding enzyme	Unwinds template DNA helix
DNA gyrase	Relaxes supercoils induced by unwinding
Primase	Primer synthesis
DNA polymerase	Synthesizes new DNA nucleotide chains
Ribonuclease	Removes primers
DNA polymerase	Fills gaps left by primer removal
DNA ligase	Seals nicks left by gap-filling

The free end created rotates around the other chain, which acts as a swivel. A single rotation relaxes the supercoil. Subsequently, the same gyrase enzyme reseals the nick by a covalent bond, leaving the template strand intact in advance of the fork. The energy required to drive the nicking-rotation-sealing reaction catalyzed by the gyrase enzyme is apparently derived from free energy released as the supercoils unwind.

Primer Synthesis and DNA Replication at the Fork

The DNA polymerase enzymes can add nucleotides only to the end of an existing nucleotide chain. Therefore, they cannot begin synthesis until short lengths of nucleotides are laid in place as primers by other enzymes.

The necessary primers are laid down by special RNA polymerase molecules called *primases*. These enzymes, in contrast to the DNA polymerases, can begin synthesis of a complementary RNA chain whether other nucleotides are already in place or not. Instead of laying down a single primer at one point on each of the two template chains, the primase enzymes synthesize a series of primers, with each primer in the series separated from the next by an open space on the template (Fig. 11-29a). The short RNA segments laid down by the primases are then used as primers by the DNA polymerases, which fill in the open spaces between the primers with DNA nucleotide chains (Fig. 11-29b). This pattern of synthesis produces a series

of short DNA lengths opposite the template chain, each with an RNA primer still attached at one end.

The final steps in replication remove the RNA primers, fill in the gaps created by their removal, and seal the new DNA chains into continuous molecules. The primers are removed by a specific RNAase enzyme (Fig. 11-29c) that removes the RNA nucleotides of the primers one at a time until the DNA part of the new chains is reached. The gaps left by primer removal are filled in by DNA polymerase enzymes, which add DNA nucleotides in the usual pattern until the remaining bases in the gaps are paired and filled in (Fig. 11-29d). Since the DNA polymerase enzymes cannot link the last nucleotide added in each segment to the end of the next segment, a nick is left open in the new DNA chain at each of these points. These remaining nicks are closed by the final enzyme active in replication, *DNA ligase*. This enzyme forms the final phosphate linkages required to seal the discontinuous segments into a single, continuous DNA molecule, at the expense of one ATP molecule for each gap closed (Fig. 11-29e). Table 11-1 summarizes the roles of the various enzymes active in DNA replication.

Replication proceeds at an overall rate of about 1000 nucleotides per second in prokaryotes and 100 per second in eukaryotes. The entire process is so rapid that the RNA primers and the gaps left by discontinuous synthesis persist for only a few seconds. As a consequence, all the enzymes and factors of replication operate only in the region of the fork. At a distance of only a few micrometers behind the fork the new DNA chains are already continuous and fully wound around their template chains into complete DNA molecules.

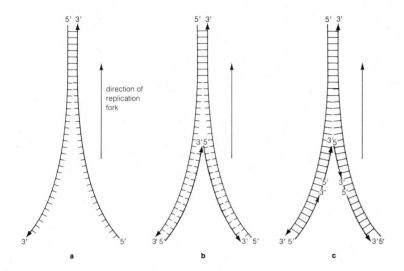

Figure 11-31 Direction of synthesis on the two template chains of a DNA molecule (see text).

Unidirectional Fork Movement in DNA Synthesis

In Chapter 9 we noted that the two chains of a DNA molecule are *antiparallel*. If they are laid out flat, as they are in Figure 9-6, it can be seen that the chains run in opposite directions. Tracing the phosphate linkages holding the left hand chain together from the bottom to the top of Figure 9-6 shows that the linkages extend from the 5' carbon of the deoxyribose sugar below to the 3' carbon of the sugar above each linkage. The bonds on the left thus run in the 5' → 3' direction from the bottom to the top of the page. On the other chain, the phosphate linkages run in the opposite direction: tracing from the bottom to the top of the figure shows that the linkages extend from the 3' carbon of the sugar below to the 5' carbon of the sugar above each linkage, or in the 3' → 5' direction.

The antiparallel arrangement of the two nucleotide chains in a DNA molecule has two important consequences for the process of DNA replication. One is that, as a DNA double helix unwinds during replication, the two nucleotide chains will run in opposite directions (Fig. 11-31a; only the 5' → 3' directions are shown in the figure). The second is that when new nucleotide chains are synthesized on the template chains, *the new chains must run in opposite directions from the template chains*. Thus, as shown in Figure 11-31b, the 5' → 3' direction of the new chain on the left will run from bottom to the top of the figure, in the direction of fork movement, and the 5' → 3' direction

of the new chain on the right will run from the top to the bottom, opposite to the direction of fork movement.

All of this has great significance for the mechanism of replication because the DNA polymerase enzymes, and the primases laying down the RNA primers can add nucleotides to a growing chain *only in the 5' → 3' direction*. This means that DNA synthesis on the two sides of a fork must move in opposite directions (as shown in the direction of the arrows in Fig. 11-31b).

If this is the case, how can replication forks move unidirectionally along a DNA molecule, starting at one end and finishing at the other? This problem is solved in nature by the mechanism that lays down the RNA primers at short intervals. This ensures that the DNA assembled on the primers is also laid down in short lengths. On one template chain, these short segments are synthesized in the direction of fork movement. On the opposite chain synthesis travels away from the fork (as shown by the direction of the arrows in Figs. 11-31a and b, and also by the direction of the arrows in Fig. 11-29). Subsequently, the short, discontinuous lengths are joined to produce single, continuous nucleotide chains (Fig. 11-31c). Since the short, discontinuous lengths are so rapidly sealed into continuous nucleotide chains, the overall movement of the fork is in one direction. (In many organisms, replication is discontinuous only on the nucleotide chain in which the direction of synthesis is opposite to the direction of fork movement.)

Many scientists contributed bits and pieces to the mechanism for DNA replication outlined in this supplement, particularly Arthur Kornberg and his associates at Stanford University. Kornberg was the first to isolate and characterize a DNA polymerase enzyme, and later filled in many of the steps in the replication process. Kornberg received the Nobel Prize in 1959 for his work with DNA polymerases and the mechanism of DNA replication.

Supplement 11-3:
Disintegration and Reorganization of the Nucleolus During Mitosis

In most organisms the nucleolus breaks down and disappears during prophase and reappears during telophase. In these organisms, nucleolar breakdown is first detected as a loosening of nucleolar structure. The nucleolus then becomes progressively smaller, and finally disappears toward the end of prophase.

During telophase, the nucleolus reorganizes on one or more chromosomes, appearing first as a small spherical body that gradually increases in size until the typical interphase condition is reestablished. During this time, ribosomal RNA synthesis, which takes place in the nucleolus (see p. 210), gradually increases until it reaches the high levels characteristic of interphase cells.

In 1934, Barbara McClintock of the Carnegie Institute demonstrated that a specific chromosomal site is necessary for reappearance of the nucleolus at telophase. McClintock, working with corn, succeeded in obtaining mutant individuals without the nucleolar site. In these individuals the nucleolus failed to reappear at telophase. Instead, numerous small, nucleoluslike bodies were distributed among the chromosomes. McClintock concluded from this that the nucleolar site, when present, collects the nucleolar material released into the nucleoplasm at prophase into a compact nucleolus again at telophase. On this basis, McClintock termed the nucleolar site the *nucleolar organizer.*

This site is readily recognized at metaphase in most species as a prominent secondary constriction in one chromosome in haploids or a chromosome pair in diploids. In diploids, the nucleolar organizers of both chromosomes of the pair containing the site are usually active. Initially, this may lead to the development of two nucleoli at telophase, but often the two sites fuse to form a single large structure. In some organisms, including humans, additional organizer sites are distributed among the chromosomes. In our own species, five pairs of chromosomes carry separate nucleolar organizer sites and form ten separate nucleoli during telophase. These usually fuse into a single nucleolus during the subsequent interphase.

Several experiments have shown that the nucleolar material reforming on the chromosomes at telophase probably includes both newly synthesized ribosomal RNA and protein and old nucleolar material released from the chromosomes during the previous prophase. In 1973 David M. and Stephanie G. Phillips of Washington University treated mitotic cells with actinomycin D, a drug that blocks RNA synthesis. Nucleoli still formed at telophase in the treated cells. Since all new RNA synthesis was blocked, the material forming the nucleoli must have been derived from "old" ribosomal RNA and proteins persisting from the nucleoli that disintegrated during prophase. Other experiments following the incorporation of radioactive nucleotides into RNA in normal cells show that ribosomal RNA synthesis also begins anew during telophase in untreated, normal cells. Thus the nucleoli reappearing at telophase include both old and new rRNA and ribosomal proteins.

These observations suggest that the nucleolar cycle visible during mitosis reflects the following molecular changes. At the beginning of prophase, the nucleolus contains rRNA percursors in association with ribosomal proteins, all at various stages of processing into ribosomal subunits. As prophase progresses, rRNA synthesis stops, and the partially processed ribosomal precursors are released from the nucleolus. These disperse into the cytoplasm, along with the enzymes associated with rRNA processing. At telophase, after division of the chromosomes, the partially processed ribosomal subunits and the processing enzymes return to the nucleolar organizer segment and collect together to reform the nucleolus at this site. Processing of this preexisting material then continues. Transcription of new rRNA also begins at this time, so that, as telophase proceeds into interphase, a progressively greater part of the nucleolus represents newly synthesized material.

Presumably, dispersal of the rRNA, ribosomal proteins, and processing enzymes during metaphase and anaphase is an adaptation reducing the chromosome mass required to be moved during anaphase. Dispersal of the nucleolus probably also reduces the chance that the chromatids carrying the nucleolar organizer regions will tangle or fail to separate during anaphase.

Supplement 11-4:
The Role of Asters and
Centrioles in Mitosis

Centriole Replication

The pattern of centriole replication was first worked out in detail by Joseph G. Gall, then at the University of Minnesota. Gall noted that at the time of DNA replication, during S, the two "parent" centrioles move apart slightly, and each produces a small, budlike *procentriole* at one end (Fig. 11-32). The budding procentrioles, which are separated from the parent centrioles by a narrow space, extend outward at right angles from the parents.

As the procentrioles take form, the typical pattern of nine triplets of microtubules becomes visible inside them, although somewhat less clearly outlined than in mature centrioles (centriole structure is discussed in detail in Supplement 8-1). The procentrioles, about 70 nanometers long when they first appear, gradually lengthen throughout S, G_2, and prophase of mitosis until the mature length of about 200 nanometers is reached.

The replicating centrioles, surrounded by the microtubules of the asters, then gradually separate and move to opposite ends of the nucleus. As they do, the developing spindle forms between them, gradually lengthening until the primary spindle is formed (see Fig. 11-10).

The Role of Centrioles and Asters in Mitosis

To investigators at the turn of the century, the movements of the centrioles and asters during prophase appeared to be related to the generation of the spindle. However, since the spindle was later discovered to form and function equally well in plant and other cells without centrioles, it became obvious that the pattern of centriole replication and separation observed in animal cells cannot be direct requirements for spindle formation.

Some of the best evidence that this is actually the case was supplied by the experiments of Roland Dietz at the Max Planck Institute in Germany. Dietz isolated living cells from a line that normally contains asters and centrioles.

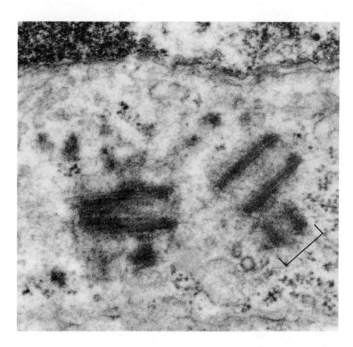

Figure 11-32 Procentriole (bracket) budding from a parent centriole in a cell of the rat. × 61,000. Courtesy of R. G. Murray, A. S. Murray, and A. Pizzo, from *Journal of Cell Biology* 26 (1965):601, by copyright permission of The Rockefeller University Press.

By flattening the cells under a microscope slide during division of the asters, Dietz succeeded in preventing migration of the asters and centrioles around the nucleus. The spindle formed nonetheless, and the subsequent mitosis and cytokinesis produced two lines of cells, one containing twice the normal number of centrioles and asters and one lacking these structures. The cells without centrioles or asters formed a spindle at the next division and divided normally, confirming that these structures are not necessary for spindle formation, even in animal cells that normally possess them.

These experiments, and the absence of centrioles and asters in many cell types under normal conditions, suggest that the complex division cycle of centrioles and asters probably serves some functional role other than generation of the spindle. Some cell biologists, Dietz among them, have suggested that the function of the asters is to separate the replicated centrioles and place them at opposite poles of the spindle. In these locations, the centrioles will be incorporated into separate daughter cells at the following mitosis and cytokinesis. Thus, instead of giving rise to the spindle, the division of the asters and centrioles is simply a mechanism ensuring that both daughter cells receive a pair of centrioles during cell division. According to this hypothesis, therefore, the spindle microtubules divide centrioles just as they do chromosomes.

If centrioles do not give rise to the spindle, what is their function in eukaryotic cells? The presence of centrioles appears to be correlated with the existence of cells that possess cilia or flagella at some time in the life cycle of the organism. In plants, for example, the species containing centrioles reproduce by means of flagellated, motile gametes. During development of flagella, a centriole gives rise to the system of microtubules that forms the axis of the flagellar shaft. This centriole usually persists as the *basal body* of the flagellum (see Supplement 8-1).

Because of the importance of flagella to reproduction and other functions in plants and animals, it is not difficult to imagine that the astral mechanism arose in evolution as an adaptation ensuring that centrioles as well as chromosomes are duplicated and equally distributed to daughter cells during cell division.

12

Meiosis and Gametogenesis: The Cellular Basis of Sexual Reproduction

The sequence of events in replication and mitosis ensures that daughter cells receive identical copies of the genetic information. During the late 1800s it became apparent that another, modified form of cell division must occur at some time in the life cycle of organisms that reproduce sexually. At this time, studies of sexual reproduction revealed that during fertilization, the nuclei of egg and sperm fuse together to form a composite nucleus. This observation made it obvious that, at some point before or after fertilization, a special division mechanism must reduce the number of chromosomes in half. Otherwise, the fusion of egg and sperm nuclei at fertilization would double the number of chromosomes each generation.

The anticipated division sequence was soon discovered. Investigators found that in animals, the number of chromosomes is reduced by one-half in a series of divisions that occurs just before the maturation of eggs and sperm. Within a few years, an equivalent series of reduction divisions was described in plants. The highly specialized process of division was called *meiosis*, derived from the Greek word *meioun*, meaning "to diminish."

In addition to reducing the chromosome number of eggs and sperm, meiosis also includes two additional processes that have great significance for the development and survival of sexually reproducing organisms. One is *recombination*, a mechanism by which segments of chromosomes are exchanged to mix gene sequences into new combinations. This process provides an almost infinite variety of new genetic types to meet the demands of a changing environment. The second additional process of meiosis, of greatest importance in animals, is synthesis of the RNA and protein molecules required for the development of eggs and sperm and at least part of the development of the embryo after fertilization.

Meiosis is followed directly or indirectly by morphological changes called *gametogenesis* that convert the cellular products of the meiotic divisions into reproductive cells, the eggs and sperm. These reproductive cells, the *gametes*, are specialized to transport or house the haploid chromosomes resulting from meiosis and bring them together in the process of fertilization. Fertilization, which involves fusion of the haploid nuclei of the gamete cells, restores the diploid chromosome number and initiates development of the new generation.

Meiotic Cell Division

Haploids, Diploids, Genes, and Alleles

Meiosis occurs only in organisms that are *diploids*, or have some higher multiple of the diploid number of chromosomes. In diploids, the nuclei of body cells contain two copies of each chromosome (see also Information Box 11-4). Thus, the chromosomes occur in pairs. For example, in humans, there are 23 different chromosomes. These occur in pairs, so that human body cells contain a total of 46 chromosomes. Meiosis reduces the chromosome number by *separating the chromosome pairs*. The sperm or egg nuclei, as a result, receive one member of each of the pairs present in body cell nuclei. Such nuclei, since they contain only half the number of chromosomes in body cells, are called *haploids* or *monoploids*. Human egg and sperm cells, for example, contain 23 chromosomes, the haploid number for our species.

Each chromosome in a diploid body cell nucleus consists of a single DNA molecule containing sequences coding for mRNA, rRNA, and tRNA molecules. Each coding sequence within a chromosome is a *gene*. The two members of each chromosome pair in a diploid contain the same genes in the same order. For this reason they are called *homologous* chromosomes. Although the two members of a homologous pair contain the same genes, there may be differences in the sequence of nucleotides for a given gene in either chromosome of a pair. These different nucleotide sequences are called the *alleles* of a gene. If mRNAs coding for a protein are transcribed from two different alleles of a gene (differences in rRNA and tRNA genes rarely occur), the mRNAs will differ in sequence. This difference, in turn, may produce different amino acid sequences in the proteins synthesized under the direction of the transcribed mRNAs.[1] Depending on the position of the changes, the biochemical properties of the two versions of the protein synthesized may differ only slightly or may be drastically different. The effect of recombination in meiosis is to mix the alleles of the two chromosomes of a homologous pair into new combinations. This is the primary source of the variability in genetic types generated by meiosis.

Meiosis: How It Works

The overall mechanism of meiosis is simple in broad outline. In mitosis, a single replication is followed by a single division of the chromosomes. In meiosis, a single replication is followed by *two* sequential divisions of the chromosomes, with the result that the chromosome number is halved in the products of the division. This is paralleled by changes in the quantity of DNA per nucleus (Fig. 12-1). If the DNA content of the eggs and sperm of a sexually reproducing organism is considered as the base or 1X level, then diploid body cells of the organism, at the G_1 stage before replication (see An Overview of the Cell Cycle, Chap. 11), contain the 2X amount of DNA. After replication, the resultant G_2 cells contain the 4X amount of DNA. If replication is followed by mitosis, the daughter nuclei contain the 2X amount at the completion of division. If the division sequence is meiotic, the two sequential divisions of meiosis reduce the DNA quantity to the 1X or haploid value. This is restored to the 2X diploid value after fusion of the haploid egg and sperm nuclei during fertilization.

---

[1] Some differences in mRNA sequence will not result in changes in the amino acid sequence of the protein coded for. This is the case if the sequences spell out different triplet codewords for the same amino acid.

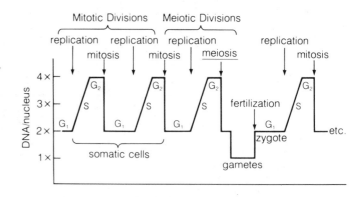

Figure 12-1 The relative amounts of DNA per nucleus during mitotic and meiotic cell cycles. The base level, 1X, represents the amount of DNA in a sperm or egg nucleus. G_1, S, and G_2 refer to the stages of interphase (see Fig. 11-1). Modified from an original courtesy of C. P. Swanson, from *Cytogenetics*, Prentice-Hall, 1967.

In the pattern most familiar to us, that of animals, meiosis directly produces eggs and sperm. Union of these reproductive cells or gametes in fertilization restores the chromosome number to the diploid level and initiates development of the new individual.

Meiosis follows the same overall pattern in all diploid organisms (for an overview of the entire process, see Information Box 12-1). After a premeiotic interphase, during which the chromosomes duplicate, meiosis proceeds through two complete chromosome divisions, each basically resembling a mitotic division:

Division I	*Division II*
Prophase I	Prophase II
Metaphase I	Metaphase II
Anaphase I	Anaphase II
Telophase I	Telophase II

There is *no DNA replication between the two meiotic divisions*.

Premeiotic Interphase Cells entering meiosis are usually the immediate products of a series of mitotic divisions. These cells enter the interphase before meiosis, which is not markedly different from an interphase preceding mitotic division. However, DNA replication during premeiotic S may be more prolonged and the nucleus may become larger during S and G_2 than in premitotic interphase. Premeiotic interphase ends and meiosis begins as the chromosomes begin to condense and become visible as threads in the nucleus.

Information Box 12-1

An Overview of Meiosis

Meiosis occurs only in organisms that have at least the diploid number of chromosomes. As in Information Box 11-2, let a single line represent all of the DNA coding for one copy of the genetic information of a cell:

Then break the line into shorter subunits representing the chromosomes:

This set of lines represents the chromosome complement of a haploid cell. Diploid cells have *two* copies of the genetic information:

Each vertical pair of lines represents a homologous pair of chromosomes. The two chromosomes of a homologous pair have the same genes in the same sequence. Different forms of a gene, however, called alleles, may be present in either of the members of a pair. During the interphase before meiosis, the chromosomes duplicate:

Each chromosome now contains two chromatids.
   In meiosis, the chromosomes condense into shorter, thicker units:

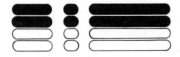

Homologous pairs then come together and pair:

While they are paired, some of the homologous chromosomes may exchange segments through a process of breakage and reunion of reciprocal parts:

(continued on next page)

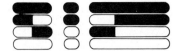

This exchange creates new combinations of alleles in the chromosomes undergoing the exchange. The paired chromosomes then line up on the spindle for the first of the two meiotic divisions. In this division, the two *chromosomes* of each homologous pair make microtubule connections leading to opposite ends of the cell:

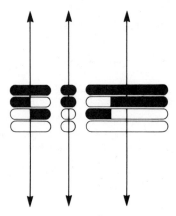

Note that both *chromatids* of any chromosome connect to microtubules leading to the same end of the cell. This division separates the chromosomes of homologous pairs:

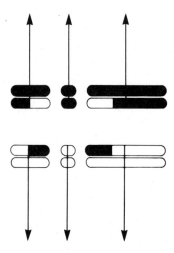

The two chromatids of each chromosome, however, are still together.

The chromosomes then line up on spindles for the second meiotic division. There is no DNA replication preceding this division. In the division, the two chromatids of each chromosome now make connections to microtubules leading to opposite ends of the cell:

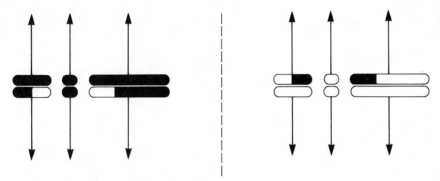

This division separates the chromatids of each chromosome:

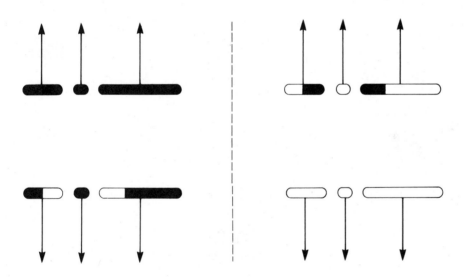

These chromosomes are then enclosed in separate nuclei or cells as the four haploid products of meiosis:

nucleus 1

nucleus 2

nucleus 3

nucleus 4

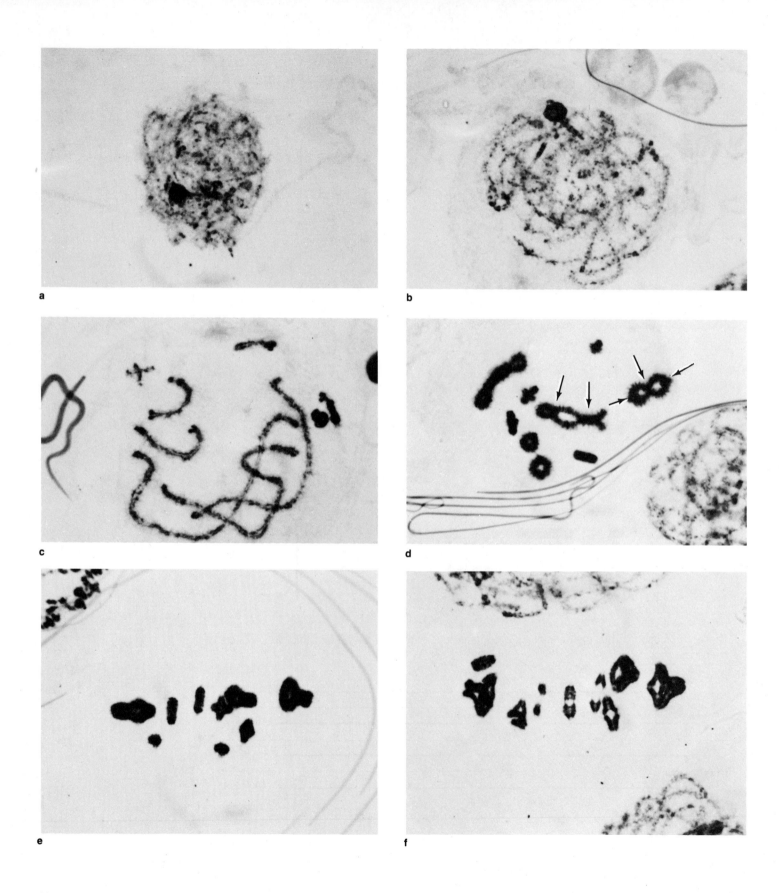

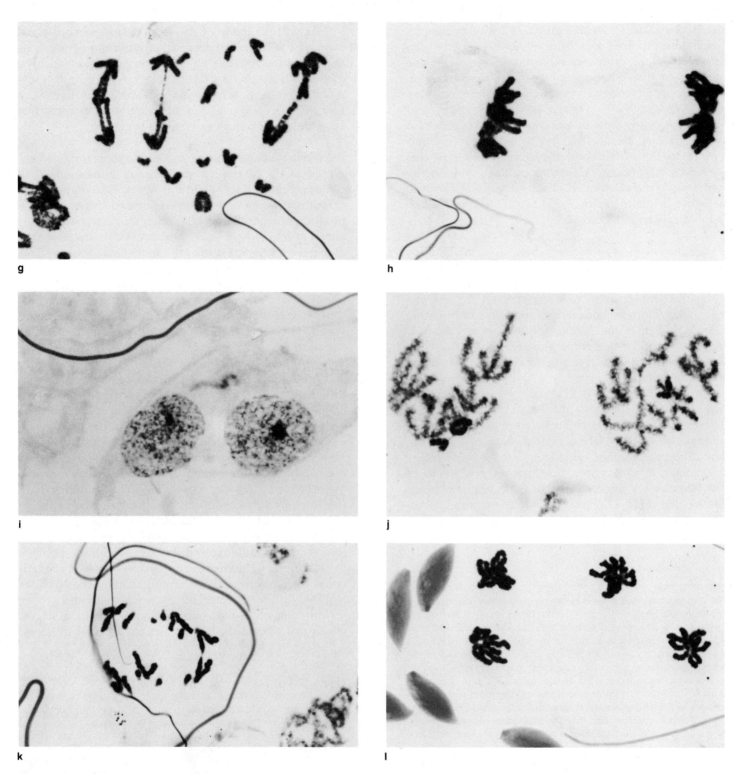

Figure 12-2 Meiosis in the grasshopper *Chorthippus*.
(**a**) Condensation stage. (**b**) Pairing. (**c**) Recombination. The narrow separation between the paired homologs is visible at most points. (**d**) Transcription. In the grasshopper, decondensation during the transcription stage is not pronounced. Crossovers are present at the points marked by the arrows. (**e**) Recondensation. (**f**) Metaphase I. (**g**) Anaphase I. (**h**) Early telophase I. (**i**) Interkinesis. (**j**) Prophase II. (**k**) Anaphase II. (**l**) Telophase II. Courtesy of J. L. Walters.

Meiotic Prophase I Meiotic prophase I is more complex than mitotic prophase and includes the chromosome rearrangements that cause recombination. This stage lasts much longer than mitotic prophase in the same organism, and may in fact extend over weeks, months, or even years.

Although it is more or less continuous, prophase I is divided for convenience into five well-defined stages according to the activities of the chromosomes: (1) condensation, (2) pairing, (3) recombination, (4) transcription, and (5) recondensation. These stages are shown in the series of photographs in Figure 12-2 (pages 294–295) and in diagrams in Figure 12-3.

Condensation Stage The condensation stage, traditionally called *leptotene*,[2] begins as the chromosomes first condense into visible threads (Figs. 12-2a and 12-3a). In the condensation mechanism, as in mitotic prophase, the chromosome fibers fold into shorter, more compact structures that become thicker as the folding becomes more extensive. Each of the condensing chromosomes contains two chromatids, which are the result of DNA replication during premeiotic interphase.

Pairing Stage During interphase and the condensation stage of prophase I, the two homologs of each pair occupy random and often widely separated locations in the nucleus. The condensation stage changes to the pairing stage (traditionally termed *zygotene*) as the two chromosomes of each homologous pair come together and begin to line up side by side (Figs. 12-2b and 12-3b). The pairing mechanism, called *synapsis*, proceeds until the homologous pairs are lined up in exact side-by-side register.

---

[2]The contemporary names used for the prophase I stages in this chapter refer to the functional activity of the chromosomes at each stage. The traditional names were derived from the morphological appearance of the chromosomes at each of the stages. *Leptotene* (*leptos* = fine or thin; *tene* = thread) refers to the threadlike appearance of the chromosomes during this stage. *Zygotene* (*zygon* = Y) refers to the Y-shaped appearance of the chromosomes during the pairing process. *Pachytene* (*pachus* = thick) describes the shorter, more compact appearance of the chromosomes at this stage. *Diplotene* (*diplos* = double) is derived from the fact that the separation between the two chromatids of each chromosome cannot be distinctly seen until this stage. The final traditional name, *diakinesis* (*dia* = across; *kinesis* = movement), meaning "movement across" refers to a change in position of the attachment points (crossovers, see p. 298) between chromatids, which are pushed toward the tips of the chromosomes by the tight condensation occurring at this stage. Frequently, the chromosomes become so compact and rounded at the recondensation or diakinesis stage that they are held together only at their tips.

As with so many of the mechanisms of mitosis and meiosis, the molecular forces that bring the homologous chromosomes together in synapsis are unknown. The questions surrounding the pairing mechanism are especially interesting because no intermolecular forces are known to operate over the relatively long distances that separate the homologs as synapsis begins.

Recombination Stage The recombination stage (or *pachytene*) begins as the pairing of homologous chromosomes becomes complete (Figs. 12-2c and 12-3c). Each of the paired chromosomes, as noted, consists of two chromatids, the products of replication during premeiotic interphase. Thus a total of four closely associated chromatids are present in each paired structure during the recombination stage. Two terms are commonly used for the paired chromosomes. When considered at the level of chromatids, the paired homologs are referred to as *tetrads* (from *tetra* = four) because four chromatids are present in each paired structure. When considered at the level of chromosomes, the same structure is called a *bivalent* because two homologous chromosomes are synapsed in each pair.

Although the synapsis of homologs is very close at this stage, the two chromosomes of each pair remain separated by a regular space about 0.15 to 0.2 micrometer wide. This space, which is just visible in the light microscope, proves under the electron microscope to contain a highly specialized structure called the *synaptonemal complex* (Fig. 12-4). Genetic recombination has been shown to take place between the four closely associated chromatids at this stage by a process of physical breakage and exchange of segments between the chromatids of homologous chromosomes (see Fig. 12-9). Supposedly, the synaptonemal complex is directly involved in these exchanges, possibly by holding the chromatids in the closely paired configuration and also by containing and aligning the enzymes required for the breakage and exchange mechanism.

In contrast to the condensation and pairing stages, which usually run their course within a matter of hours,

Figure 12-3 Meiosis in diagrammatic form. (**a**) Condensation stage. Although both chromatids are shown for diagrammatic purposes in the magnified circles of parts (a), (b), and (c) of this figure, the split between the two chromatids of a chromosome is not actually visible during these stages. (**b**) Pairing stage. Pairing is in progress at several points (arrows). (**c**) Recombination. (**d**) Transcription. The chromosomes are held in pairs only by crossover points (arrows) remaining from the recombination stage. In most organisms, all four chromatids usually become visible in the tetrads at this time. (**e**) Recondensation. (**f**) Metaphase I. (**g**) Anaphase I. (**h**) Telophase I. (**i**) Prophase II. (**j**) Metaphase II. (**k**) Anaphase II. (**l**) Telophase II.

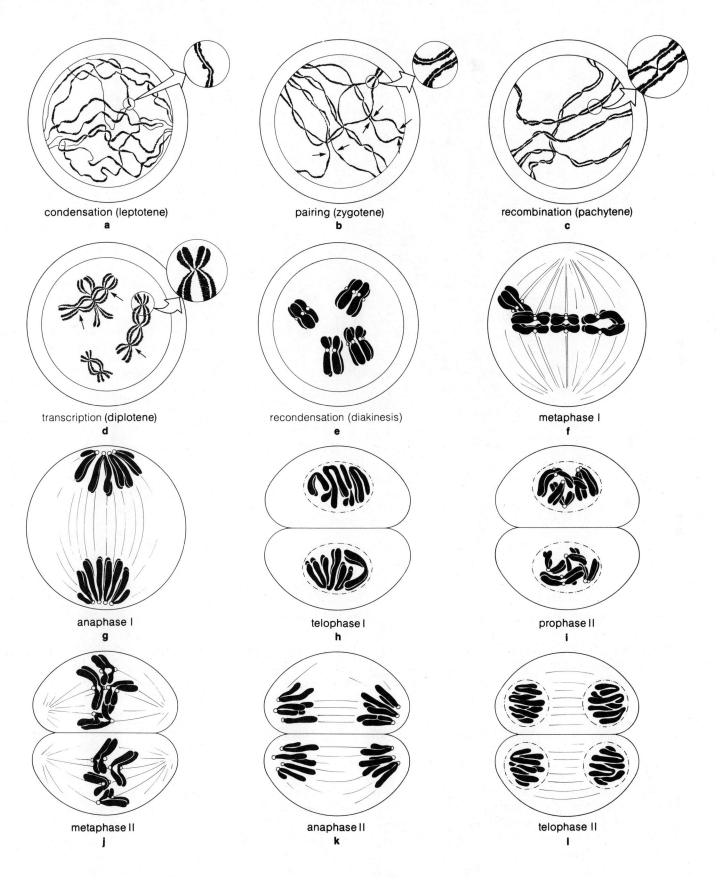

condensation (leptotene)
a

pairing (zygotene)
b

recombination (pachytene)
c

transcription (diplotene)
d

recondensation (diakinesis)
e

metaphase I
f

anaphase I
g

telophase I
h

prophase II
i

metaphase II
j

anaphase II
k

telophase II
l

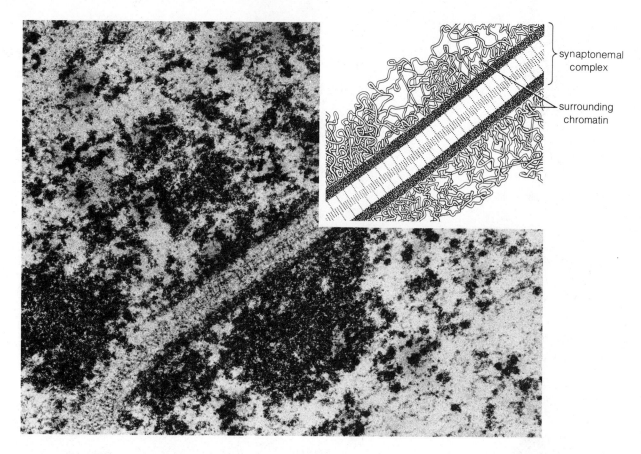

synaptonemal complex

surrounding chromatin

Figure 12-4 The paired chromosomes of a grasshopper. The narrow space between the chromosomes contains the synaptonemal complex. The inset shows the relationship of the synaptonemal complex to the chromatin of the paired chromosomes. × 74,000. Micrograph courtesy of P. B. Moens, from *Journal of Cell Biology* 40 (1969):542, by copyright permission of The Rockefeller University Press.

the recombination stage may take weeks, months, or even years. The length of this stage probably reflects the complex events associated with the breakage and exchange of chromatid segments during genetic recombination.

Transcription Stage As the recombination stage comes to a close, the homologs separate at many points (Figs. 12-2d and 12-3d). This end to close pairing signals the beginning of the transcription stage (or *diplotene*). As the homologs separate, the synaptonemal complex disappears from between the chromosomes.

The homologous chromosomes become so widely separated at this stage that they almost seem to repel each other except at scattered attachment points (arrows, Fig. 12-3d). On close inspection, these attachment points prove

to be regions in which two of the four chromatids cross over between homologous chromosomes (see also Figs. 12-5 and 12-9). These crossing places, called *crossovers* or *chiasmata* (singular *chiasma* = crosspiece), are remnants of the chromatid exchanges that took place during the recombination stage.

During the transcription stage, the chromosomes unfold to a greater or lesser extent and become active in RNA synthesis. In organisms in which RNA transcription is relatively limited at this time, the unfolding, called *decondensation*, produces only a slight fuzziness in the outline of the chromosomes. In other species, such as many insects, the chromosomes unfold almost completely and revert to an apparently interphaselike state. In some organisms, particularly in female amphibians, birds, and reptiles,

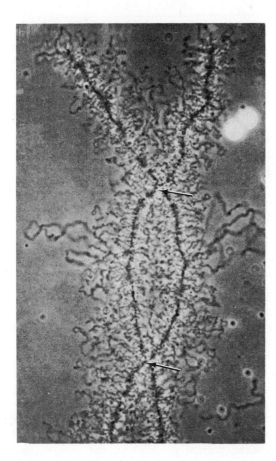

Figure 12-5 A lampbrush chromosome in the nucleus of a salamander during the transcription stage. Loops active in RNA transcription extend outward from all parts of the tetrad. The tetrad is held together by crossovers at the arrows. × 470. Courtesy of J. G. Gall.

loops extend outward from all parts of the chromatids during this stage (Fig. 12-5). The tetrads are called *lampbrush chromosomes* when in this configuration. The loops in these chromosomes are sites of intensive RNA synthesis throughout the transcription stage.

Both ribosomal and messenger RNA are synthesized in quantity during the transcription stage. Structurally, the ribosomal RNA synthesis is marked by extensive growth of the nucleolus, or production of extra nucleoli numbering in the hundreds or thousands. The entire meiotic cell also grows at this time through the synthesis of large quantities of proteins, fats, carbohydrates, and other molecules in the cytoplasm. This period of growth is most pronounced in the developing eggs (oocytes) of amphibians, birds, and reptiles, which may grow from microscopic size

to diameters of several centimeters during the transcription stage. Most of this growth is in the cytoplasm. The nucleus in these cells remains microscopic, or may become just large enough to be visible to the naked eye.

The transcription stage may last for a long time. In developing amphibian egg cells, for example, this part of meiotic prophase may last nearly a year. In humans, oocytes reach this stage in unborn females at about the fifth month of fetal life and remain arrested at this point during the remainder of prenatal life, through birth and childhood, and until the female reaches sexual maturity. Then, just before ovulation, one oocyte each month breaks arrest and continues the meiotic sequence. The time between the onset and completion of the transcription stage in human females may thus range from 12 or so to more than 50 years. Intensive RNA and protein synthesis, however, occurs only during the part of the transcription stage preceding meiotic arrest in the developing fetus.

Recondensation Stage During the recondensation stage (or *diakinesis*) the chromosomes condense again into short, tightly packed structures (Figs. 12-2e and 12-3e). If little decondensation has occurred during the transcription stage, the transition to the recondensation stage may be difficult to identify. By the end of the recondensation stage the chromosomes are packed so tightly that they may be almost spherical.

Completion of the recondensation stage marks the close of prophase I of meiosis. The various events of prophase I produce three results of great significance for the outcome of meiosis: (1) recombination, (2) synthesis of most or all of the RNA, protein, lipid, and carbohydrate molecules required for the growth of gametes and the early stages of embryonic development, and (3) condensation of the chromosomes into short rodlets. The remaining stages of meiosis are concerned primarily with division of the chromosomes to the haploid number.

The Meiotic Divisions Near the end of prophase I the spindle for the first meiotic division forms by a pattern closely similar to mitosis. The centrioles, if present, have replicated during the previous interphase. The two pairs of centrioles migrate with the asters to opposite ends of the cell and take up positions at the spindle poles just as in mitosis (see p. 263). This pattern is most typical of the meiotic divisions in male animals leading to sperm formation. In many animal eggs, the centrioles disappear and cannot be detected in developing eggs at any stage of meiosis. In these eggs, the spindle forms without the involvement of centrioles and asters essentially as it does in the higher plants. In either case, the spindle, built up from hundreds to thousands of microtubules stretching from end to end of the cell, soon fills most of the cytoplasm around the nucleus.

Metaphase I of Meiosis Just as in mitosis, the breakdown of the nuclear envelope provides a convenient reference point for the transition from prophase I to metaphase I of meiosis. The spindle moves to the position formerly occupied by the nucleus, and the tetrads, scattered by the breakdown of the nuclear envelope, make their way to the spindle midpoint (Figs. 12-2f and 12-3f). Except for the pairing of homologous chromosomes into tetrads, meiosis at this point closely matches the transition from prophase to metaphase in mitosis.

The first major divergence from the pattern observed in mitosis, and in fact the distinctive event of the meiotic divisions following prophase I, occurs as the tetrads attach to the spindle microtubules. Each of the homologous chromosomes has two kinetochores, one for each of its two chromatids (see Fig. 12-6). However, although two kinetochores are present in each homologous chromosome, both of these make connections to the *same* pole of the spindle at metaphase I of meiosis (Fig. 12-6; compare with Fig. 11-4). The two kinetochores of the other chromosome of a homologous pair both make connections to the opposite pole. Thus the two kinetochores of each homologous chromosome act as a functional unit during metaphase I of meiosis.

Anaphase I Separation of the homologous pairs by the spindle initiates anaphase I (Figs. 12-2g and 12-3g). Because of the opposite connections made by the kinetochores, the anaphase movement *separates the two chromosomes of each homologous pair and delivers them to opposite spindle poles.* At the completion of anaphase I, as a consequence, the poles receive the haploid number of chromosomes. Each chromosome, however, is still double and contains two chromatids (from the replication during the previous interphase). As a result, although the two groups of chromosomes at the spindle poles contain only one member of each homologous pair, they still contain twice the haploid amount of DNA.

The pattern of separation of the homologs during anaphase I of meiosis is as important for genetic variability as recombination during prophase I. To appreciate why, you must understand the maternal and paternal origins of the two members of each homologous pair. The haploid egg and sperm nuclei giving rise to an individual contain only one member of each homologous pair. Fusion of the

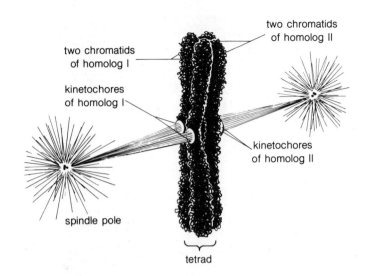

Figure 12-6 Kinetochore connections at metaphase I of meiosis. The two kinetochores of a homologous chromosome make microtubule connections leading to the same pole of the spindle (compare with Fig. 11-14).

egg and sperm nuclei in fertilization brings the homologous pairs together, creating the diploid condition. In each chromosome pair of the diploid, the chromosome originating from the sperm nucleus is called the *paternal* chromosome, and the chromosome originating from the egg is called the *maternal* chromosome. As an individual grows by cell divisions from the fertilized egg, replication and mitosis ensure that the same pairs are equally distributed, so that all of the homologous pairs in the organism contain one maternal and one paternal chromosome. Although the maternal and paternal chromosomes of a homologous pair contain the same genes in the same order, different alleles of these genes may be present in either member of the pair.

The different homologous pairs separate independently from each other at anaphase I of meiosis, so that any combination of chromosomes of maternal and paternal origin may be delivered to a spindle pole. The random combinations of maternal and paternal chromosomes delivered to the poles contribute to the total genetic variability of the products of meiosis. In humans, for example, with 23 pairs of chromosomes, there are 2^{23} possibilities for the combination of maternal and paternal chromosomes delivered to the spindle poles at anaphase I. Even without recombination, therefore, the probability that two children of the same parents will receive the same combination of maternal and paternal chromosomes would be one chance out of $(2^{23})^2$, or 1 out of 70,000,000,000,000! This variability arises from random segregation alone. The further variability introduced by recombination, which mixes chromatid segments randomly between maternal and paternal chromatids, makes it practically impossible for individuals to produce genetically identical offspring (except for identical twins, which arise from a different mechanism, by division of the fertilized egg into two separate cells that each give rise to complete embryos).

Telophase I A well-defined telophase I (Figs. 12-2h and 12-3h) does not always occur between anaphase I and the second meiotic division. All possible gradations are found in nature, from organisms with no detectable decondensation of the chromosomes at telophase I, as in some insects, to species with almost complete reversion to an interphaselike state called meiotic *interkinesis* (Fig. 12-2i). In the latter organisms, such as corn and some other plants, the chromosomes partly decondense and a nuclear envelope temporarily surrounds the polar masses of chromatin. However, in most species telophase I is a transitory stage in which meiotic cells pause only briefly before entering prophase II of meiosis. No DNA replication occurs at this time in any known organism.

During telophase I the single metaphase spindle reorganizes into two spindles that form in the regions of the telophase I spindle poles. If centrioles and asters are present, they also divide at this time, placing a single centriole at each pole of the two spindles. These events complete the cellular rearrangements for the second meiotic division.

The Second Meiotic Division The second meiotic division follows essentially the same pattern as an ordinary mitotic division. After a brief or even nonexistent interkinesis and prophase II (Figs. 12-2i and j and 12-3i), the chromosomes left at the two poles by the first meiotic division move to the midpoints of the two newly formed spindles. If decondensation has occurred during telophase I or interkinesis, the chromosomes condense back into tight rodlets as movement to the metaphase II spindles takes place. The kinetochores of the chromosomes attach to the spindle microtubules at metaphase II of meiosis (Fig. 12-3j) exactly as they would at mitotic metaphase. Each chromosome at this stage contains two chromatids; the kinetochores of these chromatids make attachments to microtubules leading to *opposite* poles of the spindle as in mitosis.

At anaphase II (Figs. 12-2k and 12-3k) the two chromatids of each chromosome, because of their connections to spindle microtubules, separate and move to opposite poles of the spindle. This separation and movement delivers the haploid number of chromatids to each pole of the spindle. As a result, each pole now contains the haploid quantity of DNA. This quantity is one-fourth of the DNA present at G_2 in the original cell entering the two sequential meiotic divisions.

At telophase II (Figs. 12-2l and 12-3l), the chromatids decondense and nuclear envelopes form around the four division products. When decondensation and nuclear envelope formation are complete, the chromatids in these nuclei are considered to be chromosomes. The four nuclei formed at this point have widely divergent fates in various species of plants and animals (see The Formation of Gametes in Animals and Plants: Gametogenesis below).

The Overall Effects of Meiosis: A Review

For simplicity we will consider meiosis in a hypothetical cell containing only two pairs of chromosomes—one long pair and one short pair, as in the review of mitosis presented in Chapter 11 (see p. 272 and Fig. 11-22). At premeiotic G_1, these uncondensed pairs are unassociated and randomly distributed in the nucleus (Fig. 12-7a). During premeiotic S, each chromosome replicates and becomes double at all points. As a result, each contains two chromatids (Fig. 12-7b). After G_2, as the condensation stage begins, the chromosomes fold into thicker threads that become visible in the light microscope (Fig. 12-7c). Synapsis at the pairing stage (Fig. 12-7d) brings homologous chromosomes together, producing two tetrads in the nucleus. Each tetrad contains two chromosomes and four chromatids. During the recombination stage, segments of the homologous chromosomes exchange (Fig. 12-7e), leading to new combinations of alleles. At the transcription stage, at least some decondensation and RNA synthesis take place. The chromosomes condense again at the final, recondensation stage of prophase I and enter metaphase I (Fig. 12-7f).

At metaphase I, the two homologous chromosomes of each pair connect to microtubules leading to opposite poles of the spindle. Anaphase I (Fig. 12-7g) therefore separates the homologs and moves the haploid number of chromosomes to each spindle pole. Thus only one long chromosome and one short chromosome are present at each pole, and pairs no longer exist. Note that at this point the chromosomes at the poles still contain two chromatids.

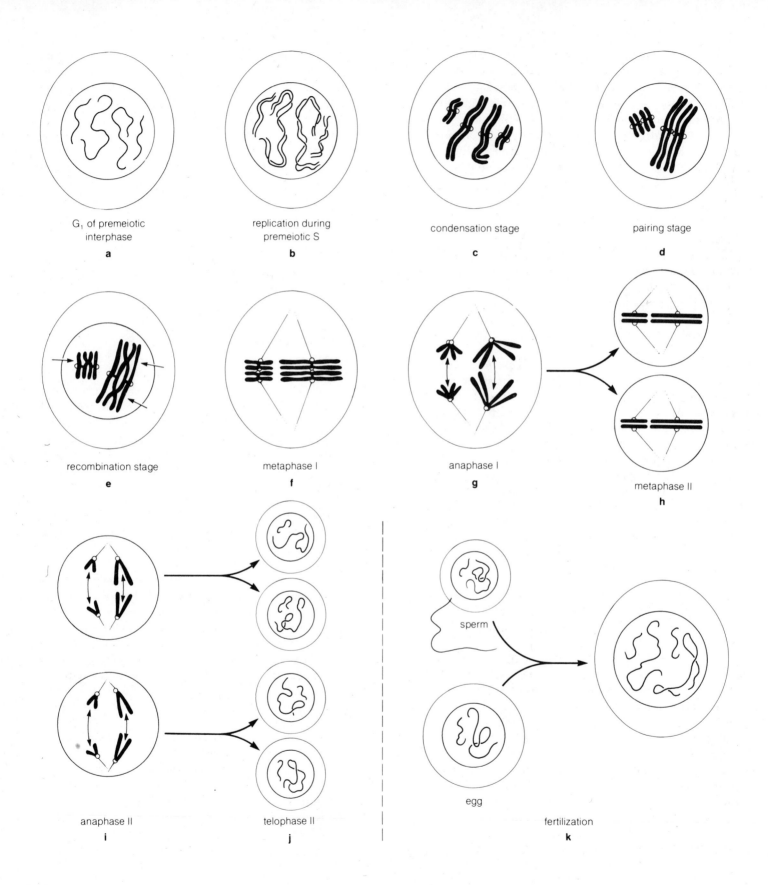

G₁ of premeiotic
interphase

a

replication during
premeiotic S

b

condensation stage

c

pairing stage

d

recombination stage

e

metaphase I

f

anaphase I

g

metaphase II

h

anaphase II

i

telophase II

j

sperm

egg

fertilization

k

During telophase I, any intervening interphaselike period, and prophase II two new spindles form at the metaphase I division poles. At metaphase II (Fig. 12-7h), the chromosomes move to the spindle midpoints and make microtubule attachments as in mitosis, in such a way that the two chromatids of each chromosome connect to microtubules leading to opposite spindle poles. Separation of the chromatids at anaphase II (Fig. 12-7i) delivers the haploid number of chromatids to the poles. The four division products at telophase II (Fig. 12-7j) each now contain the haploid quantity of DNA, and possess only two chromatids (now technically called chromosomes), one long and one short.

If, in this hypothetical organism, meiosis and gamete formation produce eggs and sperm, the end products will each contain two chromosomes, one long and one short. Fusion of an egg and sperm nucleus in fertilization (Fig. 12-7k) rejoins the pairs and returns the chromosome complement to the G_1 level of premeiotic cells.

Thus the cycle is completed and sexual reproduction, through meiosis and fertilization, is accomplished without changing the basic chromosome number. Meiosis supplies three important parts to the overall mechanism: (1) it reduces the chromosomes to the haploid number in gametes, (2) it produces genetic variability among the gametes produced by the same individual, and (3) it supplies RNA and protein molecules, synthesized during the transcription stage, needed for fertilization and the early stages of development.

The factors controlling the progress of cells through the meiotic cell cycle have proved as elusive as the controls of the mitotic cycle. In fungi and protists, essentially any cell can be induced to enter a meiotic rather than mitotic cell division by simple changes in the environment, usually in the direction of less favorable growth conditions. In higher plants and animals, the induction of meiosis probably involves an interaction between hormones and the cell surface. Various internal responses, possibly including adjustments in the concentration of substances such as the cyclic nucleotides or Ca^{2+} (see p. 255), then trigger meiosis. The observations made to date suggest, as in mitotic cycles, that a variety of control mechanisms exist in different cells and species. Ultimately, these control mechanisms probably act on a series of genes that is activated in sequence to trigger meiosis.

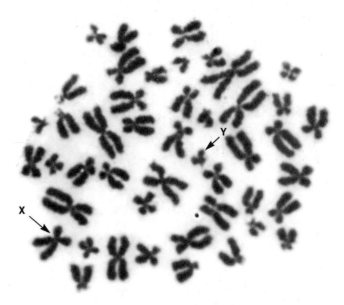

Figure 12-8 The X and Y sex chromosomes in a set isolated from a human mitotic cell at metaphase. × 2,000. Courtesy of S. Brecher.

Division of Sex Chromosomes in Meiosis

In some plants and most animals, the males and females of a species show differences in chromosome structure or number. The chromosomes regularly different in the males and females of a species are called *sex chromosomes*. In most of these species, including humans, cells in females contain a homologous pair of the sex chromosomes, and cells in males contain only one of this pair. By convention, the chromosomes present in a pair in the female are called X *chromosomes*. In the male, the single representative of this pair, the X, may exist alone ("XO" males) or may be associated with another sex chromosome, the Y, which is not found in females (Fig. 12-8). XX females and XY or XO males are the rule in most animals and the relatively few plants with sex chromosomes. In a few groups, such as the birds, butterflies, and moths, the situation is reversed, with XX males and XY females. All other chromosomes in the nucleus are found in homologous pairs in both males and females. The chromosomes that show no difference in number or morphology in either sex are called *autosomes*.

The sex chromosomes, as the name suggests, frequently have a direct developmental role in determining the sex of an individual. Also, in individuals with the XY combination (or XO; see below) unusual genetic effects

Figure 12-7 A review of meiosis (see text).

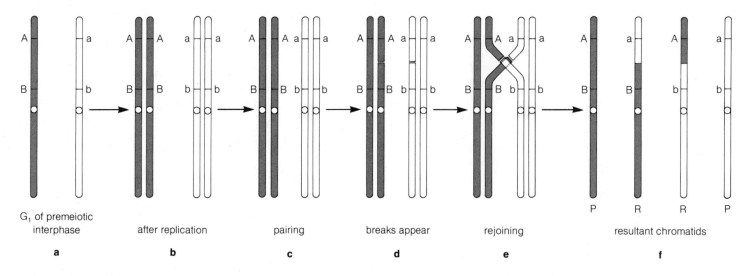

Figure 12-9 The pattern of the exchanges between chromatids occurring during recombination (see text). *P*, parental; *R*, recombinant.

G₁ of premeiotic interphase — **a** after replication — **b** pairing — **c** breaks appear — **d** rejoining — **e** resultant chromatids — **f**

may occur because the X and Y chromosomes often carry totally different genes. These unusual patterns of inheritance related to the sex chromosomes, called *sex linkage* or *sex-linked inheritance*, are important in several genetically determined diseases of humans, such as color blindness and hemophilia (see pp. 339–340).

The presence of sex chromosomes presents no mechanical difficulties during the meiotic divisions. In the sex with the XX pair (in humans, the female), the sex chromosomes go through meiosis exactly as the autosomes. Pairing and recombination occur between the two X chromosomes, and each of the products of meiosis receives a single X at the close of meiosis. All the eggs of species with XX females therefore contain a single X chromosome; the eggs of normal females never carry a Y.

In XY individuals the X and Y chromosomes may or may not contain homologous genes and pair during prophase. Whether or not pairing occurs, the X and Y line up on the spindle along with the autosomes at metaphase I of meiosis. Both the X and Y contain two chromatids at this stage as a result of replication at premeiotic interphase. The two subsequent meiotic divisions separate the four chromatids of these two chromosomes and distribute one to each of the four products of meiosis. Two receive an X chromatid, and two receive a Y chromatid. As a result, in XY individuals (such as the human male), sperm nuclei may contain either an X or Y chromosome.

In individuals without a Y chromosome (XO individuals, common in insects) both chromatids of the single X

usually go to the same pole of the spindle at anaphase I. The other pole receives no X. At anaphase II, the nucleus receiving the X chromosome divides again and distributes one X chromatid to each of the two division products. The other nucleus, which receives no X during anaphase I, divides normally, yielding two nuclei with no X chromosome. Of the four nuclei resulting from meiosis in these XO individuals, two will contain an X chromosome, and two will have no X chromosome (O nuclei).

At fertilization, the X gametes produced by an XX individual may fuse with either an X or a Y gamete produced by an XY individual (or an X or O gamete in XO individuals). The resultant diploid nuclei are therefore either XX or XY (or XX or XO), and the chromosome complements of the two sexes are reconstituted.

The Mechanism of Recombination

The genetic variability arising from meiosis results from two processes: recombination during meiotic prophase I, and the independent separation of homologous chromosomes during anaphase I. We have already considered how the kinetochore connections made by the homologs during metaphase I result in new combinations of maternal and paternal chromosomes in anaphase I, and, eventually, in eggs and sperm. Recombination generates variability through a different mechanism, by exchanging segments between homologous chromosomes.

The Genetic Consequences of Recombination

The exchanges of chromosome segments in recombination take the form shown in Figure 12-9. Consider a homologous pair in which both chromosomes contain the genes *A* and *B*. In this example, one of the two chromosomes carries the *A* form of gene A and the other has the slightly different *a* allele of the same gene. The B gene is similarly present as the two different alleles *B* and *b*. At the beginning of meiosis the chromosomes carry the alleles in the combinations shown in Figure 12-9a. One contains the *A* and *B* alleles and the other the *a* and *b* alleles of the two genes.

At G_1 of premeiotic interphase the two chromosomes of the pair are unreplicated, as shown in Figure 12-9a. After replication (Fig. 12-9b), both chromatids of the A-B chromosome will have the *A* and *B* alleles at these sites. Both chromatids of the opposite member of the homologous pair will have the alleles *a* and *b* at the corresponding locations. At this stage, these chromosomes are identical in sequence to chromosomes of the same homologous pair in any G_2 cell of the same organism. During prophase I of meiosis the homologs pair closely (Fig. 12-9c). If recombination occurs between the genes it takes place as follows. Two of the four paired chromatids break in equivalent locations, as shown in Figure 12-9d. The broken ends then cross over and rejoin with the respective broken ends in the opposite chromatid. As a result, the two chromatids exchange precisely matched segments (Fig. 12-9e). The end products after recombination consist of two chromatids of unchanged sequence, and two that have been changed. The unchanged chromatids, called *parentals*, are identical to the chromosomes of the original parent cell; that is, they contain alleles *A* and *B* together, or *a* and *b* together. Two are *recombinant*: they have the new combinations *a-B* or *A-b*. During the subsequent meiotic divisions the four chromatids separate and are enclosed in four different nuclei. Thus each of the four nuclei produced will contain one of the four chromosomes shown in Figure 12-9f.

As this example shows, any single recombination event involves only two of the four chromatids and produces two parental chromosomes and two recombinants. The location of a recombination event is random; recom-

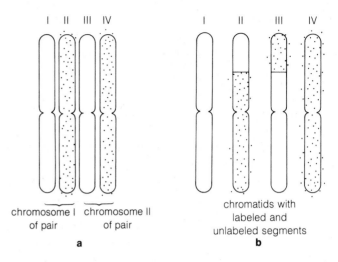

chromosome I of pair   chromosome II of pair

a

chromatids with labeled and unlabeled segments

b

Figure 12-10 The Taylor experiment demonstrating that recombination between homologous chromosomes takes place by physical breakage and exchange between homologous chromosomes (see text).

binations may take place at essentially any position along a chromosome pair and between any two of the four chromatids of a tetrad. Recombinations usually occur at a frequency of about one per chromosome "arm" (the segment of a chromosome between the kinetochore and the tip).

How Recombination Takes Place

Physical breakage and exchange between homologous chromosomes was suspected to be the mechanism underlying recombination as long ago as the first decade of this century. However, the experimental demonstration that recombination actually takes place this way was not accomplished until the mid-1960s. The critical experiment was carried out at Columbia University by J. Herbert Taylor, the scientist who first established that replication is semiconservative in eukaryotes (see Supplement 11-1).

Taylor developed a way to label one of the two chromatids of each homologous chromosome with radioactive DNA (his methods are described in Supplement 12-1). As a result, two of the four chromatids in each tetrad during meiotic prophase were labeled, and two were not (Fig. 12-10a). At the close of meiosis, Taylor was able to recover some chromatids that contained both labeled and unlabeled segments (Fig. 12-10b; see also Fig. 12-28). Chromatids of this type could be produced in meiosis only by breakage and exchange of parts between labeled and unlabeled chromatids of the paired homologs. Thus Taylor's work established that recombination actually occurs by physical breakage and exchange.

The molecular mechanisms underlying breakage and exchange in recombination, which are still not completely understood, are presently the subject of intensive investigation (a hypothesis for the mechanism, and the activity of the enzymes probably involved in recombination are described in Supplement 12-1). From results obtained by genetic crosses, it is clear that the breaks and exchanges are so precise that neither of the recombined chromatids receives any extra DNA or is missing any. As one molecular biologist put it, this is equivalent to breaking two lengthy books at the same letter of the same word and exchanging and splicing the pieces back together so perfectly that not one letter is missing from either copy! The enzymes active in recombination are probably organized on the synaptonemal complex, the framework visible between the paired homologs during the recombination stage.

The Time and Place of Meiosis in Different Organisms

There are three major variations in the time and place of meiosis in the life cycle of eukaryotic organisms (Fig. 12-11). The most familiar pattern occurs in animals, many protozoa, and a few lower plants. In this case, meiosis occurs immediately before the gametes form (Fig. 12-11a). The four haploid products of meiosis become sperm in males; in females, usually only one of the four nuclei becomes the egg nucleus.

Many plants, including all the higher plants (the angiosperms and gymnosperms) undergo meiosis at an intermediate stage in the life cycle (Fig. 12-11b). Organisms possessing this meiotic pattern alternate in each generation between haploid and diploid individuals. Fertilization produces a diploid generation, in which the individuals are called *sporophytes*. After growing to maturity by mitosis, the sporophytes produce asexual reproductive cells called *spores* by meiosis. These are diversified in genetic makeup as a result of recombination and the meiotic divisions. The spores grow directly by mitosis into haploid individuals called *gametophytes*. At some point, cells in the gametophytes differentiate into eggs and sperm following ordinary mitotic divisions. All the eggs or sperm produced from an individual gametophyte, since they arise through

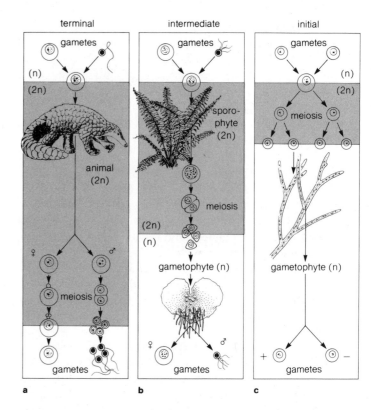

Figure 12-11 The time and place of meiosis in the life cycles of various organisms (see text). The shaded portions mark the diploid phase of the life cycles. Redrawn with permission of the Macmillan Company, from *The cell in development and heredity* by E. B. Wilson. Copyright 1925, The Macmillan Company, renewed in 1953 by Anne M. K. Wilson.

mitosis, are genetically identical. Fusion of the haploid gametes returns the cycle to the diploid, sporophyte generation.

A third major variation in the time and place of meiosis in life cycles is observed in fungi, some algae, and a few protozoa (Fig. 12-11c). This variation has proved to be of great value to researchers in cell biology. In these organisms, meiosis takes place immediately after fertilization. The two haploid gamete nuclei designated as egg and sperm (or, since they are usually undifferentiated, simply as + and −) fuse to produce a diploid nucleus. The diploid nucleus immediately enters meiosis, producing four haploid nuclei. The nuclei eventually develop into haploid spores, which grow by mitotic divisions into the haploid generation. The haploid generation thus makes up the dominant phase of the life cycle, and the diploid generation is reduced to a single nucleus. This nucleus remains diploid only during meiotic prophase I and metaphase I.

In a number of these organisms, DNA replication takes place in the haploid gamete nuclei before the gametes fuse to produce the diploid nucleus. In these organisms, therefore, replication is clearly separate in time and place from genetic recombination, which can occur only in the brief diploid phase of the life cycle during prophase I. This fact has been most useful in establishing that recombination occurs in meiotic prophase instead of during replication, as many cell biologists had previously assumed. The organisms of this group include various yeasts and the fungus *Neurospora*, which are much used in biochemical and genetic studies.

Whenever it occurs in the life cycle of sexually reproducing organisms, meiosis has the three primary results outlined in this chapter. (1) It reduces the chromosome number to the haploid level, so that the chromosome number in the species does not double at each fertilization. (2) Through the genetic rearrangements occurring in recombination and the independent separation of maternal and paternal chromatids at anaphase II, meiosis produces genetic variability in the haploid products of the division sequence. This variability is the immediate source of the variation in the offspring of sexually reproducing organisms. (3) Finally, through the RNA and protein synthesis that occurs during the transcription stage, it provides the ribosomes, enzymes, structural proteins, and raw material needed for gamete production, fertilization, and the early stages of embryonic development.

The genetic variations in the meiotic products resulting from recombination and independent separation of maternal and paternal chromosomes are responsible for most of the variability noted in the offspring of sexually reproducing organisms. The production of both blue-eyed and brown-eyed offspring from brown-eyed parents in human families, for example, results from the chromosome rearrangements occurring during meiosis.

The Formation of Gametes in Animals and Plants: Gametogenesis

Meiosis is coordinated in animals and plants with the formation of gamete cells that are specialized to carry out fertilization. The biochemical and structural changes that produce these gametes take different forms in plants and animals. In either case, the gamete cells produced function in fertilization to return the chromosome number to the diploid level and initiate embryonic development.

Gametogenesis in Animals

Meiosis in animals leads directly to the formation of gametes. In male animals, the developmental changes producing mature gametes, called *spermiogenesis*, begin after meiosis is complete. Gametogenesis in females, termed *oogenesis*, commences during the transcription stage of meiotic prophase I. In most animal species, eggs mature before meiosis is complete. Maturation of eggs in these species is accompanied by meiotic arrest, which remains in force until fertilization. At fertilization, the arrest is broken, and the remaining meiotic stages quickly run their course.

Spermiogenesis In almost all male animals the completion of anaphase II is immediately followed by development of the four products of meiosis into mature sperm cells. This development includes extensive changes in both nucleus and cytoplasm. Before spermiogenesis begins, the four products of meiosis (called *spermatids*) are not conspicuously different from ordinary somatic cells. Each contains all the major cytoplasmic organelles and structures, including mitochondria, the Golgi complex, ribosomes, endoplasmic reticulum, centrioles, and a nuclear envelope complete with pore complexes. During spermiogenesis, each of these structures undergoes extensive changes. Eventually the nucleus becomes greatly reduced in volume, and much of the cytoplasmic mass, including most of the cytoplasmic organelles, is eliminated.

Mature sperm in animals are usually greatly elongated cells, consisting typically of a *head, middle piece,* and *tail* (Figs. 12-12 and 12-13). The nucleus forms almost all the mass of the sperm head, which may be flattened, greatly elongated, hooked, or twisted into a spiral. The morphology of the head is so diverse and characteristic that many animals can be identified from the structure of the sperm head alone (see Fig. 12-13). Under the electron microscope, the mature sperm nucleus appears uniformly dense and in most species shows no evidence of chromatin fibers (Fig. 12-14). A nuclear envelope is often visible surrounding the nucleus. Usually, no pore complexes occur in the envelope in mature sperm cells.

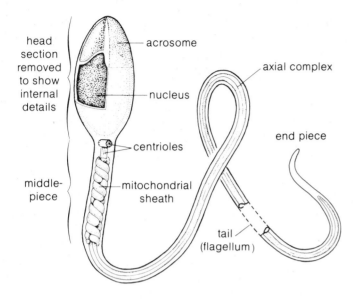

Figure 12-12 Sperm cell structure (see text).

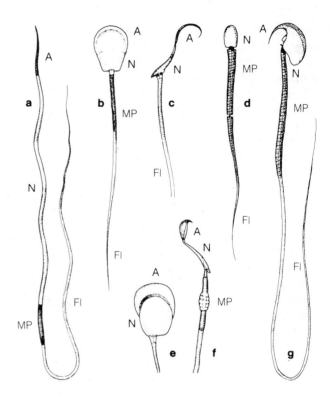

Figure 12-13 Some representative mammalian sperm cells, showing the extensive variations in head morphology. *A*, acrosome; *Fl*, flagellum; *MP*, middle piece; *N*, nucleus. (**a**) Spiny anteater; (**b**) deer; (**c**) squirrel; (**d**) bat; (**e**) and (**f**) guinea pig; (**g**) mouse.

Only a thin layer of cytoplasm remains around the nucleus of the sperm head. The usual cytoplasmic structures, such as mitochondria, ER, Golgi complex, and ribosomes, are absent from this layer. At the anterior end of the sperm head, often covering the anterior end of the nucleus like a cap, is the *acrosome,* a specialized secretion vesicle enclosed by a single, continuous membrane (see Fig. 12-14). This vesicle, which lies between the plasma membrane and the nucleus, contains enzymes and other factors that break down the external surface coats of mature eggs and promote penetration of the sperm cell to the egg plasma membrane. The acrosome forms from vesicles derived from the Golgi complex during sperm maturation.

The plasma membrane surrounds the nucleus and acrosome, and extends toward the posterior as a covering around the middle piece and tail. The middle piece (see Figs. 12-12 and 12-14) consists primarily of a sheath of mitochondria surrounding the tail axis just behind the head. Other elements of the middle piece are the centrioles and a variety of structures associated with attachment of the tail and middle piece to the base of the nucleus. Two centrioles are usually present. One, called the *proximal centriole,* lies close to the base of the nucleus just outside the nuclear envelope. The second, the *distal centriole,* is located behind the first and gives rise to the 9 + 2 microtubule complex of the tail (see Fig. 8-21).

The tail in most species is a typical flagellum and contains little more than the 9 + 2 complex of microtubules (see Chap. 8), enclosed by a sparse sheath of cytoplasm and the plasma membrane. In a few species, including mammals, additional elements called *accessory fibers* (Fig. 12-15) may also be present in the cytoplasm surrounding the axial complex. The function of these extra fibers is unknown. However, they are believed either to be motile structures that generate additional force for moving the sperm tail or structures that provide elasticity to the tail.

The basic arrangement of the sperm cell into head, middle piece, and tail is typical of almost all animals. Sperm without flagella occur in a few species of nematodes, some crustaceans, a few arachnids (mites), and some myriapods (centipedes and millipedes). Usually, absence of the tail in these sperm is correlated with the absence of the organelles of the middle piece, including mitochondria and centrioles. Multiple sperm tails are found very rarely and appear normally in only a few animal species such as the flatworms. Other less extensive

Figure 12-14 A fully condensed nucleus (*Nu*) in a sperm cell of the marmoset. The acrosome (*A*) is also visible as a caplike structure fitting over the anterior end of the nucleus. *M*, middle piece. × 14,000. Courtesy of J. B. Rattner, B. R. Brinkley, and Academic Press, Inc., from *Journal of Ultrastructure Research* 32 (1970): 316.

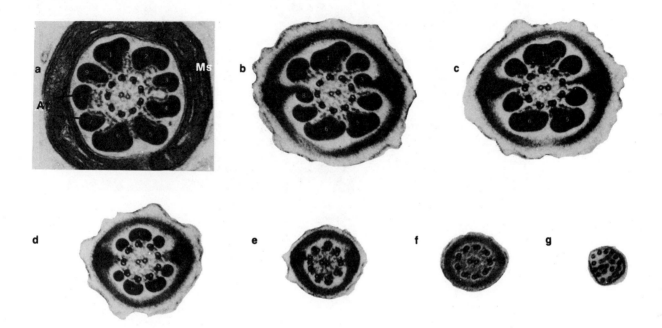

Figure 12-15 Accessory fibers (*A*, numbered from 1 to 9) in a guinea pig sperm flagellum. (**a**) to (**g**), sections made at successive points along the flagellum, *Ms*, mitochondrial sheath of middle piece. × 50,000. From *The Cell*, by D. W. Fawcett, 1966. Courtesy of D. W. Fawcett and W. B. Saunders Company.

modifications of the basic arrangement into head, middle piece, and tail are also found in a number of species among the barnacles, myriapods, and a few insects and annelids.

The organelles of mature sperm cells form and take up their positions during a series of regular rearrangements that begin at the close of meiosis. The centrioles move to the side of the nucleus that will form the rear of the sperm head. The mitochondria of the spermatid move to the same position. The Golgi complex forms a series of vesicles that fuse to form a single large secretion vesicle that later in spermiogenesis develops into the acrosome.

After secretion of the vesicle, the Golgi complex moves to the posterior end of the developing sperm cell and is eliminated near the completion of spermiogenesis along with most of the remaining cytoplasm of the spermatid.

During this time, the nucleus undergoes a progressive elongation and reduction in volume. Within the nucleus, the chromatin fibrils coalesce into platelike or granulelike aggregations that eventually fuse into a uniformly dense mass. At the close of spermiogenesis, the condensed sperm nucleus may be reduced to as little as 0.5 percent of the volume of the original cell entering spermiogenesis. These morphological changes in the nucleus are almost

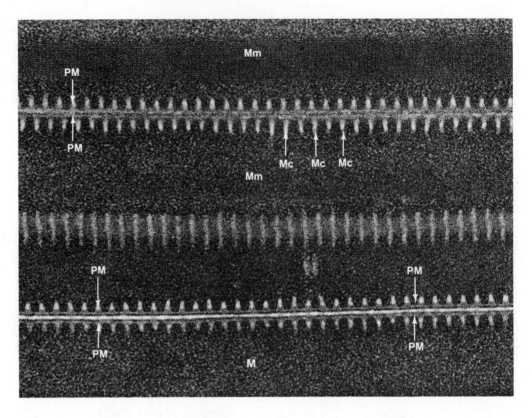

Figure 12-16 Mitochondrial derivatives in the middle pieces of three closely aligned sperm tails of the checkered beetle (Cleridae). *PM*, plasma membranes of the sperm tails; *Mc*, mitochondrial cristae; *Mm*, mitochondrial matrix. × 113,000. Courtesy of D. M. Phillips, from *Journal of Cell Biology* 44 (1970): 243, by copyright permission of The Rockefeller University Press.

always accompanied by replacement of the histone and nonhistone chromosomal proteins by more basic histones or by small, even more basic proteins called *protamines.* The end result is conversion of the nucleus into a dense, semicrystalline mass.

The distal centriole gives rise to the 9 + 2 microtubule complex of the tail. As the complex elongates, it carries with it an extension of the plasma membrane, eventually forming the tail (for details of flagellar generation, see Supplement 8-1). As these developments take place, the mitochondria take up their position near the base of the nucleus. Once reaching this position, the mitochondria extend outward in a sheath surrounding the base of the axial complex to form the middle piece. Within the sheath, the inner mitochondrial membranes may disappear or become modified, or fill with a dense, crystal-like substance. The result in these cases, termed the *mitochondrial derivative* (Fig. 12-16), may bear little resemblance to typical mitochondria. Although the significance of many of the details

of the final morphology of sperm cells is unknown, the developmental changes obviously provide motility to the mature sperm and package the nuclear contents into a form that is transportable with a minimum expenditure of energy.

Oogenesis Egg cells in animals, in contrast to sperm, are nonmotile. The structural changes occurring in oogenesis are concerned instead with the storage of information and nutrient substances required for the early growth of the embryo after fertilization and development of surface coats that protect the egg and participate in fertilization.

In animals, differentiation of the oocyte into the egg (Fig. 12-17 shows a developing mammalian egg) begins much earlier in the meiotic process than spermiogenesis. The pattern of maturation differs fundamentally from male gametogenesis in that only one of the four haploid products of meiosis becomes a functional gamete. Synthesis and storage of a variety of substances during the transcrip-

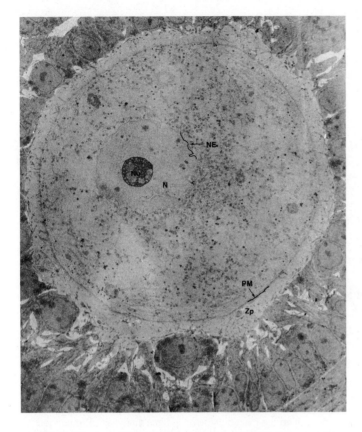

Figure 12-17 A developing mouse oocyte. *N*, nucleus; *Nu*, nucleolus; *NE*, nuclear envelope; *PM*, plasma membrane; *Zp*, zona pellucida. Immediately surrounding the oocyte is a layer of ovary cells. × 1,000. From *An atlas of ultrastructure* by J. A. G. Rhodin, 1963. Courtesy of J. A. G. Rhodin and W. B. Saunders Company.

cluding mitochondria, endoplasmic reticulum, and the Golgi complex.

In addition to the internal synthesis of yolk substances, enlargement of the egg cytoplasm may also result from the entry of proteins, lipids, and polysaccharides from outside the oocyte. Some of these materials are absorbed through the oocyte plasma membrane from the surrounding extracellular fluids. Other finished materials, including intact yolk bodies, Golgi complexes, and mitochondria may be transported into the egg cytoplasm from surrounding cells through direct cytoplasmic bridges. The final outcome of this prophase synthesis is growth of the oocyte to the largest dimensions of any cells. In some birds and reptiles oocytes reach diameters of several centimeters at maturity.

The growth period of the egg may extend from a few days to years, depending on the species. As the oocyte attains its final dimensions, synthesis ceases and the rate of metabolism falls almost to zero. The eggs of some animals are released at this stage and remain arrested in prophase I until fertilization. In others, the eggs are shed at metaphase I and remain at this stage until fertilization. Other eggs, including those of most mammals, are blocked at metaphase II of meiosis. In only a few species, such as the sea urchin, is meiosis completed before shedding of the eggs. In all cases, the mature unfertilized egg is an inert, quiescent cell, in which almost all cellular mechanisms have stopped.

No matter when meiosis arrests and resumes after fertilization, the pattern of divisions after metaphase I is essentially the same in all animals. The nucleus breaks down and the chromosomes and spindle take up an eccentric position near the plasma membrane. At anaphase and telophase I the accompanying division of the cytoplasm is unequal, with the result that one of the two prophase II nuclei is separated into a small, abortive cell known as a *polar body* (Figs. 12-18 and 12-19). The second division of cytoplasm at anaphase and telophase II is also unequal, producing a second polar body. During the second division, the first polar body in different species either disintegrates, remains quiescent, or also divides. The result, in various animals, is a single large cell containing the egg nucleus and the majority of the cytoplasm, with one, two, or three nonfunctional polar bodies attached at one side. After the final division, the remaining single egg nucleus, now called the *pronucleus*, sinks to the interior of the egg.

One or more surface coats may be secreted outside the egg plasma membrane, either by the egg itself or by cells surrounding the egg. This external covering, called the *vitelline coat* (or in mammals, the *zona pellucida*; see Fig. 12-17), differs in structure from species to species. The vi-

tion stage are reflected in a great increase in the activity and growth of the egg cytoplasm. Large numbers of ribosomes are assembled. These are for a time intensively engaged in the synthesis of proteins. Lipids and polysaccharides are also made in large quantities in the oocyte cytoplasm. Most of the protein and lipid is stored in membrane-bound structures called *yolk bodies* or *yolk platelets*. The cytoplasm of the eggs of all animals except mammals becomes closely packed with these bodies, which serve as a source of raw materials for growth of the embryo after fertilization. The activity of the synthetic machinery of the cytoplasm during the transcription stage is marked by extensive growth of the various cytoplasmic organelles, in-

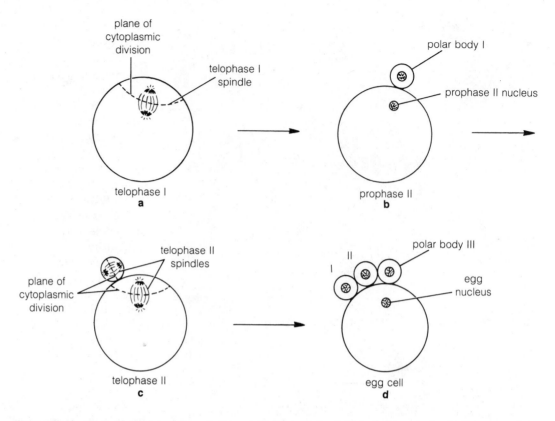

Figure 12-18 Unequal division of the cytoplasm during the mitotic divisions in female animals. The unequal division produces one large egg cell and three small, nonfunctional cells called polar bodies. In some species the first polar body does not divide during the second meiotic division. This variation in the sequence produces an egg cell with two polar bodies instead of three.

telline coat, which contains polysaccharides and glycoproteins as major constituents, probably functions to protect the egg from mechanical and chemical injury. The coat may persist after fertilization and remain until considerable development of the embryo has taken place.

Other envelopes or coats may be found outside the vitelline coat. These layers, secreted by cells surrounding the oocyte or by organs of the oviduct, also take on a great variety of forms in different species. In insects and some fish eggs an impervious layer called the *chorion* is secreted around the egg by surrounding cells. This layer, which forms the outside coat on the mature egg, contains a minute passage, the *micropyle*, through which the sperm passes during fertilization. The albumen and shell of avian and reptilian eggs are analogous layers secreted around the egg as it passes through the oviduct. In amphibians, molluscs, and echinoderms, this layer takes the form of a gelatinous mass that surrounds the freshly laid eggs.

The mature egg, with cytoplasm packed with yolk, lipid droplets, ribosomes, and inactive RNA, and arrested at various stages of meiosis, is shed from the ovary and awaits penetration of the sperm. The meeting of sperm and egg triggers a sequence of events beginning with penetration of the egg coats by the sperm and culminating in fusion of the male and female pronuclei.

Gametogenesis in Plants

Meiosis in plants is *intermediate* (see Fig. 12-11) and gives rise in the sporophyte to *spores*, which germinate and after mitotic division produce the gametophyte. After one or more mitotic divisions, cells of the gametophytes differentiate and give rise to male and female gametes.

Motile male gametes resembling animal sperm cells, called *antherozoids* or *spermatozoids*, occur in some algae

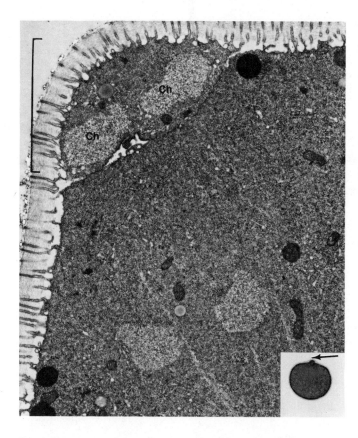

Figure 12-19 An egg with its first polar body, from the surf clam *Spisula.* A light micrograph of the egg and polar body (arrow) is shown in the inset. The main figure shows an electron micrograph of the polar body (bracket) and the underlying egg cytoplasm at higher magnification. The chromosomal material (*Ch*) in the polar body is the result of the first meiotic division. Just below the polar body, in the egg cytoplasm, the second meiotic division can be seen in progress. × 10,000; inset × 200. Courtesy of F. J. Longo, E. Anderson, and Academic Press, Inc., from *Journal of Ultrastructure Research* 33 (1970): 495.

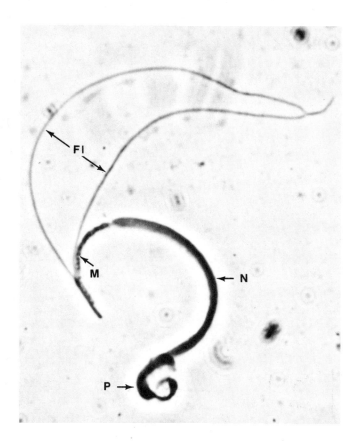

Figure 12-20 A light micrograph of a mature sperm cell of the plant *Nitella.* Fl, flagella; M, mitochondria; N, nucleus; P, plastids. × 2,700. Courtesy of F. R. Turner and The Rockefeller University Press, from *Journal of Cell Biology* 37 (1968): 370, by copyright permission of The Rockefeller University Press.

and in mosses and liverworts (Figs. 12-20 and 12-21). Motile sperm cells are also found in ferns and primitive gymnosperms such as *Zamia* and *Ginkgo.* These plant sperm are regularly multiflagellate. The flagella, usually two in number, are located at the anterior end of the cell and trail back along the side of the nucleus (as in Figs. 12-20 and 12-21). In some plants the flagella are more numerous, reaching levels of a hundred or more in some species, as in the fern *Marsilea.* During plant spermiogenesis, the flagella arise from centrioles that are similar to their animal counterparts.

The nucleus elongates and condenses during the development of motile plant sperm, usually extending into a spiral in mature sperm. During these changes, which proceed in a series of morphological steps similar to those occurring in animal spermiogenesis, the histone and nonhistone chromosomal proteins in many cases are replaced by small, protaminelike proteins as they are in animals.

In higher plants, including the angiosperms and all but the most primitive gymnosperms, the male gametes are nonmotile and arise from the gametophyte in the following pattern. Meiosis in the diploid sporophyte gives rise to four haploid products called *microspores.* In the flowering plants, these meiotic divisions occur in the anthers of flowers. Each of the microspores subsequently undergoes one mitotic division, producing the two nuclei of the pollen grain. The pollen grain (Fig. 12-22) is the male gametophyte of the higher plants. Within the pollen grain, one of the two nuclei forms the *generative nucleus* of

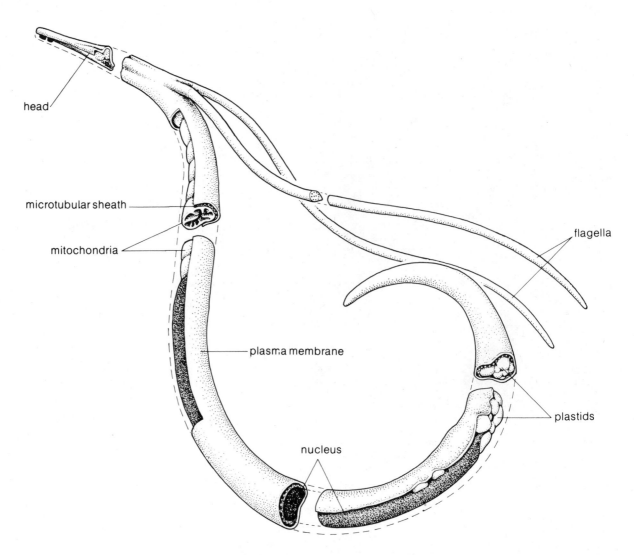

head

microtubular sheath

mitochondria

plasma membrane

nucleus

flagella

plastids

Figure 12-21 The ultrastructural features of the *Nitella* spermatozoid. Redrawn from an original courtesy of F. R. Turner, from *Journal of Cell Biology* 37 (1968): 370, by copyright permission of The Rockefeller University Press.

the male gametophyte. The second forms the *vegetative nucleus*. The two-celled gametophyte subsequently develops the hard, impermeable exterior coats characteristic of the pollen grain. The mature pollen grain enters a state of metabolic arrest, in which it remains until it comes in contact with the female parts of a flower of the same species. At this time, the pollen grain germinates, forming a pollen tube. Growth of the pollen tube is regulated by the vegetative nucleus. Within the pollen tube, the generative nucleus divides, forming two *sperm nuclei*. One of these acts as the male gamete in fertilization of the egg cell.

In lower plants, the egg cell is less complex than in animals, owing to the fact that no large quantities of yolk or other nutrients are stored within the egg cytoplasm. This is related partly to the dependence of the egg on the gametophyte for nourishment and partly to the ability of the egg to form sugars and other substances by photosynthesis after fertilization. The history of egg development in the lower plants is also simpler than in animals because the egg differentiates directly from a haploid cell following a mitotic rather than meiotic division. Thus development of an egg in the lower plants proceeds without conspicu-

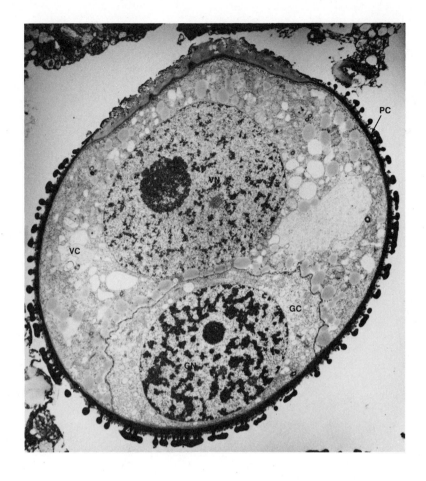

Figure 12-22 A mature pollen grain of the lily as seen in cross section. VC, vegetative cell cytoplasm; VN, generative nucleus; GC, generative cell cytoplasm; GN, generative nucleus; PC, pollen coat. Courtesy of J. M. Sanger.

ous changes, producing a quiescent cell that in most species remains awaiting fertilization within the tissues of the parent gametophyte.

In angiosperms, the female gametophyte develops from a large spore called the *megaspore* within the ovary of the flower. The megaspore originates from meiosis and is the sole survivor of the four haploid products of the meiotic divisions (Fig. 12-23a–f). The haploid megaspore immediately germinates within the ovary to form the female gametophyte and undergoes three sequential mitotic divisions, producing a total of eight nuclei (Fig. 12-23f–k). One of these is contained in a cell that differentiates into the egg (Fig. 12-24). Five of the remaining nuclei become enclosed in cells that differentiate into tissues of the gametophyte. These cells function in various ways during fertilization in different species but do not contribute to development of the embryo following fertilization. The remaining two nuclei take up a position near the egg. These nuclei, which may fuse together or remain separate but closely aligned, are called the *polar nuclei*. During fertiliz-

ation, one of the two sperm nuclei originating from the pollen grain fuses with the egg nucleus to form the zygote, returning the chromosome number to the diploid level and giving rise to the sporophyte generation of the plant cycle. The second sperm nucleus contributed by the pollen grain fuses with the two polar nuclei, producing a triploid cell that divides mitotically to form a triploid nutritive tissue called the *endosperm*.

Fertilization thus completes the cycle of events initiated by the entry of cells into meiosis and gametogenesis. Meiosis reduces the chromosome number to the haploid level, generates new combinations of genetic information through the mechanisms of recombination and independent separation of maternal and paternal chromosomes, and produces the RNA and protein molecules required for gametogenesis and part or all of embryonic development. Gametogenesis converts the products of meiosis into mature egg and sperm cells that are capable of carrying out fertilization. Fertilization restores the chromosome number to the diploid level and initiates development of the

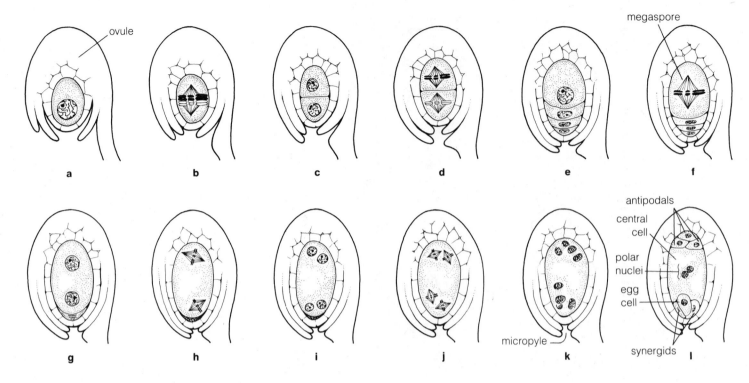

Figure 12-23 Development of the female gametophyte in a flowering plant (see text).

succeeding generation. The entire program of meiosis, gametogenesis, and fertilization generates a variety of new genetic combinations to provide the variability necessary for the survival of species in a changing environment.

Suggestions for Further Reading

Biggers, J. D. and A. W. Schuetz (eds.). 1972. *Oogenesis.* University Park Press: Baltimore.

Phillips, D. M. 1974. *Spermiogenesis.* Academic Press: New York.

Schultz-Schaeffer, J. 1980. *Cytogenetics: Plants, animals, humans.* Springer-Verlag: New York.

Stern, H. and Y. Hotta. 1973. Biochemical controls of meiosis. *Annual Review of Genetics* 7:37–66.

Westergaard, M. and D. von Wettstein. 1972. The synaptonemal complex. *Annual Review of Genetics* 6:71–110.

Wolfe, S. L. 1981. *Biology of the cell.* 2nd ed. Wadsworth: Belmont, Calif.

For Further Information

Crossing over and recombination, Chapter 13
Genes and alleles, Chapter 13
Kinetochore structure and activity, Chapter 11
Sex chromosomes and sex linkage, Chapter 13
Spindle formation and activity, Chapter 11

Questions

1. How do the results of mitosis and meiosis differ?

2. Why was meiosis suspected to occur before it was actually discovered?

3. What mechanisms produce genetic variability in meiosis?

4. Define haploid and diploid.

5. What is the difference between genes and alleles?

6. What happens during each of the stages of meiotic prophase I? What events mark the beginning and end of each stage?

7. What is the synaptonemal complex? When and where does it appear in meiosis?

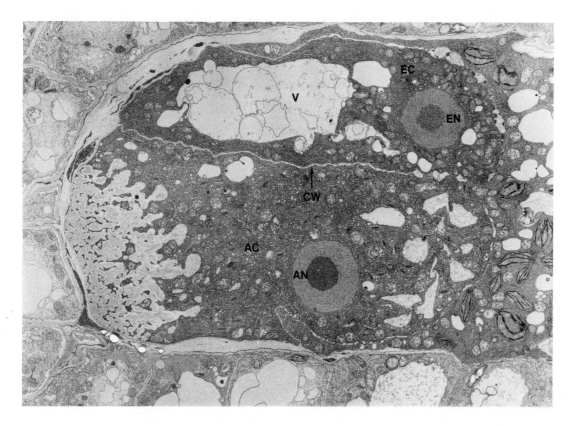

Figure 12-24 The egg cell and one of the accessory cells in the female gametophyte of the plant *Capsella* (shepherd's purse). *EN*, egg nucleus; *EC*, egg cytoplasm; *V*, vacuole; *CW*, cell wall; *AN*, accessory cell nucleus; *AC*, accessory cell cytoplasm. Courtesy of P. Schultz.

8. Define tetrad, bivalent, homolog, synapsis, and crossover.

9. Compare the pattern of kinetochore connections in the two meiotic divisions. What significance do these connections have for the outcome of meiosis?

10. What are maternal and paternal chromosomes? How are maternal and paternal chromosomes distributed in meiosis? At what point in the meiotic divisions do maternal and paternal chromosomes separate?

11. A chromosome pair has the alleles shown below at G_1 in a cell destined to enter meiosis. If a crossover occurs at the point shown by the arrow, what kinds of chromatids will appear in the products of meiosis?

12. What are paternal chromatids? Recombinant chromatids?

13. How did Taylor's experiment demonstrate that recombination takes place by an exchange of segments between homologous chromosomes?

14. How do generations alternate in the higher plants? What is the relationship of meiosis to this alternation? Is there an equivalent alternation of generations in animals?

15. The text states that meiosis in the fungi and protozoa is frequently induced by adverse environmental conditions. Does this provide any advantage for the survival of a population? Why?

16. How does the outcome of meiosis differ in male and female animals?

17. Trace the sex chromosomes through meiosis in an XX, XY, and an XO individual.

18. What are the three major results of meiotic cell division?

19. Outline the major changes that produce the head, middle piece, and tail of animal sperm cells. What organelles participate in the formation of each of these sperm regions? How do the

organelles that persist in each of these regions contribute to fertilization?

20. What major cellular developments produce a mature egg cell? How do each of these developments contribute to fertilization?

21. What are polar bodies? What is the significance of polar body formation in oogenesis?

22. Compare the developmental events that produce sperm cells in plants and animals. What structures occur in both animal and plant sperm cells? What major structures of animal sperm cells are absent in plant sperm?

23. Trace the development of male and female gametes in the flowering plants.

Supplement 12-1:
The Mechanism of Breakage and Exchange in Recombination

The Experimental Demonstration that Recombination Takes Place by Breakage and Exchange

J. Herbert Taylor's experiment showing that recombination takes place by physical breakage and exchange between homologous chromatids was carried out with meiotic cells of a grasshopper, *Romalea*. To label the chromosomes of the grasshopper, Taylor injected them with radioactive thymidine, a chemical that is incorporated in quantity only into DNA. The thymidine was injected only once, under conditions that assured that only one DNA replication would take place in the presence of the radioactive label. Taylor then followed the distribution of the label in cells undergoing meiosis.

Taylor was interested in obtaining cells with only one chromatid of each homologous chromosome labeled. This was possible, according to the experimental conditions used, in cells that picked up the label during the last premitotic interphase before meiosis. To understand why only one chromosome of each homologous pair will be labeled during the subsequent meiosis these cells must be followed through two cycles of semiconservative replication (Fig. 12-25a). As a result of semiconservative replication in the presence of labeled thymidine during the last premitotic interphase (Fig. 12-25b), each replicated DNA molecule will contain one "old" unlabeled nucleotide

chain and one "new" labeled chain. At metaphase of the mitotic division following this replication, both chromatids of each chromosome will be labeled (Fig. 12-25c).

These cells now enter the interphase before meiosis. First the cells undergo another DNA replication. Labeled thymidine has been washed from the tissues, and this replication occurs in unlabeled medium. The DNA molecules, each containing one labeled and one unlabeled nucleotide chain, unwind for replication to occur (Fig. 12-25d). The labeled nucleotide chain will serve as a template for replication; because no label is present in the medium the newly synthesized nucleotide chain copied from it will be unlabeled. These nucleotide chains, the old labeled chain and the new unlabeled chain, according to semiconservative replication, remain together to form one of the two chromatids of the replicated chromosome. This chromatid, at the subsequent metaphase, will show the presence of label. The other old nucleotide chain, the unlabeled one, will also serve as a template at this replication. This nucleotide chain, and the new chain copied from it during replication, are both unlabeled and will remain wound together to form the other chromatid of the chromosome. This chromatid will show no label at the subsequent metaphase. Therefore, all of the chromosomes at the subsequent meiotic metaphase will contain one labeled and one unlabeled chromatid (Fig. 12-25e).

At metaphase I of meiosis, these chromosomes will reveal whether recombination has occurred by breakage and exchange. If no breakage and exchange occurs, all chromatids will be either completely labeled or completely unlabeled at metaphase I (as in Fig. 12-25e). If any recombination occurs by physical breakage and exchange between labeled and unlabeled chromatids, some chromosomes should appear at metaphase with single chromatids segmented into labeled and unlabeled regions (Fig. 12-25f).

Taylor was able to find segmented label in many of the chromatids in meiotic cells of his experimental animals (Fig. 12-26). This finding demonstrated conclusively that physical breakage and exchange actually takes place between homologous chromatids during meiosis.

The Molecular Mechanism of Breakage and Exchange

While the actual molecular events underlying the breakage and exchange mechanism are unknown, it is believed to depend on complementarity in the DNA sequences of ho-

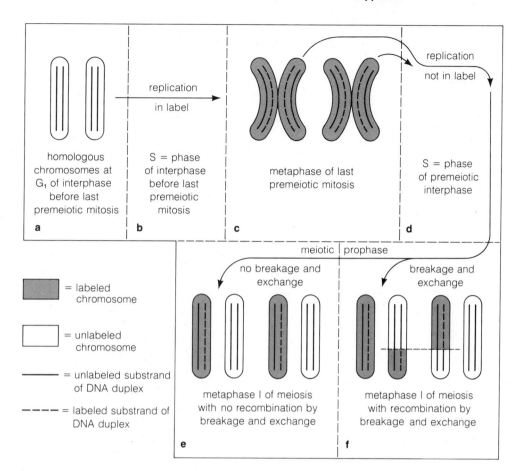

= labeled chromosome

= unlabeled chromosome

——— = unlabeled substrand of DNA duplex

- - - - = labeled substrand of DNA duplex

Figure 12-25 The Taylor experiment demonstrating that recombination occurs by breakage and exchange (see text).

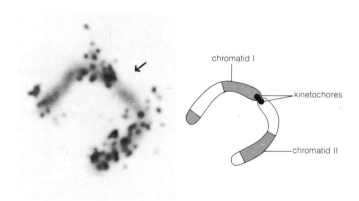

Figure 12-26 A chromosome recovered at metaphase II of meiosis in Taylor's *Romalea* experiment. The shaded regions in the tracing show the labeled segments. Reciprocal exchanges of labeled segments in this manner could take place only by physical breakage and exchange. Micrograph × 4,000. Courtesy of J. H. Taylor, from *Journal of Cell Biology* 25 (1965): 57, by copyright permission of The Rockefeller University Press.

mologous chromatids, and on enzymes that are also active in DNA replication (see Supplement 11-1). The proposed recombination mechanism proceeds as shown in Figure 12-27. The two chromatids involved in the recombination event approach and line up side by side, as they do at the pairing stage of meiotic prophase I. In the first step in recombination, a nick is made by a DNAase enzyme in one of the four nucleotide chains of the two chromatids taking part in the exchange (Fig. 12-27a). The DNA unwinds from the nick, unraveling a single nucleotide chain (Fig. 12-27b) that subsequently invades and rewinds with a complementary nucleotide chain in the opposite homologous chromatid (Fig. 12-27c). A chain with a 3′ end is a good candidate for the invading strand, since it can be extended by addition of nucleotides at its free end as it rewinds (remember that the known DNA polymerases synthesize DNA only in the 5′ → 3′ direction; see Information Box 11-1). Invasion may occur in a segment of the opposite homolog opened by a second nick, or in a segment temporarily opened by a local unwinding of the DNA. In

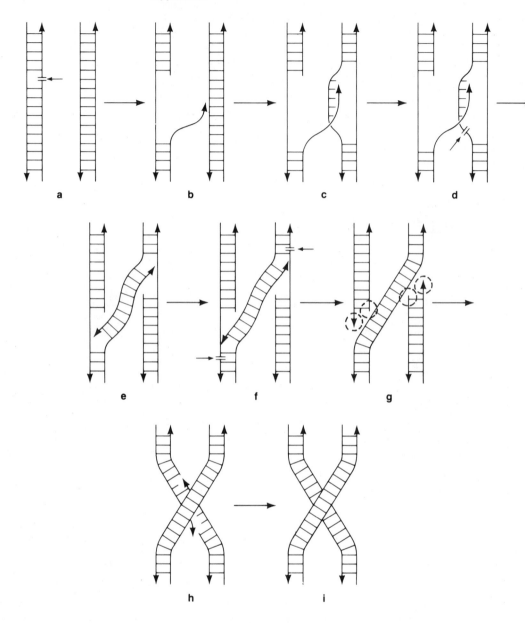

Figure 12-27 The possible steps in the molecular mechanism for recombination (see text).

either case, unwinding and rewinding are probably promoted by the same binding proteins, unwinding enzymes, and DNA gyrase enzymes active in DNA replication.

Once the invading helix is rewound, a second nick (arrow, Fig. 12-27d) frees it as a bridge extending between the homologous chromatids (Fig. 12-27e). Further nicks (Fig. 12-27f) then open the remaining chains. The nicks are then resealed with the open ends of the bridging helix,

creating a completely continuous DNA molecule (Fig. 12-27g; sealing is catalyzed by another enzyme active in DNA replication, DNA ligase—see Supplement 11-2).

These operations leave four free ends that have not as yet entered in the crossover (surrounded by dotted circles in Fig. 12-27g). These free ends then wind together to form the second bridging helix of the crossover (Fig. 12-27h). Any gaps left by the rewinding are filled in by DNA poly-

Table 12-1 Enzymes and Factors Possibly Active in Recombination

Enzyme or Factor	Possible Role in Recombination
DNAase	opens initial nicks in homologous DNA molecules
DNA binding proteins / unwinding enzyme / DNA gyrase	unwind single nucleotide chains after nicking; rewind invading nucleotide chain
DNA polymerase	extends invading nucleotide chain and fills gaps after formation of hybrid helix
DNA ligase	seals nicks after invasion and gap filling

merases, and the final remaining nicks are closed to form the second continuous DNA molecule of the crossover (Fig. 12-27i).

The molecular model for recombination fits the results of a variety of experiments, and the evidence is good that the proposed mechanism, or one very much like it, probably operates in breakage and exchange of chromosome segments during meiotic prophase I. All the enzymes required, including the DNAases opening the nicks, and the binding proteins, unwinding enzymes and DNA gyrases, DNA polymerases and ligases required in the model are known and well characterized (the proposed activity of these enzymes in recombination is summarized in Table 12-1). At least some of these enzymes have been detected in meiotic cells during the recombination stage.

The various enzymes active in recombination, as noted, are probably organized in the framework of the synaptonemal complex, the structure that appears between the paired homologs during the recombination stage. The molecular steps shown in Figure 12-27 probably occur in the middle of the complex, where they break, cross, and reseal DNA molecules that have moved from the homologous chromatids into this region.

Supplement 12-2:
Recombinant DNA and Cloning

Through a series of recently developed techniques, essentially any DNA sequence, including whole genes, can be introduced into living cells. The DNA is introduced into the recipient cells by linking it to DNA molecules normally forming a part of the host cell DNA, or to the DNA of an infecting virus. The linkage process is called "recombination" since the end result, in which DNA from two different sources is covalently linked into continuous, intact molecules, resembles meiotic recombination.

Once introduced into a host cell, the DNA is replicated along with the host cell DNA and passed on in cell division. This creates a line of descendants called a *clone*, all containing the DNA sequence introduced into the original cell.

The recombinant DNA technique depends on a group of enzymes called *restriction endonucleases* that occur normally in bacterial cells. These enzymes are capable of recognizing short, specific DNA sequences and breaking the DNA at or near the recognition sequence. Most useful of these are the enzymes that attack reverse-repeat sequences (see p. 200) in such a way that free ends containing complementary single nucleotide chains are produced. For example, one restriction endonuclease attacks the reverse-repeat sequence

$$
\begin{array}{ccccccc}
 & & & \downarrow & & & \\
---\text{G} & \text{A} & \text{A} & \text{T} & \text{T} & \text{C} & --- \\
| & | & | & | & | & | & \\
---\text{C} & \text{T} & \text{T} & \text{A} & \text{A} & \text{G} & --- \\
 & & & & & \uparrow &
\end{array}
$$

at the arrows, producing free ends with exposed, complementary single nucleotide chains:

$$
\begin{array}{cccccccccccc}
---\text{G} & & & & & & & \text{A} & \text{A} & \text{T} & \text{T} & \text{C} --- \\
| & | & | & | & | & + & | & | & | & | & | & \\
---\text{C} & \text{T} & \text{T} & \text{A} & \text{A} & & & & & & \text{G} --- &
\end{array}
$$

Since the enzyme attacks only this sequence, all of the products of the attack have complementary ends and any of the fragments produced can rewind at the ends and join together.

The sequences attacked by the endonuclease appear randomly, spaced at intervals separated by several thousand base pairs in DNA from almost any source. As a result, essentially any DNA molecule can be broken into a number of fragments, all with "sticky" ends that can pair and join with any other fragment produced by the same enzyme. To produce recombinant DNA, sticky fragments obtained from different DNA molecules by treatment with the enzyme are mixed and then exposed to DNA ligase (see p. 284) which seals any paired ends that wind together into covalently linked, hybrid DNA molecules.

The recombinant DNA molecules produced in this way are cloned in bacteria by taking advantage of *plasmids*.

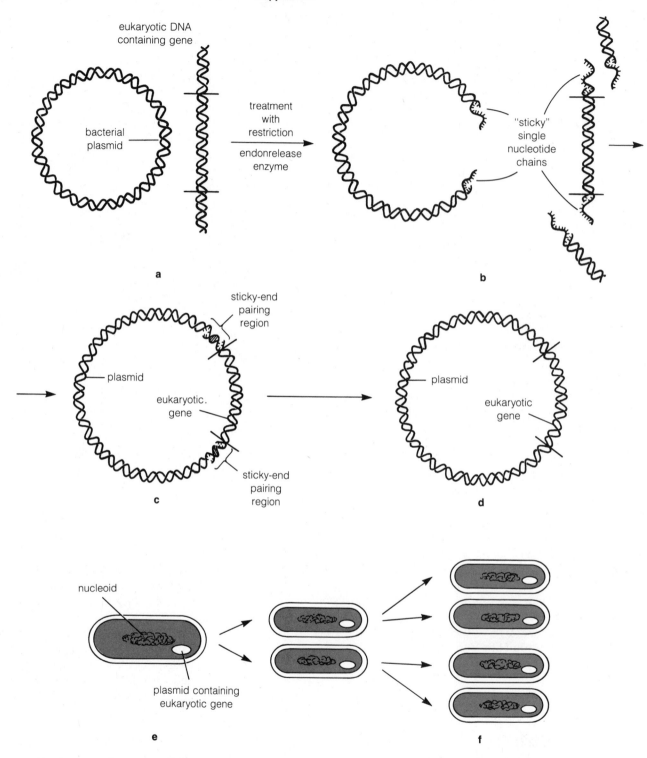

Figure 12-28 Cloning recombinant DNA produced by linking a gene to a bacterial plasmid (see text).

These are small, circular DNA molecules that exist in the cytoplasm of bacterial cells without direct connection to the major DNA circle of the bacterial nucleoid. Plasmids are replicated in bacterial cells and passed on in division as is the primary DNA circle of the bacterial nucleoid (see p. 274). To clone a sequence, both the DNA of interest and plasmids isolated from a bacterium (Fig. 12-28a) are digested with a restriction endonuclease (Fig. 12-28b). This produces sticky ends from both DNA molecules that can pair in different combinations (Fig. 12-28c). These combinations are subsequently sealed into closed circles by the DNA ligase enzyme (Fig. 12-28d). The recombinant plasmids containing the DNA of interest are then introduced into a bacterium such as *E. coli* (Fig. 12-28e; exposing the cells briefly to elevated temperatures—42°C—for a few minutes in the presence of a Ca^{2+} ions promotes uptake of the recombinant plasmids).

The recombinant plasmids are then replicated in the bacterial cytoplasm and passed on to daughter cells during division. After many rounds of replication and division, a large population of bacterial cells is produced, all containing the recombinant DNA plasmid with the DNA sequences of interest (Fig. 12-28f).

The same techniques have also been used to create clones in cells isolated from eukaryotes and grown in cultures, most frequently from the African green monkey. In this case, the restriction endonuclease enzymes are used to recombine the DNA molecules of interest with the DNA of a virus infecting the monkey cells. Green monkey cell cultures are then infected with the recombined viral DNA. In the infected cells, the inserted DNA sequences are replicated along with the viral DNA.

Cloning is not without risk, since there is danger that virulent, highly infective types may be created by the inclusion of recombinant DNA sequences in bacterial cells. However, the risk is reduced by growing the cloned cells in carefully monitored, closed environments from which escape is unlikely. The danger of escape is further minimized by the use of mutant bacteria with nutritional requirements so stringent that their chance of survival outside the laboratory is essentially zero.

The cloning technique has revolutionized DNA sequencing studies because it provides a way to increase the number of copies of a DNA sample to levels permitting biochemical analysis. Many genes have now been completely sequenced by this approach. It also provides a method for introducing genes coding for various proteins into bacterial cells, where the encoded proteins may be produced in quantity. The protein insulin, required for the survival of persons suffering from diabetes, has already been produced in bacterial cells in this way, and clones producing other proteins of human interest and benefit are presently being developed.

It is also possible that recombinant DNA techniques may eventually be developed to allow introduction of DNA sequences containing the normal forms of genes into persons suffering from hereditary diseases. This potential recombinant DNA technique, called *genetic engineering*, shows promise of revolutionizing the treatment of genetically based human diseases if workable techniques can be developed for its application.

Suggestions for Further Reading

Gilbert, W. and L. Villa-Komaroff. 1950. Useful proteins from recombinant bacteria. *Scientific American* 242:74–94 (April).

Guarente, L., T. M. Roberts, and M. Ptashne. 1980. A technique for expressing eukaryotic genes in bacteria. *Science* 209:1428–30.

13

The Genetic Consequences of Meiosis

The events of meiosis precisely recombine alleles and separate the maternal and paternal chromosomes of homologous pairs in diploid species. Following gametogenesis, the haploid egg and sperm cells resulting from meiosis unite randomly by twos to reestablish the diploid chromosome number and start development of the next generation. Since the precision of meiotic recombination and division underlies these events, the outcome of meiosis and fertilization and their effects on heredity can be followed and predicted mathematically.

The mathematical techniques used to trace the separation and distribution of genes and alleles in gametes and predicting the outcome of fertilization were developed long before anything was understood about meiosis or DNA. The groundwork for these techniques, which are still used today to follow and predict the outcome of genetic crosses, was laid down in the 1860s by an Austrian monk, Gregor Mendel. Meiosis was not fully worked out until just after 1900, and the establishment of DNA and nucleotide sequences as the physical basis for heredity took until the 1950s. Mendel's research and conclusions were so advanced for his time that many years were to pass before his findings were fully appreciated and understood.

The easiest way to understand the relationship of the events of meiosis and fertilization to heredity, and the mathematical analysis of this relationship, is to follow Mendel's original experiments. By tracing the distribution of parental traits among offspring, Mendel discovered the existence of genes and alleles, how they are distributed in the formation of gametes, and how they interact to produce the visible traits of offspring.

Mendel's Experiments

Mendel chose garden peas for his research because they could be grown easily without elaborate equipment. Many varieties of peas were known at Mendel's time, including some in which hereditary traits bred true. That is, the traits were passed on without change from one generation to the next. Mendel eventually used a total of seven varieties of peas in his experiments.

The Single-Trait Crosses

Among the inherited traits that Mendel selected for study were two that affected the shape of seeds. One variety of peas produced pods with round, smooth seeds, and another pods with wrinkled seeds. If these lines were self-pollinated, the wrinkled or round seed traits would breed true. For example, if crossed among themselves, the plants of the variety with wrinkled seeds always produced offspring with wrinkled seeds.

To test the pattern of inheritance in crosses between these two types of plants, Mendel took pollen from a plant that normally produced round seeds and placed it on the flowers of a variety that produced wrinkled seeds. The reverse cross was also carried out, in which pollen from wrinkled-seed plants was placed on the flowers of plants normally producing round seeds. Mendel noted that all the seeds produced in the pollinated plants were round and smooth, as if the trait for wrinkled seeds had disappeared or was masked. These seeds, which represent the first generation of offspring of the cross, are called the F_1 generation (F = filial, from *filius* = son). The seeds were then planted, and grown to maturity. Some of the F_1 plants were then crossed among themselves. In the seeds resulting from this cross, which formed the F_2 generation, Mendel counted 5474 of the round variety and 1850 of the wrinkled variety, in an approximate ratio of three round to one wrinkled or about 75 percent round and 25 percent wrinkled seeds. Other pairs of characteristics, seven in all, were tested in the same way (Table 13-1). In all cases, a uniform F_1 generation was obtained in which only one of the two characteristics of the pair appeared. In the second generation (the F_2 generation), both of the characteristics appeared again, with the characteristic present in the F_1 generation always representing about 75 percent of the offspring in the F_2 generation. When analyzed quantitatively, the pattern in which these traits were transmitted from parents to offspring proved to be totally unlike an even blending of parental characteristics, which might be the expected outcome. Instead, one trait of each pair was completely absent in the F_1 generation, and then reappeared in the F_2 generation in a definite proportion among the offspring.

Mendel's Hypotheses Explaining the Single-Trait Crosses

Mendel realized that the results of the cross could be explained if he assumed that the adult plants carry a *pair* of factors governing the inheritance of each trait. He proposed that these factor pairs separate as gametes are formed, so that each gamete cell receives only one of the pair. As the maternal and paternal gametes fuse, the resulting diploid nucleus (called the *zygote* nucleus) receives one factor for the trait from the male gamete and one factor for the same trait from the female gamete, reuniting the pair. Because a gamete receives only one member of each pair of factors governing a trait, Mendel termed this phenomenon the *segregation* of the factor pairs.

The observation that one of the traits, such as wrinkled seeds, disappeared in the F_1 generation and reappeared in the F_2 was interpreted by Mendel to mean that, although undetected in the F_1, the factor was still present and merely masked in some way by the "stronger" factor. Mendel called this effect *dominance* and assumed that when the two different factors occurred together in an organism only the factor for roundness in seeds had effect. The factor for wrinkled seeds was assumed to be *recessive*, and to be expressed only when both members of the pair in an individual were the factor for wrinkling.

With no knowledge of DNA or meiosis, Mendel had discovered genes and alleles, their manner of distribution to gametes, and their reunion in fertilization. His "factors" are genes; the alternate forms of these factors are alleles. The occurrence of genes in pairs reflects the fact that chromosomes occur in pairs in diploid individuals: each of the chromosomes of a pair carries one copy of the gene. Segregation of the gene pairs reflects the two divisions of meiosis, in which the chromosome pairs carrying the genes are separated and placed singly in the haploid gametes. The reunion and interaction of the gene pairs in

Table 13-1 Results of Mendel's Crosses Using Seven Different Characteristics in Peas

Characteristic	F_1	F_2
Round × wrinkled seeds	All round	5474 round; 1850 wrinkled
Yellow × green seeds	All yellow	6022 yellow; 2001 green
Red × white flowers	All red	705 red; 224 white
Green × yellow pods	All green	428 green; 152 yellow
Large × small pods	All large	882 large; 299 small
Tall × short plants	All tall	787 tall; 277 short
Flowers along stem × flowers at ends of stems	All flowers along stems	651 along stems; 207 at ends of stems

Figure 13-1 The result of Mendel's cross between pea plants with round seeds and wrinkled seeds in the F₁ generation (see text).

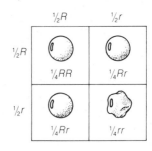

Figure 13-2 The results of Mendel's cross between plants with round and wrinkled seeds in the F₂ generation (see text).

offspring results from the fusion of egg and sperm nuclei in fertilization, where homologous chromosomes are brought together again into pairs.

Mendel's assumptions about genes[1] and their segregation in gametes and reunion in fertilization allowed him to explain the results of his crosses (Fig. 13-1). He assumed that both members of the gene pair controlling seed shape in the original parent plant with round seeds were of the same form, or allele (remember that alleles are alternate forms of a gene that differ in DNA sequence; see p. 274). In the round-seed parent, both of the alleles of the gene for seed shape caused seeds to develop the round form. This allele was given the symbol *R*, so in this parent, the genetic constitution of the pair was *RR*. An individual that carries a pair of alleles of the same type is now called a *homozygote*. In the production of gametes the two members of the pair separate and, because only *R* alleles are present in the pair, all gametes receive one *R* allele (the left heading in Fig. 13-1). In the original parent with wrinkled seeds, both alleles of the gene governing seed shape are also assumed to be the same. This allele was given the symbol *r*, and the wrinkled parent, also a homozygote, has the genetic constitution *rr*. These *rr* alleles are also separated and distributed singly to each gamete. All gametes from this parent thus receive one *r* allele (the top heading in Fig. 13-1). When these gametes unite to produce the F₁ generation, they combine at random. However, the results of the combination are always the same in the F₁. All zygotes produced by fertilization between gametes from the round and wrinkled plants will produce the combination *Rr* (shown in the square in Fig. 13-1). An individual of this type, in which the two alleles of a gene pair are different, is now called a *heterozygote*. Because *R* is dominant over *r*, all the offspring will have round seeds even though the

allele for wrinkled seeds is present in all individuals of the F₁ generation.

These assumptions also explain the results of Mendel's F₁ crosses. According to his explanation, all F₁ plants produce two kinds of gametes. Because the *Rr* gene pairs separate in the production of gametes, one-half receive the *R* factor and one-half the *r* factor. (These gametes are entered in both the horizontal and vertical headings in Fig. 13-2; the squares show the resulting combinations.) Combining two gametes that carry the *R* factor produces an *RR* F₂ plant; combining *R* from one parent and *r* from the other produces an *Rr* F₂ plant. Combining *r* from both F₁ parents will produce an *rr* F₂ plant. The *RR* and *Rr* plants in the F₂ generation will have round seeds; however, the *rr* offspring will have wrinkled seeds. Thus Mendel's assumptions explain the reappearance of the wrinkled trait in the F₂ generation. These assumptions were that (1) the genes governing a trait occur in pairs, (2) one allele is dominant over the other if different alleles are present in a gene pair, and (3) the alleles of a gene pair are separated and delivered singly to gametes during their formation.

Mendel's assumptions also successfully explained the *proportions* of offspring obtained in the F₂ generation. In the crosses between round and wrinkled-seed plants, according to Mendel's principle of segregation of alleles in gamete formation, one-half of the gametes of the F₁ generation will contain the *R* factor and one-half will contain the *r* factor (see Fig. 13-2). To produce an *RR* zygote, two *R* gametes must be selected from among the gametes and combined. Similarly, to produce an *rr* zygote, two *r* gametes must be selected from the gametes and combined. The chance of accomplishing this is predicted by a simple law of probability, which states that the chance of two independent choices occurring together is determined by multiplying their individual probabilities. Since *R* and *r*

---

[1]Mendel called genes and alleles *factors*; the modern terminology will be used here.

gametes occur in equal numbers, the chance of selecting either an *R* or *r* gamete from the gamete pool is one out of two or ½. The chance of selecting two *R* gametes in a row to produce an *RR* zygote is then equal to the chance of selecting one *R* gamete (½) times the chance of selecting a second *R* gamete, or ½ × ½ = ¼. Therefore, one-fourth of the F_2 offspring of the F_1 cross *Rr* × *Rr* is expected to be *RR*, and to have round, smooth seeds. By the same line of reasoning, one-fourth of the F_2 offspring will be *rr*, with wrinkled seeds. If one-fourth of the F_2 is *RR* and one-fourth is *rr*, it follows that the remaining one-half of the F_2 generation must be *Rr*. Therefore, three-fourths of the F_2 offspring will be expected to have round, smooth seeds (¼*RR* + ½*Rr*) and one-fourth wrinkled seeds, in the ratio 3:1. This, as shown in Table 13-1, is the approximate ratio Mendel actually obtained in his cross; the same ratios hold for all other gene pairs used in his experiments. Thus Mendel's three assumptions about hereditary factors successfully explained both the types of offspring and their proportions in the F_1 and F_2 generations.

Figure 13-2 illustrates a simplified method for determining the expected proportions of *RR*, *Rr*, and *rr* plants in the F_2 generation. The chance of obtaining each type of gamete from the "male" parent is entered with the gametes on the horizontal heading to the diagram, and the gametes from the "female" parent are entered on the vertical heading to the diagram. The various combinations possible in the offspring can be obtained by combining the factors carried by the gametes in the squares opposite the headings. The frequency of the combination in each square is calculated by multiplying the frequencies of the gametes used to produce it. In the diagram, two squares, each with a frequency of one-fourth, contain the combination *Rr*. Added together, these two squares provide the total chance of obtaining the *Rr* class in the F_2 generation: ¼ + ¼ = ½. This device, called a *Punnett square* after R. C. Punnett, an early geneticist, is useful for determining the expected outcome of a genetic cross (for an algebraic method, see Supplement 13-1).

Mendel's Experimental Proofs of His Hypotheses

Mendel noted that the validity of his assumptions could be proved by observing how closely his assumptions could predict the outcome of a cross different from any tried so far. To carry out this test, Mendel crossed an F_1 plant with round seeds, assumed to have the gene pair *Rr*, with a

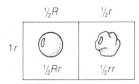

Figure 13-3 The results of Mendel's testcross proving the hypothesis (see text).

wrinkled-seed plant of the original parental type, assumed to have the pair *rr*. The outcome of this cross (*Rr* × *rr*) can be predicted by constructing a Punnett square. According to Mendel's assumptions, all of the gametes of the *rr* plant should contain a single *r* factor. Therefore, the probability of selecting an *r* gamete from this parent is 1. The gamete and its probability are entered in the left heading of the Punnett square in Figure 13-3. The *Rr* parent should produce two types of gametes, one-half containing the *R* factor and one-half containing the *r* factor. These are entered along the top heading of the square. Filling in the possible combinations in Figure 13-3 gives the two classes *Rr* and *rr*, both with the expected frequency of one-half. Therefore, the cross *Rr* × *rr* is expected to produce one-half round-seed plants and one-half wrinkled-seed plants. Mendel's actual results in one experiment testing this cross were 49 round-seed plants and 47 wrinkled-seed plants, closely approximating the expected 1:1 ratio. (A contemporary statistical technique for evaluating how closely actual results match expectations in experimental tests of this type, called the *chi square* method, is described in Supplement 13-1).

The cross used by Mendel to test his hypotheses, in which the F_1 generation is crossed with the homozygous recessive parent, is still used in genetics to determine whether individuals carrying a dominant trait are homozygotes or heterozygotes for that trait. This question often arises because when one allele is dominant the homozygote (*RR* in the case of round versus wrinkled seeds) is externally indistinguishable from the heterozygote (*Rr* in the case of round versus wrinkled seeds). In the cross used to determine whether an individual is a homozygous or heterozygous dominant, called a *testcross*, the individual in question is always crossed with a homozygous recessive. If the offspring of the cross are of two types, with one-half displaying the recessive trait and one-half the dominant trait, as in Mendel's testcross, then the individual in question must be a heterozygote (*Rr* in the case of round versus wrinkled seeds). If all of the offspring display the dominant trait, the individual in question must be a homozygote (RR in the case of round versus wrinkled seeds; the cross producing these results is $RR \times rr$).

Dihybrid Crosses

All of Mendel's initial crosses, outlined in Table 13-1, involved single gene pairs determining the same characteristics such as seed form. Crosses of this type are called *monohybrid* crosses. Mendel was also curious about the effects of crossing parental stocks with differences in two sets of hereditary characteristics. For a cross of this type, called a *dihybrid* cross, Mendel chose seed shape and a second characteristic, seed color (green versus yellow).

In Mendel's dihybrid cross, plants with round, yellow seeds were crossed with plants with wrinkled, green seeds. The F_1 generation produced from this cross consisted entirely of round, yellow-seed offspring. Mendel then crossed individuals of the F_1 generation among themselves and obtained, in the F_2 offspring of these crosses, 315 round, yellow-seed plants; 101 wrinkled, yellow-seed plants; 108 round, green-seed plants; and 32 wrinkled, green-seed plants. Mendel noted that in numbers these offspring approximate a 9:3:3:1 ratio.

The ratios that Mendel obtained were consistent with his previous findings if one additional assumption was made: that the two sets of factors separate independently during formation of gametes. That is, separation of the pair of genes controlling seed shape has no effect on the separation of the pair of factors for seed color when gametes are produced. Mendel termed this assumption *independent assortment*.

To understand the effect of independent assortment in the cross, assume that the parental types are $RRYY$ (round, yellow seeds) and $rryy$ (wrinkled, green seeds). The round, yellow parent will produce only RY gametes, and the wrinkled, green parent will produce only ry gametes. In the F_1 generation all the possible combinations of these gametes will produce only one class of offspring, $RrYy$. Assuming that the allele for yellow is dominant over green, all of the F_1 will be, as observed, round, yellow-seed plants.

If the genes for shape and color separate independently in gamete formation, each of the F_1 pea plants will produce four types of gametes (indicated on the horizontal and vertical headings in the Punnett square in Fig. 13-4). The R allele for seed shape can combine with either the Y

Figure 13-4 The results of crossing F_1 $RrYy$ plants (see text).

or y allele of the pair for color. Similarly, the r allele can be delivered to a gamete with either Y or y. This independent assortment of $RrYy$ alleles thus would be expected to produce four types of gametes, appearing with equal frequency: $\frac{1}{4}RY$, $\frac{1}{4}Ry$, $\frac{1}{4}rY$, and $\frac{1}{4}ry$. These gametes, and their expected frequencies, are entered for both parents in the headings to Figure 13-4.

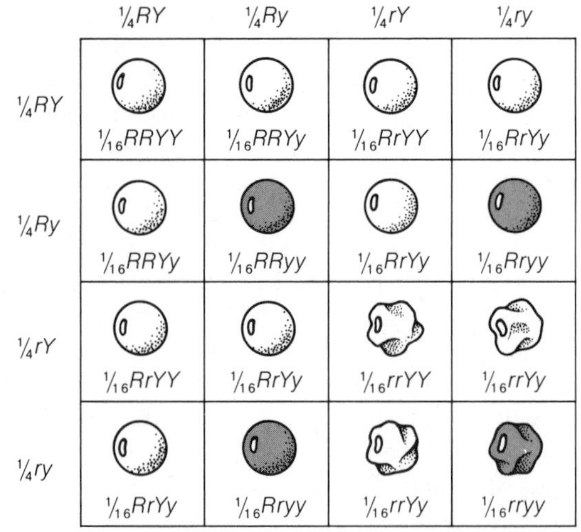

The results of all possible combinations of these gametes can be obtained by filling in the squares of the diagram. Sixteen classes are obtained, all with an expected frequency of 1 out of 16 offspring. Of these, the classes $RRYY$, $RRYy$, $RrYY$, and $RrYy$ will have the same appearance, with round, yellow seeds. These combinations occur in 9 out of the 16 squares in the diagram and therefore offspring with round, yellow seeds are expected in a total frequency of 9/16. The combinations producing wrinkled, yellow seeds, $rrYY$ and $rrYy$, are found in three squares for a total expected frequency of 3/16. Similarly, the combinations $RRyy$ and $Rryy$, yielding round, green-seed plants, occur in three of the squares for a total expected frequency of 3/16 of the offspring. Finally, the combination $rryy$, producing wrinkled, green-seed plants, occurs in only one square and thus is expected in 1/16 of the offspring. These expected frequencies of round yellow:wrinkled yellow:round green:wrinkled green, in a ratio of 9:3:3:1, closely approximate the actual results of 315:101:108:32 obtained by Mendel (a chi square test of this experiment is given in Supplement 13-1). Thus Mendel's hypotheses, with the added assumption that different gene pairs separate independently, explains the observed results. Mendel's tests of the added hypothesis completely confirmed it: the testcross *RrYy* × *rryy*, for example, gave the expected 1:1:1:1 ratio in the offspring. Once again, with no knowledge of the details of meiosis, Mendel had detected a part of the meiotic mechanism. The independent assortment of gene pairs reflects the fact that the distribution of the maternal and paternal chromosomes of one homologous pair during anaphase I of meiosis is completely independent of any other pair (see p. 300). Because one chromosome pair separates independently of other chromosome pairs during anaphase I, the genes carried on the different chromosome pairs will also separate independently (Fig. 13-5).

Mendel's Results: A Summary

Mendel's hypotheses concerning the inheritance of traits in sexually reproducing organisms were thus confirmed by his experiments: (1) the genes controlling a single characteristic occur in pairs in an individual; (2) the two genes of a pair are separated and occur singly in gametes (now called the *principle of segregation*); (3) one allele of the pair is dominant in its effects over the other if different alleles of the gene are present in the gene pair of a single individual (now called the *principle of dominance*); and (4) any one pair of genes separates independently of other gene pairs (now called the *principle of independent assortment*). Later research was to show that gene pairs separate independently from each other only if they are carried on different chromosome pairs (see The Discovery of Linkage and Recombination, below).

Mendel also discovered that the genetic makeup of an individual cannot be determined from its morphological appearance alone. Individuals may look the same and still be different genetically, as in the experiments demonstrating that *RR* and *Rr* plants, although different genetically, still have the same outward appearance. In modern terminology, the genetic makeup of an organism is termed the *genotype* to distinguish it from its outward appearance, called the *phenotype* (*phainein* = to show). According to this terminology, the two different genotypes *RR* and *Rr* produce an identical phenotype with round seeds.

Mendel's results were published in 1866. Unfortunately, they were not understood by the scientists of his day, and they lay unnoticed until the early 1900s, when three investigators—Hugo de Vries, Carl Correns, and Erich von Tschermak—independently repeated the same line of reasoning and experimentation used earlier by Mendel and came up with the same results. To their surprise, they found that they had been "scooped" 34 years earlier by an Austrian monk. Mendel died in 1884, sixteen years before his research and results were rediscovered and finally understood.

Genes, Chromosomes, and DNA

The Chromosome Theory of Heredity

By the time Mendel's results were rediscovered in 1900, the details of meiotic cell division had been worked out and it was not long until the similarities in behavior between chromosomes and genes were pointed out. In an historic paper published in 1903, Walter Sutton, then a graduate student at Columbia University, drew all the necessary parallels. In his paper, Sutton drew attention to the fact that chromosomes occur in pairs in sexually reproducing organisms, as do genes. The two chromosomes of each pair are separated and delivered singly to gametes, as are genes. Further, the separation of any one pair of chromosomes in gamete formation is independent of the separation of any other pair, as in the behavior of different pairs of genes in Mendel's dihybrid crosses. Finally one member of each chromosome pair is derived in fertilization from the male parent and one from the female parent, in an exact parallel to the two members of a gene pair. On the basis of this total coincidence in behavior, Sutton concluded that Mendel's factors, the genes, are carried on the chromosomes. In the intervening years Sutton's hypothesis has become established as the *chromosome theory of heredity*.

Genes, Alleles, and DNA

When DNA and the genetic code were discovered in the 1950s, it became apparent that the alleles studied by Mendel and his successors are different versions of the sequence of nucleotides coding for a given mRNA molecule (differences in tRNA or rRNA sequences rarely occur). Thus the dominant form *B* of a gene coding for blue flower color might have the sequence

   . . . ATAGAGATTGCATTAGACATAGGC . . .

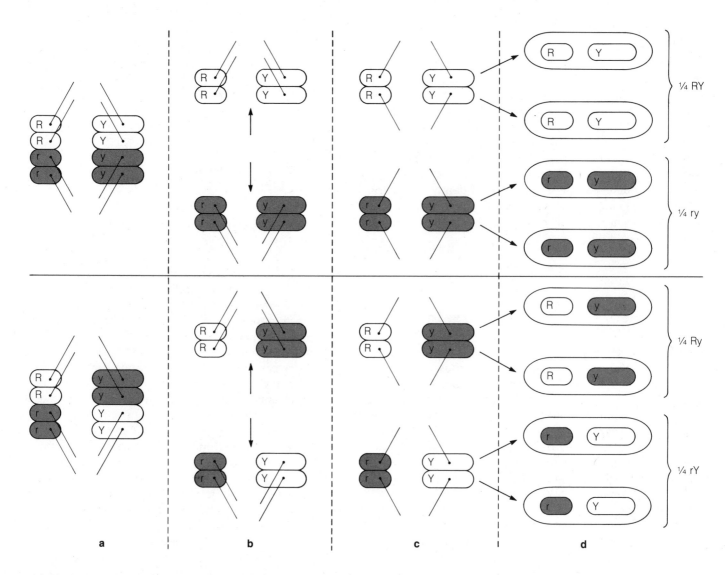

Figure 13-5 The independent assortment of alleles carried on different chromosome pairs in meiosis. (**a**) There are two possible arrangements of the two chromosome pairs on the spindle at metaphase I. These two possibilities are shown as the top and bottom of part (**a**). The chromosomes carrying the recessive alleles are shaded. These two possible arrangements would deliver the combinations of alleles shown in the top and bottom of (**b**) to the poles at anaphase I of meiosis. Note that among the four possibilities the *R*-bearing chromosome can be delivered to a pole with either the *Y*- or *y*-bearing chromosome of the other pair; the *r*-bearing chromosome can also be delivered to a pole with either the *Y*- or the *y*-bearing chromosome of the other pair. Thus the chromosomes of the short pair are delivered to the poles independently of the long pair. Formation of the spindles at metaphase II (**c**) and separation of chromatids into gamete nuclei at anaphase II and metaphase II (**d**) produce the four possible combinations *RY*, *ry*, *Ry*, and *rY*. These combinations, all expected in an equal frequency of ¼ if the total possibilities of the top and bottom rows are combined, are the same as the four possible combinations entered in the vertical and horizontal headings of Figure 13-4. The shaded chromosomes in Figure 13-5 could represent either the maternal or paternal chromosome in either the long or short pair.

in one region within its boundaries on the DNA (the brackets indicate the coding triplets). The recessive *b* allele of this gene might have the same sequence except for a base substitution at one point:

$$... \overline{ATA}\,\overline{GAG}\,\overline{ATT}\,\overline{GCA}\,\overline{TTA}\,\overline{GTC}\,\overline{ATA}\,\overline{GGC}\,...$$

These alternate forms of the gene are copied into mRNA molecules by the transcription mechanism:

$$... \overline{UAU}\,\overline{CUC}\,\overline{UAA}\,\overline{CGU}\,\overline{AAU}\,\overline{CUG}\,\overline{UAU}\,\overline{CCG}\,...\ (B\ form)$$

or

$$... \overline{UAU}\,\overline{CUC}\,\overline{UAA}\,\overline{CGU}\,\overline{AAU}\,\overline{CAG}\,\overline{UAU}\,\overline{CCG}\,...\ (b\ form)$$

The mRNA molecules then attach to ribosomes and direct the synthesis of protein molecules. The proteins produced will be identical except for the substitution of a single amino acid. The CUG codon of the mRNA transcribed from the *B* allele will direct the insertion of leucine at the corresponding point in the amino acid sequence of its protein; the CAG codon of the *b* allele mRNA will substitute glutamine at this point.

The different protein molecules produced by the *B* and *b* alleles of the gene might produce different phenotypes by the following pattern. Assume that the protein coded by the B gene is an enzyme catalyzing a step in the pathway leading to the production of blue pigment in the cells of flower petals:

$$\text{substance A}\ \xrightarrow{\text{enzyme 1}}\ \text{substance B}\ \xrightarrow{\text{enzyme 2}}$$

$$\text{substance C}\ \xrightarrow{\text{enzyme 3}}\ \text{blue pigment}$$

The protein coded by the B gene is enzyme 2 in the pathway. In the form coded by the *B* allele, the presence of leucine allows the amino acid chain of the enzyme to fold into a form in which it is fully active. All the steps in the pathway then run in sequence, and the blue flower pigment is produced. However, the presence of glutamine at the same position alters the folding conformation of the enzyme at the active site, completely destroying its ability to catalyze the conversion of substance B to substance C. As a result, the pathway is blocked and blue pigment is not produced.

BB individuals, with the *B* allele on both of the chromosomes of the pair carrying this gene, produce blue pigment and therefore have blue flowers. Although *Bb* individuals carry one of the recessive alleles, the *B* allele on the other chromosome of the pair codes for the active

form of the enzyme; this form is produced in sufficient quantity to keep the pathway going. Therefore, *Bb* individuals will also have blue flowers, and the *B* allele is said to be dominant to the *b* allele. Individuals with the *b* allele on both chromosomes of the pair (*bb* individuals) will produce no pigment; this condition is seen as white flowers in the plant. (Supplement 13-2 outlines the origins of the different alleles of genes in spontaneously occurring changes in DNA sequence called *mutations*).

The Genetic Effects of Recombination in Meiosis

Mendel studied seven pairs of factors, all of which separated independently in the formation of gametes. By chance or good management the total number of pairs Mendel chose for his experiments matches the number of chromosome pairs in his experimental organism: there are seven pairs of chromosomes in peas. If Mendel had extended his study to eight or more pairs, his hypothesis that different gene pairs always separate independently would have required modification.

This fact was first pointed out shortly after Sutton published his hypothesis of the equivalence of behavior between genes and chromosomes. Several people, including Sutton, drew attention to the fact that completely independent separation of gene pairs is not always expected if the number of gene pairs exceeds the number of chromosome pairs in an organism. Some of the multiple gene pairs would be carried on the same chromosome, and would therefore segregate together as a linked group during meiosis. Because it was obvious that there are at least hundreds or thousands of genes in an organism (there are probably tens of thousands or more), it seemed likely that many sets of genes would be found to be linked together on the same chromosomes and transmitted together rather than independently to offspring.

Further genetic research in the early part of this century led to the discovery of many examples of genetic *linkage* of this type. Work with linkage soon led to a second discovery: that the mechanisms of meiosis can rearrange linked alleles into new combinations. This meiotic mechanism was termed *recombination* when it was discovered in the early 1900s. It is now known that recombination occurs during meiotic prophase by a process of breakage and exchange of chromatid segments (for details, see Supplement 12-1).

The Discovery of Linkage and Recombination

Linkage was first detected in peas by the English scientists W. Bateson and R. C. Punnett (for whom the Punnett square is named) in 1906. These investigators found that two pairs of genes, one controlling flower color (purple versus red) and one the length of pollen grains (long versus short) were transmitted to offspring in the same combinations and rarely separated independently. Instead, if one parent had purple flowers and long pollen grains, these tended to be transmitted to offspring together, as if flower color and pollen length were controlled by a single gene. It was obvious that the two characteristics could not be controlled by a single gene, however, because other plants could be found in which purple flower color was combined with short pollen grains. In this case, the purple color and short pollen length were also usually transmitted together as a unit. Occasionally, offspring of a parent with the linked characteristics showed the opposite arrangement. For example, a parent with purple flowers and long pollen grains might sometimes produce offspring with the opposite arrangement—purple flowers and short pollen grains.

This pattern of inheritance was successfully explained in 1910 by Thomas H. Morgan and his coworkers at Columbia University. Morgan noted that linkage could be explained if the different gene pairs usually inherited together are carried on the same homologous pair of chromosomes. At the same time, he also offered the correct explanation for the occasional changes in the combinations of genes carried on the same chromosome. He proposed that this recombination arises through the *crossovers* or *chiasmata* observed between homologous chromosomes during prophase I of meiosis (see p. 298).

A classical experiment with corn plants by C. B. Hutchison of Cornell University in the 1920s illustrates how linkage and recombination are detected in genetic crosses. Hutchison followed two pairs of genes in his experiment. One pair, designated *S*, controlled the shape of kernels in the ears; these were either round (*S*) or shrunken (*s*). The second pair of genes, designated *C*, affected the color of the kernels, which were either colored (*C*) or colorless (*c*). Crosses between plants that had round, colored kernels with plants that had shrunken, colorless kernels produced offspring with round, colored kernels. This result is similar to Mendel's dihybrid crosses and can be explained by assuming that the parent with round, colored kernels possessed the gene pairs *CCSS*, and that the

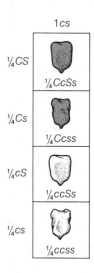

Figure 13-6 Expected results of the testcross *CcSs* × *ccss* if the *C* and *S* genes separate independently (see text).

parent with shrunken, colorless kernels had the genetic constitution *ccss*. The F_1 offspring, with round, colored kernels, would then have the pairs *CcSs* and would all show the round, colored traits because these are dominant over the shrunken, colorless condition.

Hutchison then backcrossed a number of *CcSs* F_1 plants with one of the parent varieties, the *ccss* type with shrunken, colorless kernels. Figure 13-6 shows what would happen if these gene pairs separated independently as in Mendel's dihybrid crosses. The gametes expected in the *CcSs* parent are shown in the vertical headings in the Punnett square. The *ccss* parent can produce only one type of gamete, with the constitution *cs* (entered in the top heading of the Punnett square).

Filling in the squares of the diagram reveals that four classes of offspring would be expected from this cross if the two pairs of genes separated independently. Each of these classes, *CcSs*, *Ccss*, *ccSs*, and *ccss*, occurs with an expected frequency of one-fourth. Therefore, if separation of the *C* and *S* gene pairs is independent, the cross would yield offspring with colored, round:colored, shrunken:colorless, round:colorless, shrunken kernels in a 1:1:1:1 ratio.

In Hutchison's experiment these expected ratios were not even remotely approximated. Of the 8368 offspring, almost all were like the parents: 4032 with colored, round kernels (*CcSs*) and 4035 with colorless, shrunken kernels (*ccss*). These parental types amounted to 8067/8368 × 100 or about 96.4 percent of the total offspring. The very few

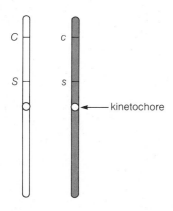

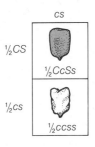

Figure 13-8 Results of a cross between *CcSs* and *ccss* parents if the *C* and *S* genes are located on the same chromosome (see text).

Figure 13-7 Arrangement of the alleles *Cc* and *Ss* on the two chromosomes of a homologous pair (see text).

remaining offspring fit into the other two classes. Only 149 colored, shrunken (*Ccss*) and 152 colorless, round (*ccSs*) individuals were counted, amounting to only 301/8368 × 100 or about 3.6 percent of the total.

The preponderance of combinations resembling the parental types is explained by assuming that the genes are linked on the same chromosome. In Hutchison's experiment, assume that the two gene pairs *Cc* and *Ss* in the colored, round parent are located at different points on the same pair of chromosomes (Fig. 13-7). One chromosome of the pair carries the two alleles *C* and *S*, and the other has the two alleles *c* and *s* at corresponding locations. As the chromosomes separate in meiosis, gametes receive either the chromosome bearing the *C* and *S* alleles together or the chromosome carrying *c* and *s*. Thus only two types of gametes, *CS* or *cs*, are expected with an equal frequency of one-half. The other parent used in the cross, *ccss*, possesses two identical chromosomes with respect to the gene pair studied: both contain the *c* and *s* alleles. These gametes are entered in the Punnett square in Figure 13-8. Filling in the squares in the diagram reveals that only two classes of offspring are expected, *CcSs* and *ccss*, in a 1:1 ratio. Thus the assumption of linkage explains the predominance of parental combinations of alleles in the offspring.

The small percentage of the new *Ccss* and *ccSs* combinations observed with a frequency of about 3 in 100 is explained by assuming that recombination occasionally occurs between the two gene pairs controlling kernel shape and color. As Morgan first realized, the new combinations, called *recombinants*, are related to the crossovers or chiasmata that appear between chromatids of the tetrads in prophase I of meiosis. These crossovers produce the

recombinants noted by Hutchison by the mechanism shown in Figure 13-9. In most meiotic cells of the *CcSs* parent entering meiosis, no crossovers formed between the points on the tetrad bearing the alleles for seed shape and color (Fig. 13-9a). These four chromatids separated and entered the gametes singly, producing the two gamete types shown in Figure 13-9b. But in a small percentage of the cells entering meiosis in the *CcSs* parent, a crossover formed between the two genes (Fig. 13-9c). This crossover, created by breakage and exchange of segments between two of the four chromatids, brought together the new combinations *C* and *s* in one recombinant chromatid and *c* and *S* in the other. (Note that two of the four chromatids of the tetrad remained unrecombined; these are called *parental* chromatids.) These four chromatids separated during the two meiotic divisions and entered separate gametes, giving rise to four different types: two parentals, with the *CS* or *cs* combination (from the two chromatids not entering into the crossover), and two recombinants, with the new *cS* and *Cs* combinations (Fig. 13-9d; see also Fig. 12-9). When combined in fertilization with the *cs* gametes from the *ccss* parent, the recombinants produced the two additional *ccSs* and *Ccss* classes noted in 3 percent of the offspring.

Figures 13-7 through 13-9 assume that the dominant alleles *C* and *S* are carried together on one chromosome, and the two recessive alleles *c* and *s* are carried together on the other chromosome of the homologous pair of the *CcSs* parent used in the cross. One chromosome is thus *CS*, and the other is *cs* as shown in Figure 13-7. This condition, when the dominant alleles are carried together on the same chromosome of a homologous pair, is called the *cis* arrangement of alleles (from *cis* = same side). In genetic shorthand, the two chromosomes of a *CcSs* individual having this arrangement are written as *CS/cs*, in which the alleles carried on one chromosome of the pair are written together in front of the slash, and the alleles carried

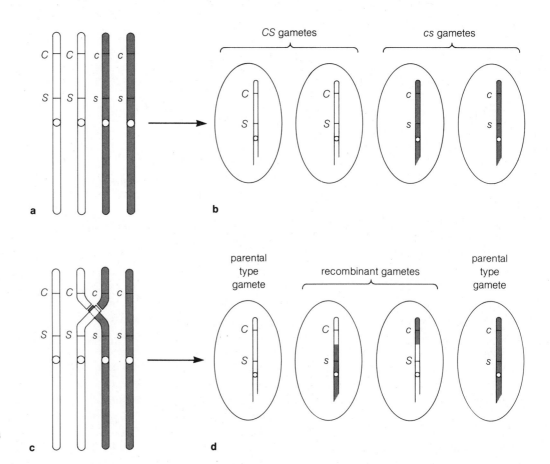

Figure 13-9 How crossing over in meiosis produces recombination between alleles (see text).

together on the second chromosome are written together following the slash.

The alleles of a *CcSs* parent can also be distributed in the opposite pattern, so that one chromosome of a homologous pair has the dominant *C* linked to a recessive *s*, and the second chromosome has the recessive *c* allele linked to a dominant *S*. In contrast to the linkage arrangement shown in Figure 13-7, the two chromosomes of this *CcSs* parent would show the combination:

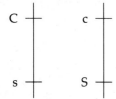

In genetic shorthand, this arrangement is written as *Cs/cS*. This condition, when the dominant alleles of the linked genes under study are carried on *opposite* members of a homologous pair, is called the *trans* arrangement of alleles (from *trans* = across).

The parental linkage combinations *Cs* and *cS* also dominate the offspring when a CcSs parent having the *trans* linkage arrangement *Cs/cS* is crossed with a *ccss* plant having colorless, shrunken kernels (or, in genetic shorthand, a *cs/cs* individual). Hutchison actually carried out a cross of this type in his original experiments. Of a total of 44,595 offspring, colorless, round-kernel plants (*cS/cs*, numbering 21,906) and colored, shrunken-kernel plants (*Cs/cs*, numbering 21,379) formed 43,285 of the total, or 43,285/44,595 × 100 = 97 percent. These offspring obviously received chromosomes from the *CcSs* parent carrying the parental combinations *Cs* and *cS*. Recombination in the *CcSs* parent would produce chromosomes with the opposite *CS* and *cs* arrangement (see if you can diagram the crossover producing these recombinant types). Gametes carrying these recombinant chromosomes, when combined with *cs* gametes from the *cs/cs* parent, produced colored, round-kernel offspring (*CS/cs*, 638 actually counted) and colorless, shrunken-kernel offspring (*cs/cs*, 672 actually counted). The recombinants together numbered only 1310/44,595 × 100 = 3 percent of the total. These results closely match the percentages obtained in the first of Hutchison's crosses; about 97 percent of the combinations match the parental arrangement of alleles, and only about 3 percent fit the alternate arrangement.

Morgan's hypothesis of linkage and recombination was thus able to reconcile the new observations with Mendel's hypothesis that different pairs of genes separate independently when gametes are formed. It was now realized that completely independent assortment, as proposed by Mendel, applies only to genes carried on separate chromosomes. Linked genes, carried on the same chromosome, are inherited according to Mendel's pattern for single genes except for rearrangement of the linkage due to recombination.

Chromosome Mapping by Recombination Studies

As a part of his research with recombination, Morgan noted that the percentage of cells undergoing recombination varies depending on the particular genes being studied. Some linked genes are almost always found in the parental combinations, with recombinants appearing in very low frequency, sometimes in less than 1 percent of the total gametes. Other linked genes show a much higher rate of recombination, so high that it is difficult to detect that the genes are actually linked together. In other words, the alleles of two genes, even though known by other crosses to be carried on the same chromosome, segregate in an approximate 1:1:1:1 ratio if the basic *AaBb* × *aabb* testcross used in Hutchison's experiment is carried out. From observations of this kind, Morgan and one of his students, Alfred M. Sturtevant, proposed that the amount of recombination observed between any two genes located on the same chromosome pair is a reflection of the distance between them. The greater this distance, the greater the chance that a crossover can form between the genes and the greater the recombination frequency. In more formal terms, Morgan and Sturtevant's hypothesis states that the frequency of recombination between two genes is proportional to the distance between them on the chromosome.

It was quickly realized that if the recombination frequency is proportional to the distance between linked genes on a chromosome as Morgan and Sturtevant proposed, these frequencies could be used to *map* the chromosome and assign the genes to relative locations. For example, assume that three genes, *A*, *B*, and *C*, are known to be linked together on the same chromosome. In genetic crosses, 10 percent recombination is detected between genes *A* and *B*. Crosses between *A* and *C* reveal 8 percent recombination, and crosses between *B* and *C* show 2 percent recombination. These frequencies are compatible with only one possible arrangement of genes on the chromosome:

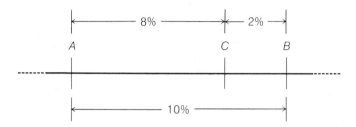

Note that both the order of the genes and the relative distance between them can be estimated from their recombination frequencies.

It is important to distinguish between the percentage of meiotic cells undergoing recombination and the percentage of recombinants showing up in gametes. For example, the observation that 3 percent of the offspring in Hutchison's experiment were recombinants means that 3 percent of the gametes originating from the *CcSs* parent contained recombined chromosomes, or that for every 100 gametes formed, 3 contained recombinants. To produce an average of 3 percent recombinants among the gametes, 6 percent of the cells in meiosis must undergo a recombination between the *C* and *S* genes. This is because any cell undergoing a single recombination or crossover between the *C* and *S* genes to produce two recombined chromosomes also yields two unrecombined chromosomes (see Fig. 13-9d). In male meiosis, for example, the 6 cells out of every hundred undergoing recombination between the *C* and *S* genes produce $6 \times 4 = 24$ haploid gametes. Out of every 4 chromatids in the tetrad containing the *C* and *S* genes, 2 are recombinants and 2 are parentals. Therefore, recombination produces $6 \times 2 = 12$ recombinants and $6 \times 2 = 12$ parental chromatids. Each of these chromatids enters a separate gamete, giving the total of 24. The remaining 94 cells out of every 100 completing meiosis without recombination between the two genes will produce $94 \times 4 = 376$ gametes, all with the parental combination of *C* and *S* alleles. The total number of gametes with the parental combination of *C* and *S* alleles will then be $376 + 12 = 388$. The total number of recombinant gametes will be 12, giving a percentage of 12/400 or 3 percent. In general, then, the percentage of cells entering recombination is always twice the percentage of gametes containing recombinant chromosomes.

Using the mapping method most of the known genes

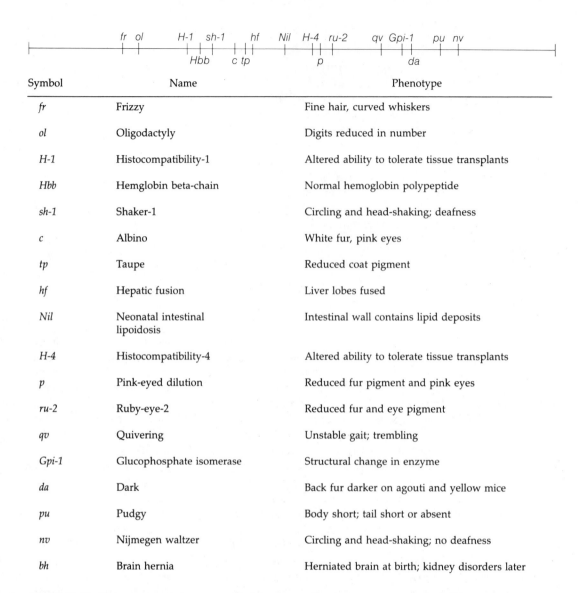

Symbol	Name	Phenotype
fr	Frizzy	Fine hair, curved whiskers
ol	Oligodactyly	Digits reduced in number
H-1	Histocompatibility-1	Altered ability to tolerate tissue transplants
Hbb	Hemglobin beta-chain	Normal hemoglobin polypeptide
sh-1	Shaker-1	Circling and head-shaking; deafness
c	Albino	White fur, pink eyes
tp	Taupe	Reduced coat pigment
hf	Hepatic fusion	Liver lobes fused
Nil	Neonatal intestinal lipoidosis	Intestinal wall contains lipid deposits
H-4	Histocompatibility-4	Altered ability to tolerate tissue transplants
p	Pink-eyed dilution	Reduced fur pigment and pink eyes
ru-2	Ruby-eye-2	Reduced fur and eye pigment
qv	Quivering	Unstable gait; trembling
Gpi-1	Glucophosphate isomerase	Structural change in enzyme
da	Dark	Back fur darker on agouti and yellow mice
pu	Pudgy	Body short; tail short or absent
nv	Nijmegen waltzer	Circling and head-shaking; no deafness
bh	Brain hernia	Herniated brain at birth; kidney disorders later

Figure 13-10 Linkage map of chromosome I of the mouse. From data compiled by M. C. Green, in *Handbook of Biochemistry*. Ed. H. A. Sober. Chemical Rubber Company, Copyright The CRC Press, Inc. Reprinted by permission. 1970.

of the best-studied organisms, *Drosophila*, corn, *Neurospora*, *Escherichia coli*, and various other species, have been assigned positions on the chromosomes of these species (Fig. 13-10 shows the linkage map of one of the chromosomes of a mouse). Note that these positions are only relative; the positions of the genes are indicated only by recombination percentage distances (called *crossover* or *map units*) and not by physical distances in micrometers or nanometers.

The Effects of Genes Carried on Sex Chromosomes: Sex Linkage

Sex chromosomes, as noted in Chapter 12, are chromosomes that are different in males and females of the same species (see p. 303). In organisms such as our own species, cells in females carry a homologous pair of the sex chromosomes, and males carry only one of the pair. The

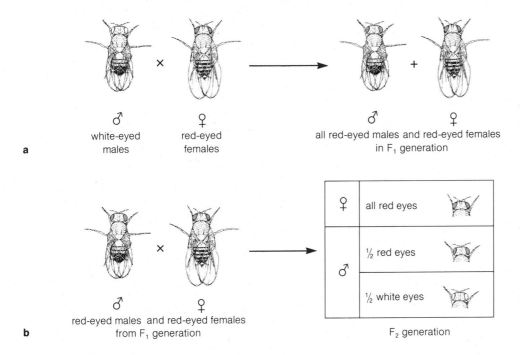

Figure 13-11 Sex linkage in a cross with *Drosophila* (see text). (**a**) A cross between white-eyed males and red-eyed females. (**b**) The F$_1$ cross between the red-eyed males and red-eyed females resulting from (**a**).

chromosomes occurring in a pair in females are called *X chromosomes*; therefore females, since they carry a pair of X chromosomes, are *XX*. In males, the single representative of this pair may exist alone (*XO* males), or may be associated with another chromosome, the *Y chromosome*, that does not occur in females. Males of this type are therefore *XY*. Since the chromatids of these chromosomes are separated in meiosis and placed singly in gametes, females produce eggs containing only X chromosomes, and males produce sperm cells that contain either an X or a Y. In humans and *Drosophila*, union of two gametes containing X chromosomes reunites the XX pair and leads to development of a female; union of a Y-bearing sperm with an X-bearing egg produces an XY male.

In 1910 Morgan noted a curious pattern of inheritance in *Drosophila* in which eye color seemed to vary according to the sex of the fly. All the genetic traits studied up to this time had been transmitted equally between male and female offspring, and the sex of the individuals could be ignored in analyzing the frequency of various classes. Morgan's discovery of an exceptional pattern of inheritance dependent on the sex of the offspring, called *sex linkage*, was to have great significance for the later understanding of several hereditary diseases in humans that are transmitted as sex-linked traits.

Morgan first detected sex linkage in a line of flies that had white eyes instead of the usual red. If white-eyed males of this line were crossed with red-eyed females, all the F$_1$ generation, both males and females, had red eyes (Fig. 13-11a). This suggested that red and white eye color is controlled by two alleles of a single gene. Red is dominant over white if both alleles are present in the same individual. In the offspring of a cross between individuals of the F$_1$ generation, these alleles would then be expected to segregate without respect to sex according to a 3:1 ratio, with red eyes as the most frequent class if the gene for eye color followed the usual pattern of Mendelian inheritance. Surprisingly, Morgan found that all the females in the F$_2$ generation were red-eyed and, among the males, half were red-eyed and half white-eyed (Fig. 13-11b).

In the reciprocal cross, in which white-eyed females were crossed with red-eyed males, different results were obtained. In the F$_1$ offspring of this cross, all the females were red-eyed and all the males white-eyed. Crosses between F$_1$ individuals produced two classes of eye color in each sex of the F$_2$ generation: one-half of all males and females were red-eyed, and one-half were white-eyed (Fig. 13-12).

Morgan realized that this unexpected pattern of inheritance paralleled the division of sex chromosomes in meiosis and their distribution to gametes (see p. 304). He reasoned that the segregation of eye color between males

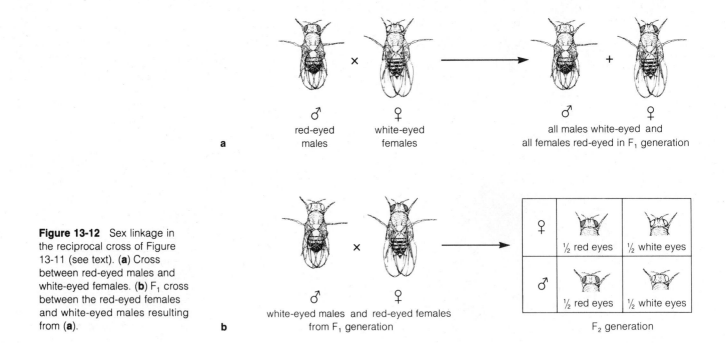

Figure 13-12 Sex linkage in the reciprocal cross of Figure 13-11 (see text). **(a)** Cross between red-eyed males and white-eyed females. **(b)** F₁ cross between the red-eyed females and white-eyed males resulting from **(a)**.

(a)

♂ red-eyed males × ♀ white-eyed females → ♂ all males white-eyed and ♀ all females red-eyed in F₁ generation

(b)

♂ white-eyed males and red-eyed females from F₁ generation

F₂ generation

♀	½ red eyes	½ white eyes
♂	½ red eyes	½ white eyes

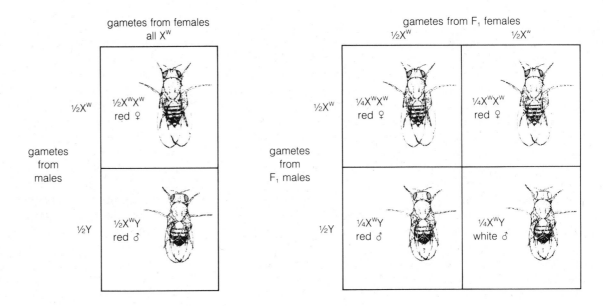

Figure 13-13 The explanation of Morgan's results (see text).

and females could be explained if the gene for this trait is carried on the X chromosome but not on the Y (the Y chromosome in *Drosophila* and many other species is almost inert genetically and carries few active genes). Inheritance of eye color would then follow the pattern shown in Figure 13-13. In the first cross carried out by Morgan (see Fig. 13-11), red-eyed females were crossed with white-eyed males. He assumed that the two X chromosomes of the females both carried the dominant allele W for red color, and that the single X in the male carried the recessive allele w for white. These females produced one kind of egg with respect to eye color, all carrying an X chromosome with the W allele for red eyes. The males produced two kinds of sperm cells, one carrying an X with the recessive w allele for white eyes and one with the Y, which does not carry a gene for eye color. These two types of sperm cells occurred with an equal frequency of one-half. The egg and sperm cells are entered in the Punnett square in Figure 13-13a; note in the results of the cross that all F_1 females are Ww, with red eyes. All of the males received an X carrying the W form of the allele from their mothers and were also red-eyed.

Crossing these F_1 individuals gave the results shown in Figure 13-13b. The Ww females of the F_1 generation produced two types of eggs with an equal frequency of one-half, one with an X chromosome carrying the W allele for red eyes, and one with an X carrying the w allele for white eyes. The F_1 males produced two types of sperm cells in equal numbers. One type carried the X chromosome with the W allele for red eyes, and the second type the inert Y. These egg and sperm cells and their frequencies are entered in the headings to the Punnett square. The filled-in squares of the diagram predict that one-half of the females (one-fourth of the total offspring) would receive an X chromosome with the W allele from both the egg and sperm and thus would have red eyes. The other half of the females would receive an X with the recessive w allele for white eyes from the egg and the dominant W-bearing X from the sperm and would also be red-eyed. Among the males, one-half would receive an X through the egg with the dominant W allele for red eyes, and one-half would receive an X with the recessive allele w for white eyes. Thus one-half of the males would be expected to be red-eyed and one-half white-eyed.

These predictions matched the results actually obtained by Morgan. Since Morgan's first discovery of sex-linked inheritance, more than 100 genes with a similar pattern of inheritance have been described in *Drosophila*.

Sex linkage can also be readily detected in humans. For example, certain kinds of color blindness and *hemophilia*, a disease in which the blood fails to clot in injuries, are transmitted in the same pattern as the red-white eye color distribution noted in *Drosophila* by Morgan. The inheritance of human red-green color blindness shows how sex linkage works and how it is detected in humans. Although testcrosses cannot be carried out with humans, a similar analysis can often be made by looking into family records and constructing a chart called a *pedigree*. A pedigree shows all marriages and offspring for as many generations

as possible, the sex of individuals in the different generations, and the presence or absence of the trait of interest.

A pedigree of a family with a history of color blindness is shown in Figure 13-14. Females are designated by the symbol ♀ and males by ♂; solid black circles in the pedigree indicate the presence of the trait, and white circles the absence of the trait. At the very top of the pedigree (generation 1) the earliest recorded marriage in the family involved a color-blind male and a normal female. The single child of this marriage (generation 2), a daughter, had normal vision and married a normal male. The trait reappeared, however, in five of the seven sons of this marriage (generation 3). The remaining child of this marriage, a daughter, had normal vision.

This pattern of inheritance and the rest of the pedigree can be explained if it is assumed that the allele for color blindness is recessive and carried on the X chromosome. According to this assumption, the color-blind male in the first marriage of the pedigree passed his X chromosome carrying the recessive allele to his daughter. The daughter also received an X carrying the normal, dominant form of the allele controlling color vision from her mother and thus had normal vision. However, the daughter was a *carrier* of the trait for color blindness because one of her X chromosomes carried the recessive allele. Five of her seven sons received the X chromosome carrying the recessive allele and were color blind. The remaining sons received the X chromosome from their mother carrying the allele for normal vision; the daughter also had normal vision because she had an X chromosome carrying the dominant allele for normal vision from her father (all daughters of this marriage, in fact, would be expected to have normal vision since one of their two X chromosomes would carry the normal allele from the father).

Three of the color-blind males of generation 3 married normal females. From the pattern of inheritance of the offspring in the marriages of the color-blind sons (generation 4), it seems probable that none of their wives carried a recessive allele for color blindness because none of the sons or daughters in the fourth generation showed the trait. All the daughters of color-blind fathers were carriers of the recessive gene for the trait, however, because each received the X chromosome carrying the recessive allele from her father. As a result, the trait showed up again in sons in the next generation (generation 5) who received the X with the recessive allele from their mothers. Note the presence and absence of the trait alternates from generation to generation in the males. This is because a father does not pass on his X chromosome, with either recessive or dominant traits, directly to his sons; the X chromosome received by a male always comes from his mother. As a

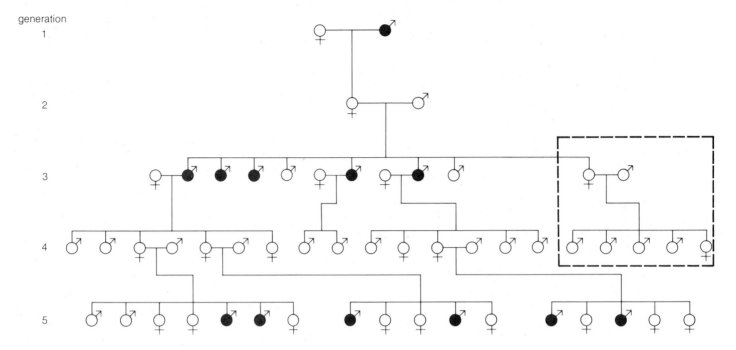

generation
1
2
3
4
5

Figure 13-14 Pedigree of a family with a history of color blindness. Females are designated by the symbol ♀ and the males by ♂; solid black circles indicate the presence of color blindness. White circles indicate the absence of the trait. Marriages are indicated by a horizontal bar between two circles. An alternate possibility for the segment enclosed by dashed lines is shown in Figure 13-15.

result, alternation or "skipping" of a trait in a human pedigree suggests that the allele under study is recessive and is carried on the X chromosome.

None of the females in the pedigree in Figure 13-14 developed color blindness. But this would have been possible if a female with normal vision, who carried a recessive allele for color blindness on one chromosome, had married a color-blind male. A possible outcome of a marriage of this type is shown in Figure 13-15, in which the marriage enclosed in dotted lines in Figure 13-14 is considered as if the daughter of the line had married a color-blind male. Some of the sons would be likely to receive the dominant allele for normal vision from the mother, and some would receive the X with the recessive allele for color blindness. Thus both color-blind and normal sons could arise from this marriage. The same possibilities hold true for the daughters. The father's X chromosome brings one recessive allele to each daughter. This allele, if combined with the mother's X bearing the recessive allele, would produce a color-blind daughter. If combined with the mother's X carrying the dominant allele for normal vision, the daughter would have normal color vision but would be a carrier of the trait.

Lack of Dominance and Multiple Alleles

Detailed study of the distribution of hereditary traits among offspring has revealed two additional patterns of inheritance that, while not anticipated by Mendel, still rest on his principles and the mechanisms of meiosis. These patterns are *lack of dominance*, in which the effects of both alleles can be detected in heterozygotes, and *multiple alleles*, a condition in which more than two alternate alleles of a gene can be detected among the members of a population.

Lack of Dominance

Mendel's principle of dominance assumed that if two different alleles of a gene are present in the same individual, one masks the effects of the other and the individual displays only the trait of the dominant allele. Although this assumption was true for the traits Mendel studied in his

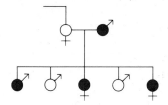

Figure 13-15 A possible outcome of the marriage enclosed in dotted lines in Figure 13-14 if the daughter carrying the trait married a color-blind male. Both color-blind and normal sons and daughters could arise from this marriage.

garden peas, further work revealed that the effects of recessive alleles are not always completely masked by dominant alleles.

The inheritance of feather color in blue Andalusian chickens follows a pattern showing lack of dominance. Blue Andalusians, if crossed among themselves, always produce offspring with two new colors, black and speckled white in addition to blue, in the ratio of 1 black:2 blue:1 speckled white. Black mated with black always breed true, giving only black offspring. Speckled white chickens, if mated among themselves, also maintain the speckled white color. Black and speckled white birds, however, give rise to all blue Andalusian offspring if mated.

These results can be explained if it is assumed that the three colors come from different combinations of two alleles of the gene for feather color. These alleles are C for black and c for speckled white. Black does not completely dominate white; as a result, in heterozygous Cc individuals, the mixture of effects produces the Andalusian blue color. A cross between black (CC) and speckled white (cc) yields all Cc (blue Andalusian chickens) in the F_1 generation, as actually observed. Crossing these individuals produces the combinations CC, Cc, and cc in the F_2 offspring in the ratio 1:2:1. These combinations, which produce black (CC), blue Andalusian (Cc), and speckled white birds (cc), thus explain the results obtained in crosses between blue Andalusian birds.

The assumption that the three colors come from the CC, Cc, and cc combinations of the two alleles can be tested by backcrossing individuals of the F_2 generation with the original black and speckled white parents. The expectations of these crosses, if the assumption is correct, are:

black parent $\times$ blue F_2 ($CC \times Cc$) =
$$\tfrac{1}{2} \text{ black} + \tfrac{1}{2} \text{ blue offspring}$$
speckled white parent $\times$ blue F_2 ($cc \times Cc$) =
$$\tfrac{1}{2} \text{ blue} + \tfrac{1}{2} \text{ speckled white offspring}$$
These results, in the predicted ratios, are actually obtained from such crosses.

In lack of dominance, the different genotypes produced by a cross each result in different phenotypes, that is, in different outward appearances in the individuals obtained in crosses. In other words, each genotype produces a distinct and recognizable genotype.

Multiple Alleles

Mendel assumed that each hereditary trait is controlled by pairs of factors, the alternate alleles of each gene. Although the alleles of each gene in any one diploid individual do occur in pairs, as Mendel proposed, more than two alleles of a gene may be detected if several individuals of a species are compared.

Coat colors in rabbits are inherited in this pattern. Wild rabbits usually have brown fur that is fairly uniform in color over the whole animal; this color, since it occurs in the majority of individuals in nature, is called the *wild type* for the trait. At the other end of the spectrum of coat colors is the all-white fur of the *albino* domestic rabbit. Crosses between wild-type and albino rabbits reveal that the wild-type and white colors are controlled by two alleles of a single gene. A dominant C allele of this gene produces the brown fur color of wild-type rabbits; individuals homozygous for the recessive c allele are albinos. Thus CC or Cc rabbits have brown fur, and cc rabbits are albinos; crosses between Cc parents produce brown and albino offspring in a 3:1 ratio (albinism in higher animals is frequently the result of a homozygous recessive gene).

A number of other fur colors have been discovered that also give a 3:1 segregation in the F_2 generation in crosses with wild-type rabbits. *Chinchilla* rabbits are silver-colored instead of brown. If chinchillas are crossed with wild-type rabbits, the offspring or F_1 generation are brown; crosses among the F_2 give three-fourths brown and one-fourth chinchilla, indicating that chinchilla is also recessive to the wild-type. However, crosses between chinchilla and albino reveal that the chinchilla allele is dominant to albino. One other allele of interest, called *Himalayan*, produces white rabbits with pigmented nose, ears, tail, and legs. This coat is also due to a recessive form of the gene for fur color; Himalayan rabbits crossed with the wild-type

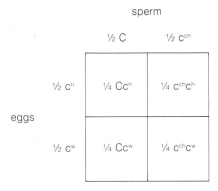

sperm

	½ C	½ c^ch
½ c^h	¼ Cc^h	¼ c^ch c^h
½ c^w	¼ Cc^w	¼ c^ch c^w

eggs

Figure 13-16 Expected outcome of the cross $Cc^{ch} \times c^h c^w$ in rabbits (see text).

also give a 3:1 ratio of wild-type to Himalayan in the F_2 generation.

Crosses between different combinations of the Himalayan, chinchilla, and albino alleles reveal that Himalayan is recessive to chinchilla but dominant to albino. These alleles are identified in genetic shorthand as C = wild type, c^{ch} = chinchilla, c^h = Himalayan, and c^w = albino. The dominance relationships of the alleles are summarized as follows:

C with C, c^{ch}, c^h, or c^w = wild-type brown fur (CC, Cc^{ch}, Cc^h, Cc^w)
c^{ch} with c^{ch}, c^h, or c^w = chinchilla ($c^{ch}c^{ch}$, $c^{ch}c^h$, $c^{ch}c^w$)
c^h with c^h or c^w = Himalayan (c^hc^h, c^hc^w)
c^w and c^w = albino (c^wc^w)

The coat color produced in any individual depends on which pair of the four alleles for coat color is received from the parents: although four alleles exist in the population as a whole, any individual has only two, one on each of the two chromosomes of the homologous pair carrying the gene controlling coat color. For example, in the cross between a Cc^{ch} father and a c^hc^w mother, four distinct alleles of the gene for color are present in the two parents. The father, however, has only two, and produces sperm cells containing either the C allele or the c^{ch} allele in equal frequencies of ½ (Fig. 13-16). The mother similarly has only two alleles and produces c^h and c^w eggs with equal frequencies of ½. Filling in the Punnett square in Figure 13-16 indicates that these gametes produce four possible combinations when joined by twos in fertilization. Two of

A allele	...ATGCAGATACCGATTACAGACCATAGG...
a_1 allele	...ATGCAGA<u>G</u>ACCGATTACAGACCATAGG...
a_2 allele	...ATGCAGAT<u>G</u>CCGATTACAGACCATAGG...
a_3 allele	...ATGCAGATACCGATTACAG<u>G</u>CCATAGG...

Figure 13-17 The multiple alleles of a gene consist of small differences in the nucleotide sequence at one or more points (underlined).

these combinations, Cc^{ch} and Cc^w, produce wild-type rabbits; the other two combinations, $c^{ch}c^h$ and $c^{ch}c^w$, both produce chinchilla. Therefore the offspring of this cross are expected to be ½ wild-type and ½ chinchilla.

Most genes exist as a series of multiple alleles in this way rather than in only two alternate forms. In terms of DNA molecules, the different alleles of a gene represent differences at one or more points in the sequence of nucleotides making up the gene (Fig. 13-17). In any one individual, the separation of the two alleles present follows the usual pattern of separation and placement in separate gamete cells through the mechanism of meiosis, and produces the same genetic outcomes predicted by Mendel's principles.

Gene Interactions

Two frequently observed patterns of inheritance follow pathways that at first seem to be unrelated to Mendel's principles and the mechanisms of meiosis underlying these principles. However, these patterns, termed *polygenic inheritance* and *epistasis*, prove on closer examination to depend on genes that are carried in pairs in individuals and separated and distributed to gametes by the usual mechanisms of meiosis.

Polygenic Inheritance

Some inherited traits, such as skin color or height in humans, follow a pattern of inheritance that at first seems to involve an even blending of parental characteristics rather than the distinct segregation ratios predicted by Mendelian genetics. The offspring of parents of differing skin color, for example, often seem to have skin pigment of intermediate values, rather than a distribution of pigments among children approximating 3:1, 9:3:3:1, or other ratios. Careful analysis of such offspring, however, shows that the even blending of parental traits is only apparent, and regular segregation ratios actually do exist among the products of a cross. These regular ratios are not detected on casual inspection because there are many possible classes of offspring that differ only slightly. Moreover, the most frequent classes expected are those that fall at intermediate values between the characteristics of the parents. Inheritance along this pattern proves to be the result of several different gene pairs that affect the same trait; hence the name polygenic (poly = many) inheritance. The inheritance of skin color in humans, first investigated by C. B. Davenport in studies carried out with mulattoes in Bermuda and Jamaica, follows the polygenic pattern.

Davenport's analysis confirmed that apparent blending of parental skin color to produce offspring with intermediate color was the most frequent, but not the only outcome of matings between mulattoes. In his investigations of mulatto families, Davenport found that such matings can also produce children that are either darker or lighter than the parents with a low but regular frequency. In fact, matings between mulattoes can produce a spectrum of skin pigmentation ranging from apparently pure black to white. However, in this distribution, intermediate colors are most common.

Davenport saw that this distribution of skin colors could be explained if pigmentation in humans is controlled by more than one gene pair. Although it is probably an oversimplification, he proposed that two different genes, P_1 and P_2, control human skin color. Although there are at least two alleles of each gene, P and p, no dominance exists between any of the alleles. Pure African Negroes, with no white ancestors, were assumed to carry the four alleles $P_1P_1P_2P_2$ of the two gene pairs. Persons of pure Caucasian ancestry were assumed to carry the four alleles $p_1p_1p_2p_2$. Matings between pure black and pure white individuals would produce the genotype $P_1p_1P_2p_2$, which is assumed to be an offspring of an intermediate brown color. Because no dominance is exerted between the P and p alleles of either gene, $P_1p_1P_2p_2$ individuals can be distinguished in external appearance, or phenotype, from either the $P_1P_1P_2P_2$ or the $p_1p_1p_2p_2$ parent.

A mating between two $P_1p_1P_2p_2$ individuals with the same intermediate degree of pigmentation would produce very different results, some of them unexpected if even blending were to occur. The possible combinations of gametes from two such individuals are shown in the Punnett square in Figure 13-18. Both persons will produce the four gametes shown in equal frequencies of one-fourth. These gametes, when combined in the squares of the figure, produce a range of pigment combinations varying more or less evenly from pure black, $P_1P_1P_2P_2$, through a range of intermediate values (indicated by the shading in Fig. 13-18) to white, $p_1p_1p_2p_2$. Either of the two extremes, pure white or pure black, would be expected in 1 out of 16 births if pigmentation is controlled by two genes as Davenport suggested. This range of pigments would not be expected if the inheritance of skin color depended simply on the blending of parental pigmentation; if blending actually occurred, two parents of intermediate color would always produce children of the same intermediate color.

Analysis of pigmentation in the offspring of parents of intermediate color shows that the children of such parents actually show the expected range of color from black to white, with intermediate values arising more frequently than either extreme. The extremes, in fact, are much rarer than would be expected if skin pigmentation in humans is controlled by only two genes. The low frequency with which pure black or white offspring are encountered in matings between individuals of intermediate color suggests that skin pigmentation may be controlled by as many as five genes, each with at least two alleles P and p. In that case, crosses between $P_1p_1P_2p_2P_3p_3P_4p_4P_5p_5$ individuals, with intermediate skin color, would be expected to produce either pure black or pure white skin color in approximately 1 out of every 1000 children, an expectation that is close to the results actually observed.

Plotting the distribution among individuals of a trait that is controlled by polygenic inheritance, such as skin color, typically produces a bell-shaped curve. Figure 13-19 shows a plot of the frequencies of various body heights in men; the bell-shaped curve obtained indicates that this human trait is also controlled by polygenic inheritance. Other organisms show equivalent patterns for certain characteristics, such as ear length in corn, seed color in wheat, and color spotting in mice. Thus the inheritance of such traits in which several genes interact to affect the same trait can still be explained by paired genes that are separated and distributed singly to gametes by meiosis.

Epistasis

Unusual patterns of offspring that at first seem to violate Mendelian genetics are also produced when the alleles of one gene can override the effects of a different gene. This pattern is detected in the inheritance of coat color in guinea pigs and many other mammals. In guinea pigs, the

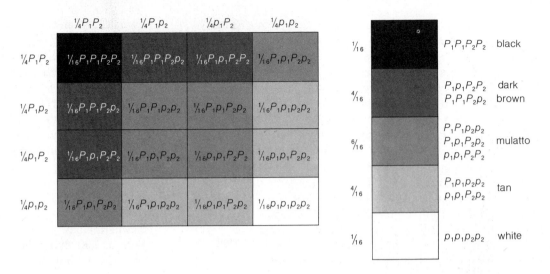

	$\frac{1}{4}P_1P_2$	$\frac{1}{4}P_1p_2$	$\frac{1}{4}p_1P_2$	$\frac{1}{4}p_1p_2$
$\frac{1}{4}P_1P_2$	$\frac{1}{16}P_1P_1P_2P_2$	$\frac{1}{16}P_1P_1P_2p_2$	$\frac{1}{16}P_1p_1P_2P_2$	$\frac{1}{16}P_1p_1P_2p_2$
$\frac{1}{4}P_1p_2$	$\frac{1}{16}P_1P_1P_2p_2$	$\frac{1}{16}P_1P_1p_2p_2$	$\frac{1}{16}P_1p_1P_2p_2$	$\frac{1}{16}P_1p_1p_2p_2$
$\frac{1}{4}p_1P_2$	$\frac{1}{16}P_1p_1P_2P_2$	$\frac{1}{16}P_1p_1P_2p_2$	$\frac{1}{16}p_1p_1P_2P_2$	$\frac{1}{16}p_1p_1P_2p_2$
$\frac{1}{4}p_1p_2$	$\frac{1}{16}P_1p_1P_2p_2$	$\frac{1}{16}P_1p_1p_2p_2$	$\frac{1}{16}p_1p_1P_2p_2$	$\frac{1}{16}p_1p_1p_2p_2$

$\frac{1}{16}$	$P_1P_1P_2P_2$	black
$\frac{4}{16}$	$P_1p_1P_2P_2$ $P_1P_1P_2p_2$	dark brown
$\frac{6}{16}$	$P_1P_1p_2p_2$ $P_1p_1P_2p_2$ $p_1p_1P_2P_2$	mulatto
$\frac{4}{16}$	$P_1p_1p_2p_2$ $p_1p_1P_2p_2$	tan
$\frac{1}{16}$	$p_1p_1p_2p_2$	white

Figure 13-18 The proposed inheritance of skin color in the offspring of parents with the $P_1p_1P_2p_2$ genotype (see text).

gene *B* determines whether coat color is black or brown. The dominant allele *B* produces black color in *BB* and *Bb* individuals; *bb* individuals are brown. However, the expression of the *B* gene is controlled by a different gene *C* for color. In this gene, the *C* allele is dominant, and both *CC* and *Cc* guinea pigs develop the black or brown fur color determined by the *B* allele. However, homozygous recessive *cc* individuals develop no fur color and are albinos no matter what alleles are present at the *B* gene. This type of interaction, in which one gene overrides the effects of another, is called *epistasis* (from *epi* = over and *stasis* = standing or stopping).

Epistasis has an easily understood molecular basis. In such cases, the genes involved usually code for different enzymes that catalyze steps in a common pathway leading to production of a substance such as a skin, hair, or eye pigment:

$$\text{substance A} \xrightarrow{\text{enzyme 1}} \text{substance B} \xrightarrow{\text{enzyme 2}}$$
$$\text{substance C} \xrightarrow{\text{enzyme 3}} \text{pigment}$$

Suppose that in an animal the *B* gene codes for enzyme 1 in the pathway. In its dominant *B* allele the gene codes for a highly efficient form of enzyme 1, leading to intensive activity of the pathway and production of large quantities of a fur pigment. The recessive *b* allele codes for a modified form of the enzyme that catalyzes the first step in the pathway only very slowly, leading to greatly reduced pig-

ment production by the remainder of the pathway when the *b* allele is present on both chromosomes carrying the gene. This condition could lead to black (*BB* or *Bb*) or brown (*bb*) fur or skin colors, since the two colors are produced by different concentrations of the same black-brown pigment, *melanin*.

Production of the pigment, however, depends on a different gene that codes for another enzyme catalyzing a subsequent step in the pathway. Assume that this is the enzyme catalyzing step 3. In the active form coded by the dominant *C* allele of this gene, step 3 proceeds at a rate sufficient to convert all available molecules of substance C into the pigment, whether the individual has the genotype *CC* or *Cc*. However, the recessive *c* allele codes for a completely inactive form of the enzyme. Individuals homozygous for this allele (*cc* individuals) produce no active molecules of enzyme 3, and the pathway is blocked at step 3. As a consequence, these individuals are albinos and have white fur no matter which alleles of the *B* gene are present.

The two gene pairs involved in an epistatic interaction may be on the same or different chromosome pairs. When on different chromosome pairs, as is often the case, the ratios produced resemble the products of a dihybrid cross, with some of the expected categories lumped together. In guinea pigs, for example, epistasis by the *C* gene lumps the *ccBB*, *ccBb*, and *ccbb* genotypes together as albinos, producing the following distribution of phenotypes:

$\frac{9}{16}$ black + $\frac{3}{16}$ brown + $\frac{4}{16}$ albino

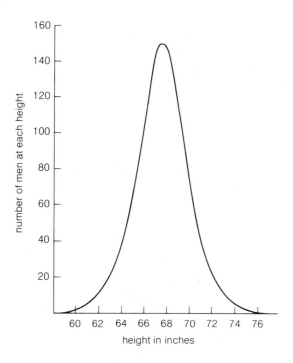

Figure 13-19 A graph of numbers of men of different heights ranging from 58 to 77 inches, with the height in inches plotted against the frequency of each height. The distribution of frequencies produces a typical bell-shaped curve, indicating that height is controlled by more than one gene pair.

instead of the 9:3:3:1 ratio expected in a dihybrid cross. There are many additional examples of epistatic interactions between different genes that lead to the same 9:3:4 or other ratios.

The Changing Gene Concept

The results of experimentation in heredity described in this chapter—monohybrid and dihybrid crosses, linkage and recombination, sex linkage, incomplete dominance, polygenic inheritance, and epistasis—are usually grouped together as the findings of classical genetics. All these patterns are completely compatible with genes that are carried in pairs in individuals, and recombined, separated, and distributed to gametes by the mechanisms of meiosis.

The research in classical genetics, and the later investigations into the molecular characteristics of DNA and its role in information storage, have led to a gradually emerging concept of the gene. Classical genetics established that genes control more or less clearly defined characteristics of organisms. This functional definition was first translated into molecular terms by the biochemical research of the 1950s and 1960s. The biochemical work led to the discovery of DNA structure, the genetic code, RNA transcription, and the mechanism of protein synthesis. From this work, genes were initially proposed to be coding units that spell out the directions for making enzymes. As a result, a "one gene, one enzyme" concept arose and held favor for a time. Gradually it became apparent that not all proteins are enzymes; some, such as the protein subunits of microtubules, serve motile or structural roles in cells. Others are hormones or membrane receptor proteins. In addition, some genes were found to code for polypeptide subunits of complex, multichain proteins rather than complete proteins. As a consequence of these findings, the gene concept was modified to assert that each gene codes for a polypeptide, or "one gene, one polypeptide."

By the mid- to late 1960s, it became obvious that even this concept was too narrow to define genes adequately. At this time, research into the kinds of RNA transcribed in the nucleus revealed that genes code for ribosomal RNA and tRNA as well as polypeptides. Regulatory genes or regions, such as the operators of prokaryotic DNA (see p. 236), were also discovered; these DNA regions are never transcribed into RNA of any kind.

These more recent findings have led to the concept that a gene is a segment of a DNA molecule that codes for a unit of function. The function may be to code for the entire polypeptide chain of a simple protein, or a polypeptide subunit of a complex protein with several polypeptide subunits. The proteins may be enzymes, structural units, peptide hormones, recognition and receptor proteins, or other protein-based cellular molecules. Alternatively, the unit of function may code for ribosomal or transfer RNA, or it may be involved in regulating the activity of other functional units within the DNA. (Contributions from bacterial and viral genetics to the gene concept are discussed in Supplement 13-3.)

Suggestions for Further Reading

Corwin, H. O. and J. B. Jenkins. 1975. *Foundations of modern genetics.* Houghton Mifflin: Boston.

Goodenough, V. and R. P. Levine. 1974. *Genetics.* Holt, Rinehart & Winston: New York.

Herskowitz, I. H. 1979. *Elements of genetics.* Macmillan: New York.

Strickberger, M. W. 1976. *Genetics.* 2nd ed. Macmillan: New York.

Sturtevant, A. H. 1965. *A history of genetics.* Harper & Row: New York.

Watson, J. D. 1976. *Molecular biology of the gene.* 3rd ed. W. A. Benjamin: Menlo Park, Calif.

Woeller, B. R. 1968. *The chromosome theory in inheritance.* Appleton-Century-Crofts: New York.

Wolfe, S. L. 1981. *Biology of the cell.* 2nd ed. Wadsworth: Belmont, Calif.

For Further Information

DNA structure, Chapter 9
Genes in mitochondria and chloroplasts, Supplement 10-2
Meiosis, Chapter 12
Protein synthesis, Chapter 10
Recombination, Chapter 12 and Supplement 12-1

Questions

1. What are dominant and recessive traits?

2. What is a homozygote? A heterozygote?

3. What were Mendel's basic hypotheses about the hereditary factors (genes) and their pattern of transmission to offspring?

4. What is a monohybrid cross? A dihybrid cross?

5. What is a testcross? How did Mendel use a testcross to test his hypotheses concerning monohybrid crosses?

6. Explain what independent assortment means. How are independent assortment and segregation related to meiosis?

7. What is a genotype? A phenotype?

8. Explain alleles in terms of DNA sequences.

9. What is linkage? How was linkage discovered?

10. How does recombination affect linkage?

11. Why are the recombinant chromosomes produced by a single recombination event found in only one-half of the gametes?

12. Explain recombination in terms of chromosomes and chromatids.

13. What is the relationship between the frequency of recombination between genes and their positions on chromosomes?

14. Why are the "maps" developed from recombination frequencies laid out in crossover units instead of in nanometers?

15. What is sex linkage? How does it differ from ordinary linkage? What are sex chromosomes?

16. Explain sex-linked inheritance in terms of the meiotic division sequence.

17. What observations led to the discovery of sex linkage?

18. What is a pedigree? How are pedigrees used in studies of human inheritance?

19. What is a carrier? Explain why a human male cannot be a carrier of color blindness or hemophilia.

20. What is incomplete dominance?

21. Explain the difference between the number of possible alleles of a gene in individuals and in populations of individuals.

22. What is polygenic inheritance? How does it differ from the inheritance of traits governed by a single gene? In what way is the fact that mulattoes can have children with unpigmented skin related to the reappearance of recessive traits in the F_2 generation of Mendel's crosses?

23. What is epistasis?

24. What is a gene? Why were the "one gene, one enzyme" and "one gene, one polypeptide" hypotheses inadequate? Explain the difference between the alleles of a gene in terms of DNA sequences.

Problems*

1. The C allele of a gene controlling color in corn produces kernels with color; plants homozygous for a recessive c allele of this gene have colorless or white kernels. What kinds of gametes, in what proportions, would be produced by the parent plants in the following crosses? What seed color would be expected in the offspring of the crosses? In what proportions?

$$CC \times Cc \qquad Cc \times Cc \qquad Cc \times cc$$

2. In peas, the allele T produces tall plants, and the allele t produces dwarfs. The T allele is dominant to t. If a tall plant is crossed with a dwarf, the offspring are distributed about equally between tall and dwarf plants. What are the genotypes of the parents?

3. Assume that eye color in humans is controlled by a pair of alleles of a single gene, with brown dominant to blue. Can two brown-eyed parents have a blue-eyed child? Can blue-eyed parents have a brown-eyed child? A brown-eyed couple, both of whom had one blue-eyed parent and one brown-eyed parent, are expecting their first child. What is the chance that the child will have blue eyes? Suppose the first child actually has blue eyes. What is the chance that the second child of the same couple will have blue eyes?

4. Two brown-eyed parents have four blue-eyed children. What is the chance that their next child will have brown eyes? Blue eyes?

5. In four-o'clock flowers, red plants are homozygous for the dominant allele R of a gene for flower color. White plants are homozygous for the r allele of the same gene. Heterozygotes, with one R and one r allele, have pink flowers. What offspring and in what proportions would be expected from the following crosses?

$$RR \times Rr \qquad RR \times rr \qquad Rr \times Rr \qquad Rr \times rr$$

6. In guinea pigs, an allele for rough fur (R) is dominant over an allele for smooth fur (r); an allele for black coat (B) is dominant over that for white (b). You have an animal with rough, black fur. What cross would you use to determine whether the animal is homozygous for these characteristics? What results would you accept as proof?

7. You cross a lima bean plant breeding true for green pods with another lima bean that breeds true for yellow pods. You note that all the F_1 plants have green pods. These green-pod F_1 plants, when crossed, produce 675 plants with green pods, and 217 with yellow pods. How many genes probably control pod color? What are the alleles? Which one is dominant?

8. Some people can roll their tongue into a complete circle, and some cannot. This ability is inherited as if it is controlled by two alleles of a single gene, with the ability to roll the tongue as the dominant trait. Two tongue-rolling parents have a child who cannot roll his tongue. Can you explain this? Write the genotypes and phenotypes for the three individuals.

9. Some recessive alleles have such a detrimental effect that they are lethal when present in both chromosomes of a pair. That is, the homozygous recessive cannot survive and dies at some point during embryonic development or early life. Suppose that the allele r is lethal in the homozygous rr condition. What ratios of offspring would you expect from the following crosses?

$$RR \times Rr \qquad Rr \times Rr$$

10. In garden peas, green pods (GG or Gg) are dominant over yellow (gg); tall plants (TT or Tt) are dominant over dwarfs (tt); and round seeds (RR or Rr) are dominant over wrinkled (rr). What offspring are expected in the F_1 generation if a true-breeding tall plant with green pods and round seeds is crossed with a true-breeding dwarf plant with yellow pods and wrinkled seeds? What offspring are expected if F_1 individuals are crossed? (Assume no linkage.)

11. Suppose a man has the following combination of alleles for the genes A, B, C, D, E, F, and G:

$$AaBBCcddEeFfGg$$

How many different kinds of gametes can be produced with respect to these genes? (Assume no linkage.)

12. Persons with the dominant allele A can taste the chemical phenylthiocarbamide (PTC); those who are homozygous for the alternate a allele of this gene cannot taste PTC. Suppose a brown-eyed taster marries a blue-eyed nontaster. What kinds of children, in what proportions, could this couple expect if one of the brown-eyed taster's parents was a blue-eyed nontaster? What kinds of children in what proportions could a pair of brown-eyed, taster parents expect if each had a blue-eyed, nontaster parent? (Assume no linkage, and that eye color in humans is controlled by a pair of alleles of a single gene, with brown dominant to blue.)

13. In humans, red-green color blindness is a sex-linked recessive trait. If a man with normal vision and a color-blind woman have a son, what is the chance that the son will be color-blind? What is the chance that a daughter will be color-blind?

14. The following pedigree shows the pattern of inheritance of color blindness in a family (persons with the trait are indicated by black circles):

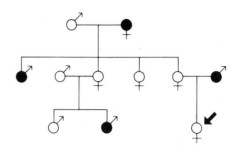

What is the chance that a son of the third-generation female indicated by the arrow will be color blind if she marries a normal male? A color-blind male?

15. Persons affected by a condition known as *polydactyly* have extra fingers or toes. The trait was present (black circles) in the following members of one family:

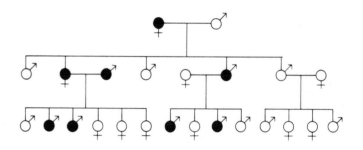

From the pedigree, can you tell if polydactyly comes from a dominant or a recessive allele? Is the trait sex-linked? As far as you can determine, what is the genotype of each person in the pedigree with respect to the trait?

16. In chickens, feathered legs are produced by a dominant allele F. Another allele f of the same gene produces featherless legs. The dominant allele P of another gene produces pea comb; a recessive allele p of this gene causes single comb. A breeder makes the following crosses with birds 1, 2, 3, and 4; all parents have feathered legs and pea comb (assume no linkage):

1 × 2: all feathered, all pea comb
1 × 3: ¾ feathered, ¼ featherless, all pea comb
1 × 4: ⁹⁄₁₆ feathered, pea comb; ³⁄₁₆ featherless, pea comb; ³⁄₁₆ feathered, single comb; ¹⁄₁₆ featherless, single comb

What are the genotypes of the four birds?

17. A brown rabbit that is crossed with a chinchilla produces three brown rabbits and one white. What are the genotypes of the parents?

18. In humans, the blood groups M and N are determined by two alleles of a single gene without dominance: M individuals are homozygous MM, MN individuals are heterozygous Mm, and N

individuals are homozygous *mm*. The ABO blood groups are controlled by three alleles of the gene I: I^A, I^B, and I^o. The dominance relationships of the three alleles are:

$I^A I^A$ = type A blood $I^A I^B$ = type AB blood
$I^B I^B$ = type B blood $I^o I^o$ = type O blood

Although the I^A and I^B alleles are both dominant over the I^o allele, there is a lack of dominance between the I^A and I^B alleles. A mix-up in a hospital ward convinced a mother with type O and MN blood that a baby given to her really belonged to someone else. Tests in the hospital revealed that the doubting mother was able to taste PTC. The baby given to her had type O and MN blood and could not taste PTC. The mother has four other children with the following blood types and tasting abilities:

type A and MN blood, taster
type B and N blood, nontaster
type A and M blood, taster
type A and N blood, taster

Without knowing the father's blood types and tasting ability, can you determine whether the child is really hers? (Assume that all of her children have the same father.)

19. A number of genes carried on the same chromosome are tested and show the following crossover frequencies. What is their sequence in the map of the gene?

Genes	Crossover frequency between them
C and A	7%
B and D	3%
B and A	4%
C and D	6%
C and B	3%

20. In *Drosophila*, the recessive genes for black body color and purple eyes are carried on the same chromosome. You make a cross between a fly with normal eye and body color and a black-bodied fly with purple eyes. Among the offspring, about half have normal eye and body color, and half have purple eyes and black bodies. A small percentage have (1) normal eye color and black bodies, and (2) purple eyes with normal body color. What alleles are carried together on the chromosomes in each of the flies used to make the cross? What alleles are carried together on the chromosomes of the F_1 flies with normal eye color and black bodies, and purple eyes with normal body color?

21. Using the Chi square method, determine how closely the following proportions of offspring fit a 9:3:3:1 ratio (see Supplement 13-1):

a.	74	33	38	1
b.	880	310	330	100
c.	1810	597	603	198
d.	807	410	400	205

22. You carry out a cross in *Drosophila* that produces only half as many males as females in the offspring. What might you suspect as a cause? (Hint: Review problem 9.) Show the genetic basis for your answer.

23. In cats the genotype *AA* produces the tabby fur color; *Aa* is also a tabby, and *aa* is black. Another gene pair is epistatic to the gene for fur color. When present in its dominant *W* form (*WW* or *Ww*) this gene blocks the formation of fur color and all the offspring are white; *ww* individuals develop normal fur color. What fur colors, in what proportions, would you expect from the cross:

$$AaWw \times AaWw$$

(Assume no linkage.)

*Answers to genetics problems begin on p. 399.

Supplement 13-1:
Some Methods for Determining Expected Ratios and Testing the Fit of Observed to Expected Results

An Algebraic Method for Determining Expected Ratios in Offspring

Constructing a Punnett square to determine expected ratios of alleles in offspring is convenient for monohybrid or dihybrid crosses. But if used for more than two pairs of genes, the technique is cumbersome and time-consuming. However, the outcome of such crosses can be easily calculated by a simple algebraic method instead of a Punnett square.

The method works this way in monohybrid crosses. In the first of Mendel's experiments described in this chapter, two heterozygous (*Rr*) parents, both carrying the *R* allele for smooth seeds and the *r* allele for wrinkled seeds, were crossed. In algebraic terms, the gametes expected are ($\frac{1}{2}R + \frac{1}{2}r$), produced by the "male" parent, and ($\frac{1}{2}R + \frac{1}{2}r$), produced by the "female" parent. Multiplying these individual alleles and probabilities algebraically to obtain the combined outcome and probabilities gives:

$$(\tfrac{1}{2}R + \tfrac{1}{2}r)(\tfrac{1}{2}R + \tfrac{1}{2}r) = \tfrac{1}{4}RR + \tfrac{1}{2}Rr + \tfrac{1}{4}rr \quad (13\text{-}1)$$

the expected genotypes and their ratios in the offspring. Because the *RR* and *Rr* classes are phenotypically identical in the offspring of this cross, the outcome can also be designated:

$$(\tfrac{1}{2}R + \tfrac{1}{2}r)\,(\tfrac{1}{2}R + \tfrac{1}{2}r) = \tfrac{3}{4}R\_\_ + \tfrac{1}{4}rr \qquad (13\text{-}2)$$

where $R\_\_$ indicates either RR or Rr.

The algebraic method is most useful when the cross under study includes more than a single pair of genes. In Mendel's cross testing the inheritance of seed shape and color, the two parents in the F_1 cross had the genotype $RrYy$, with round, yellow seeds. The Rr and Yy pairs in the cross $RrYy \times RrYy$ are assumed to be on different chromosome pairs and therefore completely independent of each other. Since the two pairs are independent, the outcome of the cross with respect to the Rr pair is $\tfrac{3}{4}R\_\_ + \tfrac{1}{4}rr$, where $R\_\_$ can be either RR or Rr (from Equation 13-2). Similarly, the outcome of the cross with respect to the Yy pair is $\tfrac{3}{4}Y\_\_ + \tfrac{1}{4}yy$. The combined probability of these events occurring independently and simultaneously is the product of their individual probabilities, or

$$(\tfrac{3}{4}R\_\_ + \tfrac{1}{4}rr)\,(\tfrac{3}{4}Y\_\_ + \tfrac{1}{4}yy)$$
$$= \tfrac{9}{16}R\_\_Y\_\_ + \tfrac{3}{16}\,rrY\_\_ + \tfrac{3}{16}R\_\_yy + \tfrac{1}{16}rryy \qquad (13\text{-}3)$$

giving the expected 9:3:3:1 ratio in the offspring (redo problems 10 and 23 using the algebraic method).

Testing the Fit of Observed to Expected Results

In experimental genetic crosses it is rare that the ratios of offspring exactly match the expected frequencies. For example, in Mendel's dihybrid cross outlined in this chapter, the expected 9:3:3:1 ratio was approximated but not precisely matched by the observed distribution of different classes in the offspring. The numbers actually obtained, 315:101:108:32, were obviously close to the expected ratios but did not exactly duplicate them. How far can observed frequencies depart from expected results and still be considered to support a hypothesis? To answer this question it must be determined whether the departure from expected results is due to chance rather than to mistakes in the hypothesis.

To make this determination a statistical technique called the *chi square* (χ^2) *method* is often used to decide whether the outcome is close enough to expectations to support the hypothesis. The method will be applied to the Mendel experiment described on page 328 to illustrate its use.

In the experiment, 556 offspring were obtained with differences in seed form and color distributed among four classes. The offspring from the cross ($RrYy \times RrYY$) were expected in a 9:3:3:1 ratio, or 313 round, yellow seeds ($\tfrac{9}{16} \times 556$); 104 wrinkled, yellow seeds ($\tfrac{3}{16} \times 556$); 104 round, green seeds ($\tfrac{3}{16} \times 556$); and 35 wrinkled, green seeds ($\tfrac{1}{16} \times 556$). As noted, the numbers actually obtained were 315:101:108:32.

These results are evaluated by the chi square method according to the equation:

$$\chi^2 = \Sigma\,(d^2/e) \qquad (13\text{-}4)$$

where χ^2 is chi square, d is the difference between observed and expected numbers for each class, and e is the expected number for each class. The symbol Σ means "the sum of." Calculation of chi square from this equation using the data from the Mendel experiment is shown in Table 13-2. The observed results for each class of offspring are entered in the top line of the table, and the expected results are listed beneath them. The difference between observed and expected numbers of offspring is found (line 3) and squared (line 4). Then, for each column, line 4 is divided by line 2 (this operation carries out d^2/e), and the result is entered in line 5. Finally, the chi square value for the experiment is calculated by adding the results from each column together [$\Sigma(d^2/e)$; line 6].

The figure obtained is then located on a table of chi square values (Table 13-3). To use the table, the *degrees of freedom* for the experiment must first be determined. Usually, the degrees of freedom are equivalent to one less than the total number of classes in the experiment, or

$$\text{degrees of freedom} = n - 1 \qquad (13\text{-}5)$$

where n is the number of classes. In the experiment entered in Table 13-2, there are four expected classes, and thus the degrees of freedom are $4 - 1 = 3$. Locating three degrees of freedom in Table 13-3 and following across the line to the figures closest to the value for chi square obtained in our calculation shows that the value falls between columns 3 and 4 of Table 13-3, or between $P = 0.95$ and $P = 0.90$, closer to 0.95. By definition, values for P greater than 0.05 indicate that chance variations are responsible for the observed deviations from expected results, and that the hypothesis is acceptable. The P value obtained for Mendel's experiment, falling between 0.95 and 0.90, is well above this arbitrary limit. If lower values for P, between 0.05 and 0.01, are obtained the differences between expected and observed ratios are considered significant and indicate that further experiments should be done before the hypothesis is accepted or rejected. Values

Table 13-2 Calculation of χ^2

	Column 1	Column 2	Column 3	Column 4
	Round, Yellow	Wrinkled, Yellow	Round, Green	Wrinkled, Green
1. Observed results	315	101	108	32
2. Expected results (e)	313	104	104	35
3. d (observed − expected)	2	−3	+4	−3
4. d^2	4	9	16	9
5. d^2/e	0.01	0.08	0.15	0.2

6. $\chi^2 = \Sigma(d^2/e) = (0.01 + 0.08 + 0.15 + 0.2) = 0.44$

Table 13-3 Chi Square Values

Degrees of Freedom	P = .99	P = .98	P = .95	P = .90	P = .80	P = .70	P = .50	P = .30	P = .20	P = .10	P = .05	P = .02	P = .01
1	0.00006	0.00063	0.0039	0.016	0.064	0.148	0.455	1.074	1.642	2.706	3.841	5.412	6.635
2	0.0201	0.0404	0.103	0.211	0.446	0.713	1.386	2.408	3.219	4.605	5.991	7.824	9.210
3	0.115	0.185	0.352	0.584	1.004	1.424	2.366	3.665	4.642	6.251	7.815	9.837	11.341
4	0.297	0.429	0.711	1.064	1.649	2.195	3.357	4.878	5.989	7.779	9.488	11.668	13.277
5	0.554	0.752	1.154	1.610	2.343	3.00	4.351	6.064	7.289	9.236	11.070	13.388	15.086

for P less than 0.01 indicate that the hypothesis is probably wrong.

Supplement 13-2: Mutations

The ultimate source of the different alleles of each gene is through *mutations*. A mutation may be defined as any change in the DNA sequence of a gene. For a mutation to be significant in genetics and heredity, it must occur in the DNA of reproductive cells; that is, in cells that give rise to new individuals. In animals and plants that reproduce sexually, the important cells are those that give rise to eggs and sperm. Mutations in the DNA of these cells are incorporated into the nucleus of the zygote at fertilization and, through the mitotic divisions taking place during development, are duplicated and passed on to all the body cells of the new organisms.

Gene Mutations

Mutations that occur within the length of DNA coding for a single function such as an mRNA or rRNA molecule are called *gene mutations*. Most common are changes in single bases of a DNA sequence, involving the substitution of one base for another or the addition or deletion of single bases. Mutations of this type may cause the substitution of one amino acid for another in the protein encoded in the mutated gene (Figs. 13-20a and b), or may change a triplet coding for an amino acid to a terminator codon.

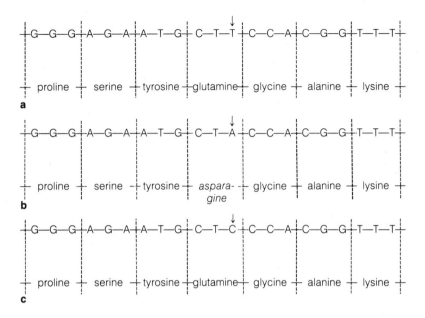

Figure 13-20 The effects of mutations on proteins encoded in a gene. (**a**) The gene sequence and the corresponding sequence of amino acids in a protein. A mutation changing thymine to adenine at the position marked by the arrow would cause substitution of asparagine for glutamine in the encoded protein (**b**). However, a change from thymine to cytosine at the same position (**c**) would lead to no change in the amino acid sequence, since the codewords CTT and CTC both stand for glutamine.

However, not all mutations involving single base changes cause amino acid substitutions because many of the amino acids are specified by more than one three-letter codeword in DNA. For example, mutation of the codeword CTT to CTC would cause no change in the amino acid chain of protein coded for by a gene because both CTT and CTC are codewords for the amino acid glutamine (see Figs. 13-20a and c). If an amino acid substitution does occur, the mutation creates a new allele of the gene.

Gene mutations that involve single nucleotide changes are sometimes called *point mutations*. Changes in single nucleotides may occur in either of two ways. One, called a *transition*, occurs when one purine is replaced by another (a replacement of adenine by guanine or guanine by adenine) or when one pyrimidine is replaced by another (cytosine by thymine, or thymine by cytosine; see Fig. 13-21a). Transitions have the effect of substituting a different base pair for the original one. *Transversions*, the second type of nucleotide change, take place when a purine is replaced by a pyrimidine, or a pyrimidine by a purine (the replacement of either cytosine or thymine by guanine or adenine, or vice versa; Fig. 13-21b). In transversions, the same base pair is retained, but the purine and pyrimidine bases of the pair change sides in the double helix.

These changes may arise from a variety of sources. Rarely, one of the nucleotide bases may take up an altered distribution of its atoms in space that permits unusual base pairs to form (Fig. 13-22a). These changes, for example, would permit adenine to pair with cytosine instead of its normal pairing partner, thymine (Fig. 13-22b), during replication. Changes of this type, called *tautomeric shifts*, occur naturally in the nucleotide bases at a low frequency.

Other changes may be induced in the nucleotide bases of DNA by the action of chemicals such as nitrous acid, hydroxylamine, and mustard gas. Nitrous acid, for example, converts adenine to a modified base, *hypoxanthine*, by removing an amino group (Fig. 13-23a). At a subsequent replication, instead of pairing with thymine, as the adenine normally would, hypoxanthine pairs with cytosine (Fig. 13-23b). As a result, C is substituted for the T that would normally be placed in this position in the newly synthesized chain, and an A-C base pair is converted to a G-C pair in the DNA in subsequent generations. Hydroxylamine has the reverse effect, and induces chemical alterations that change a G-C base pair to an A-C pair.

Tautomeric shifts and chemical alterations of the DNA bases can also be induced by *ionizing radiations* such as x

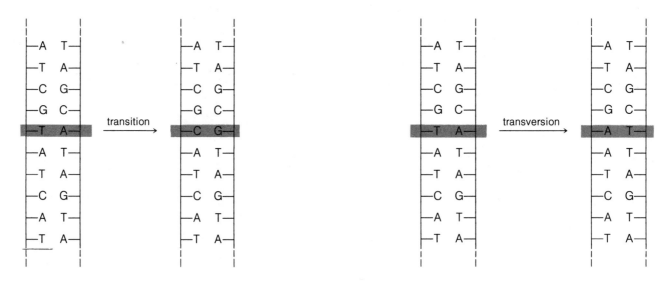

Figure 13-21 (a) A transition, a mutation that has the effect of substituting another base pair for the correct one. (b) A transversion, a mutation in which the correct base pair is retained but the purine and pyrimidine bases of the pair change places in the double helix.

rays, gamma rays, and neutrons. These forms of radiation are called "ionizing" because they cause ejection of an electron from one of the DNA bases or from molecules in the medium surrounding the DNA. This ejection creates a chemically reactive ionic group that can alter the bases or cause breaks in the DNA backbone chain. Ultraviolet light, a nonionizing form of radiation, can also alter DNA, in this case by inducing chemical crosslinks to form between pyrimidine bases in the DNA chain. These links, by causing the bases to pair incorrectly, can lead to both transitions and transversions.

Depending on their position in the DNA code for a protein or polypeptide, these sequence alterations may change a coding triplet from one amino acid to another, or may change a coding triplet into a terminator codon. These changes may have various effects on the polypeptide coded for by the transcription unit. If a change in amino acid sequence occurs in the active site of an enzyme, the protein may be altered so extensively that its catalytic activity is destroyed. Changes in amino acids in other parts of the polypeptide chain of an enzyme may have lesser or no significant effects. Changes from one codon to another specifying the same amino acid would have no effect on the protein encoded in a gene.

Gene mutations also arise through additions or deletions of one or more nucleotide bases in a DNA chain (Fig. 13-24). These alterations may be caused by acridine dyes,

for example, which can remove from one to 20 or more adjacent nucleotides from a chain. Such deletions and insertions cause a change in *reading frame* of the DNA code. From the position of the insertion or deletion to the end of the gene, the second or third base of each triplet will now be read as the first base of each triplet (Fig. 13-25). Single deletions or insertions, especially if located near the beginning of the gene, cause extreme changes in the amino acid sequence of the encoded protein, usually with complete loss of activity. Insertions or deletions that occur near the end of the gene, or additional insertions or deletions that combine to reestablish the reading frame within a few triplets will have less drastic effects.

Mutations noted in the genes coding for hemoglobin in humans give an idea of the extent of such changes. Of the 169 mutations reported in different regions of the two genes coding for the polypeptides of this protein, 161 can be accounted for by single base changes; one would require a change of two bases within the coding triplet affected. One mutation involves a change in a terminator codon that results in the addition of 31 extra amino acids to the polypeptide. The rest of the changes are deletions that have various effects ranging from amino acid changes to removal of 1 to 5 amino acids from the polypeptide chains. The effects of these mutations range from unnoticeable to seriously debilitating for the persons carrying them.

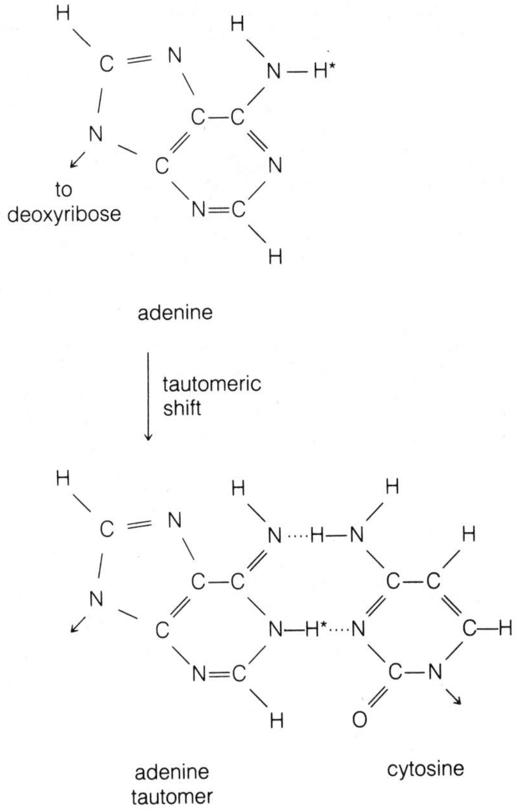

Figure 13-22 A tautomeric shift in adenine (**a**) leading to pairing with cytosine instead of its normal pairing partner, thymine (**b**).

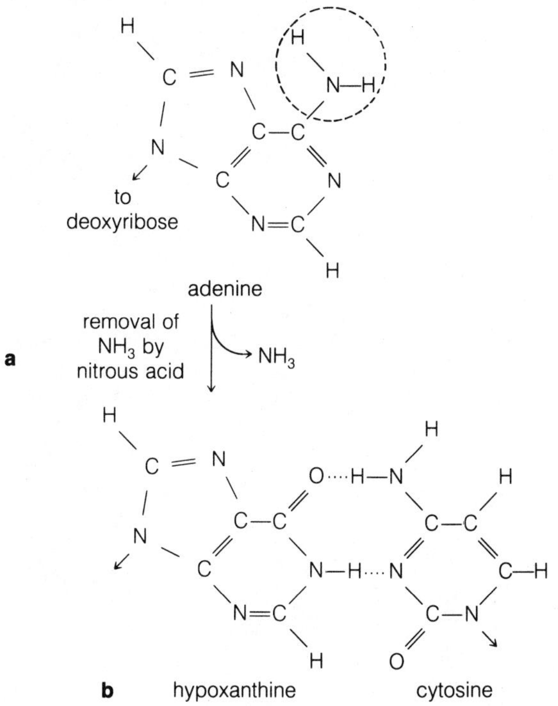

Figure 13-23 Conversion of adenine to hypoxanthine by nitrous acid by removal of an amino group (dotted lines) (**a**); hypoxanthine pairs with cytosine instead of thymine (**b**). This change would produce a transition after DNA replication.

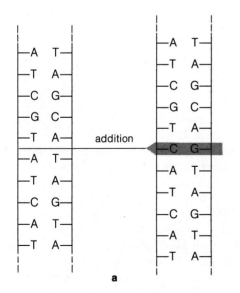

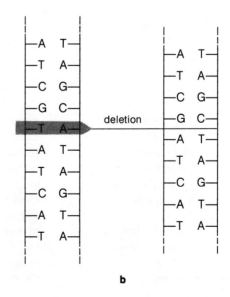

Figure 13-24 Mutations through additions (**a**) and deletions (**b**) of single base pairs.

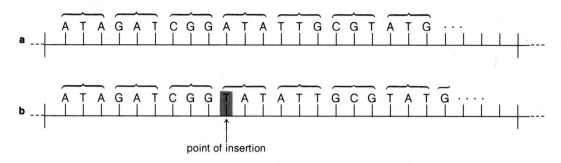

Figure 13-25 A change in reading frame of a DNA sequence because of an addition. (**a**) The unmutated gene; triplets are read in sets of three bases as shown. (**b**) An insertion of a base at the point shown causes a change in reading frame. All the triplets are read incorrectly after the point of insertion. Deletion of a single base would have the same effect.

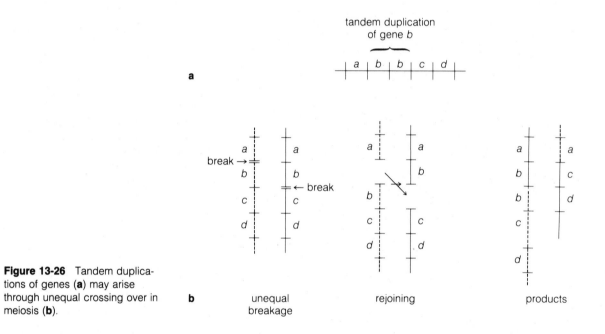

Figure 13-26 Tandem duplications of genes (**a**) may arise through unequal crossing over in meiosis (**b**).

Occasionally, gene mutations involve duplications of an entire gene, so that offspring receive two copies of the gene that are repeatedly tandemly in the chromosome (Fig. 13-26a). Such changes may arise through rare mistakes in the breakage and exchange mechanism producing recombination in meiosis (see Fig. 12-9). In such cases, the breakage and exchange is unequal, so that one chromosome receives extra DNA (Fig. 13-26b). Gene duplication by this or other mechanisms has probably been an important source of genetic changes in evolution. For example, the hemoglobin molecules of lower vertebrates are rela-

tively simple proteins consisting of polypeptides transcribed from a single gene. The more complex and efficient hemoglobin molecules of higher vertebrates contain polypeptides transcribed from two different genes. These genes, which are similar in sequence and arranged tandemly on the chromosome, probably arose through a duplication resulting from unequal crossing over during meiosis.

Gene mutations of all kinds occur at low but steady and predictable rates, from the rarest at 1 out of 1,000,000,000 to the most frequent at 1 out of 10,000 copies of a gene. Average rates fall in the vicinity of about 1 in 100,000 gene copies. This means that for each gene with this average mutation rate in a diploid species, about 1 out of every 50,000 individuals would carry a newly mutated form of that gene (the number is 50,000 instead of 100,000 since diploids carry two copies of each gene). While these mutation rates seem low, the large number of genes in individuals and the large number of individuals in many species assure that new mutations will appear frequently. For example, assuming that humans have about 30,000 different genes, the average rate of 1 mutated gene out of 100,000 copies indicates that each person is expected to carry about $30,000 \times 2 \times 1/100,000 = 0.6$ new mutations (30,000 is multiplied by 2 since we are diploid and carry 2 copies of each gene). In other words, each of us has about 1 chance in 2 of harboring a mutation not found in our parents. Over the human population as a whole, which includes about 4,000,000,000 people, $0.6 \times 4,000,000,000 = 2,400,000,000$ mutations are expected to appear each generation.

Most gene mutations, whether they involve a single or many nucleotides, are deleterious or, at least, bring no advantage to the individual carrying them. This is because most mutations ultimately show up as changes in structural and enzymatic proteins. Changing these molecules at random, as mutations do, is likely to upset the fine and complex balance that produces the coordinated biochemical activity of an organism. A good analogy would be to note the effects of inducing "mutations" in a complex electronic device such as a computer by changing the values of resistors, capacitators, and other components in the instrument at random. Most of the random changes would interfere with performance or, at best, would have no effect. However, a very few mutations out of thousands or millions might increase the speed, efficiency, or capacity of the computer, or, in the case of mutations in nature, the ability of organisms to survive and reproduce.

Chromosome Mutations

More extensive changes involving the DNA of several genes are called *chromosome mutations*. Most of these changes result from breaks introduced into the sugar-phosphate backbone chains of DNA (Fig. 13-27): These breaks separate a DNA molecule into two or more pieces, which may subsequently reattach in the same position (Fig. 13-27a),

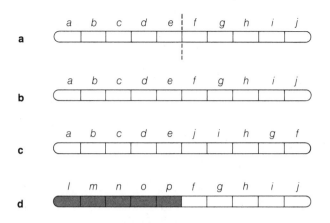

Figure 13-27 Chromosome mutations. If the DNA of a chromosome breaks at the position shown by the dotted lines (**a**) the broken segment may attach (**b**) in the same position, (**c**) in reversed position, or (**d**) to the end of a different chromosome.

attach in reversed position (Fig. 13-27b), attach to the DNA molecule of a different chromosome (Fig. 13-27c), or be lost entirely. All of these changes, which take place in nature, can be induced by various chemicals found in the environment or by ionizing radiation such as x rays.

Deletions of chromosome segments are usually lethal to any offspring carrying them. However, reversed attachment of a broken segment, called an *insertion*, or the attachment of a broken segment to the end of a different chromosome, called a *translocation*, may have less drastic effects. Among these may be prevention of recombination of alleles within reversed or translocated segments through interference with chromosome pairing in meiosis. Or, such rearrangements might affect the mechanisms regulating the transcription of genes within the altered segments.

Insertions and translocations are frequently encountered in nature. Chromosome rearrangements are especially common in plants; some animals, such as scorpions, grasshoppers, cockroaches, and various species of flies also show a large proportion of individuals with altered chromosomes. Chromosome rearrangements also occur in humans. Down's syndrome, a mutation with a deleterious effect in humans, involves addition of an entire extra chromosome. In a few animal and plant species, chromosome mutations are so frequent that it is difficult to decide which chromosome arrangement is the normal or wild-type for the species.

Supplement 13-3:
The Genetics of Bacteria and Viruses

During the 1940s and 1950s, geneticists working in several laboratories established that genetic recombination takes place in the DNA of bacteria and viruses as well as in eukaryotes. This was a most significant discovery, because bacteria and viruses can be grown with ease by the millions or even billions in generations that pass in a matter of minutes. In comparison, a generation of *Drosophila* takes two weeks to grow from zygote to sexual maturity, and geneticists working with corn are lucky to get two generations a year.

The rapid generation times of bacteria and viruses allowed crosses and their outcomes to be traced much more quickly. In addition, the millions or billions of offspring obtained from a cross permitted rare events, such as recombination between very closely spaced points along DNA sequences, to be detected. This, in turn, allowed mapping to be extended to points within genes, sometimes to distances as short as individual nucleotides. Even more important to the study of genes and their activity was the fact that the effects of different alleles could be defined in terms of biochemical effects rather than morphological differences in offspring. Because bacteria can be grown in large quantities in culture solutions that are completely defined chemically, what goes into bacterial cells, what chemical changes occur inside, and what comes out can in many cases be detected and closely followed. Studies of this type provided much of the information revealing how DNA replication, RNA transcription, protein synthesis, and other biochemical processes take place in cells, and led to a more complete and accurate definition of genes and alleles and their activities at the biochemical and molecular level.

Genetic Studies with Bacteria

Sexual reproduction, with recombination between alleles and independent segregation of maternal and paternal chromosomes in meiosis, occurs in diploids in eukaryotes. Although bacteria are normally haploids and do not undergo meiosis, the processes of recombination and segregation of alleles can still occur in individual cells tem-

porarily converted to the diploid condition by the entry of DNA from another bacterial cell. The extra DNA may include only a segment of the DNA circle of another bacterium; cells carrying segments of this type in addition to their own DNA are called *partial diploids*. In this case, which is the most frequent condition observed in bacteria, recombination and segregation of alleles occurs only in the segment of the host cell DNA that is homologous to the piece of DNA entering from outside. More rarely, a complete DNA circle may enter from another cell, creating a complete diploid. Three routes have been recognized by which DNA from one bacterial cell may enter another to create partial or complete diploids: *conjugation, transformation,* and *transduction*.

In conjugation, which resembles sexual reproduction in diploid eukaryotes, bacterial cells of the same species make contact and fuse together at one point, forming a cytoplasmic bridge that directly connects the two cells (Fig. 13-28). Once the bridge forms, a copy of part or all of the donor cell's DNA may cross the bridge and enter the recipient cell.

In transformation, pieces of DNA released from fragmenting bacterial cells are absorbed from the surrounding solution by healthy cells. For the DNA pieces to be absorbed, they must originate from the same species as the recipient cell. Although transformation is best known from laboratory studies in which recipient cells are exposed to DNA artificially extracted and purified from donor cells, the process probably also occurs under natural conditions. The molecular basis for penetration of the absorbed DNA through the wall and plasma membrane of the recipient cell is unknown.

Transduction occurs through a transfer of DNA fragments from one cell to another in virus particles. The process depends on the cycle of infection and multiplication of viral particles inside bacteria (see p. 21). Occasionally, a fragment of the DNA of a bacterial host cell is packed into a virus particle along with the viral DNA. The viral particles mistakenly containing bacterial DNA, which are released along with the normally constituted particles, are infective and can attach to another cell of the same species. When attachment occurs, the fragment of bacterial DNA contained inside the viral particle is injected into the recipient cell just as the viral DNA would be.

Recombination in Partial Diploids

All three processes, conjugation, transformation, and transduction, result in the insertion of an extra piece of

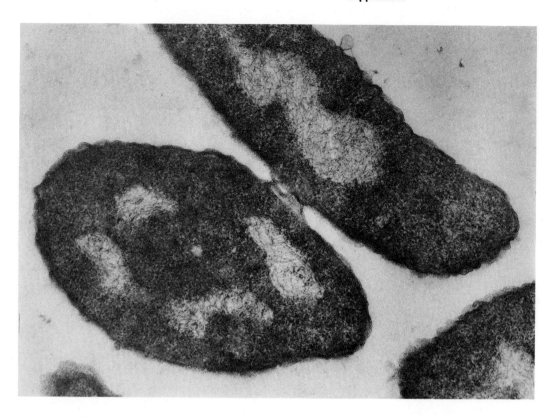

Figure 13-28 Conjugating cells of *E. coli*. A cytoplasmic bridge has formed between the cells in the region of contact. × 68,000. From L. G. Caro, *Journal of Molecular Biology* 16 (1966):269, courtesy of L. G. Caro and Academic Press, Inc. (London) Ltd.

DNA of the same species inside a bacterial cell, creating a partial or complete diploid. Once the extra piece of DNA is inside, it can pair with the homologous sequences of the DNA circle of the bacterial cell (Fig. 13-29a) and enter into recombination. During recombination, which occurs by essentially the same molecular mechanisms as in eukaryotes, homologous segments break and exchange between the bacterial DNA circle and the extra piece of DNA. The result, if different alleles of the genes in the diploid region are present in the two molecules entering into the exchange, is a new combination of alleles in the bacterial DNA circle (Fig. 13-29b). Following recombination, the bacterial DNA replicates and the cell divides normally, producing a cell line with the new combination of alleles.

The progeny of a single bacterium, since they are haploids, usually contain only one of the four possible outcomes of a recombination event (the two parentals, and the two recombinants—see Fig. 13-9). However, because

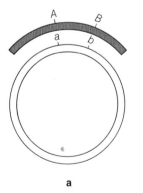
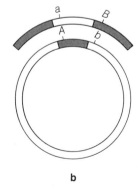

a   b

Figure 13-29 Recombination of a region in which a bacterial cell has been converted to a partial diploid. (**a**) Pairing between the extra DNA segment and the DNA circle of the recipient cell; (**b**) the result of recombination.

so many cells can be used for a cross between parental types (millions or billions), the chance is good that each of the four possible outcomes of recombination can be recovered among the many offspring. Thus, although meiosis does not occur in bacteria, the outcome of sexual reproduction and recombination can be traced by noting the percentages of parental and recombinant types among the offspring as in eukaryotes.

Hereditary traits in bacteria are detected through the changes they cause in phenotypes, just as they are in eukaryotes. However, most of the bacterial phenotypes studied in genetic crosses involve biochemical characteristics such as the chemical requirements of the growth medium, differences in cell processes such as protein synthesis, transcription, or replication, or changes in products secreted by the cells.

One of the experiments that first detected recombination in bacteria, conducted by Joshua Lederberg and Edward L. Tatum at Yale University, illustrates how bacterial crosses and recombination studies are carried out. Lederberg and Tatum worked with two types of *Escherichia coli* that had lost their ability to grow on the *minimal medium* required by normal, wild-type cells (wild-type *E. coli* can grow if supplied only with a solution of inorganic salts and an energy source such as glucose). One type could grow only if the vitamin *biotin* and the amino acid *methionine* were added to the medium. The second type had no requirement for biotin or methionine, but could grow only if the amino acids threonine and leucine were added to the growth medium. These strains can be represented as:

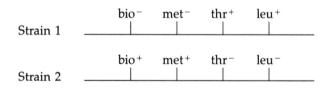

where *bio+* indicates the wild-type allele providing cells with the ability to synthesize biotin for themselves, and *bio−* the allele producing cells that cannot synthesize biotin. Similarly, *met+*, *met−*; *thr+*, *thr−*; and *leu+* and *leu−* are the respective wild-type and deficient alleles for methionine, threonine, and leucine synthesis.

Lederberg and Tatum mixed about 100,000,000 cells of the two deficient strains together and placed them on a minimal medium. Most of the cells were unable to survive, but a few hundred viable colonies of bacteria were formed containing cells descended from single individuals that were able to grow on the minimal medium. Since these cells could grow, they must have had the

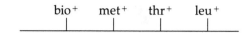

combination of alleles and were therefore completely wild-type bacteria.

It was obvious from this that recombination had taken place between DNA containing the different alleles to produce the wild-type strain (Fig. 13-30). In this case, the recombination probably occurred in partial diploids created by conjugation. The results of such studies can be used to construct maps of bacterial chromosomes through comparisons of crossover frequencies as in eukaryotes.

Work with DNA ligase, one of the enzymes active in DNA replication (see Supplement 11-2), illustrates how the ability to detect alleles and carry out crosses with bacteria has allowed biochemists and geneticists to trace out the steps in biochemical pathways. Through genetic crosses, bacteria were found that had an allele coding for a defective form of DNA ligase. Bacteria with the wild-type allele for DNA ligase are able to carry out DNA replication normally. However, the newly synthesized DNA was left in short, disconnected pieces in bacteria with the allele coding for the defective DNA ligase. This finding showed that DNA ligase probably catalyzes the last step in DNA replication, in which the short pieces assembled in earlier steps are joined into long, continuous molecules (see Fig. 11-29d and e). Other steps in DNA replication were traced by a similar approach.

Genetic experiments with bacteria have also been of fundamental importance in research unraveling the molecular mechanism of recombination. Some of this research has been concerned with the enzymes and other factors active in recombination, and the sequence of molecular changes that produce breaks and exchanges between homologous DNA molecules. This investigation has been approached by the same techniques used in studies of replication, transcription, and protein synthesis: the steps in the mechanism are traced by noting the effects of alleles coding for deficient forms of enzymes and other factors suspected to be active in recombination.

Genetic Crosses and Recombination in Viruses

Recombination can also occur between DNA molecules originating from different virus particles if two or more viruses simultaneously infect the same cell. At some point

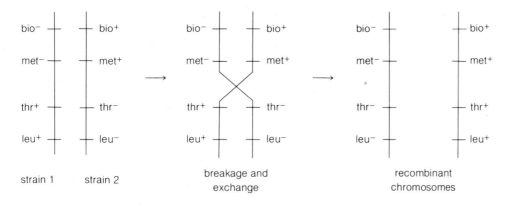

Figure 13-30 Recombination between the bacterial strains used in the Lederberg and Tatum experiment (see text).

during the growth of the new virus particles within the infected cell, viral DNA molecules from the different parental types may pair and cross over by breakage and exchange. Any recombinant viral chromosomes produced from this breakage and exchange are then packed into protein coats and, upon rupture of the host cell, are released to the medium to infect further hosts.

Viral recombinants are detected primarily through changes in the biochemistry of viral DNA replication, recombination, cell infection, or the molecular structure of virus particles produced by the infection. For example, in the viruses infecting *E. coli*, various alleles cause differences in enzymes replicating or recombining the viral DNA, or in the proteins of the viral coats. These phenotypes, which can be detected by biochemical tests or electron microscopy, have also revealed much new information about the molecular and biochemical steps occurring in DNA replication, recombination, RNA transcription, protein synthesis, and other cellular mechanisms.

Suggestions for Further Reading

Goodenough, V. and R. P. Levine. 1974. *Genetics*. Holt, Rinehart & Winston: New York.

Watson, J. D. 1976. *Molecular biology of the gene*. 3rd ed. W. A. Benjamin: Menlo Park, Calif.

SEVEN

Cell Origins

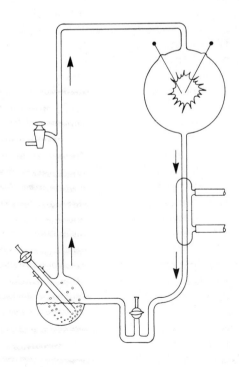

14

The Origins of Cellular Life

The most ancient rocks of our solar system have been dated as 4.5 billion years old. The earth is estimated to be of the same age. At this time, 4.5 billion years ago, our planet condensed out of the primordial matter and began its long transition into the environment we know today. The oldest known fossils of bacterialike cells, discovered in deposits at North Pole, a region in remote northern Australia, are estimated to be approximately 3.5 billion years old. Therefore life must have originated on our planet at some time during the first billion years of its existence.

Very few fossils remain to tell us of the characteristics of the cellular life of 3.5 billion years ago; there is just enough information to indicate that the earliest cells were probably bacterialike prokaryotes (Fig. 14-1). Nothing at all exists to inform us about the earlier period, between 4.5 and 3.5 billion years ago, when the transition from nonliving to living matter took place. No intermediate forms survive, and there is no readable fossil record of this period. Thus we are left with hypothesis, speculation, and conjecture to unravel the events of this time.

Hypotheses explaining the origin of life have had a long and interesting history. The biblical explanation that life was placed here by divine intervention was concerned primarily with human and other higher or more prominent forms of life. Lesser creatures like insects and worms were commonly believed as late as the mid-nineteenth century to arise spontaneously and more or less continuously from rotting organic matter. Beliefs of this kind have always been a part of folklore, and even scientists have at times been convinced that lesser life could arise by spontaneous generation.

These ideas were refuted among scientists by a series of experiments that culminated in Louis Pasteur's famous demonstration in 1862 that nutrient fluids sterilized and sealed to protect them from contamination could be kept indefinitely without generation of microbial or other life. Pasteur's refutation of spontaneous generation was so effective that scientists generally came to reject the idea that life on earth could have arisen spontaneously at any time. As a consequence, theories about the possible origins of life were not considered seriously by anyone in the world of science for nearly a hundred years after Pasteur's experiments. This situation began to change in the 1920s, following the publication of a major hypothesis for the origin of life advanced independently by A. I. Oparin, a Russian, and J. B. S. Haldane, an Englishman. Oparin and Haldane agreed that spontaneous generation of complex cellular life under present-day conditions is impossible. They argued, however, that the earth's surface and atmosphere were radically different from today's conditions during the first millions of years of its existence. Haldane in particular proposed that the earth's initial atmosphere probably contained reduced substances such as methane (CH_4), ammonia (NH_3), and water vapor instead of high concentrations of oxygen as in today's atmosphere.

According to Oparin and Haldane, the action of sunlight, heat from the earth, and lightning on the initial atmosphere produced significant quantities of organic molecules. These molecules persisted and accumulated in the lakes and seas of the primitive earth because oxidation and microbial decay, the two main routes by which organic molecules are broken down in the present-day environment, could not take place. Over millions of years, these organic molecules became more concentrated and interacted spontaneously to produce still more complex substances such as nucleic acids and proteins. Haldane thought that the concentration of these molecules in the seas might even reach the consistency of a "hot, dilute soup."

According to Oparin and Haldane these organic substances, through constant mixing and interactions, eventually collected by chance into assemblies that could carry

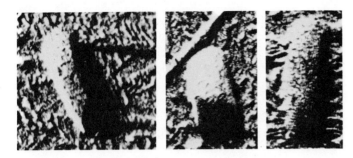

Figure 14-1 Fossils of bacterialike prokaryotes exposed in three-billion-year-old rocks by polishing. Courtesy of E. S. Barghoorn, from *Science* 152 (1966):758. Copyright 1966 by the American Association for the Advancement of Science.

out some of the reactions of life. As an energy source, these first assemblies used the organic molecules that were present in abundance in the environment.

The molecular assemblies that were most successful in competing for space and raw materials persisted and accumulated. Eventually the most successful of these aggregations developed the full qualities of life. As the chemical activity of this emerging life increased, the store of organic molecules used as an energy source was gradually exhausted. Life persisted, however, through the development of photosynthesis as the ultimate source of energy for the replenishment of organic matter. The release of oxygen by more advanced photosynthesizers, and the depletion of organic molecules in the environment changed the earth and atmosphere so drastically that the spontaneous generation of life was no longer possible. At this point, only life arising from preexisting life could exist, as on the present-day earth.

Oparin and Haldane's hypothesis stirred some interest among scientists and others in the 1920s, and it was admitted that their proposal seemed plausible. However, the entire idea was considered little more than an interesting intellectual exercise for another 30 years, until major discoveries in two areas changed these attitudes and led to a burst of new interest and investigation in the possible origins of life. One was the demonstration in 1953 by Stanley Miller at the University of Chicago that the action of electrical discharges on reduced gases could actually produce a variety of organic compounds. The second was the observation that other planets of our solar system, most notably Jupiter and Saturn, still retain atmospheres of reduced gases of the type envisioned by Oparin and Hal-

dane. Vast clouds of frozen particles containing these gases and even organic molecules were also detected in interstellar space. Thus the initial conditions and developments proposed for the origin of life by the two theorists were shown to be likely events in the history of our solar system and earth.

Most scientists now agree that life could have originated spontaneously on the primitive earth under the environmental conditions thought to exist at that time. On this assumption, a series of stages based on Oparin and Haldane's original hypothesis has been proposed as the most likely course of events in the origin of life.

Before we consider these stages we must define the point at which a collection of interacting molecules takes on the quality of life. Probably, the first spark of life appeared in molecular aggregates much simpler than even the most simple cells we know today. Present-day cells at a minimum have a boundary membrane separating the cell interior from the exterior environment, one or more nucleic acid coding molecules, a translation system capable of using the coded information to produce biological molecules, and a metabolic system providing the energy to carry out the cell's activities. Because these systems are so complex, it is hardly conceivable that life appeared at this level without being preceded by less organized forms. Therefore, the transition from nonliving matter to the first cells was probably gradual, and no sudden event caused cellular life in all its present-day complexity to appear at one instant.

The problem, then, is to decide what minimum level of organization and activity is required for a collection of interacting molecules to be considered alive. Some investigators claim that very simple groups of molecules are alive if they can use an energy source to carry out a single, continuous energy-requiring reaction. Others insist that life at a minimum must include the ability to grow and reproduce in kind.

Because the latter, more complex requirements are closer to our present-day concept of life, we will use this level of organization to define precellular life. Therefore, collections of molecules will be considered to have the first spark of life if they can (1) use either light or chemical energy to drive internal reactions requiring energy, (2) increase in mass by controlled synthesis, and (3) reproduce into additional collections of matter of the same kind. The ability to reproduce in kind is considered to carry with it a requirement for an information-coding system and a system for translating the coded information into finished molecules. At this minimum level, life is relatively complex but still much simpler than the simplest cells we know of.

Stages in the Evolution of Life

Although probably a continuous process, the first evolution of life can be broken for convenience into four successive stages. The first was the formation of the earth and its initial atmosphere. The first stage provided inorganic raw materials for the evolution of life and set up the conditions under which they could interact. In the second stage, complex organic molecules were produced by inanimate forces such as lightning or ultraviolet radiation acting on the inorganic chemicals in the environment. This hypothetical stage has been successfully duplicated in the laboratory, with results indicating how this step probably took place. In the third stage, the newly produced organic molecules collected by chance into assemblies capable of chemical interaction with the environment. Experiments testing ideas dealing with this stage have also met with some success. In the fourth and final stage, some of these assemblies were successful in converting the energy of complex molecules absorbed from the environment into useful chemical energy; some of this energy was used to synthesize other complex molecules such as proteins and nucleic acids. Gradually, in this stage, the coding function of the nucleic acids became established, and the relationship between the nucleic acid sequences and the sequences of amino acids in proteins became fixed into the genetic code. This development provided direction to synthesis and reproduction in the primitive molecular assemblies. At this level of organization, life appeared in the assemblies. Evolution then changed from chemical to organic, with natural selection of favorable mutations in the coding system as the basis for further evolutionary development and change. Although the development of the coding relationship between nucleic acids and proteins is the most difficult part of the entire process of evolution to deal with experimentally, it has also been partially reconstructed and tested with some success.

Once life began in the molecular assemblies and the basis for organic evolution was established, natural selection led to the appearance of cells equivalent to the most primitive prokaryotes known today. Presumably, this level was reached at or before the time the North Pole deposits were laid down in Australia some 3.5 billion years ago. Interactions between these early prokaryotes and further selection eventually established the lines leading to eukaryotic cells, a development that may have required another 1 to 1.5 billion years.

To study the evolution of life by the scientific method we must assume that all the processes occurring in these successive stages were inanimate and took place by chance.[1] Given similar conditions and sufficient time, it is probable that a similar process could occur again and that life has evolved or is evolving at other locations in the universe.

The First Stage: The Origin of the Earth and Its Primitive Atmosphere

The contemporary view of the earth's origins is that the sun and planets of our solar system and all the stars and other bodies of the universe condensed out of cosmic clouds of gas and dust particles. The composition of these clouds, which still persist in the universe today (Fig. 14-2), has been studied by analyzing the light transmitted or reflected by them. Most of the matter of the clouds is hydrogen gas at extremely low concentrations; lesser amounts of nitrogen, helium, and neon are also present. Other elements and compounds are also suspended in the clouds as solid particles. These include metallic iron and nickel; the silicates, oxides, sulfides, and carbides of these and other metals; inorganic and organic carbon compounds; ammonia; and frozen water (see Table 14-1).

According to the condensation hypothesis, stars, suns, and planetary systems are continually forming from cosmic clouds of gas and dust and disintegrating into dust again. Our own solar system condensed from one of these large clouds of dust. Most of the cloud condensed rapidly around a single center, causing high pressure and heat to develop in the interior and setting off a thermonuclear release of energy; this release created the sun. Other, smaller centers of condensation produced the planets.

As the condensing planets formed, their temperatures increased because of the effects of solar heating, gravitation, and internal pressure. Although never reaching the temperature of the sun, the heat inside the condensing earth became high enough to melt the collected substances. The heavy, metallic elements settled to form the core of the earth. The lighter materials, such as the silicates and carbides of these metals, rose to the surface,

---

[1]Some theoreticians feel that the chemical processes giving rise to the first life did not occur by chance. According to their view, the particular collection of inorganic molecules assembled together when the earth first appeared, the conditions of pressure, temperature, and so on, existing at the time, and the types of energy input could lead only to the chemical results actually obtained. Thus, the results were not chance or random, but were fixed as chemical probabilities when the earth first condensed.

Figure 14-2 A cosmic cloud of gas and dust (the horsehead nebula in Orion) in space some 1300 light years from the earth. Palomar Observatory photograph.

Table 14-1 Atoms, Molecules, or Chemical Groups Detected in Cosmic Clouds or Outer Space

Atom, Molecule, or Radical	Symbol
Hydrogen atom	H
Hydroxyl radical	OH
Ammonia	NH_3
Water	H_2O
Formaldehyde	HCHO
Carbon monoxide	CO
Cyanogen radical	CN
Hydrogen cyanide	HCN
Cyanoacetylene	HC_2CN
Methyl alcohol	CH_3OH
Formic acid	HCOOH
Carbon monosulfide	CS
Formamide	$HCONH_2$
Silicon oxide	SiO
Carbonyl sulfide	OCS
Acetonitrile	CH_3CN
Isocyanic acid	HNCO
Hydrogen isocyanide	HNC
Methylacetylene	CH_3C_2H
Acetaldehyde	CH_3CHO
Thioformaldehyde	HCHS
Hydrogen sulfide	H_2S
Methylene imine	H_2CNH

Adapted from S. W. Fox, *Molecular and Cellular Biochemistry* 3 (1974):129, courtesy of S. W. Fox and *MCB*.

where they cooled and solidified to form the rocks and particles of the surface crust.

The original atmosphere of the earth probably contained large quantities of hydrogen, nitrogen, and water vapor originating from the cosmic cloud. As the atmosphere and surface cooled, much of the water vapor condensed into droplets and rained down on the dust and rocks of the crust. Eventually, after years of torrential rains, water collected into the rivers, lakes, and seas of the primitive earth.

Some of the water was retained as vapor in the atmosphere. Other gases such as hydrogen, nitrogen, and carbon dioxide interacted with each other and with the carbides, nitrides, and sulfides in the crust to produce methane, ammonia, and hydrogen sulfide. Carbon dioxide may also have been expelled from the interior of the earth along with other gases by erupting volcanoes. Free oxygen is not believed to have been present in the atmosphere in significant amounts because it would have reacted quickly with particles and rocks of the crust to form oxides.

The fact that the oldest rocks of the earth's crust, now deeply buried but originally exposed to the atmosphere, contain reduced, rather than oxidized substances supports

Table 14-2 Natural Sources of Energy on the Earth

Source	Energy (cal/sq. cm/year)
Sun (total radiation including ultraviolet)	260,000
Ultraviolet light	4,000
Electrical discharges	4
Shock waves	1.1
Radioactivity (to 1 km depth)	0.8
Volcanoes	0.13
Cosmic rays	0.0015

Adapted from S. L. Miller, H. C. Urey, and J. Oro. *Journal of Molecular Evolution* 9 (1976):59.

these ideas about the characteristics of the earth's primitive atmosphere. In addition, analysis of the light transmitted and reflected by the atmospheres of the largest planets, Jupiter and Saturn, which are believed to have retained their primitive complement of gases almost intact, reveals the same strongly reduced character. Free oxygen is absent, and ammonia, methane, and water vapor are prominent among the gases present on these planets.

The absence of oxygen and the presence of hydrogen, methane, ammonia, and water vapor gave the atmosphere of the primitive earth a nonoxidizing rather than an oxidizing quality. This nonoxidizing quality was fundamental to the next stage of evolution: the appearance of complex organic molecules through the action of natural energy sources on the inorganic matter of the crust and atmosphere.

The Second Stage: The Spontaneous Production of Organic Molecules on the Primitive Earth

The hydrogen, nitrogen, methane, ammonia, water vapor, and other gases of the primordial atmosphere, and the same gases dissolved into bodies of water, were exposed to continual inputs of energy from a number of natural sources (Table 14-2). One source, then as now, was sunlight. Besides the visible light from the sun, greater quantities of ultraviolet light reached the lower atmosphere and acted on surface chemicals and waters than in the present-day environment (at present, most of the ultraviolet light approaching the earth is absorbed by oxygen and ozone in the outer atmosphere and never reaches lower levels or the surface). Another energy source was provided by heat from absorbed light and volcanic activity. Electrical discharges during the violent rainstorms of the period also supplied energy to the atmosphere and surface of the earth. According to Oparin and Haldane's ideas, chemical activation of the gases in the primitive atmosphere by these energy sources led to the spontaneous production and accumulation of a wide variety of complex organic molecules.

This proposal received direct support in 1953, when Stanley Miller developed his apparatus to test the effect of electrical discharges on a simulated primitive atmosphere. In the Miller apparatus (Fig. 14-3), water vapor, methane, ammonia, and hydrogen flowed continuously through a chamber subjected to repeated sparking from electrodes. Below the chamber, the water vapor and any organic chemicals produced cooled, condensed, and became trapped at a low point in the tubing. Operating the apparatus for one week yielded a surprising variety of organic chemicals, including urea, several amino acids, and lactic, formic, and acetic acids. Thus organic compounds can actually be produced by the action of an energy source on the gases of the primitive atmosphere as Oparin and Haldane had proposed.

After Miller's pioneering experiments showed the way, a great many additional experiments demonstrated that other organic molecules can be synthesized by varying the energy source and the gases present in the starting mixture. Adding hydrogen cyanide, which is readily produced by interactions between methane and nitrogen, produced additional amino acids and the purine and pyrimidine building blocks of the nucleic acids. Adding formaldehyde, another gas readily produced by reactions between the gases of the primitive atmosphere, led to the production of sugars, including the ribose and deoxyribose sugars of DNA and RNA. Other variations produced the subunits of lipids. A key feature in these experiments is the absence of oxygen in the simulated atmosphere. If oxygen is added to the mixture of gases used, very few or no organic molecules are produced.

Thus it has been possible to demonstrate that the building blocks of all the major kinds of biological molecules could have been synthesized on the primitive earth. The results of the analysis of meteorites and interstellar dust clouds also support the hypothesis that organic molecules could have been synthesized spontaneously. Organic molecules found in a meteorite impacting in 1969 near Murcheson, Australia include 6 of 20 amino acids found in proteins, and 12 additional nonbiological amino acids. Purines and pyrimidines have also been found in trace quantities among the organic molecules of the meteorite. These organic substances, and the organic molecules detected in interstellar dust clouds were presumably synthesized by the same inanimate, spontaneous mechanisms proposed in the Oparin-Haldane hypothesis.

Even proteinlike chains of amino acids have been synthesized in laboratory experiments approximating the primitive environment. Sidney W. Fox of the University of Miami produced proteinlike molecules by heating dried mixtures of amino acids to 160 to 210°C for several hours. Fox termed these proteinlike molecules *proteinoids*. Fox and others have also formed nucleic acids by heating mixtures of nucleotides and phosphates to about 60°C and holding them at this temperature for some time.

All these experiments establish that a wide variety of complex organic substances could have arisen spontaneously over the millions of years between the formation of the earth and the first appearance of living matter. Accumulation of these compounds, which probably included all the major molecules found in living organisms, provided the raw materials for the third stage in the evolution of life, the collection of organic matter into assemblies capable of interacting with each other and the environment.

The Third Stage: The Spontaneous Collection of Organic Molecules into Functional Assemblies

As the concentration of organic substances increased in the environment, one or more types of molecules assembled spontaneously into aggregates of various kinds. As a preliminary to this assembly, they may have been concentrated somewhat by the evaporation of water from lakes, inland seas, and tidal flats. These more concentrated solutions of organic matter may then have followed one or more of several possible routes of assembly into functional units. Some of these routes have been shown to be possible under conditions imitating the primitive environment.

Oparin proposed that a mechanism known as *coacervate* formation was important in the assembly process. Coacervates form when protein molecules in solution separate into discrete, concentrated droplets as a result of attraction between charged and polar groups on the surfaces of different polypeptide chains. If many of these polar groups face the surfaces of the protein droplet, the sur-

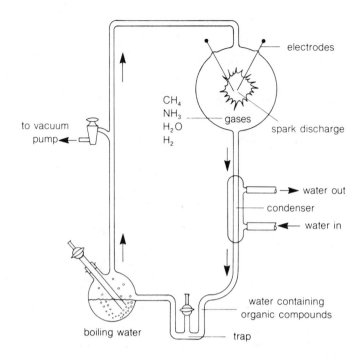

Figure 14-3 The Miller apparatus demonstrating that organic molecules can be produced spontaneously in a primitive atmosphere (see text). Courtesy of S. L. Miller. Copyright 1955 by the American Chemical Society.

rounding water molecules tend to form an ordered film or boundary layer several molecules in thickness around the coacervate. The film of water molecules gives the boundary between the droplet and the surrounding medium many of the properties of a membrane. As a result, coacervates may concentrate molecules from the surrounding medium, or shrink or swell in response to changes in inside-outside concentrations as do living cells.

A similar process of droplet formation has been studied by Fox and his colleagues. They showed that if solutions of proteinoids are heated in water and then allowed to cool, small spherical particles containing the proteinoids separate out of solution. These small particles (Fig. 14-4), termed *microspheres* by Fox, are similar to coacervates in activity and can absorb various substances from the surrounding medium. Some of the microspheres in Fox's experiments also had weak catalytic activity and were able to speed the rate of various biochemical reactions including the hydrolysis of ATP to ADP (see Table 14-3). Other microspheres were observed to bud, split, or fragment in a process superficially similar to cell division (see Fig. 14-4b).

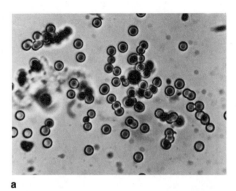

a

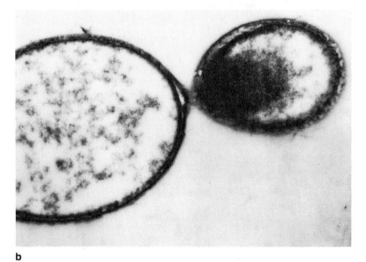

b

Figure 14-4 Fox's microspheres. (**a**) A collection of microspheres made by cooling a heated solution of proteinoids (light micrograph). × 360. (**b**) Microspheres fragmenting in a process superficially similar to cell division (electron micrograph). × 11,000. Courtesy of S. W. Fox, from *Molecular evolution and the origin of life* by S. W. Fox and K. Dose, Freeman, San Francisco, 1972.

Table 14-3 Catalytic and Related Activities of Microspheres

Type of Reaction	Substrate Broken Down by Microspheres
Hydrolysis	ATP Nitrophenyl acetate Nitrophenyl phosphate
Decarboxylation	Glucaronic acid Pyruvic acid Oxaloacetic acid
Amination	Glutamic acid
Reduction-oxidation	H_2O_2
Synthesis (with added ATP)	Nucleic acids Peptides

Adapted from S. W. Fox, *Molecular and Cellular Biochemistry* 3 (1974):129, courtesy of S. W. Fox and *MCB*.

Other assembly mechanisms may also have been important. J. D. Bernal of the University of London has proposed that molecules may have collected into aggregates by absorption on clay particles in mud around tidal flats and river mouths. Clays are built up from alternating layers of inorganic ions and water molecules arranged in a highly ordered molecule lattice. This structure provides internal surfaces that strongly absorb organic molecules and promote chemical interactions between them. This absorption on clays can enhance reactions that link building-block substances into biological macromolecules. Other reactions that release energy can also be promoted by clays, especially if the clays contain metallic ions in abundance. In this way, clays may have served as catalysts for early life reactions in addition to providing surfaces for absorption and aggregation. The mud flats containing these clays could have reached extended size, so that the areas available for spontaneous generation of life might have been widely distributed on the primitive earth.

Another possible route of aggregation is the formation of films or particles by many types of lipid molecules. When placed in water, these lipid molecules spontaneously form bilayers that have many of the properties of cell membranes (see Fig. 2-14 and p. 38). Often lipid bilayers suspended in water round up into closed vesicles consisting of a saclike continuous "membrane" enclosing a central space (Fig. 14-5). Bilayers can absorb various substances, including proteins or polypeptides, from the surrounding medium. Substances absorbed by bilayers, or trapped inside a bilayer vesicle, can be significantly altered in chemical properties. This is particularly true of the bilayer interior, which excludes water and could have promoted a variety of chemical reactions that take place more favorably in a nonwatery environment. Thus bilayers, like coacervates, microspheres, and clays, could have acted both as catalysts and points of aggregation for organic substances.

Supposedly, one or more of these aggregation processes took place repeatedly in the primitive environment. Sooner or later, within some of the collections of protein, nucleic acid, lipid, and carbohydrate molecules, life appeared.

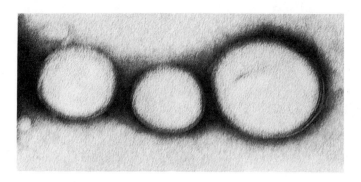

Figure 14-5 Vesicles produced from phospholipids synthesized under primitive earth conditions, as seen in an electron micrograph of a negatively stained preparation. Courtesy of W. R. Hargreaves, from *Nature* 266 (1977):78.

The Fourth Stage: The Development of Life in the Primitive Assemblies

We have defined life as the stage at which the primitive molecular assemblies could (1) use an energy source to drive energy-requiring reactions to completion, (2) increase in mass by controlled synthesis, and (3) reproduce into additional assemblies like themselves. These steps toward life have proved to be much more difficult to reconstruct and test in the laboratory. Nevertheless, some ideas about them have been developed, and a few have even been tested experimentally.

In contemporary cells reactions that release and require energy are linked through adenosine triphosphate, the ATP molecule. Reactions that release energy are used to add a phosphate to adenosine diphosphate or ADP to produce ATP. The energy stored in ATP is then released to energy-requiring reactions by removal of the added phosphate, converting ATP back to ADP (see p. 55). Since ATP is used in all present-day cells as the molecule linking energy-releasing and energy-requiring reactions, and since it is relatively easily synthesized under primitive earth conditions, it seems reasonable to assume that ATP was important in the earliest reactions leading toward life.

In present-day cells, the primary energy-releasing reactions are oxidations, in which high-energy electrons are removed from organic molecules. The energy of these electrons is tapped off gradually by passing them along a chain of carrier molecules called an *electron transport chain* (Fig. 14-6 and p. 148). As the electrons pass from one carrier to the next, they release some of their energy until, at the end of the chain, much of their energy has been given up. In contemporary cells, this released energy is used to convert ADP to ATP, creating an energy store in the form of phosphate groups linked into ATP.

Oxidations were also likely to have been among the primary energy-releasing reactions of the primitive molecular aggregates. At first, the energy released by oxidations would have been utilized in single steps, through direct transfer of high-energy electrons from oxidized substances to other molecules that were converted in the process to more complex forms. However, the advantages and efficiency of energy release in smaller steps would have favored the development of intermediate carriers, opening the pathway for the appearance of the first electron transport systems.

The porphyrins (Fig. 14-7) are prime candidates for the intermediate carriers of the first electron transport chains. These molecular groups, which could have been readily synthesized in the primitive environment, gain or lose electrons with relative ease. In this capacity, they serve as the active groups transferring electrons in the cytochromes, which occur as carrier molecules in the electron transport systems of both respiration and photosynthesis in present day cells (see Information Box 6-2).

ATP may have made its first entry into the system simply as one of the many organic molecules absorbed and broken down as an energy source. Since ATP breakdown is reversible, it is likely that some of the energy released in the transport of electrons along the first transport chains drove the reaction in the opposite direction, toward synthesis of ATP from ADP. Later, because of the efficiency and versatility of energy transfer by removing phosphate groups from ATP, or adding them to ADP, ATP gradually became established as the primary molecule linking energy-releasing and energy-requiring reactions.

As these systems developed and gradually specialized in the use of ATP as an energy-carrying intermediate, synthetic pathways also increased in complexity. N. H. Horowitz has suggested a pattern by which complex, sequential synthetic pathways might have appeared. Suppose that a modern pathway in cells makes a required substance, such as an amino acid, starting with a simple organic or inorganic substance A and leading through steps producing substances B, C, D, and finally the amino acid. Initially the amino acid was abundant in the environment, and was absorbed directly for use in the molecular cluster. Later, as the amino acid became scarce because of its use, chemical selection favored clusters that could make the amino acid from substance D, another slightly less complex organic molecule still found in abundance in the

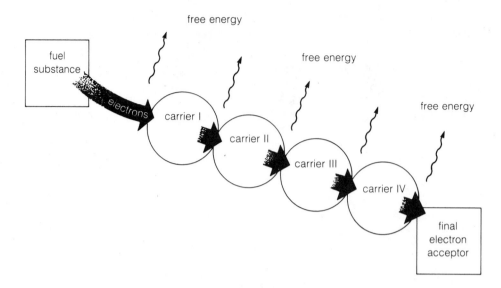

Figure 14-6 Oxidation and movement of electrons along a chain of electron carriers. Free energy is released as electrons pass from one carrier to the next. At some points, enough energy is released to drive the synthesis of ATP.

environment. As D became exhausted, selection favored assemblies developing a pathway in which the even simpler substance C could be absorbed and used as a precursor to make D. This process continued until the entire pathway leading from the simple inorganic substance A to the amino acid was established.

All these developing reaction systems yielding and utilizing energy were probably initially speeded by the nonspecific, general catalytic activities of coacervates, lipid vesicles, microspheres, or the clays absorbing organic molecules. Gradually, catalytic functions were taken over by more specialized molecules with increased specificity, leading eventually to enzymes.

Some ideas about the possible steps in the appearance of enzymes have been developed through the study of the catalytic properties of substances like iron in pure form and in various compounds. Metallic iron can act as a catalyst to increase the breakdown of hydrogen peroxide (H_2O_2) to water and oxygen. Combining iron into iron oxide further increases the rate of the reaction. If iron is bound into a porphyrin ring its catalytic ability is increased about a thousand times. Finally, if combined with the correct protein, the catalytic properties of iron are increased millions of times. Thus biological catalysts probably first evolved from small inorganic and organic molecules, and later increased in activity by combination with more complex organic molecules and polypeptides.

The first polypeptides included in the early catalytic systems may have simply provided a stable framework anchoring an inorganic catalyst such as a metal ion to the primitive molecular assemblies. However, since many amino acids, particularly those with acidic or basic properties, also have catalytic activity, some of the catalyst-polypeptide complexes would have been more efficient in speeding reactions than the inorganic catalyst alone. With the development of a coding system directing the synthesis of proteins, the catalytic polypeptides took on greater specificity and reproducibility until the first enzymes appeared.

Development and reproduction of any of these reaction systems to levels in which more than two or three coordinated steps took place would not be likely without an information system to direct synthesis and reproduction of the primitive enzymes. Therefore, it is likely that a mechanism for storing, reproducing, and translating information had to develop in parallel with the systems releasing and utilizing chemical energy.

The beginnings of the information system may have developed as an offshoot from the use of the nucleotides such as ATP as energy sources and carriers. Occasionally, these nucleotides probably assembled spontaneously into nucleic acids, as in the experiments simulating primitive earth conditions. Some combinations of nucleotides in the nucleic acids formed in this way may have favored the

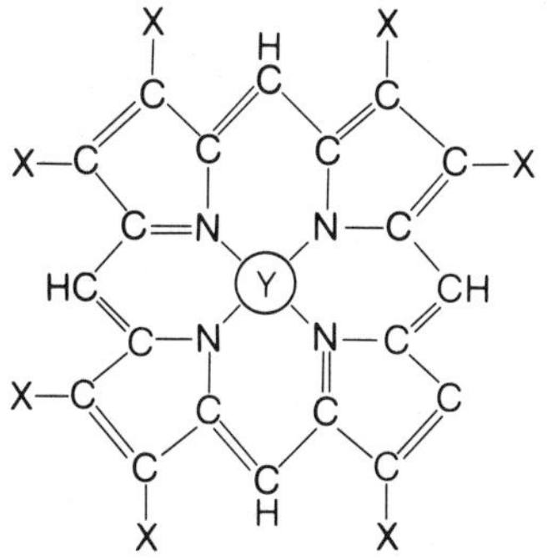

Figure 14-7 A porphyrin ring, a pigment molecule that could have been synthesized in quantity under primitive earth conditions. In present-day organisms, porphyrin rings form the active chemical groups of chlorophyll, hemoglobin, and the cytochromes. These porphyrins differ in the substitution of chemical groups at the positions marked with an X, and the central ion marked by a Y (Y is a magnesium ion in chlorophyll and an iron ion in hemoglobin and the cytochromes).

absorption of particular amino acids into the assemblies. As a result, the sequence of amino acids in the polypeptides made in the assemblies became less random, and the first vestiges of directed synthesis appeared. With time, and the appearance of additional spontaneous interactions between nucleic acids and proteins, enough specificity developed to lay down the beginnings of the genetic code.

Sidney Fox has reported that microspheres containing both proteinoids and nucleic acids are actually able to carry out the first steps in this process. In Fox's experiments some correlation was detected between the type of nucleic acid included in a microsphere and the type of amino acid absorbed. Including a nucleic acid molecule containing only one type of base, adenine, favored the absorption of the amino acid lysine into the microspheres. Adding a nucleic acid molecule containing only cytosine bases favored the absorption of proline into the microspheres. Similarly, guanine-containing nucleic acid molecules favored the uptake of glycine, and a uridine nucleic acid favored phenylalanine. These results are of special interest because AAA is the codeword for lysine, CCC is the codeword for proline, GGG is the codeword for glycine, and UUU is the codeword for phenylalanine in the genetic code!

Once the coding system appeared and the relationships between the nucleic acid codewords and the amino acids became fixed, the way was open for the change from inanimate chemical evolution to organic evolution. Mutations in the sequence of the coding nucleic acids would then cause changes in the proteins of the assemblies. Some of these changes would be favorable and would increase the ability of the assembly to compete for organic molecules in the medium. Other favorable mutations would allow the assemblies to manufacture more of their required organic substances for themselves. By this stage, the assemblies would probably have acquired surface membranes. These could have arisen either during the initial aggregate formation, or later as an adaptation that improved the ability of the assemblies to survive and compete. At this stage, these first living beings would have taken on the important characteristics of cellular life, including a coding system, a system capable of translating the coded information into specific proteins, and systems coupling energy-yielding to energy-requiring reactions, all surrounded by a limiting surface membrane. These cells, in structural complexity and biochemistry, would approach the most simple and primitive modern-day prokaryotes.

The Events Leading from Prokaryotes to Eukaryotic Cells

Interactions between the primitive prokaryotic cells and their environment led to more advanced prokaryotes such as the blue-green algae, and set up conditions for the evolution of eukaryotic cells. Three sequential events occurring at this time were highly critical for the later appearance of eukaryotes. One was the development of photosynthetic reactions using water as a raw material and liberating oxygen to the atmosphere as a by-product. This development set up conditions for the second event critical to the later evolution of more advanced cells: the conversion of the atmosphere, through the oxygen released from photosynthesis, from a nonoxidizing to an oxidizing character. This change made possible the third event, the appearance of cells using oxygen as the final electron acceptor for their energy-releasing reactions.

The Appearance of Photosynthesis, Oxygen, and Aerobes

While the functions of life were developing, the primitive molecular assemblies were dependent on absorbed organic substances for their energy source. Because there was no free oxygen in the environment, the assemblies probably used inorganic compounds such as sulfates or nitrates as final acceptors for the electrons removed in their oxidations and were thus *anaerobic* (see Fig. 14-6). As the anaerobes became more widely distributed, the supply of organic fuel substances dwindled until the lakes and seas more closely resembled those of today. The dwindling fuel supply would certainly have led to extinction of the new life except for the appearance of photosynthesis among the early cells. The appearance of photosynthesis allowed some of the primitive cells to use light instead of absorbed organic substances as their energy source. It thereby liberated them, and the cells feeding on them, from dependence on the supply of fuel molecules in the environment.

Photosynthesis as we now know it has a number of significant steps (Fig. 14-8). In the first step, electrons are released by a donor substance (Fig. 14-8a) and passed to a pigment molecule. The pigment molecule then absorbs light energy; this energy is used to raise the electrons derived from the donor substance to a high-energy form (Fig. 14-8b). The electrons, raised in energy level by this mechanism, then power the chemical work of the cell. Thus, in photosynthesis, the energy of the electrons entering chemical reactions comes from sunlight rather than from complex organic molecules.

The earliest photosynthetic reactions may have involved pigment molecules such as porphyrins that were synthesized spontaneously and included by chance in molecular aggregates. As with the other biochemical activities of the emerging life, light absorption and the excitation of electrons to high energy levels became regulated as enzymes appeared in the molecular assemblies.

The earliest photosynthetic pathways were forced to use an initial donor substance such as H_2S that releases electrons at relatively high levels, as the photosynthetic bacteria still do today. Eventually, mutations appeared that allowed water, which releases very low-energy electrons, to be used as the electron donor for photosynthesis. This established the pathway for photosynthesis used in the modern blue-green algae and the eukaryotic plants, in which water is split as a source of electrons and oxygen is released to the atmosphere as a by-product.

As a consequence of this new adaptation, which appeared in ancestral prokaryotes giving rise to the blue-green algae, oxygen was released to the atmosphere in ever-increasing quantities. This release gradually changed the atmosphere from a nonoxidizing to an oxidizing character and led to the balance of gases present in our environment today. Release of oxygen into the atmosphere also formed an ozone layer in the outer reaches of the atmosphere, effectively protecting the cellular life developing on the earth's surface from the injurious mutations and other chemical effects of ultraviolet light.

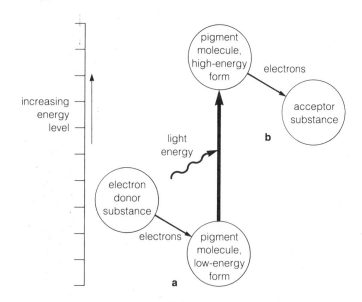

Figure 14-8 The overall steps of photosynthesis. (**a**) Electrons released by a donor substance are picked up by a pigment molecule, which absorbs light and raises the electrons to higher energy levels. (**b**) These high-energy electrons are then released from the pigment molecule and used to reduce an acceptor substance, which is converted into a high-energy, more complex molecule in the process. In eukaryotic plants, the acceptor substance is CO_2, which is converted into units of carbohydrate by the reaction.

The gradual increase in the levels of oxygen in the atmosphere set the stage for the third biochemical development that was to be of critical importance for the later evolution of more complex cells. This was a series of mutations that allowed the newly abundant oxygen to be used as the final acceptor for the electrons removed in oxidative reactions. This development was highly advantageous, since oxygen accepts electrons at very low energy levels. Consequently, much more energy could be tapped from the electrons removed from fuel substances and used to power cellular activities (Fig. 14-9). The prokaryotic cells using this complete oxidative pathway were the first *aerobes* (see p. 144) to appear on the earth.

Thus the evolutionary developments of this period include three major events that were of supreme importance for the later development of eukaryotes: (1) appearance of photosynthesis and adaptation of some of the primitive photosynthetic prokaryotes to the use of H_2O as an electron source, giving rise to the forerunners of the blue-

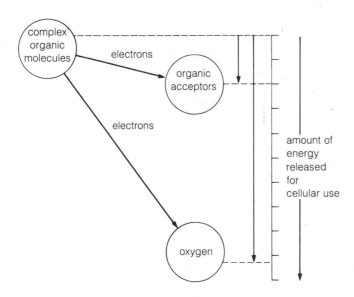

Figure 14-9 The advantage of using oxygen as final electron acceptor for electrons removed during cellular oxidations. Oxygen accepts electrons at very low energy levels, and permits more of the energy of the electrons to be tapped off for cellular use.

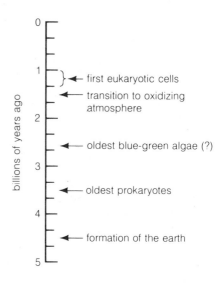

Figure 14-10 The time scale of cellular evolution (see text).

green algae; (2) release of oxygen by these organisms to the atmosphere, converting its character from reducing to oxidizing; and (3) evolution of prokaryotes using oxygen as final acceptor for the electrons removed during cellular oxidations. By this time, the change in character of the atmosphere and the nearly total depletion of organic molecules in the rivers, lakes, and seas had altered the earth to conditions approaching the present-day environment.

The fossil and geological record gives some indications of the probable time scale of these events (Fig. 14-10). The oldest bacterialike cells yet discovered date back about 3.5 billion years. Evolution of blue-green algae from these early bacteria may have required a minimum of another 500 million years. This conclusion is based on discoveries of limestone-containing deposits called *stromatolites* in rocks laid down 1.6 to 2.7 billion years ago (Fig. 14-11). Stromatolites are good indicators of the presence of blue-green algae since they are still formed by certain species of these organisms today. The time required for oxygen to appear in abundance in the atmosphere has been estimated from the degree of oxidation of iron-con-

taining sediments. Sediments in nonoxidized form persist in layers deposited as recently as 1.8 billion years ago. These finally give way to oxidized "red beds" about 1.5 billion years old. This indicates that the transition to an oxidizing atmosphere, begun nearly 3 billion years in the past, probably occupied 1.5 billion years and became complete not much more than 1.5 billion years ago.

The Appearance of Eukaryotic Cells

Eukaryotic cells undoubtedly developed from one or more of the different kinds of prokaryotic cells that became abundant some 1.5 billion years ago. This process, which included evolution of new structural features and organelles such as mitochondria, chloroplasts, and the nuclear envelope, may have taken place by one or more of several possible routes. One possibility is that many of the membranous structures characteristic of eukaryotes may have arisen through invaginations of the surface or plasma

a b

Figure 14-11 (**a**) Fossil stromatolites (arrows) embedded in rock. Courtesy of Biology Media. (**b**) Living blue-green algae depositing stromatolites (arrows); photographed under the ice in an Antarctic lake. Photograph by F. G. Love and L. Hoare; courtesy of B. C. Parker and G. M. Simmons, from *Trends in Biochemical Sciences* 6 (1981).

membranes of prokaryotic cells (Fig. 14-12). In prokaryotes, the enzymes and biochemical activities associated with oxidation and photosynthesis are linked to the plasma membrane (see pp. 136 and 160). Possibly, through invaginations of the plasma membrane (Fig. 14-12a), portions of the membrane containing these enzymes and biochemical activities extended into the cytoplasm and pinched off (Fig. 14-12b), giving rise to membranous organelles that gradually evolved into mitochondria and chloroplasts.

The nuclear envelope membranes may have also developed by the same process, through invaginations of the plasma membrane that extended inward and gradually surrounded the nucleus (Figs. 14-12c and d). Similar invaginations may also have given rise to the membranes of the rough and smooth endoplasmic reticulum.

A different hypothesis about the possible origins of mitochondria and chloroplasts is based on the obvious resemblances between these organelles and contemporary free-living prokaryotes. According to this idea, advanced in its most recent and complete form by Lynn Margulis of Boston University, mitochondria and chloroplasts evolved from ancient prokaryotes that were originally eaten as food particles. Instead of breaking down, the ingested cells persisted as functional units in the cytoplasm of the feeding cells. Mutations increasing the interdependence between the ingested prokaryotes and their host cells eventually led to their conversion into chloroplasts and mitochondria.

Margulis proposes that in the development of mitochondria, the evolution of prokaryotes proceeded to the point where complete photosynthesis was common and oxygen was present in large quantities in the atmosphere. Among these prokaryotes were some nonphotosynthetic types that had developed the capacity to engulf other cells through invaginations of the plasma membrane. These cells were thus feeding types, and lived by oxidizing the organic molecules picked up by this means. Some of these nonphotosynthetic feeding cells were capable of using oxygen as the final electron acceptor, and were thus aerobic. Others were still limited in their oxidative reactions to using organic substances that accept electrons at intermediate energy levels and were thus anaerobic.

Interactions among these cells led to the evolution of mitochondria, which began when groups of the nonphotosynthetic, anaerobic cells ingested nonphotosynthetic aerobic cells in large numbers. Instead of breaking down, some of the ingested aerobes persisted intact in the cytoplasm of the capturing cells and continued to respire aerobically. As a result, the cytoplasm of the host anaerobes, formerly limited to the use of organic molecules as final electron acceptors, became the place of residence of aerobes capable of carrying out the much more efficient transfer of electrons to oxygen.

The new association would have brought advantages to both the host cell and the ingested prokaryote. Some of the chemical energy produced by the ingested aerobe

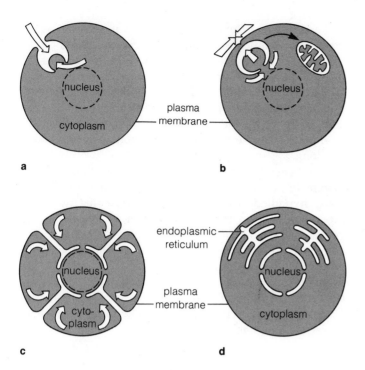

Figure 14-12 A hypothetical route by which mitochondria may have formed, through invagination of the plasma membrane of a pro-karyote. (**a**) Invagination of the plasma membrane, which contains oxidative enzymes in prokaryotes, to form a cytoplasmic vesicle. (**b**) Separation and gradual conversion of the vesicle into a mitochon-drion. (**c** and **d**) The hypothetical origin of the nuclear envelope of eukaryotic cells through invaginations of the plasma membrane. Fragments destined to form endoplasmic reticulum could have originated through the same route.

would diffuse into the cytoplasm of the anaerobe, thereby benefiting the host cell. In turn, the host cell, which was probably highly efficient in the capture and ingestion of organic matter, would supply the aerobe living in its cy-toplasm with the foodstuffs needed for survival. As this relationship developed, mutations gradually increased the interdependence between the host and ingested cells. Among these mutations were the loss of many motile and synthetic functions of the ingested aerobes because these activities could be supplied by the host. The ingested aerobes then became more and more specialized as en-ergy-converting organelles, and eventually completed their transition into mitochondria. All that remains today of the nucleus of the ancestral parasites that gave rise to mito-chondria are small circles of DNA that code for a few lim-ited and essential functions that evidently cannot be supplied by their "hosts" (see Supplement 10-2).

This new cell type, a former anaerobe containing an aerobe in its cytoplasm, gave rise to the eukaryotic cell line. Other features of eukaryotic cells, including the nu-clear envelope and endoplasmic reticulum, gradually ap-peared in the developing eukaryotes, possibly through invaginations of the plasma membrane as proposed in the alternate hypothesis.

Some time after the evolution of eukaryotic cells to this level, Margulis proposes, a second interaction of the same type led to the development of chloroplasts. In this case, some of the feeding cells already containing mito-chondria ingested prokaryotes capable of photosynthesis. Some of the photosynthetic cells persisted in the cyto-plasm of the ingesting cells and supplied their hosts with the ability to use light as a source of energy. As this rela-tionship developed, the ingested photosynthetic prokar-yotes gradually evolved into chloroplasts by the same process of gradually increasing interdependence between the host cells and their cytoplasmic residents. These prim-itive eukaryotes, containing both mitochondria and chlo-roplasts, founded the cell lines leading to the eukaryotic algae and plants.

Some modern examples of very recently established relationships between ingested and host cells provide evi-dence that this route of origin could actually have taken place. Many kinds of animal cells, including representa-tives of eight major phyla, ingest algal cells or chloroplasts and retain them as beneficial cytoplasmic residents. One of the most interesting examples, described by Robert Trench of the University of California at Santa Barbara, involves a group of marine snails that contain chloroplasts derived from algae used as food. Initially, the snails hatch and develop as embryos without chloroplasts. As soon as they reach a larval feeding stage, chloroplasts from algae eaten by the young snails are taken up by cells lining the gut. These chloroplasts persist in the gut cells as the snails grow to adult form, and continue to carry out photosyn-thesis in their new location (Fig. 14-13). Trench has shown by the use of radioactive labels that molecules synthesized by the ingested chloroplasts diffuse into the surrounding cytoplasm and are used as fuel substances by the host snail. Individual chloroplasts may remain active in the gut cells for months. Thus the routes envisioned by Margulis for the origin of mitochondria and chloroplasts can still be demonstrated today.

As these structures developed, additional major ad-aptations completed the conversion of the primitive eu-karyotes into contemporary forms. These adaptations were increases in the complexity of chromosomes and the de-velopment of microtubules to regulate cell division and motility. Later evolution led to the aggregation of cells into

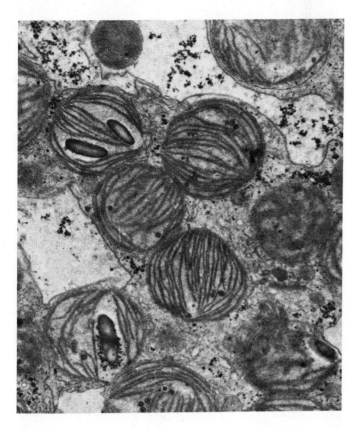

Figure 14-13 Active chloroplasts in the cytoplasm of a digestive cell of the snail *Elysia*. Courtesy of R. K. Trench, from *Proceedings of the Royal Society*, Series B, 184 (1973):63.

prokaryotes date from 3.5 billion years in the past would mean that as much as 2 billion years was required for the evolution of eukaryotes from prokaryotes (see Fig. 14-10). If so, this interval is greater than the total period required for the entire evolution of prokaryotes from the earth's origin. Considering the complexities of eukaryotic cells, and the advances they represent over the prokaryotes, it is entirely possible that their evolution may have taken this long.

The chemical and biological events leading from the inanimate earth to the first eukaryotic cells contain so many hypothetical and speculative steps that the entire process may seem unlikely. However, the total time involved, some 3.5 billion years, is so great that even events of very low probability are likely to have happened more than once. As George Wald of Harvard University has put it, given the time span of these events, " . . . the impossible becomes possible, the possible probable, and the probable virtually certain. One has only to wait: time itself performs the miracles."[2]

Suggestions for Further Reading

Bernal, J. D. 1967. *The origin of life*. World: New York.

Calvin, M. 1969. *Chemical evolution*. Oxford University Press: New York.

Day, W. 1979. *Genesis on planet earth*. House of Talos: East Lansing, Mich.

Dickerson, R. E. 1978. Chemical evolution and the origin of life. *Scientific American* 239:70–86 (September).

Fox, S. W. and K. Dose. 1977. *Molecular evolution and the origin of life*. 2nd ed. Marcel Dekker: New York.

Keosian, J. 1964. *The origin of life*. Reinhold: New York.

Margulis, L. 1970. *Origins of eukaryotic cells*. Yale University Press: New Haven, Conn.

Ponnamperuma, C. 1972. *The origins of life*. Thames and Hudson: London.

Schopf, J. W. 1978. The evolution of the earliest cells. *Scientific American* 239:111–38 (September).

Wolfe, S. L. 1981. *Biology of the cell*. 2nd ed. Wadsworth: Belmont, Calif.

colonies, and eventually to the appearance of many-celled organisms of greater and greater complexity, including our own form of life.

The total time required for complete evolution of the first fully eukaryotic cells from prokaryotes may have extended over as much as two billion years. Fossils in rocks as old as 1.4 billion years have been claimed to be eukaryotic cells on the basis of apparent nuclei and division stages resembling those of eukaryotic algae. These conclusions are controversial, however; others have claimed that the apparent nuclei in these ancient cells resulted from precipitation of cell structures during fossilization and that eukaryotic cells are not likely to have been on the earth much longer than 1 billion years. Assuming that the first fully eukaryotic cells appeared at some time between these limits of 1.4 and 1 billion years ago, and that the oldest

Questions

1. Outline the Oparin-Haldane hypothesis for the origin of life.

---

[2]George Wald. The origin of life. *Scientific American* 191:45.

2. How would you define life as applied to a bacterium? A eukaryotic cell? A human? A tree?

3. Review the probable stages in the evolution of life.

4. What is the condensation hypothesis for the origin of the sun and planets of our solar system?

5. Why was the absence of oxygen among the gases of the primitive atmosphere important for the evolution of life? What evidence indicates that the atmosphere of the primitive earth was actually nonoxidizing?

6. What energy sources are believed to have acted on the chemicals of the primitive environment to produce complex organic molecules? What evidence indicates that complex substances may actually have been produced in this way? What kinds of molecules have been synthesized in systems that simulate the primitive environment?

7. What are coacervates and microspheres? What roles might clays and lipids have played in the assembly of molecules in the primitive earth? What is a proteinoid?

8. Outline the possible steps in the development of oxidative reactions. What roles did ATP probably play in these reactions?

9. Outline the possible steps in the development of complex synthetic pathways. How did enzymes catalyzing steps in these pathways possibly develop?

10. What importance did molecules such as ATP, GTP, UTP, TTP, and CTP have in the evolution of life?

11. Which is probably older, anaerobic or aerobic life?

12. How might photosynthesis have developed? What was the importance of photosynthesis in the evolution of life?

13. Outline the two hypotheses that account for the origins of mitochondria and chloroplasts.

14. What environmental conditions were necessary for the appearance of eukaryotes?

15. Could eukaryotic cells have survived if they had been introduced into the primitive earth from a spaceship before life had evolved here?

16. What modern evidence supports Margulis's hypothesis that mitochondria and chloroplasts might have arisen from ingested prokaryotic cells?

17. Outline the time scale for the major events in the evolution of life.

Appendix: Techniques of Cell Biology

This appendix provides a brief description of the major techniques used in cell biology research. The techniques included form the basis for experiments supplying the observations and tests on which many of the conclusions presented throughout this book are based.

Light and Electron Microscopy

Both light and electron microscopes are used extensively in biological research today. Prokaryotic cells and nucleoids, and the larger organelles of eukaryotic cells are readily visible in the light microscope; objects as small as single protein molecules or even single atoms can be seen in the electron microscope. Many of the techniques used for preparing specimens for viewing in either instrument allow individual molecular types to be located within these cell structures.

The Brightfield Light Microscope

In the standard instrument known as a *brightfield light microscope* (Fig. A-1), light rays from an illumination source are focused on the specimen by a condenser lens. The rays leaving the specimen are focused into a magnified image of the specimen by two lenses placed at either end of a tube. The lens nearest the specimen is termed the *objective* lens. The lens at the opposite end of the tube is called the *ocular* lens. Each of these lenses, in order to correct faults in the image, is actually constructed from a series of lenses placed close together, usually as many as 8 to 10 for the objective lens and 2 to 3 for the ocular lens. Since the individual elements in these lenses are placed close together, their net effect is to act as a single, highly corrected lens.

The image is observed by looking directly into the ocular lens. Coarse and fine controls are provided for movement of the specimen stage or lens tube to place the specimen in the correct position for focusing by the objective lens. The position of the condenser lens can also be adjusted so the light from this lens converges on the specimen and spreads, after leaving the specimen, into a cone of light that completely fills the objective lens.

The extent to which a microscope can distinguish fine details in the specimen as separate, distinct image points is termed its *resolution*. Resolution in light microscopes is described by the equation:

$$\text{resolution} = d = \frac{0.61\,\lambda}{n\,\sin\,\alpha} \tag{A-1}$$

in which λ is the wavelength of the light used for illumination in the microscope, n is the refractive index[1] of the transmitting medium surrounding the specimen and filling the space between the specimen and the objective lens, and α is the half-angle of the cone of light entering the objective lens from the specimen (Fig. A-2). The quantity 0.61 is a constant describing the degree to which image points can overlap and still be recognized as separate points by an observer. Since resolution is a measure of the ability of a microscope to image fine details, the quantity d becomes smaller as resolution improves. Therefore, for best resolution, 0.61 λ should take on the smallest possible value, and $n\,\sin\,\alpha$, the largest possible value.

---

[1]The refractive index is a measure of the degree to which a transparent substance slows and bends (refracts) a beam of light rays. A vacuum is considered to have a refractive index of 1; all other transparent substances have higher values.

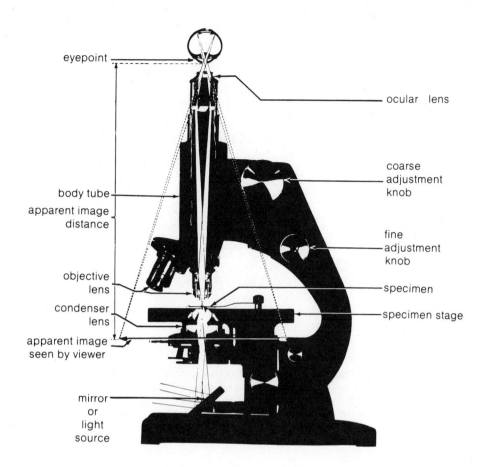

eyepoint

ocular lens

coarse adjustment knob

body tube

apparent image distance

fine adjustment knob

objective lens

specimen

condenser lens

specimen stage

apparent image seen by viewer

mirror or light source

Figure A-1 Construction of the brightfield light microscope (see text). Objectives of different magnifying power can be selected by rotating the nose-piece. Redrawn from an original courtesy of the American Optical Company.

In a good light microscope operated under the best conditions, all the quantities in Equation A-1 take on fixed values. The value for n can be pushed to its maximum by placing a drop of immersion oil (refractive index approximately 1.5) in the space between the objective lens and the specimen. In some microscopes, immersion oil can also be placed between the condenser lens and specimen to provide optimum conditions for observation. The half-angle of the cone of light entering the objective lens in the best microscopes is about 70°, giving a maximum value for sin α of about 0.94. The quantity $n \sin \alpha$, called the *numerical aperture* of the objective lens, therefore takes on a maximum value of about 1.4.

This leaves λ, the wavelength of the light used as an illumination source in the light microscope, as a variable in the equation. To keep d as small as possible, λ must take on a minimum value. The lower limit of this value is fixed by the shortest wavelength of visible light usable for illumination, lying in the blue range at wavelengths of about 450 nanometers. Substituting this value for $n \sin \alpha$ = 1.4 in Equation A-1 reveals that the best resolution pos-

sible with microscopes using visible light approaches about 0.2 micrometer, or about 200 nanometers. Using ultraviolet light as an illumination source (λ = 250 nanometers) pushes this value down to about 0.1 micrometer.

Dark Field Light Microscopy

One method employed to improve the visibility and apparent resolution of specimen points in the light microscope is *dark field* illumination. In this technique, an opaque disc is placed in the center of the condenser so that a hollow cone of light strikes the specimen. The hollow cone of light is adjusted to an angle wide enough so that no light directly transmitted through the specimen can enter the objective lens (Fig. A-3). However, any light scattered by specimen points to smaller angles will enter the objective lens, making these points appear bright against a dark background. The method apparently improves resolution, because objects too small to be directly

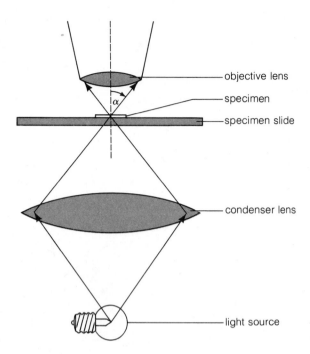

Figure A-2 Light path from the condenser to the objective lens. For maximum resolution, the half-angle (α) of the cone of light entering the objective lens should take on as large a value as possible.

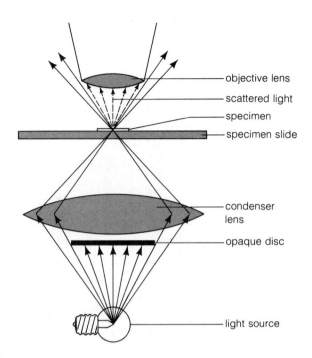

Figure A-3 Light path from the condenser to the objective lens in dark field illumination. An opaque disc blocks the central region of the condenser so that no rays from the light source directly enter the objective lens. Objects in the specimen scattering light into the objective lens appear bright against a dark background (see text).

resolved will still scatter and divert the path of light waves. By this technique, objects as small as individual microtubules, with a diameter of approximately 25 nanometers, can be seen in the light microscope (see Fig. 8-5).

Phase Contrast Light Microscopy

Most cell structures are transparent and colorless and show little contrast if living material is observed in the ordinary light microscope. The visibility of structures in living cells is greatly improved by the *phase contrast microscope*, a viewing system that makes regions of differing refractive index appear as regions of differing brightness in the image.

The phase contrast microscope takes advantage of the fact that light rays are both delayed and bent from their path (refracted) as they pass through transparent objects of differing refractive index. Many components of the cell, although practically transparent to light, have relatively high refractive indexes. The light waves passing through these components are delayed by about one-quarter wavelength, and are bent from the paths followed by waves passing through less refractive components. Although the delayed rays follow a different path through the lenses of the ordinary light microscope, they are focused into the image plane along with the unrefracted rays. Since the two groups of rays differ in phase by only one-quarter wavelength on the average, they do not interfere sufficiently in the image to cause visible alterations in the brightness of image points.

The phase contrast microscope increases the difference in wavelength of refracted and unrefracted light enough to produce maximum interference in the image. In the microscope, an annulus similar to the device used to produce a hollow beam of light for dark field illumination is placed below the specimen (Fig. A-4). The annulus restricts the unrefracted light to a path that passes through a corresponding ring-shaped depression in a glass *phase plate* placed just above the objective lens. The refracted

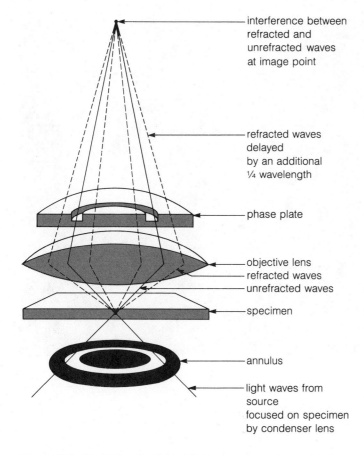

interference between
refracted and
unrefracted waves
at image point

refracted waves
delayed
by an additional
¼ wavelength

phase plate

objective lens
refracted waves
unrefracted waves

specimen

annulus

light waves from
source
focused on specimen
by condenser lens

Figure A-4 The delay of refracted light waves by a phase plate in a phase contrast light microscope (see text).

waves are deflected widely enough by specimen points to pass through other, thicker regions of the phase plate. The phase plate is constructed so that the difference in thickness between the two regions of the plate is just enough to delay the refracted waves by an additional one-quarter wavelength (the delay of light by glass is proportional to its thickness). The refracted and unrefracted light waves are then focused together into the image. At image points corresponding to points of higher refractive index in the specimen, the crests of the refracted waves, since they now differ in phase by one-half wavelength, coincide in the image with the troughs of the unrefracted waves. The two waves therefore cancel or interfere, producing regions of minimum brightness at these points. The effect over the entire image is to create differences in brightness corresponding to differences in refractive index in the specimen. By this imaging system, many structures in living

cells can be clearly seen without staining or alterations of any kind (Fig. A-5).

Electron Microscopy: The Transmission Electron Microscope (TEM)

In overall design, electron microscopes are similar to light microscopes, with some important differences. The primary differences are in the illumination and type of lenses used. Electron microscopes use a beam of electrons rather than light as an illumination source. The electron beam is focused by magnetic or electrostatic fields to produce a focused image. In the most commonly employed system, magnetic focusing fields are generated by massive coils of wire through which precisely controlled electric currents are passed. The magnetic fields generated are shaped into the correct three-dimensional shape required for focusing electrons by *pole pieces* (Fig. A-6), iron inserts placed into the bore of the electromagnets. By adjusting the current applied to the magnetic lenses, their focus and magnification can be changed at will.

Another difference between light and electron microscopes is that all the spaces traversed by electrons inside the electron microscope must be kept under a high vacuum. Otherwise, the electrons of the beam, which have relatively poor penetrating power, would be completely scattered and absorbed by gas molecules in the microscope. The requirement for a vacuum inside the microscope, and the poor penetrating power of electrons places special restrictions on the specimen, which must be dry and very thin.

These special operating conditions are successfully met in the *transmission electron microscope* (*TEM*), so called because the electrons used to form the image pass through the specimen. In construction, a TEM resembles an inverted light microscope (Fig. A-7; Fig. A-8 shows a typical high performance TEM). The central column houses the magnetic lenses and their lens coils and provides shielding for the operator from x rays that arise when metal surfaces are struck by the electron beam. At the top of the column is the electron "gun," consisting of a filament and anode. The filament, a thin tungsten wire, is heated to high temperature by an electric current, releasing a cloud of electrons from its surface. The filament and its holder, which are electrically insulated from the rest of the column, are maintained at −50,000 to −100,000 volts. The anode is grounded and thus is strongly positive with respect to the

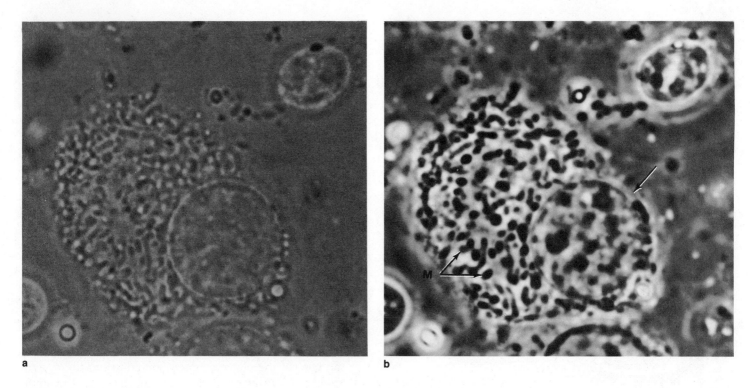

a

b

Figure A-5 (**a**) Brightfield image of a living cell from frog liver. (**b**) The same cell viewed with phase contrast optics. The nucleus is marked by an arrow; numerous bodies including mitochondria (*M*) are visible in the cytoplasm. × 2300. Courtesy of B. R. Zirkin.

filament. Electrons from the cloud around the filament are attracted to the anode. As they move through the space between the filament and anode, the electrons are accelerated to high velocity, with wavelengths in the region 0.005 to 0.003 nanometer.

Electrons from the gun are focused on the specimen by the condenser lens. In modern instruments, two condenser lenses are placed in tandem to allow focusing of a very small, intense spot of electrons. The electrons passing through the specimen are focused by a system of lenses usually including an objective, intermediate, and projector lens (see Fig. A-7). The lenses are arranged so that successively magnified images are formed by each lens, with total magnification ranging between about 3000 and 300,000 times. For most specimens, values between 5000 and 50,-000 are routinely used. Since the lenses are fixed in position, focusing is carried out by adjusting the current passing through the lens coils rather than moving the lenses as in the light microscope.

The projector lens focuses the magnified image onto a fluorescent screen at the bottom of the column. This screen, similar to the screen of a television tube, is coated with a layer of crystals that responds to electrons by emitting visible light. In this way, the electron image is converted to a visual image. Permanent records of the image are made by exposing a photographic plate to the electron beam at the level of the screen. Ordinary films and plates can be used for this purpose because the response of photographic emulsions to electrons and light is essentially the same. Interlocks are provided so that specimens and photographic plates can be exchanged without disturbing the microscope vacuum.

The wavelength of electrons in an electron beam is inversely proportional to the accelerating voltage. The higher the voltage, the shorter the wavelength. At an operating voltage of 50,000 volts, routinely achieved in transmission electron microscopes, the wavelength of the electrons in the beam is 0.005 nanometer. This wavelength, which is about 100,000 times shorter than the wavelength of blue light, greatly improves the resolution of the TEM as compared to light microscopes. Under the operating conditions used in electron microscopy, this

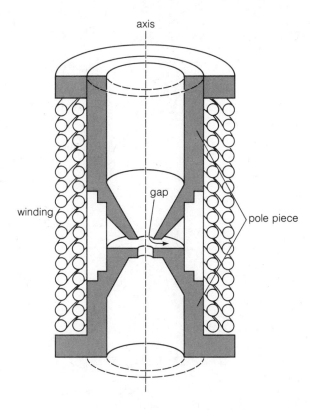

Figure A-6 A magnetic lens. The magnetic field, generated by a current passed through the winding, is shaped into a focusing field by the pole piece (see text). Redrawn from R. B. Setlow and E. Pollard, *Molecular biophysics*, Addison-Wesley, 1962. Courtesy of R. B. Setlow.

wavelength allows objects with diameters as small as 0.6 to 0.7 nanometer to be resolved. At these levels, electron microscopes can easily "see" objects as small as protein molecules. Images of individual heavy metal atoms have even been produced under special operating conditions.

Specimen Preparation for Light and Electron Microscopy

Light Microscope Preparations Although living specimens can be observed directly in the light microscope with either ordinary or phase contrast optics, the best viewing conditions are frequently obtained in material that has been stabilized by chemical fixation, cut into sections, and stained by dyes that produce contrasting colors in cell components.

The chemical fixatives used for light microscopy are reactive substances such as formaldehyde, glutaraldehyde, acetic acid, and alcohol. These substances fix cells by precipitating or coagulating cell structures or by introducing chemical crosslinks that anchor specimen molecules and structures in place. After fixation, specimens are embedded in a supportive material, usually following dehydration by exposure to successively higher concentrations of alcohol or acetone. For light microscopy, the embedding material of preference is often paraffin wax. The dehydrated tissues are impregnated with melted paraffin; after impregnation, the preparation is cooled, producing a solid block that is cut into sections about 5 to 10 micrometers in thickness. The sections are then placed on a glass slide and the paraffin is replaced with a transparent material, usually a resin of high refractive index.

Cells may be stained at any point in the process from living material to finished sections. A wide variety of organic dyes are employed as stains in light microscopy. These can be used in procedures that specifically color molecules such as RNA, DNA, and various proteins and polysaccharides. Enzymes are also used as probes for specific molecules by testing the susceptibility of cellular substances to enzymatic breakdown. For example, if a cell structure is destroyed or extensively altered after exposure of tissue to an enzyme breaking down RNA (ribonuclease) it can be assumed that RNA forms a part of the structure. Alternatively, the presence of enzymes as parts of cell structures can be tested by adding a substance attacked by the enzymes, in combination with a chemical that produces a specific color in the presence of a product of the reaction.

Specimen Preparation for Electron Microscopy Because electrons have very low penetrating power and are easily scattered, specimens prepared for electron microscopy must be dried and made as thin as possible, much thinner than even the smallest cells. These requirements for dehydration and thinness have frustrated attempts to view living material in the electron microscope. Instead, specimens for electron microscopy are prepared by one of several methods: *sectioning, shadowing, negative staining,* or *freeze-fracture.*

Sectioning Producing sections for electron microscopy includes several steps. As in specimen preparation for sectioning in light microscopy, the tissues to be observed are stabilized by placing them in chemical fixatives. Solutions of osmium tetroxide, formaldehyde, or glutaraldehyde, either singly or in combination, are most frequently used

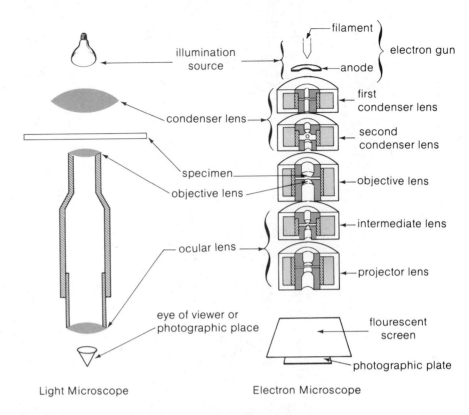

Light Microscope   Electron Microscope

Figure A-7 The arrangement of the illumination source, lenses, specimen, and viewing screen in a transmission electron microscope (right) resembles an inverted light microscope (left).

for this purpose. The tissue is then impregnated with liquid plastic, hardened, and sliced into sections as thin as 40 to 60 nanometers with diamond or glass knives.

Sections of unstained cells have little contrast when viewed under the electron microscope because the atoms occurring most commonly in tissues—carbon, hydrogen, oxygen, and nitrogen—all scatter electrons to about the same extent. To improve the contrast, salts of various heavy metals are added to the tissue as stains (substances with high atomic numbers, such as the heavy metals, deflect electrons to the greatest extent). In electron micrographs of stained sections (as in Fig. 1-6) the dense areas are regions in which many atoms of the heavy metal stain have accumulated. These regions appear dark in micrographs because the electrons passing through the heavy metal deposits are either absorbed or deflected so strongly that they are lost from the electron beam and are not focused into the image by the electron lenses (Fig. A-9). Solutions of lead citrate or uranyl acetate are commonly

used as stains in electron microscopy; osmium tetroxide is used as both a fixative and a stain. Enzymes are also used as chemical probes in electron microscope preparations as they are in light microscopy.

Shadowing and Negative Staining Cell parts and molecules are frequently isolated and examined in this form rather than in sections. The contrast of the isolated material, dried down on thin plastic support films, is usually increased by one of two techniques, *shadowing* or *negative staining,* for viewing in the electron microscope.

In the shadowing technique, isolated cell organelles or molecules dried on the plastic supporting film are coated under vacuum by a heavy metal such as platinum, evaporated from a source located to one side (Fig. A-10). Atoms of the metal, evaporated by electrically heating a small quantity of the metal to the boiling point, travel in straight lines and deposit on raised surfaces of the specimen facing the source. The raised surfaces, because of

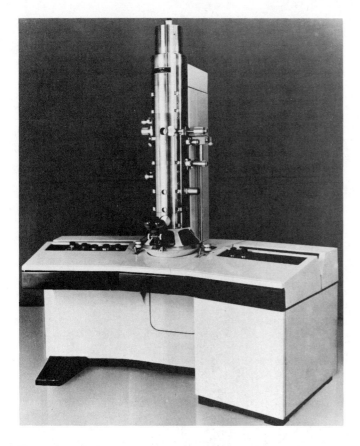

Figure A-8 A high performance transmission electron microscope. The electron lenses are housed in the central column, which is maintained under a high vacuum. To the right and left of the column are banks of controls by which lens current (focus) can be regulated. The image, formed on a fluorescent screen at the base of the column, is viewed through the windows with the binocular microscope. Photographic plates for permanent records of the image can be exposed at the level of the screen. Courtesy of the Perkin-Elmer Corporation.

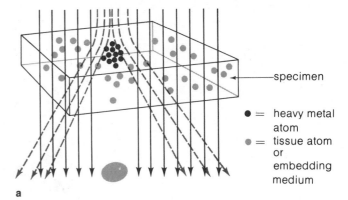

specimen

● = heavy metal atom

● = tissue atom or embedding medium

a

Figure A-9 Scattering of electrons by an object point, represented by a cluster of atoms of high density or atomic number in the specimen. As a result of scattering, fewer electrons fall on the corresponding region of the focused image, producing a "shadow" of the object point at this level.

their coating of heavy metal atoms, have high contrast in the electron microscope. Depressions in the specimen, located in the shadow of higher points, are not coated by heavy metal atoms and appear transparent in the electron microscope. The effect in the electron microscope is as if a strong light is directed toward the specimen from one side, placing surface depressions in deep shadow (as in Fig. 8-9).

In negative staining, a heavy-metal stain is allowed to dry around the outside surfaces of an isolated cell particle or molecule. The stain molecules deposit into surface crevices in the specimen during the drying process, often out-

lining details with remarkable clarity. The technique typically produces a "ghost" image, such as that shown in Figure 8-17, in which the specimen appears light against a dark background.

Freeze-Fracture Preparations The freeze-fracture technique employs cells that have been quick-frozen by plunging them into liquid nitrogen. At the temperature of liquid nitrogen (−196°C), cells freeze so rapidly that the molecules of membranes and other cell structures are immobilized in the positions they occupy in living state. The method induces so little change in cell structure that living cells frozen in liquid nitrogen after treatment with glycerol (to reduce the size of ice crystals formed inside the cells) can be stored for long periods and later thawed and revived without apparent damage. Tissue culture cells and spermatozoa are routinely frozen and stored for later use in this way.

Once frozen, the specimen is placed under vacuum and fractured by a knife edge. The fracture travels through the specimen and exposes membranes and other internal surfaces. The specimen is then frequently "etched" briefly by allowing water to dry off from the fractured surface (which is still frozen and under vacuum). The surface is then shadowed by evaporating a heavy metal from a source located to one side, just as in the shadowing technique. The shadowed surface is next coated with a layer of carbon evaporated from directly above; this carbon layer is transparent to electrons and is strong enough to support the specimen in the electron microscope. The tissue block

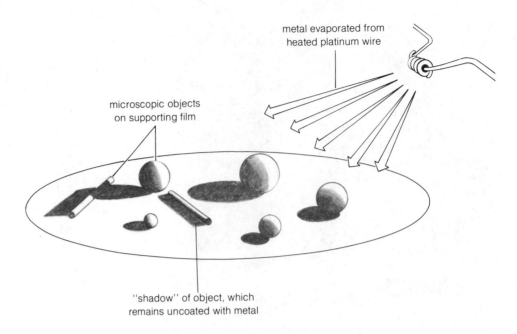

metal evaporated from
heated platinum wire

microscopic objects
on supporting film

"shadow" of object, which
remains uncoated with metal

Figure A-10 Shadowed preparations are made for electron microscopy by exposing specimens to metal evaporated from one side (see text).

is then thawed and dissolved away chemically, leaving only the shadowing metal backed by carbon. This "replica" of the fractured and etched cell surface is then ready for viewing in the electron microscope. Figure 4-2 shows a segment of a cell prepared for electron microscopy by this technique.

Radioactive Labeling and Autoradiography The radioactive labeling technique uses isotopes to locate or identify molecules within cells or cell extracts. The radioactive isotopes most used in this work, ^{3}H, ^{14}C, ^{32}P, ^{35}S, ^{125}I, and ^{131}I, are unstable and tend to decay into other elements, releasing energy in the form of radiation as they break down. For example, the ^{14}C isotope of carbon slowly breaks down to produce nitrogen, expelling a high-energy electron called a *beta particle* at the same time (see p. 27).

The beta particles ejected during radioactive decay have considerable energy and can be detected by various means. One of the most frequently used is based on the fact that beta particles can expose crystals in a photographic emulsion, producing an "image" corresponding to a radioactive point.

In application of this method, called *autoradiography*, living cells are exposed to a chemical substance containing one of the radioactive isotopes. Ideally, the substance should be a building-block molecule used by the cell only to make the molecules of interest. For example, most cells exposed to thymidine, made radioactive by the inclusion of ^{3}H (tritium) atoms in its structure, use this substance as a building block only in the synthesis of DNA. By exposing cells to radioactive thymidine, the DNA can be labeled by radioactivity. Similarly, RNA can be labeled by exposing cells to radioactive uridine, or proteins by exposure to radioactive amino acids.

After exposure to the radioactive precursor, cells are washed in unlabeled medium to remove any unincorporated radioactivity, and fixed, embedded, and sectioned for either light or electron microscopy (Fig. A-11). The sections are then placed on a specimen support and coated in a darkroom with a photographic emulsion, usually by dipping them in an emulsion that has been melted by heating. After coating, the slides are stored in the dark for a period of several days or weeks. During the storage period, radioactive decay at labeled sites exposes crystals in the photographic emulsion (see Step 5 in Fig. A-11). The

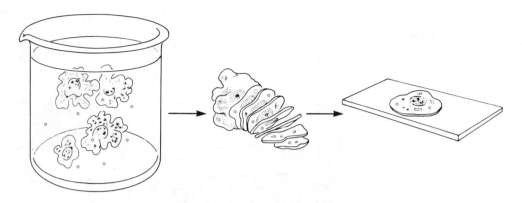

1. Living cells are exposed to radioactive label.

2. Cells are washed to remove any unincorporated label and then fixed and sectioned.

3. Sections are placed on microscope slides.

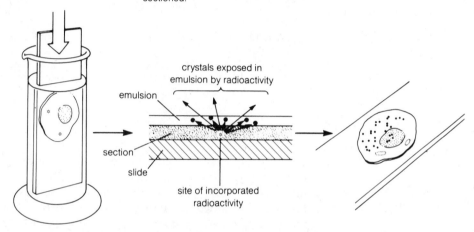

crystals exposed in
emulsion by radioactivity

emulsion

section

slide

site of incorporated
radioactivity

4. A slide containing a section is dipped into melted photographic emulsion in the darkroom.

5. Slides are dried and stored in the dark. During storage, crystals in the photographic emulsion are exposed over radioactive sites in the section.

6. The emulsion is developed by standard photographic techniques and slides are examined under a light microscope. Grains exposed in the emulsion mark sites of radioactivity.

Figure A-11 The preparative techniques used in autoradiography.

emulsion is then developed by standard photographic techniques.

After development, the sections are examined in the microscope. Radioactive areas in the sections are marked by silver grains developed over these regions (Fig. A-12). The developed pattern of grains, called an *autoradiograph*, marks the sites containing the molecules of interest.

Methods Using Antibodies Specific molecules and molecular classes are also identified for light or electron microscopy by reacting them with *antibodies*. Antibodies are

protein molecules made by animals in response to invasion by foreign substances, primarily proteins or protein fragments and some polysaccharides, originating from another individual or species. The foreign substance causing the antibody reaction is called the *antigen*. Antibodies are highly specific and usually react and bind to only the molecule used as antigen.

Antibodies are prepared by injecting a suitable animal with the purified antigen of interest. A rabbit, horse, or goat is frequently used for antibody production. Two or three injections of the antigen are given over a period of one to two months. At the end of this time, an antibody specific in its reaction to the antigen can be detected in the bloodstream of the injected animal. Extraction of a relatively small quantity of blood usually yields large quantities of the antibody, which is easily purified from the blood proteins.

The extracted antibody is then used as a probe for the molecule used as antigen. To accomplish this, the antibody is usually linked chemically to a "marker" of some type so that its reactions in cells or cell extracts can be followed. The marker used may be a radioactive group. In this case the reaction and the binding of the antibody in cells, indicating the presence of the antigen, can be followed and localized by autoradiography. Another marker frequently bound to the antibody is *ferritin*, a large iron-protein complex that can be resolved in the electron microscope. Attachment of a ferritin marker makes each antibody, and its site of reaction with its antigen, visible in tissues prepared for electron microscopy (as in Fig. 4-17). The antibody reaction is so sensitive that proteins with sequences differing by only one or two amino acids can often be separately detected and identified.

For analytical work with the light microscope, antibodies are frequently attached to a fluorescent dye. Fluorescent dyes, when exposed to ultraviolet or visible light, absorb at certain wavelengths and take on an "excited" state. Return to the unexcited or "ground" state occurs by release of some of the absorbed energy as visible light (see also p. 112, Chap. 6). The wavelength of the released light is always longer than the wavelength of the light absorbed. Because of the light release, the fluorescent dyes appear to glow strongly when exposed to light. Since the wavelengths absorbed by the dyes used in fluorescence light microscopy are in the ultraviolet or blue range, and the light released is green, red, orange, or yellow, the cell structures marked by fluorescent dyes stand out strongly against the cellular background (as in Fig. 1-13). Dark field illumination is often used to enhance the contrast of the fluorescent sites.

The Scanning Electron Microscope (SEM)

The scanning electron microscope (*SEM*) is useful for viewing the surfaces of cells and small organisms such as mites and insects. Although the SEM borrows some basic operating principles and lens systems from the TEM, its theory of operation is radically different. In the SEM (Fig.

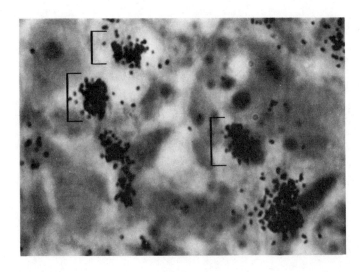

Figure A-12 Light microscope autoradiograph showing masses of grains (brackets) exposed over a section. The cells in the section do not show clearly in this micrograph because the microscope is focused on the grains in the emulsion above the section. × 600.

A-13) an electron gun and condenser lens produce and focus an electron beam into an intense spot on the specimen surface. This spot is then moved rapidly back and forth or *scanned* over the specimen surface by beam deflectors (charged plates that attract or repel the electron beam) placed below the condenser lenses. The intense spot of electrons scanning the specimen surface excites specimen molecules to high energy levels. This energy is released by the excited molecules in several forms, including high-energy electrons called *secondary electrons*.

The number of secondary electrons released depends on the angle of specimen points with respect to the scanning beam. Surface features perpendicular to the beam are struck fully and release a maximum burst of secondary electrons; surfaces at greater angles release proportionately fewer electrons. Thus the number of secondary electrons released depends on the three-dimensional shape of the specimen surface. The pulses of electrons leaving the specimen surface are picked up by a *detector* placed to one side of the specimen, and are converted into pulses of electrical current. The electrical pulses are fed into the circuitry of a television tube, where they modify the brightness of a spot scanning the front of the tube at the same rate and direction as the spot scanning the specimen. The effect at the front of the screen is to reconstruct an image of the specimen surface, with bright regions correspond-

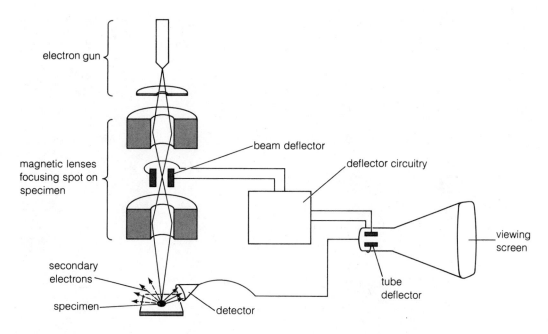

Figure A-13 Construction of the scanning electron microscope (see text).

ing to ridges in the specimen surface, and darker regions corresponding to the valleys (Fig. A-14). The spot scanning the television tube, and the corresponding spot scanning the specimen, move so rapidly that it is seen by the viewer as an instantaneous, total image with no detectable movement.

Because the secondary electrons emitted from the specimen surface are easily scattered by gas molecules, the SEM column housing the lens system, specimen, and detector must be kept at high vacuum. This operating restriction requires that the specimen must be dry, or must release gas or water molecules so slowly that the vacuum is not significantly disturbed. To meet this requirement, most biological objects, such as cells and tissues, must be fixed and dehydrated before viewing as in the TEM and cannot be examined in the living state. However, a few organisms have been placed in the SEM alive, and have survived the vacuum and bombardment by the scanning beam without extensive damage.

The SEM produces excellent images of object surfaces ranging in size from whole cells up to small insects. Since its resolution is about 20 times better than the light microscope, the surfaces of objects in this size range are imaged with significantly greater fidelity. The limited resolution of the SEM, as compared to the TEM, however, makes it less useful for observing details inside cells.

X-Ray Diffraction

X rays, like electron beams, are radiated in waves much shorter than the wavelengths of visible light. In contrast to electrons, x rays pass through most biological objects so rapidly that they are not scattered or deflected significantly unless the object is comparatively thick. As a consequence, x rays have not proved to be practical as a source of illumination for microscopy. However, they have been used successfully to work out detailed internal arrangements of atoms in some biological molecules. The principle of the method, called *x-ray diffraction*, is related to the wave interference used to produce patterns of image brightness in phase contrast microscopy.

When a beam of x rays strikes an object with considerable internal order, such as a crystal, some of the waves are reflected from atoms in the object. Since the atoms in crystals occur in regularly repeated layers, the reflected waves from one layer may interfere with waves reflected from deeper or more shallow layers.

Figure A-15 illustrates how this interference occurs. Consider a wave *A* in the x-ray beam reflected from an atom in the top layer of atoms in a crystal. Another wave *B* reflected from an atom in the next level is reflected at the same angle. However, wave *B* must travel a longer

Figure A-14 Scanning electron micrograph of the surface cuticle of a spider mite. Courtesy of J. Mais.

distance to reach an atom in the second level (dotted segment of wave B in Fig. A-15). If wave B follows the same path as wave A leaving the crystal, the two waves may interfere, depending on the extra distance traveled by wave B. If this distance is equivalent to exactly one wavelength, the two waves will leave the crystal in phase. In other words, the troughs and crests of the two waves will coincide and reinforce. If the extra distance traveled by wave B is equivalent to one-half wavelength, the two waves will be exactly out of phase, and will interfere. In this case, the troughs of one wave will correspond to the crests of the other, and the two waves will cancel. Whether the waves leaving the crystal reinforce or cancel depends on the angle at which they strike the specimen, their wavelength, and the distance between the layers of atoms in the crystal.

The regular, three-dimensional arrangement of atoms in crystals produces multiple reflective layers running at

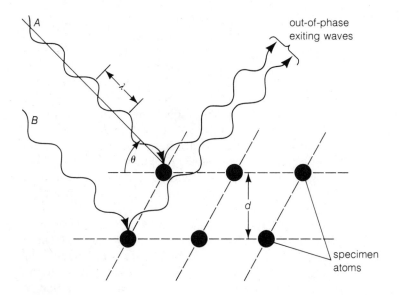

Figure A-15 Reflection of x rays from atomic planes in a crystal (see text).

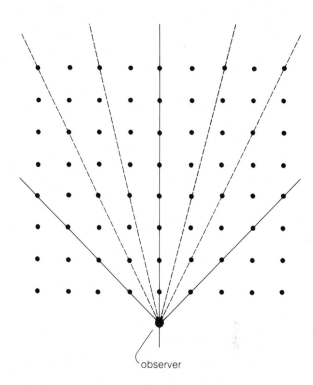

Figure A-16 The tree trunks visible from a single point in a regularly planted orchard line up in rows running in different directions. Some of the rows (solid lines) contain more trees than others (dotted lines), and are more apparent to the observer.

different angles and directions within the crystal. This is equivalent, in two dimensions, to the different angles at which trees in an orchard appear to line up when viewed from a single point (Fig. A-16). As a result, the waves in a beam of x rays directed at a crystal will be reflected at different angles, with each individual angle related to the distances between the different layers of atoms within the crystal. Since some of the layers will contain more or fewer reflecting atoms (equivalent to the different numbers of trees in the rows visible from a single point in Fig. A-16), beams reflected from the layers will be reinforced to a greater or lesser extent and will, as a result, have greater or lesser intensity. The net effect, if a photographic plate is placed in the path of the reflected beams, is the production of a pattern of spots with different spacing and intensity. From the position and intensity of the spots in the pattern, and their angles with respect to the crystal, the distances between the layers of atoms in the crystal and their arrangement can be deduced.

Some biological molecules, such as the tRNAs and some proteins, can be purified and cast into crystals. Other molecules, such as DNA, contain regularly repeated arrangements of atoms as a part of their native structure. Although the x-ray diffraction patterns produced by the internal order in these crystals or molecules are much more complex than the patterns produced by simple inorganic crystals, it has been possible in some cases to use them to reconstruct the arrangement of atoms of the biological molecule. Preeminent among the examples of successful applications of x-ray diffraction to biological molecules is the work leading to the discovery of the internal structure of DNA (see p. 216), which stands among the most significant findings of biology.

Finding Molecular Weights With Gel Electrophoresis

In this technique, cellular molecules are separated according to their rate of movement through a gel in response to an electrical field. The basic apparatus used in the technique, called *gel electrophoresis* (Fig. A-17), consists of the

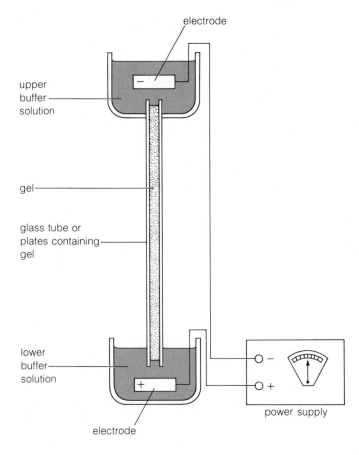

electrode

upper buffer solution

gel

glass tube or plates containing gel

lower buffer solution

electrode

power supply

Figure A-17 Apparatus for gel electrophoresis (see text).

gel, usually formed from polyacrylamide or starch, enclosed in a glass tube or sandwiched as a slab between glass plates. The tube or slab is suspended so that its ends are placed in contact with separate water solutions of a salt that conducts an electric current. The conducting solutions at either end of the tube are connected to the + and − electrodes of an electrical power source. When connected in this way, current flows from one solution to the other through the tube or slab. In this way, one end of the tube is made electrically positive or negative with respect to the other. Any molecules added to the tube will migrate through the gel in accordance with their own electrical charge.

The rate of migration through the gel depends primarily on two factors. One is the size and shape of the migrating molecules. The gels take the form of an open molecular network, with spaces large enough to allow passage of molecules such as proteins and nucleic acids along with the salt solutions used in the apparatus. The rate of migration with respect to size and shape reflects the ability of different molecules to thread through the gel network. The second factor is the *charge density* of the migrating molecules; that is, the relative numbers of + or − charges per unit of surface area. The higher the charge density, the more rapidly a molecular species will migrate toward the end of the gel with opposite charge, subject to the restrictions imposed by size and shape. Since the charge of many biological molecules varies with pH, adjustment and maintenance of pH is critical to the rate of migration through electrophoretic gels.

In application of the method a mixture of the molecules to be separated is carefully layered on the gel at the top of the slab or tube, usually by means of a hypodermic syringe. The salt solutions at the ends of the tube are then connected to the power source, in such a way that the charge of the solution at the bottom of the tube is opposite to that of the molecules under study. The layered molecules then move downward through the gel in response to the attractive charge. Since their relative rates of movement depend on size, shape, and charge density, the different molecular types in the mixture will gradually separate into a series of distinct bands. Smaller, more highly charged types will move faster and become distributed in bands closer to the bottom of the tube, and larger, less charged types will move more slowly and remain closer to the top of the tube.

Once separated into distinct bands, the molecules in the gels may be further treated for examination in various ways. Frequently, if the bands consist of proteins or other molecules that take up organic dyes, the gels are stained to reveal their distribution (Fig. A-18 shows a gel prepared in this way). Bands may also be identified by fluorescence, radioactivity, or reaction with antibodies (the bands in the gel shown in Fig. A-21 have been identified by radioactivity).

Since migration through gels reflects the size of different molecules, the positions taken by migrating bands provides a measure of molecular weight if the charge density and shape of the molecules under study is reasonably uniform. Nucleic acids satisfy these requirements and will separate into distinct bands in gels primarily according to molecular weight. Proteins, however, vary so extensively in shape and charge that accurate molecular weight approximations cannot usually be made by gel electrophoresis in the native state. This problem is circumvented by reacting them with the detergent *sodium dodecyl sulfate* (*SDS*). The SDS molecules, which carry a negative charge, coat the surfaces of protein molecules and give them a more or less uniform negative charge. This effectively eliminates the charge differences of proteins in their native state as a source of variation in the rate of migration through gels.

The coating of SDS molecules also unwinds the amino acid chain of most proteins into a random coil (see p. 44). The size of the coil in this condition primarily reflects the length of the amino acid chain. As a consequence of this conformation, and the uniform charge density imposed by the SDS coat, different proteins move through electrophoretic gels toward the positive electrode primarily according to the length of their amino acid chains, that is, according to molecular weight.

In practice, molecular weights are determined by mixing the unknowns with several molecules of known molecular weight, and running them together through the gel. The knowns and unknowns move in bands through the gel, distributed in such a way that the relative distances traveled by the knowns and unknowns give a reasonably accurate estimate of the molecular weights.

Centrifugation

Centrifugation is widely used in both the isolation and purification of cell organelles and molecules and in the analysis of molecular weight. Basically, a centrifuge (see Fig. 11-24) consists of a rotor driven at high speeds by an electric motor. The rotor contains holders for tubes containing the materials to be centrifuged. For ultra-high speed centrifugation, the centrifuge is enclosed in an armored chamber that can be cooled and held at a vacuum to reduce heat and friction due to air resistance and turbulence around the rotor. The centrifugal forces generated by the spinning rotor greatly increase the tendency of molecules or cell parts to sediment in the solution in the centrifuge tubes.

The tendency of molecules to sediment is related to their density and shape, and the density and viscosity of the surrounding solution. If a molecule of interest is denser than the surrounding solution, it will move downward in the tube; if it is less dense than the surrounding solution, it will be displaced by the molecules of the solution and will tend to move in the opposite direction, toward the top of the tube. The speed of the upward or downward movement is modified by the shape of the particle. Generally, the less spherical a molecule, the more slowly it will move through the tube in response to centrifugation.

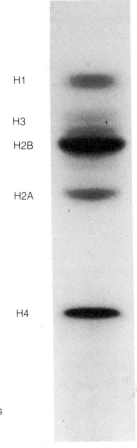

H1

H3

H2B

H2A

H4

Figure A-18 The bands produced by gel electrophoresis of the histone chromosomal proteins.

Density Gradient Centrifugation

For precise separation of molecules or molecular weight determinations, a technique known as *density gradient centrifugation* is used. In the most widely applied version of this technique, a plastic centrifuge tube is filled with a cesium chloride (CsCl) or sucrose solution, mixed in gradually more dilute proportions as the tube is filled. This produces an even gradient of concentration, with the solution most concentrated and therefore most dense at the bottom of the tube. A solution of the molecules to be separated is carefully layered at the top of the tube. During centrifugation, the different molecules in the tube travel downward through the tube in separate bands according to their density and shape. Maintenance of the molecules in distinct, sharp bands as they move downward depends on the solution gradient in the tube, from less dense at the top to more dense at the bottom. The molecules at the advancing front of a band will be slowed slightly in their progress, because they encounter a more dense medium than the molecules at the trailing portion of the band. Similarly, the trailing molecules will tend to move slightly faster than the molecules at the front of the band, since they encounter a less dense medium. The tendency of the leading molecules constantly to slow slightly, and the trailing molecules to move at slightly faster rates concentrates the molecules in a sharp band and maintains the band as it moves downward through the gradient.

The gradient may be preformed, as in the technique described above. Alternately, when a dense substance such as CsCl is used to form the centrifuging solution, the centrifugal forces may form the gradient. In this case, the CsCl molecules tend to sediment and pack more closely together toward the bottom of the tube, producing an even gradient in density from top to bottom.

Centrifugation is stopped before any of the bands reach the bottom of the tube. The final positions of the bands are usually determined by punching a hole in the bottom of the tube and collecting each band separately as it drains off. The relative positions of the bands in the tube can be used to determine the molecular weights of the molecules in the bands, as in the gel electrophoresis technique.

In centrifugation studies the rates at which molecules descend in a centrifuge are often described in terms of S units (S stands for *Svedberg*, one of the pioneers of centrifugation techniques). In general, the higher the S value, the higher the molecular weight of a substance. Ribosomal RNAs, in particular, are frequently identified in terms of their S values (see Information Box 10-1).

Buoyant Density Centrifugation

For highly critical work, molecules may be separated according to density by a variation of the basic density gradient technique called *buoyant density centrifugation*. The technique is best understood by considering how it is applied in nucleic acid research, a purpose for which it is often used. The centrifuge tube is filled with a solution, usually of CsCl, made up to a concentration with density closely approximating the molecules under study. The nucleic acid sample is layered on the solution, and the tube is spun at very high speed, sometimes for several days. As the tube spins in the centrifuge, the CsCl molecules pack more tightly toward the bottom of the tube, produc-

ing a gradient in solution density from the top to the bottom. Because the initial, uniformly dispersed CsCl solution matched the density of the nucleic acid sample, the final gradient produced is less dense than the nucleic acid at the top of the tube and more dense at the bottom. As a result, the nucleic acid molecules in the tube will descend until they reach a level at which the CsCl gradient exactly matches their density. The nucleic acid molecules will tend to remain in a sharp band at this level. The method is so sensitive that it can distinguish between "lighter" DNA molecules containing the ^{14}N isotope of nitrogen and "heavier" DNA containing the ^{15}N isotope (see Supplement 11-1).

Cell Fractionation Using the Centrifuge

Since most cell structures, such as nuclei, mitochondria, chloroplasts, and ribosomes, differ in density and are reasonably resistant to disruption, they can often be separately purified by centrifugation. This approach forms the basis of a variety of separatory techniques generally classified as *cell fractionation*. Usually, in application of these techniques, cells are first disrupted by breaking the plasma membrane in a solution held at cellular pH. This produces a mixed suspension of cell organelles and structures. Several centrifugations, at successively higher speeds, are then used to separate the structures. At lower speeds, large, dense bodies such as nuclei are driven down and concentrated. Low speed centrifugation can also be used to remove debris from a preparation. Centrifugation of the remaining solution at successively higher speeds next drives down and concentrates structures of intermediate size and density, such as mitochondria and chloroplasts. Final centrifugation at very high speeds concentrates ribosomes and larger molecules such as proteins and nucleic acids.

Nucleic Acid and Protein Sequencing

Nucleic Acid Sequencing

One of the most frequently employed methods for sequencing nucleic acids uses reagents that break DNA molecules at specific bases. The lengths obtained are then run

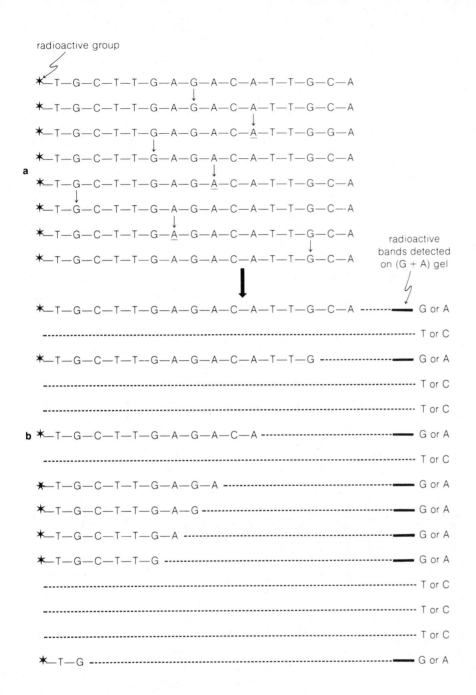

Figure A-19 Nucleic acid sequencing (see text). To produce the (G + A) gel, a uniform class of DNA fragments (**a**) is broken randomly whenever G or A occurs in the sequence (arrows). The fragments are then run on an electrophoretic gel (**b**). The bands on the gel indicate points where a G or A occurs in the sequence; empty spaces (dotted lines in the gel) show where T or C occurs in the sequence.

on electrophoretic gels, and the sequence is read directly from the distribution of DNA fragments on the gels.

In application of the technique, the purified DNA molecule to be sequenced is first broken into manageable, uniform fragment classes by restriction endonucleases (see Supplement 12-3). The DNA lengths in a given fragment class are then labeled at one end by attachment of a radio-

active phosphate group. The fragment class is subsequently treated by reagents that break the nucleic acid chain at sites where specific bases appear. For example, dimethyl sulfate modifies guanines and adenines in the DNA fragments, making the DNA chain unstable at these points. The chain is then easily broken at the modified points by treatment with alkali at 90°C. Treatment with

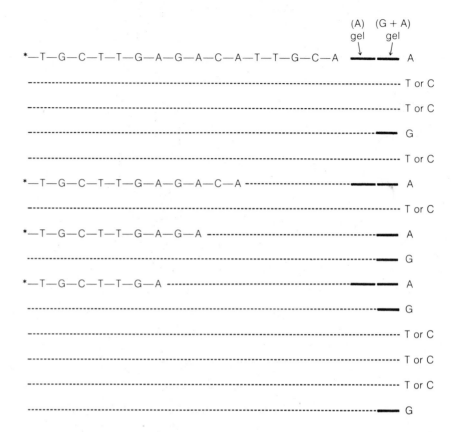

Figure A-20 Comparison of the (G + A) and (A) gels to sort out G and A in the sequence. To produce the A gel, the DNA fragment class shown in Figure A-19 is treated to break the sequence at A's (underlined points in Fig. A-19a). The fragments obtained are run on a gel and compared with the (G + A) gel. A occurs in the sequence at levels where both gels show a band. Bands present in the (G + A) but not (A) gel indicate that a G occurs at these points in the sequence. Blank levels in both gels (dotted lines) indicate a T or a C. The T and C bases are sorted out by the same approach, by attacking the sequence at (T + C) and (C) sites and comparing the gels obtained.

dimethyl sulfate is terminated before all of the guanine and adenine bases in a given fragment class are modified. The resulting positions of the modified and unmodified guanines and adenines are random, so that from all of the chains in a given fragment class, a family of lengths is obtained, each one broken at a guanine or adenine at different points in the sequence (Fig. A-19a, page 395). The segments are then run on electrophoretic gels, and the positions of individual bands detected by autoradiography. Since the only segments detected as bands will be those that contain the labeled end, the fragments will be sorted out in order of the position of the adenine or guanine from the labeled end, with successively shorter lengths toward the bottom of the gel (Fig. A-19b). This approach gives the relative positions of guanines and adenines in the sequence and is termed the (G + A) gel.

A modification of the technique gives segments broken only at random adenines. Running the fragments obtained sorts them out according to length, with successively shorter lengths broken at adenines placed toward the bottom of the gel. Noting the bands missing in the (A) gel,

when placed side by side with the (G + A) gel, allows each band of the (G + A) gel to be identified as a segment terminating in either a G or an A (Fig. A-20). Similar techniques give (C + T) and (C) gels, with fragments sorted out in order of the positions of C and T bases with respect to the labeled end. Placing the (A), (G + A), (C), and (C + T) gels side by side allows the entire sequence of the fragment to be read in order from the bottom of the gel to the top (Fig. A-21). Combining the sequences of all the restriction endonuclease fragment classes obtained from the original molecule thus allows the entire sequence of the DNA molecule to be reconstructed.

Protein Sequencing

Before sequencing, any internal disulfide linkages (see p. 46) in a protein are broken. In the most widely applied sequencing method, the protein is then reacted with a chemical "marker" that attaches to the amino acid at the

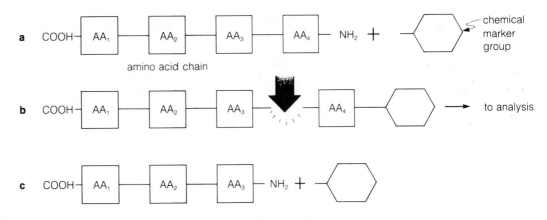

Figure A-22 Sequencing a polypeptide chain (see text).

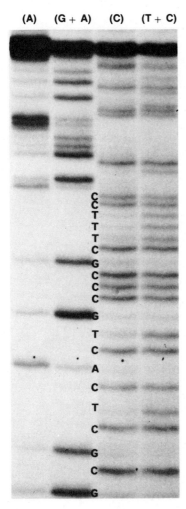

Figure A-21 The (A), (G + A), (C), and (T + C) gels showing the sequence of a DNA fragment from a bacterial cell. The sequence is read directly from bottom to top. Courtesy of A. M. Maxam and W. Gilbert, from *Proceedings of the National Academy of Sciences* 74 (1977):560.

end of the polypeptide chain with an exposed —NH_2 group (Fig. A-22a). Treatment with acid then splits off the marked amino acid (Fig. A-22b), which is then identified by standard chemical analytical techniques.

The procedure leaves the remainder of the amino acid chain intact, with another —NH_2 group exposed at one end (see Fig. A-22c). The amino acid at this end of the remaining chain is then identified by the same procedure. The process is repeated, one amino acid at a time, until the end of the polypeptide chain is reached. The technique, which is extended and laborious, has been automated by a device called a *sequinator*, which automatically carries out the sequential removal and identification of the amino acids at the —NH_2 end of the polypeptide chains. Since the automated sequinators can sequence only short polypeptides containing a limit of 20 amino acids or so, the polypeptide chain or chains of a protein must first be broken into fragment classes by protein-digesting enzymes or other means. The sequences of the individual fragment classes are then combined to reconstruct the complete amino acid sequence of the original protein or polypeptide chain.

Suggestions for Further Reading

Bradbury, S. 1967. *The evolution of the microscope.* Pergamon: New York.

Edman, P. and G. Begg. 1967. A protein sequinator. *European Journal of Biochemistry* 1:80–91.

Everhart, T. E. and T. C. Hayes. 1972. The scanning electron microscope. *Scientific American* 226:54–69 (January).

Maxam, A. M. and W. Gilbert. 1977. A new method for sequencing DNA. *Proceedings of the National Academy of Sciences* 74:560–64.

Meek, G. A. 1976. *Practical electron microscopy for biologists.* 2nd ed. Wiley: New York.

Schachman, H. K. 1959. *Ultracentrifugation in biochemistry.* Academic Press: New York.

Wischnitzer, S. 1970. *Introduction to electron microscopy.* 2nd ed. Pergamon: New York.

Work, T. S. and E. Work. 1969. *Laboratory techniques in biochemistry and molecular biology.* Vol. 1. North-Holland: London.

Answers to Genetics Problems

1. In the $CC \times Cc$ cross, the CC parent produces all C gametes, and the Cc parent produces $\frac{1}{2}C$ and $\frac{1}{2}c$ gametes. All offspring would have colored seeds—one-half homozygous ($\frac{1}{2}CC$) and one-half heterozygous ($\frac{1}{2}Cc$).

In the $Cc \times Cc$ cross, both parents produce $\frac{1}{2}C$ and $\frac{1}{2}c$ gametes. Of the offspring, three-fourths would have colored seeds ($\frac{1}{4}CC + \frac{2}{4}Cc$) and one-fourth colorless ($\frac{1}{4}cc$).

In the $Cc \times cc$ cross, the Cc parent roduces $\frac{1}{2}C$ and $\frac{1}{2}c$ gametes, and the cc parent produces all c gametes. One-half of the offspring are colored ($\frac{1}{2}Cc$) and one-half are colorless ($\frac{1}{2}cc$).

2. The genotypes of the parents are Tt and tt.

3. Yes, the brown-eyed parents can have a blue-eyed child, but the blue-eyed parents *cannot* have a brown-eyed child. The chance of a blue-eyed child being born to the brown-eyed couple described is $\frac{1}{4}$. Since each combination of gametes is an independent event, the chance of the second child (and any child) having blue eyes is also $\frac{1}{4}$.

4. The chance that the fifth child will have brown eyes is $\frac{3}{4}$ and that it will have blue eyes is $\frac{1}{4}$.

5. In the $RR \times Rr$ cross, one-half of the plants will have red flowers ($\frac{1}{2}RR$) and one-half will have pink flowers ($\frac{1}{2}Rr$).

In the $RR \times rr$ cross, all of the plants will have pink flowers (Rr).

In the $Rr \times Rr$ cross, one-fourth of the plants will have red flowers ($\frac{1}{4}RR$), one-half will have pink flowers ($\frac{1}{2}Rr$), and one-fourth will have white flowers ($\frac{1}{4}rr$).

6. Use a backcross; that is, cross the guinea pig having rough black fur with a double recessive individual, *rrbb* (smooth white fur). If your animal is homozygous ($RRBB$) you would expect all of the offspring to have rough, black fur.

7. One gene probably controls fur color, with the alleles G (green pods) and g (yellow pods). The G allele for green pods is dominant to the g allele.

8. The tongue-rolling parents are both heterozygotes; that is, they are both carriers of the recessive trait. If T is the dominant allele, and t is the recessive, then both parents are Tt and the child is the homozygous recessive tt.

9. The cross $RR \times RR$ will produce $\frac{1}{2}RR$ and $\frac{1}{2}Rr$ offspring. The cross $Rr \times Rr$ will produce $\frac{1}{3}RR$ and $\frac{2}{3}Rr$ offspring, since rr is lethal and does not appear among the progeny.

10. The parental cross is $GGTTRR \times ggttrr$. All offspring of this cross are expected to be tall plants with green pods and round seeds, or $GgTtRr$. This heterozygous F_1 generation, when crossed, is expected to produce eight types of offspring, green-tall-round : green-dwarf-round : yellow-tall-round : green-tall-wrinkled : yellow-dwarf-round : green-dwarf-wrinkled : yellow-tall-wrinkled : yellow-dwarf-wrinkled, in a 27:9:9:9:3:3:3:1 ratio.

11. Thirty-two different kinds of gametes can be produced.

12. In the first instance, the children would be expected to be: $\frac{1}{4}$ brown-eyed tasters, $\frac{1}{4}$ brown-eyed nontasters, $\frac{1}{4}$ blue-eyed tasters, and $\frac{1}{4}$ blue-eyed nontasters.

In the second instance, the children would be expected to be: $\frac{9}{16}$ brown-eyed tasters, $\frac{3}{16}$ brown-eyed nontasters, $\frac{3}{16}$ blue-eyed tasters, and $\frac{1}{16}$ blue-eyed nontasters.

13. All sons would be color-blind (100 percent chance), but no daughters would be color-blind.

14. If the woman marries a normal male, the chance that her son would be color-blind is $\frac{1}{2}$. If she marries a color-blind male, the chance that her son would be color-blind is also $\frac{1}{2}$.

15. Polydactyly is caused by a dominant allele, and the trait is not sex-linked. Thus,

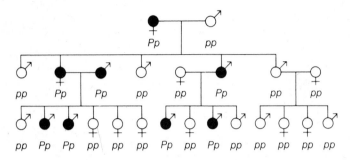

16. The genotypes are: bird 1, *FfPp*; bird 2, *FFPP*; bird 3, *FfPP*; bird 4, *FfPp*.

17. The genotype of the brown rabbit is Cc^w; the genotype of the chinchilla rabbit is $c^{ch}c^w$.

18. Yes. The child cannot be hers.

19. The sequence of the genes is ADBC.

20. Let the allele for normal body color = *B*, and the allele for black body = *b*. Let the allele for normal eye color = *P*, and the allele for purple eyes = *p*. Then the parents are

$$
\begin{array}{cc}
\dfrac{+B}{+P} & \dfrac{+b}{+p}
\end{array}
\;\times\;
\begin{array}{cc}
\dfrac{+b}{+p} & \dfrac{+b}{+p}
\end{array}
$$

The F$_1$ flies with normal eye color and black bodies are

$$
\begin{array}{cc}
\dfrac{+b}{+P} & \dfrac{+b}{+p}
\end{array}
$$

and the F$_1$ flies with purple eyes and normal body color are

$$
\begin{array}{cc}
\dfrac{+B}{+p} & \dfrac{+b}{+p}
\end{array}
$$

21. Examples *a*, *b*, and *c* can be accepted as proving the hypothesis. Example *d* does not match results closely enough, and the hypothesis that it follows a 9:3:3:1 ratio is probably wrong.

22. A recessive allele *l* carried on one of the two X chromosomes of the female parental type used in the cross is lethal when present in males. Thus in the cross $X^L X^l \times X^L Y = \frac{1}{4} X^L X^L + \frac{1}{4} X^L X^l + \frac{1}{4} X^L Y + \frac{1}{4} X^l Y$ half the males are lethals and die.

23. This cross is expected to produce white, tabby, and black kittens in a 12:3:1 ratio.

Index